Livestock and Wealth Creation

Improving the husbandry of animals kept by resource-poor people in developing countries

Cover photos:

Top: Hafiza and her son observing their precious goats in Chariswardia village near Mymensingh, Bangladesh. Hafiza is a member of a group of landless people participating in action research of DFID-LPP Project R8109 to study livestock keeping as a pathway out of poverty (courtesy of LPP Project R8109)

Middle: Poultry in Chariswardia village near Mymensingh, Bangladesh. In DFID-LPP Project R8109 poultry keeping by landless people is being studied as a pathway out of poverty (courtesy of LPP Project R8109)

Bottom: Transporting water to neighbouring farms in Bulawayo, Zimbabwe, using 'donkey power' (courtesy of T. Smith)

Livestock and Wealth Creation

Improving the husbandry of animals kept by resource-poor people in developing countries

Edited by:

E Owen

Formerly of Department of Agriculture, School of Agriculture, Policy and Development, University of Reading, Earley Gate, P.O. Box 237, Reading RG6 6AR, UK

A Kitalyi

Regional Land Management Unit (RELMA) at World Agroforestry Centre (ICRAF), P.O. Box 30677, Nairobi 00100, Kenya

N Jayasuriya

National Science Foundation, 47/5, Maitland Place, Colombo 07, Sri Lanka

T Smith

Formerly of Department of Agriculture, School of Agriculture, Policy and Development, University of Reading, Earley Gate, P.O. Box 237, Reading RG6 6AR, UK

Commissioned by:

Nottingham University Press
Manor Farm, Main Street, Thrumpton
Nottingham NG11 0AX, United Kingdom
www.nup.com

NOTTINGHAM

First published 2005

British Library Cataloguing in Publication Data
Livestock and Wealth Creation
Improving the husbandry of animals kept by resource-poor people in developing countries
E Owen, A Kitalyi, N Jayasuriya and T Smith

ISBN 1-904761-32-1

Disclaimer

Every reasonable effort has been made to ensure that the material in this book is true, correct, complete and appropriate at the time of writing. Nevertheless, the publishers and authors do not accept responsibility for any omission or error, or for any injury, damage, loss or financial consequences arising from the use of the book.

Typeset by Nottingham University Press, Nottingham
Printed and bound by Hobbs the Printers, Hampshire, England

Foreword

This new book on *Livesto* ... is one contribution to the
Millennium Developmen ... incentive for
commissioning th
by the increase in
in the major urb
increases in dem
offers opportuni
increased produ
the ability of dev
need look no fur
of milk, most o

Developed and
Creation is th
contributors fr
– 37 from 9 Af
countries; 33

Livestock and
coincides wi
Commission
economic de ... t
institutional ... d
Technical O ... e
(AU-DREA ... or
Africa's De ... nd
hunger-red ... es
for improv ... elp
to empov ... to
contribut ... ger
and pover

We trust that readers will appreciate the unique qualities of this book, and the relevance of the information to all involved in promoting improved and affordable livestock husbandry practices to the resource-poor, throughout the developing world. We feel confident that prudent adoption of improved practices will lead to wealth creation for the many poor people who keep livestock and provide them with an opportunity to escape from poverty.

The Rt. Hon. Hilary Benn, MP
Secretary of State for International Development, United Kindom

HE Mme Rosebud Kurwijila
Commissioner for Rural Economy and Agriculture, African Union Commission

Professor Richard Mkandawire
Agricultural Advisor for African Union's New Partnership for Africa's Development (NEPAD)

Editors' preface

Livestock and Wealth Creation – improving the husbandry of animals kept by resource-poor people in developing countries was conceived during a workshop in Tanzania in early 2002. Participatory consultation with academics and other stakeholders (researchers, NGOs, consultants, policy makers) in developing and developed countries during 2002-2003 confirmed that a book such as this was much needed.

Current textbooks on animal husbandry/animal science for lecturers and students in developing countries lack focus on the opportunities and problems faced by resource-poor livestock keepers; rather they are based mainly on more commercial high-input livestock production systems under temperate conditions. In view of the critical importance of livestock to the livelihoods of the resource-poor, both in rural and urban situations, it was deemed imperative that relevant information on more market-oriented livestock production systems be available for teaching so that appropriate advice is given to smallholder farmers. Most of the texts currently available have been written by developed-country authors, often with insufficient understanding of the cultural impact on livestock keeping. Furthermore, texts currently available are usually too expensive for developing-country purchasers and are also becoming out of date.

Part 1 of the book sets the agenda on key issues and principles in livestock development and poverty alleviation, and on cross-cutting issues which need to be understood before embarking on improving output from a given species. As well as answering the key question 'Why keep livestock if you are poor?', chapters in Part 1 include addressing issues on livestock systems, poverty assessment methods, livestock development and poverty, knowledge – key to empowerment, livestock products and improvement, marketing to promote development, livestock and the environment, response to nutrient supply, feeds and feeding to improve productivity, sustainable breeding strategies and improving livestock health.

Part 2 considers species individually, with emphasis on how to improve productivity (with examples) to achieve sustainable livelihoods for livestock keepers. There are chapters on bees, giant African snails, poultry, small mammals (grasscutters, guinea pigs and rabbits), pigs, goats, sheep, camels, cattle, buffalo, yak, equines and wildlife.

A unique feature is the cameos preceding most chapters. Each cameo involves people relating how their livelihoods were changed for the better by adopting a given improvement in animal husbandry.

The book concludes with a chapter considering the lessons learned and the way ahead. Improving the survival and production of livestock kept by the resource-poor is vital for security of livelihoods and transformation from poverty to relative prosperity. Providing appropriate information and an enabling environment are key elements in facilitating this process.

The information in this text book is relevant to smallholder livestock keepers in all developing countries.

Livestock and Wealth Creation is an output from the Livestock Production Programme (LPP) of the UK Department for International Development (DFID) for the benefit of developing countries. The views expressed are not necessarily those of DFID.

Acknowledgements

As mentioned in the Preface, this book was conceived at a Department for International Development (DFID) Livestock Production Programme (LPP) workshop in Tanzania, in January 2002, when Dr Wyn Richards, Manager of the LPP, requested we consider the idea. We are therefore very grateful to Wyn, not only for initiating the 'book project' but also for orchestrating the project's approval and funding by the DFID's LPP. Both Wyn Richards and Ms Sarah Godfrey, Assistant Manager of LPP, have given us much support and encouragement, throughout the project; we are very grateful to them and their colleagues at NR International.

We are delighted to have The Rt. Hon. Hilary Benn, MP, HE Madam Rosebud Kurwijila and Professor Richard Mkandawire contribute the Foreword, thereby giving the book such a high stamp of approval.

We are very grateful to the many participants in this 'book project' - the stakeholders who contributed so freely with ideas and opinions during the year-long participatory consultation in 2002-03, the principal authors, co-authors and consulting authors who contributed chapters and responded to the suggestions and requests of the editors, the livestock keepers and others in developing countries who generously shared their personal experiences in the cameos and case studies, the contributors of photos and those photographed for permission to publish their photos - without their generous help and participation, this book would not have been achieved.

We owe a particular debt of gratitude to our proof reader, Dr John Clatworthy. John not only corrected proofs but also corrected (and edited) manuscripts prior to submission to the publisher. Emyr Owen is also very grateful to his ever-patient wife, Jane, who acted as a voluntary proof reader and general helper.

Finally, we wish to thank Ms Sarah Keeling, Production Manager of Nottingham University Press for being so accommodating and helpful.

The Editors

Contents

1

Introduction – the need to change the 'mind-set'

Emyr Owen[*], *John Best*[†], *Canagasaby Devendra*[†], *Juan Ku-Vera*[†], *Louis Mtenga*[†], *Wyn Richards*[†], *Tim Smith*[†], *Martin Upton*[†]

Introduction

This chapter aims to explain why a textbook, such as this, is considered necessary at the start of the New Millennium. The chapter will also explain the rationale for the structure, content and authorship of the book.

Why a new livestock text-book?

Need for a 'mind-set' change

As indicated in the Preface, the editors and LPP - DFID (Livestock Production Programme of the UK's Department for International Development) undertook a participatory consultation with livestock-development specialists during 2002-03. This involved discussions and e-mail correspondence with academics, researchers, non-government organisations (NGOs), practitioners, consultants and policy makers in both developing and developed countries. The consultation yielded clear conclusions:

- There was a need for a new livestock textbook which illustrated how improving the husbandry of animals kept by resource-poor people in developing countries could increase their income and reduce poverty. No textbook currently exists which deals with this issue in a comprehensive manner. There was a need to promote a change in mind-set, especially amongst academics, and thereby their students, to acknowledge:
    - the importance of livestock (especially small stock) in the livelihoods of resource-poor people in developing countries

---

[*] Principal author
[†] Co-author

- that the resource-poor are generally small-scale (smallholder) or landless farmers.

- The mind-set change needed to include giving 'academic respectability' to the study of how to improve livestock husbandry to create wealth and hence alleviate poverty of resource-poor people in developing countries.
- The mind-set change needed to give greater emphasis to the 'participatory approach' to research and development, i.e. involving livestock-keepers and other stakeholders from the outset (and throughout) in development projects.
- The mind-set change needed to give greater emphasis to the 'holistic approach', i.e. appreciating that livestock-keeping is part of a farming system and that changes in husbandry need to be considered in the context of the system as a whole.

Millennium Development Goals

This book was conceived as a contribution towards meeting the Millennium Development Goal of reducing poverty. A series of United Nations (UN) conferences in the 1990s agreed on International Development Targets. These were endorsed by 149 Heads of State at the UN Millennium Summit in New York in September 2000. As listed by the UK Government White Paper (UK Government, 2000), they are:

- A reduction by one-half in the proportion of people living in extreme poverty by 2015.
- Universal primary education in all countries by 2015.
- Demonstrated progress towards gender equality and the empowerment of women by eliminating gender disparity in primary and secondary education by 2005.
- A reduction by two-thirds in the mortality rates for infants and children under the age of five and a reduction by three-fourths in maternal mortality – all by 2015.
- Access through primary health-care systems to reproductive health services for all individuals of appropriate ages as soon as possible, and no later than the year 2015.
- The implementation of national strategies for sustainable development in all countries by 2005, so as to ensure that current trends in the loss of environmental resources are effectively reversed at both global and national levels by 2015.

These International Development Targets are more often now referred to as 8 Millennium Development Goals (MDGs) and were listed by UK DFID (DFID, 2005) as:

- Eradicate extreme poverty and hunger.
- Achieve universal primary education.
- Promote gender equality and empower women.
- Reduce child mortality.
- Improve maternal health.
- Combat HIV/AIDS, malaria and other diseases.
- Ensure environmental sustainability.
- Develop a Global Partnership for Development.

The question of where and who are the poor people of the world is discussed in Chapter 2. It is unquestionable that almost all of the poor are in developing countries. Gryseels *et al.* (1997) estimated that 76 per cent of the poor in developing countries were rural people, and LID (1999) estimated that two-thirds of the rural poor were livestock-keepers. These estimates highlight the case for a new text book focusing on livestock kept by the poor.

It is also reasonable to hypothesise that, providing there is an increased demand for livestock products, improving the production and productivity of livestock kept by resource-poor people would improve livelihoods and alleviate poverty. The new opportunities offered by the fast growth in consumption of livestock products in the developing countries, known as 'the Livestock Revolution' (Delgado *et al.*, 1999), discussed in the next section, appear to support this hypothesis.

The Livestock Revolution

The rapid growth of demand for livestock products, which has occurred over the last quarter century, is largely driven by increases in per capita incomes, population growth and urbanisation of the developing countries. As Figure 1.1 shows, while consumption per capita of livestock products is much greater in the developed countries, it has fallen slightly over the last decade. In the developing countries, although per capita consumption starts from a much lower base, substantial growth has occurred. In these countries average per capita consumption of milk, eggs and meat has grown by two, four and six per cent respectively per year. It is noteworthy that average per capita consumption of cereals has fallen slightly in both groups of countries. About 20 per cent of total cereal consumption in the developing countries is as animal feed. This proportion has grown along with the expansion in intensive livestock production.

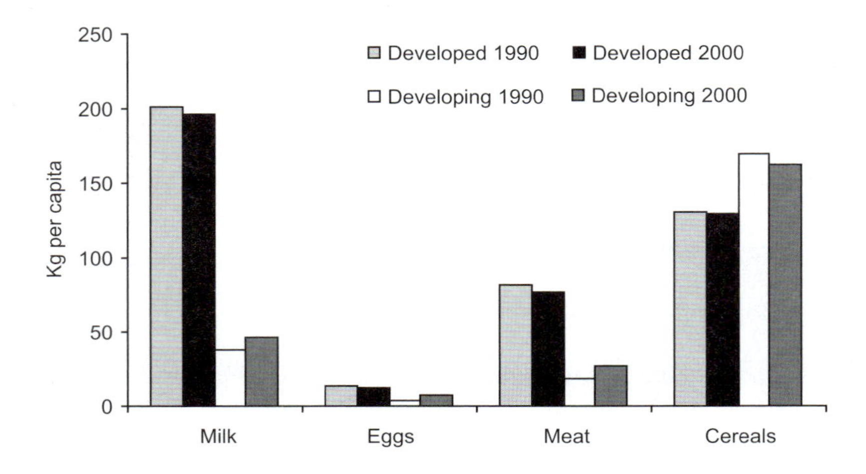

Figure 1.1 Per capita annual consumption of livestock products and cereals, in 1990 and 2000, in developed and developing countries (FAOSTAT, 2003). Note: data recorded as 'supply' assumed to be the amounts consumed

The contributions of different animal species to the average developed and developing country intake of meat and their growth are shown in Figure 1.2. In the group of developing countries, consumption of meat from intensive livestock production systems has grown fastest. For instance, the per capita intake of pig meat has increased by a half over the decade (by four per cent annually), while that of poultry meat has more than doubled (over seven per cent annual growth) from 1990 to 2000. These, so-called, 'landless production systems' are largely responsible for the rapid growth in average meat supply per person in the developing countries. The quantities of pig and poultry meat in the average diet of people in developing countries now exceed the quantity of bovine meat. It should be noted that the very high estimated growth rate and current level of pig-meat consumption in China have a major influence on the averages for all developing countries and doubts have been raised regarding their reliability (Bruinsma, 2003).

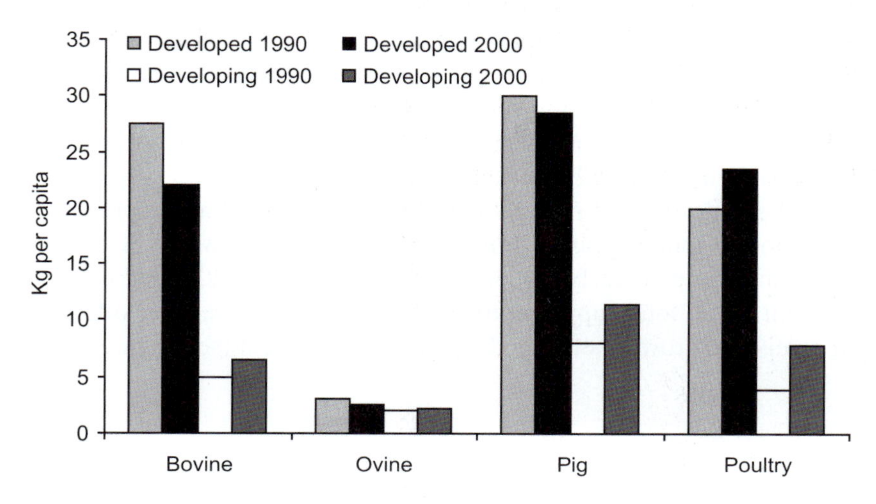

Figure 1.2 Types of meat consumed annually, per capita in 1990 and 2000, in developed and developing countries (source: FAOSTAT, 2003)

Much of the increase in supply, needed to meet the growth in consumption per head and that of the population, is derived from increased imports of these products from the developed countries. The growth of imports is illustrated in Figure 1.3 in which the total height of each column represents the gross imports, this being the sum of exports and net imports. Here then is an opportunity, and a challenge, for developing-country livestock-producers to meet more of the growing demand for livestock products and stem the rapid growth of imports, thus to conserve precious foreign exchange.

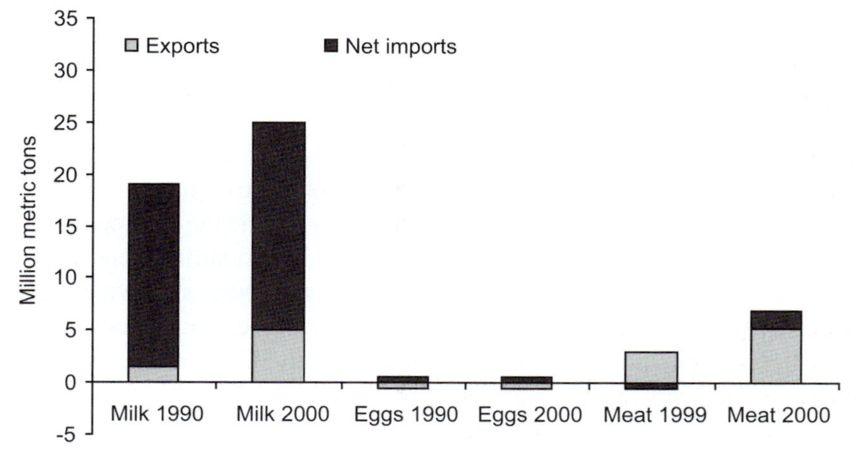

Figure 1.3 Imports of livestock products to developing countries in 1990 and 2000 (source: FAOSTAT, 2003)

Dairy products are by far the most important type of livestock product imported into developing countries, both quantitatively and in value terms. Dairy products make up 54 per cent of the total dollar value of net imports of livestock and their products, or 71 per cent if non-food items such as hides, skins and wool are excluded. Imports of dairy products have grown at 2.4 per cent annually, and in 2000 represented nearly 12 per cent of total supply of these products in the developing countries. Net imports of eggs have declined and represent a very small fraction of total supplies. Imports of meat have grown by two and a half times over the decade (nearly 10 per cent annually) and in 2000 represented over five per cent of the total supply of meat. However there are large differences in import levels and growth between different types of meat (see Figure 1.4).

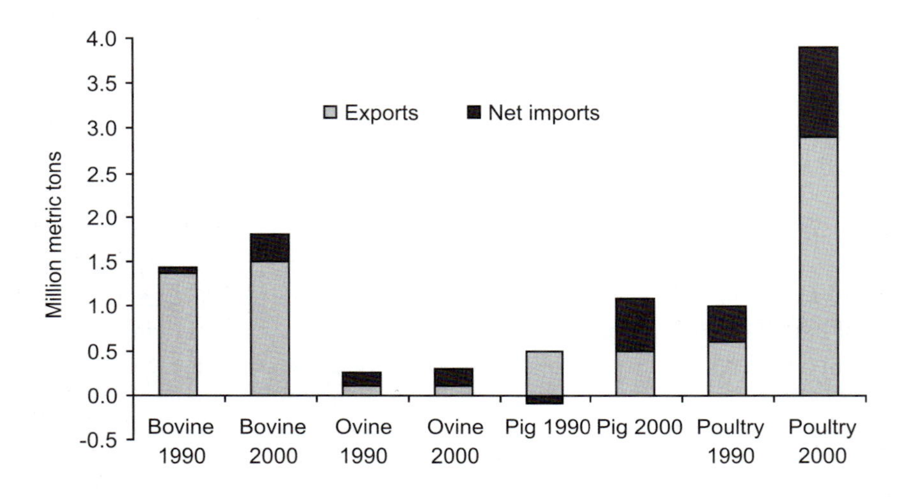

Figure 1.4 Imports of different types of meat to developing countries in 1990 and 2000 (source: FAOSTAT, 2003)

While trade between developed and developing countries in ovine meat has stagnated over the last decade, imports of bovine, pig and poultry meat to the developing countries have grown rapidly. For bovine meat, the developing countries were initially net exporters to the developed countries. Subsequently, imports of bovine meat have increased by nearly 50 per cent and now account for six per cent of total developing country supplies. Imports of pig meat have tripled (nearly 12 per cent growth annually), but still contribute only two per cent of supply. Imports of poultry meat have increased by four and a half times (by nearly 16 per cent annually), make up 13.5 per cent of total supply, and exceed imports of all other types of meat put together.

Increasingly, over the next two decades, the Livestock Revolution will create opportunities in developing countries for import substitution and greater production of livestock products. The key question is: will resource-poor livestock-keepers be able to take advantage of these opportunities and thereby move out of poverty? This question is addressed in Chapter 8 and other chapters.

The book's structure

Contributors

This book has been developed and written using a participatory approach (greatly enabled by e-mail), and has involved over 100 contributors from 26 countries. Nearly two-thirds of the contributors are from developing countries – 37 from nine African countries, 20 from seven Asian countries and eight from four Latin American countries. Most of the developed-country contributors are from the UK. Almost half of the contributors (26 from developing countries, 22 from the UK) have participated in LPP - DFID research projects in developing countries during the last decade, where the emphasis was on demand-driven research with a poverty-alleviation focus. Over half (53 per cent) of the contributors are university teachers; the remainder are researchers (16 per cent), livestock specialists in NGOs (11 per cent), policy makers (10 per cent) and consultants, etc. (10 per cent).

The editors have tried to ensure that the contributors of given chapters are from different countries, and if possible different regions of the developing world. In addition to the principal author and co-authors, each chapter has one or more 'consulting authors'. This novel approach was to ensure authenticity and widen the participation.

Another novel approach is the inclusion of one or more cameos at the beginning of most chapters. The purpose of the cameos is to illustrate to the reader how the livelihoods of resource-poor livestock-keepers were changed for the better by adopting a given improvement in animal husbandry.

Cross cutting issues, species and lessons learned

Chapters 1 to 13 set the agenda on key issues and principles in livestock development and poverty alleviation, and on cross-cutting issues which need to be understood before embarking on improving output from a given species. These chapters are components of the 'need for a mind-set change', referred to earlier.

They include: considering why poor people keep livestock, livestock systems, poverty assessment methods, knowledge dissemination, products and marketing, livestock and the environment, and improvement strategies concerning the nutrition, breeding and health of livestock.

Chapters 14 to 26 consider species individually, with emphasis on how to improve productivity (with examples) to achieve sustainable livelihoods for livestock-keepers. Readers are asked to note that the order of considering the species in Chapters 14 to 26 is, more or less, in order of increasing physical size, commencing with bees. However, it could be argued that this order also reflects the order of importance, or potential importance, of the species to the livelihoods of the very poor. Part of the 'mind-set change' needed is to recognise the importance of small stock (Chapters 14 to 20) to resource-poor people and recognise that improving the productivity of small stock is likely to make a large contribution to alleviating poverty.

Chapter 27 considers the lessons learned and the way ahead. As indicated in the Preface, an obvious conclusion is that improving the survival and production of livestock kept by the resource-poor is vital for security of livelihoods and transformation from poverty to relative prosperity. The provision of appropriate information and an enabling environment are key elements in facilitating this process.

Other species

Fish

Fish are not included in the species considered in Chapters 14 to 26. This is not to disregard the importance of fisheries to the livelihoods of poor people in developing countries. Rather the editors decided that to incorporate material on fisheries would extend the scope of the publication beyond manageable limits since the contribution of fisheries to rural livelihoods is made via so many different types of system, e.g. capture fisheries (marine and freshwater), managed open-water fisheries, and cage culture, as well as pond culture (which may or may not involve integration with livestock production). Readers should note that there is a large literature on aquaculture and fisheries (e.g. Pillay, 1993; Stickney, 2000; World Fish Center, www.worldfishcenter.org). Also there are three DFID-funded research programmes (parallel to LPP) covering different aspects of fisheries and exploring roles for aquaculture and fisheries in providing sustainable benefits to poor communities.

These are: the Aquaculture and Fish Genetics Research Programme (www.dfid.stir.ac.uk), the Fisheries Management Science Programme (www.fmsp.org.uk), and the Post-harvest Fisheries Programme (www.phfp.uk.com).

Others

Silkworms are also not included in Chapters 14 to 26; it could be argued they should be in view of their importance to silk production, particularly in China.

The editors excluded consideration of ostriches as a separate chapter on the grounds that ostrich farming is capital-intensive and not normally directly relevant to resource-poor people. However, a development in Zimbabwe, some time ago, involved resource-poor smallholders being paid by commercial ostrich farmers to incubate ostrich eggs using broody hens. Ostrich products are also briefly discussed in Chapter 7.

Other species not included are lizards, frogs, turtles and bats. Although important historically as food sources for humans, these are now endangered species, except for frogs.

Animal welfare

Farm animal welfare is a subject of increasing interest in developed-country livestock husbandry with implications for farmers and researchers regarding husbandry practices (e.g. Ewbank *et al.*, 1999; Peterson and Palmer, 1991). The topic is alluded to in several chapters (e.g. Chapters 13 and 25) in the present book, and will assume increasing importance in the future, particularly in relation to international trade (Chapter 8) and the conduct and publication of experiments with farm livestock.

Other literature

The list of 'Further reading' at the end of each chapter provides sources of other relevant literature. Readers will appreciate that each chapter is often a brief review of a large topic which has had whole books devoted to it, albeit not from the perspective of poverty alleviation in developing countries.

Readers are also reminded that *An introduction to animal husbandry in the tropics* (Payne, 1990; Payne and Wilson, 1999) remains a classical general text on the subject.

For two recent companions to the present book, readers are referred to *Developing smallholder agriculture – a global perspective* (Tinsley, 2004) which has a broader perspective on smallholder agriculture, and *Participatory livestock research: A guide* (Conroy, 2005) which provides a comprehensive guide to modern approaches to livestock development research.

Two other classics which should be read in conjunction with the present book are *Reality in rural development aid, with emphasis on livestock* (Ørskov, 1993) and *Strategy for sustainable livestock production in the tropics* (Preston and Murgueitio, 1992).

Further reading

Conroy, C. 2005. *Participatory livestock research: A guide*. ITDG Publishing, London, UK.

Ørskov, E.R. 1993. *Reality in rural development aid, with emphasis on livestock*. Rowett Research Services Ltd., Bucksburn, Aberdeen AB2 9SB, UK.

Payne, W.J.A. 1990. *An introduction to animal husbandry in the tropics. Fourth edition*. Longman Scientific and Technical, Harlow, Essex CM20 2JE, UK. Co-published in the United States by John Wiley & Sons, New York, USA.

Payne, W.J.A. and Wilson, R.T. 1999. *An introduction to animal husbandry in the tropics. Fifth edition*. Blackwell Science Ltd., London, UK.

Preston, T.R. and Murgeitio, E. 1992. *Strategy for sustainable livestock production in the tropics*. Centro para la Investigacion en Sistemas Sostenibles de Produccion Agropecuaria (CIPAV), AA20591 Cali, Colombia; Swedish Agency for Research Cooperation with Developing Countries (SAREC), P.O. Box 16140, S-103 23, Stockholm, Sweden.

Tinsley, R.L. 2004. *Developing smallholder agriculture – a global perspective*. AgBe Publishing, 1050 Brussels, Belgium.

Author addresses

Emyr Owen, Formerly of Department of Agriculture, School of Agriculture, Policy and Development, University of Reading, Earley Gate P.O. Box 237, Reading RG6 6AR, UK

John Best, Department of International and Rural Development, School of Agriculture, Policy and Development, University of Reading, Earley Gate

P.O. Box 237, Reading RG6 6AR, UK
Canagasaby Devendra, Formerly of the International Livestock Research Institute (ILRI), P.O. Box 30709, Nairobi, Kenya. Current address: 130A Jalan Awan Jawa, 58200 Kuala Lumpur, Malaysia
Juan Ku-Vera, Facultad de Medicina Veterinaria y Zootecnia, Universidad Autonoma de Yucatán, C.P. 97100 Merida, Yucatan, Mexico
Louis Mtenga, Department of Animal Science and Production, Sokoine University of Agriculture, P.O. Box 3004, Morogoro, Tanzania
Wyn Richards, NR International, Park House, Bradbourne Lane, Aylesford, Kent ME20 6SN, UK
Tim Smith, Formerly of Department of Agriculture, School of Agriculture, Policy and Development, University of Reading, Earley Gate P.O. Box 237, Reading RG6 6AR, UK
Martin Upton, Department of Agricultural and Food Economics, School of Agriculture, Policy and Development, University of Reading, Earley Gate P.O. Box 237, Reading RG6 6AR, UK

2

Why keep livestock if you are poor?

Aichi Kitalyi[*], *Louis Mtenga*[†], *John Morton*[†], *Anni McLeod*[‡], *Philip Thornton*[‡], *Andrew Dorward*[‡], *M. Saadullah*[‡]

Cameo - The impact of acquiring a cow

Ms Batisheba Enock of Gezaulole of Dar es Salaam, Tanzania, had this to say:

"Through a Heifer-in-Trust project, I acquired a cow at a time when I was a desperate divorcee with a plot of less than 0.1 hectare. I raised this cow on forage from open-access land and crop residues. In seven years the herd grew to seven. I gave one heifer to another farmer, ploughed back into the project proceeds from the sale of one offspring, and sold another for TZS 200,000 (about US$ 200) to pay school fees. Next, after selling one cow, I will buy a farm to produce my own forage. My two milk cows provide 15 litre daily, enough for my household needs and for sale. The manure fertilises my small vegetable garden and pawpaw trees, whose fruit we mainly consume but occasionally sell. I sometimes sell the manure, but I also give it free to neighbours. I have learned from seminars and other farmers that manure can produce gas fuel for cooking, but this needs investment. In future I will have a small plant to provide gas for cooking and lighting. That cow I acquired in 1996 transformed my life by giving me food, income and employment. It is everything to me."

Introduction

This chapter looks at why the poor keep livestock. An attempt is made to characterise the poor and relate that to livestock keeping. This is followed by a brief description of key socio-economic concepts to understand livestock keeping by the poor in developing countries. The last section of the chapter looks at livestock roles and functions, with examples from different livestock production systems.

---

[*] Principal author
[†] Co-author
[‡] Consulting author

Who and where are the poor?

We all think that we know what is meant by the concepts 'poor' and 'poverty', but defining them is complex although they have probably been the most common terms in human development discussions in the last decade. A measure based on consumption or income data obtained through household surveys has received the most attention, and the World Bank has been using it since the early 1990s. Using purchasing power parity prices estimated by the International Comparison Program, a benchmark consumption-based poverty line for the low-income countries was selected in 1993 by World Bank (World Bank, 2001). In 1999 the lower poverty line was set at about US$ 1 a day, and the upper one at US$ 2. These poverty lines are generally used as indicators of global progress in poverty alleviation, and more meaningful poverty lines are those developed by each country for its own conditions. For example, the poverty line for Ethiopia between 1996 and 1999 was 1075 birr a year, or less than US$ 0.50 per day. The World Development Report 2000/01 (World Bank, 2001) and the UNDP Human Development Report (UNDP, 1997) give detailed accounts of global estimation of poverty, and country-specific poverty lines for some developing countries can be obtained from their poverty reduction strategy papers (PRSPs).

The rich perceive poverty as deprivation of materials for well-being. However the poor perceive poverty as a more multidimensional social phenomenon: it ranges from food and material deprivation to the psychological experience of multiple deprivations (World Bank, 2001).

The multiple dimensions of poverty as described by the poor underscore the importance to them of livestock. To understand the link between livestock keeping by the poor and the interlocking dimensions of poverty, we will refer to the Sustainable Livelihood (SL) approach advocated by the UK Department for International Development (DFID). To quote DFID's Sustainable Livelihood Guidance Sheet (DFID, 1999),

> "... the framework views people as operating in a context of vulnerability. Within this context, they have access to certain assets or poverty-reducing factors. These gain their meaning and value through the prevailing social, institutional and organisational environment. This environment also influences the livelihood strategies - ways of combining and using assets - that are open to people in pursuit of beneficial livelihood outcomes that meet their own livelihood objectives."

The SL framework is based on five types of capital: natural, human, financial, physical and social. The World Bank Report, Voices of the Poor Initiative, suggests that the poor focus on assets rather than income and therefore link their lack of assets (physical, human, social and environmental capital) to their vulnerability and exposure to risk (Narayan *et al.*, 2000). Livestock do not have a clear position in either framework of assets, and Morton and Meadows (2000) describe livestock under natural, financial and social capital. This could explain the under-representation of livestock in poverty-reduction strategy papers observed in a recent study commissioned by FAO using the Pro-poor Livestock Policy Initiative (Blench *et al.*, 2003).

Within the SL framework, Dorward and Anderson (2002) have extended the debate by bringing in the importance of relating asset functions and livelihood strategies. Discussing this concept in relation to livestock, three livelihood strategies are suggested: 'hanging in', when livestock play buffering and insurance roles; 'stepping up' where livestock go beyond subsistence and contribute to building up other assets; and 'stepping out' where productivity of livestock is less important than their holding value as savings (Table 2.1). This concept has been used in investigating the roles of livestock in livelihoods of poor families in the Yucatan, South-East Mexico, where clear differences were found among different wealth strata in husbandry practices and reasons for keeping pigs.

Table 2.1 Principal roles of livestock by livelihood strategy

Livelihood strategy	*Principal livestock roles*
'Hanging in'	Subsistence
	Complementary production
	Buffering
	Insurance
'Stepping up'	Accumulation
	Complementary production
	Market production/income
'Stepping out'	Accumulation

Source: Dorward and Anderson (2002).

In the last decade, much work has been devoted to defining the poor and in mapping poverty in an endeavour better to target development programmes (LID, 1999; Thornton *et al.*, 2002). Table 2.2 shows the extent of rural poverty in the six regions of developing countries. Using data on the distribution of people by livestock production systems and agro-ecological zones, together with poverty statistics from the Human Development Report and studies on livestock ownership patterns, LID (1999) estimated that 678 million poor were livestock-keepers (Table 2.3). These are estimates, and more studies are needed.

Table 2.2 Extent of rural poverty in developing country regions, 1994–95

Region	*Number of rural poor (millions)*	*Share of rural to total (%)*
East Asia	114	81
South Asia	417	81
South-East Asia	121	83
Latin America and the Caribbean	76	42
West Asia and North Africa	40	50
Sub-Saharan Africa	248	88
Total	1016	76

Source: Gryseels *et al.* (1997).

Table 2.3 Number and location of resource-poor livestock keepers

Agro-ecological zone	*Resource-poor livestock keepers in different livestock production systems (millions)*		
	Extensive grazers	*Mixed rain fed*	*Landless*
Arid or semi-arid	63	213	
Temperate, including tropical highlands	72	85	
Humid, sub-humid and subtropical	-	89	
Total	135	387	156

Source: LID (1999).

The ground-breaking work of Thornton *et al.* (2002), which mapped poverty and livestock in developing countries, indicated that high densities of poverty-stricken livestock-keepers appear mainly in mixed irrigated systems in parts of South Asia, the mixed rain-fed systems of India and most of Sub-Saharan Africa. They show that in East Africa higher poverty rates occur in the arid and semi-arid lands, where over 60 per cent of the households fall below the poverty line. Poverty among livestock-keepers in this ecological zone is exacerbated by conflict, drought and disasters, such as the 1995-97 drought and conflict in Somalia that decreased the national herd of cattle by 70 per cent and that of goats by 60 per cent (Ndikumana *et al.*, 2000).

Poverty mapping is still in its infancy. Thornton *et al.* (2002) pointed out the limitations of the data used, which mainly stem from lack of detailed data for the different parameters at the country level. Furthermore, the general poverty maps may inadequately represent the distribution of poor livestock-keepers, particularly if the multidimensional nature of poverty is taken into account. Another aspect to consider in future poverty mapping of livestock-keepers is the difference in importance of livestock in household income, which is a function of various socio-economic factors. Delgado *et al.* (1999) cite about twenty case studies from Africa, Asia and parts of Latin America which show that the poor and the landless derive a greater share of their household income from livestock sources than do the relatively better-off in the same rural community. While broad-scale poverty mapping can provide useful aggregated information, little is said about the causes of poverty and its dynamics in particular places. For this, studies need to be carried out at the local level that can identify the reasons why people fall into and climb out of poverty, and the roles that livestock play in these processes. Work that links high-resolution poverty mapping with community surveys is starting to throw light on these complex issues.

Key issues and concepts

In studying livestock keeping by the poor and the marginalised, we have to be conscious of various key socio-economic concepts that are not necessarily relevant in commercial livestock production.

Ownership, control and access to benefits

It is not always easy in traditional small-scale livestock production systems to decide who is the 'owner' of an animal; 'ownership' is not a simple or indivisible concept, but a 'bundle of rights'. Good examples can be found in various pastoral societies where male heads of a family assign 'ownership' of particular animals to wives or sons but still make the important decisions regarding herding. The head of the family, the assigned owner, or both may have to be involved in decisions regarding slaughter or sale. The position may be further complicated if other people who have given the stock in some form of traditional arrangement retain some sort of rights or are associated with the animal in people's minds or speech. Stock loans further complicate ownership, as the loaned status of an animal may not be revealed without careful consideration. Neither ownership nor control necessarily relates to access to benefits - a much wider pool of people might expect to share in benefits such as milk and meat or the proceeds when the animal is sold.

Labour

The people who look after an animal (often known as livestock 'keepers') are not necessarily those who own or control it or have access to its benefits. The division of labour by age and gender determines who looks after an animal. In many traditional livestock-keeping societies, management and care of large species (cattle, buffalo and camels) are the preserve of adult men, whereas women or children take a larger role in herding small stock. Women are often associated with milking and processing of dairy products but, in some traditional pastoral societies in North-East Africa, women are prohibited from milking. However, such customs are liable to change with external trends such as commercialisation, increasing education and male labour migration. There are also traditional, evolving or new institutions for hiring labour, such as wage-herding contracts paid in kind or cash. Labour is a key factor in livestock development, particularly in Sub-Saharan Africa, mainly because many of the technologies developed for improving livestock feeding are more labour-intensive than those they replace.

Land

Land tenure interacts with livestock production in complex ways. Livestock-keepers may rely primarily on their private land, as do many farmers in mixed systems or specialist livestock producers around the world. Where this is the case, the evolution of the land-tenure system has a major effect on livestock production. Shrinkage of landholdings through complex inheritance laws or increasing population may drive producers to intensify their livestock production by more efficient use of resources, to specialise further in livestock production, or to leave it altogether. Many livestock-keepers rely on some rights to use community or state land, or ownerless land. This could include tiny patches of land such as roadside verges or small village commons in landscapes dominated by private tenure (as in much of India), or large-scale open rangelands like those typical in pastoralism in arid and semi-arid lands. There is major debate on the management of such lands. Some consider these lands 'common' and subject to the so-called 'tragedy of the commons', in which livestock owners inevitably overgraze and degrade common rangelands.

Others point out that true commons, as distinct from open-access resources, have their own governing mechanisms and institutions, and it is the neglect or erosion of these institutions that causes degradation. Still others note that real-world situations often involve changing mixtures of private, common and individual tenure.

Knowledge

Knowledge of livestock and of the resources that sustain them is a key factor in livestock production. There is much important knowledge in communities that have specialised over generations in livestock production and this is often referred to as indigenous technical knowledge (ITK). Livestock production ITK has received less research attention than animal health ITK. Those involved in assisting poverty-stricken livestock-keepers need to respect these ITK traditions. But livestock systems change as production intensifies and crop and livestock production become integrated, and needs for knowledge change accordingly. These new needs are often satisfied by livestock-keepers' own evolving knowledge and it is therefore important not to underestimate the extent to which livestock-keepers can experiment with new techniques and learn from successes and failures. In many cases, however, changing systems require more thorough dissemination of external knowledge, which is the role of extension backed by research. It is to be hoped that extension and research systems can respond to these needs in ways that foster participation of poverty-stricken livestock-keepers, to ensure their needs are heard and their knowledge respected. Chapter 6 provides more discussion of knowledge and empowerment in livestock keeping.

Markets

Chapter 8 will examine markets and marketing in more detail, but we should note that:

- Markets affect all livestock-keepers: all sell livestock or produce, and buy non-livestock based food and other necessities.

- Markets are now global. Even if livestock-keepers sell only locally and do not use purchased inputs, their prices are still affected by global prices of meat and feed, which are in turn influenced by changing demand around the world.

- Livestock products such as milk and meat are highly perishable, and livestock themselves have unique practical constraints on how they are marketed. Therefore local market conditions matter very much, e.g. where the market is, how livestock are taken there, and the relationships between buyers and sellers.

- Intermediaries in livestock and livestock-product trading, as in other food trading, have a reputation for being exploitative. These perceptions are sometimes justified, but many traders do considerable work in matching sellers to end-markets and take considerable risks, and so 'exploitation' needs to be examined critically.

Gender

Gender roles are the socially constructed expectations for women and men. Although often considered 'natural', these roles vary greatly among societies. As a general rule, women do not herd large livestock, but among the Rashaida Arabs of Sudan - in other respects a very 'conservative' Muslim society - one sees veiled women and girls herding on camelback (Morton and Meadows, 2000). Gender roles are important in the way livestock-keepers (and those who assist them) think about livestock ownership and control, the division of livestock-related labour, and the transmission and validity of livestock knowledge. We need to understand both the importance and the cultural nature of these roles.

Definitions of poverty and its causes vary by gender, age, culture and other social and economic attributes. Understanding the issues surrounding ownership of livestock, control of and access to resources, income and expenditure, and their implications for gender roles and equity, is important for understanding the vulnerability of some of the gender categories in developing countries. In the World Bank-led Participatory Poverty Assessment (PPA) exercises, the most vulnerable groups were found to be female-headed households, single mothers, orphans, men with large families, unemployed youths, adolescent mothers, casual workers and women married to irresponsible or alcoholic husbands (World Bank, 2001).

Livestock keeping, food security and rural livelihoods

The human-animal relationship is ancient. One of the main reasons for animal domestication, which started at the end of the Pleistocene age, (12,000 years ago), was to address the problem of unpredictability of food supply associated with unpredictable weather. The earliest livestock species to be domesticated for food were pigs, sheep, goats and cattle. Over the millennia, the number of animal species domesticated has increased as the mode of production has changed with increased intensification in response to increasing demand for animal products.

Food insecurity is one of the dimensions of poverty and it encompasses food production, stability of supply and access to quality food. 'Food poverty' is a common term in the current discussion on poverty reduction. Food poverty is described as the inability to afford the basket of food items typically considered as adequate to provide minimum energy requirements. Livestock production adds quality and quantity to that basket.

The nature of the contribution of livestock to household food security varies from place to place. Where markets exist, poor livestock-keepers generally tend to sell the animals or their products for money to buy cereal and leguminous grains, as these are cheaper than animal products and provide much more energy and protein to the household. A DFID Livestock Production Project (R8109) with landless livestock-keepers in the Terai, Nepal, showed that goats are a source of occasional cash, not food security (McLeod *et al.*, 2002).

Poverty-stricken livestock-keepers use the smaller mammals and poultry more for food than the larger species. Smaller animals are more prolific, have lower requirements in terms of capital and maintenance costs and are less risky to keep. They are also easier to sell or dispose of where there are no means of preservation or easy access to markets. In Bolivia, with very simple improvements in husbandry, small animals such as chickens, ducks, pigs, hair-sheep and guinea pigs increased the annual income of a poor household by over 25 per cent (Paterson and Rojas, 2002).

It is also evident that the poor benefit greatly from increased consumption of relatively small quantities of livestock protein. Animal products such as milk, meat and eggs raise the quality of the mainly cereal-based diets of poverty-stricken livestock-keepers, because they provide readily digestible, high-quality protein and energy, as well as essential micronutrients such as calcium, iron, zinc, and vitamins A (retinol), B6 (pyridoxine) and B12 (cyanocobalamin) (see Chapter 7). Again, poor people who rear livestock tend to consume a greater quantity of livestock products than do those without livestock. Improving the husbandry of livestock kept by the poor may therefore increase consumption of animal products.

The sale of live animals and their products, and leasing of draught animals, generates income for livestock-keepers. Many pastoral societies are involved in the sale of live animals, milk, and hides and skins for income. Wool, hair and manure are other products contributing to the income of livestock-keepers. Landless people in some parts of India graze their flocks of sheep and goats on state-owned and open-access land. After crop harvest, farmers hire these animals to graze (or be kraaled) on their crop fields at night to provide manure for soil fertility. The livestock-keepers accrue an income of over US$ 200 per flock in two months. Use of animals for commercial transport (camels, camelids, cattle, buffalo, donkeys, horses) is another growing income-generating activity for poverty-stricken livestock-keepers.

Photo 2.1 Eunice Wambui, a mother of 13 and a widow, living on an 800 m sq. plot, says the cow has taken her out of poverty (courtesy of Heifer Project International, Kenya)

Contribution to increased agricultural production

Draught animals are an important asset for the poor in developing countries and are highly valued by many communities in arid and semi-arid lands, and hilly and mountainous areas. Although not much is documented on selection of animals for draught purposes, special breeds or types have been conserved for draught in many regions in developing countries. Draught animals permit more land to be cultivated in a timely manner and with less human drudgery. In some areas they also have important uses in crop processing and water lifting.

The tradition of animal traction is ancient in much of Asia, the Middle East and Highland Ethiopia. In Latin America and parts of Southern and East Africa, it was a colonial introduction that has taken root. Animal traction is much needed in most of Sub-Saharan Africa following the failure of government support to rural mechanisation programmes. Furthermore, the increasing miniaturisation of landholdings in the mixed farming system limits use of machinery. A new approach to animal traction, with appropriate equipment for most farm operations, is playing a significant role in agricultural intensification.

Pack and transport animals are of great importance, especially in drier and more mountainous areas, and among pastoral communities. The donkey is extensively used as a pack animal in many livestock keeping communities. Among the Maasai of East Africa, the best present for a newly married woman is a donkey, which is used to fetch water and fuel wood and in transportation during pastoral migration. Large numbers of donkeys are found in China, Ethiopia, Pakistan, Mexico and India. Other common transport animals include the mule, ox, yak, water buffalo, camel and llama; most of these animals are also used for draught. In addition, over 60 per cent of the different zebu cattle types are primarily used for draught.

In mixed farming systems, households without draught animals rely on a number of customary arrangements that help them acquire their use, such as draught-animal sharing, hiring, and exchange of draught power for human labour or a share of the crop or crop residues. In Ethiopia, not having draught animals means delayed planting, low crop yields and higher costs of production, since they have to be hired, often at inconvenient times. Use of animals for draught is also considered in other chapters (21, 22, 23, 24 and 25).

Livestock are important as a source of nutrients for crop production in all mixed farming systems. In West Africa, customs exist that encourage pastoralists who do not grow crops to use their animals to manure crop farmers' fields for cash, goods or permission to graze the stubble. Poverty-stricken farmers cannot afford chemical fertilisers, and manure may be the only source of nutrients for their fields. Research on integrated land management has generated manure management techniques (such as composting and use of liquid manure and effluent from biogas production) that have increased the value of manure in soil fertility. Increased integration between crops and livestock in many smallholder, mixed systems has been shown to be one of the major pathways out of poverty. Lekasi *et al.* (2001) give good scientific evidence for this, showing how improved husbandry and manure management practices can improve land productivity. Another economic benefit linked to crop-livestock integration in one homestead is the reduction in transaction costs for each enterprise.

Livestock as investment, insurance and tokens of relationship

Beyond their role of providing food and inputs for agriculture, livestock are important as 'savings' or 'investments' for the poor, and provide security or insurance through various ways in different production systems.

- Livestock allow what economists call 'consumption smoothing', because they provide food (dairy products) for almost the whole year and because they can be sold to buy food and other necessities at any time of year; unlike crops, which are highly seasonal. Crop harvests do not necessarily coincide with needs for cash.

- Livestock are an excellent way of accumulating wealth over the years or even over generations in systems where other investment opportunities may be few or untrustworthy. Many remote areas populated by livestock-keepers have no banking systems, and in many countries livestock have been a better investment than bank accounts in unstable and depreciating national currencies.

- Livestock act as an insurance against droughts that plague many dry-land areas, although livestock themselves are extremely vulnerable to drought. Much literature exists on this and related questions such as:

  - For agro-pastoralists and dry-land mixed farmers, whether livestock really are an important 'buffer stock' (Fafchamps *et al*., 1998).

  - For pastoralists, whether livestock losses from drought are proportionately greater for poor than wealthy livestock-keepers.

  - For livestock-keepers, what are the best strategies in the face of looming drought (e.g. selling or retaining livestock) and how are these strategies affected by real-world variations in marketing opportunities and external action (Morton and Barton, 2002).

- Livestock are a means of creating and maintaining social relationships, through marriage payments (such as lobola, which is widespread in Southern Africa), as an allotment to children or wives, and as traditional forms of livestock loans. Livestock loans can be made to social equals, where they cement relationships but also spread risk, as the loaned livestock may be herded in more than one microenvironment. Livestock loans are also made to the poor and the destitute as a form of charity or for social solidarity with the poor - though often there is also some advantage to the lender in spreading risk or attracting labour. The terms of such loans vary. The recipient generally has rights to the produce (milk, eggs, manure or draught power), but may also have rights to a portion of the offspring. Where this is the case, these customs can become the basis for indigenous restocking strategies, although they have severe limitations where drought or some other disaster strikes the wider community.

Livestock in climatically marginal environments

Livestock are of particular importance for people who live in marginal environments - who are often among the poorest - both because of the difficulties they face eking out a livelihood from such environments, and because they are often far from power centres and markets. Marginal environments are principally high mountain areas and arid zones (broadly defined by average annual rainfall of below 400 mm, but the shortness of the growing season, the variability of rainfall, and the lack of surface water are also important). Such areas constitute significant proportions of the world, particularly in developing countries.

Livestock are primarily important in these areas because of their mobility - they can make use of water and forage resources scattered in low concentrations across the landscape, and also make use of growing seasons too short for food crops. These environments favour particular species and breeds of livestock - goats, desert-adapted sheep, camels, donkeys and some breeds of zebu cattle. This mobility has given rise to the traditional ways of life referred to as pastoralism, agro-pastoralism or transhumance, where some or all of the livestock-keeping household moves with the livestock, and where traditional knowledge and flexible forms of property, such as common-property regimes, are important. Such people are sometimes wealthy in numbers of stock but many live at or just above the subsistence level and are also vulnerable to the vagaries of climate and misconceived government policies.

One feature of the use of marginal environments by livestock-keepers is that often they need to use the less marginal environments at certain times of the year. Dry land pastoralists need more humid refuges for the dry season, and montane pastoralists need lowland refuges for winter. Failure to recognise this fact, as demonstrated by the encroachment of arable cropping on these refuges, not only further undermines the livelihoods of poor and vulnerable people, but also prevents any productive use of the most marginal areas.

What is the future of livestock keeping by the poor in developing countries?

Livestock keeping contributes to on-farm diversification and intensification, which can be a strategy for poor households to escape poverty. However, the vulnerability of the poor in terms of access to economic and social assets and services calls for multisectoral support to transform them from subsistence to market producers. An important rural development dimension closely linked to livestock keeping by the poor focuses on the effects of policy and legal and institutional aspects on farm production. Without favourable policies to promote access to markets, to market information and to financial services, it may be difficult for the poor to penetrate the competitive global market. Sub-Saharan Africa has much to learn from South Asia, where huge investments have been allocated to vertical integration, i.e. linking farm production to agro-processing, value adding and marketing. Availability of tools for feed conservation and product processing, and infrastructure improvement, are all factors that can support livestock keeping by the poor.

Further reading

Cole, H.H. and Garrett, W.N. (ed.) 1980. *Animal agriculture. The biology, husbandry and use of domestic animals*. W.H. Freeman and Company, San Francisco, USA.

Delgado, C., Rosegrant, M., Steinfeld, H., Ehui, S. and Courbois, C. 1999. *Livestock to 2020: the next food revolution. Food, Agriculture, and the Environment Discussion Paper No. 28*, International Food Policy Research Institute (IFPRI), Washington D.C., USA.

Dorward, A.R. and Anderson, S. 2002. Understanding small stock as livelihood assets: indicators for facilitating technology development and dissemination, *Report on review and planning workshop 12th to 14th August 2002*, pp. 4-7, Imperial College, Wye, UK.

FAO (Food and Agriculture Organisation of the United Nations). 2003. *Pro-poor Livestock Policy Initiative*. http://www.fao.org/ag/againfo/projects/en/pplp/home.html

LID (Livestock in Development). 1999. *Livestock in poverty focussed development*. Livestock in Development, Crewkerne, UK.

Peacock, C.P. 1996. *Improving goat production in the tropics. A manual for development workers*. Oxfam/FARM-Africa, Oxford, UK.

Thornton, P.K., Kruska, R.L., Henninger, N., Kristjanson, P.M., Reid, R.S., Atieno, F., Odero, A.N. and Ndegwa, T. 2002. *Mapping poverty and livestock in the developing world*. International Livestock Research Institute (ILRI), Nairobi, Kenya.

UNDP (United Nations Development Programme). 1997. *Human development report 1997: poverty from a human development perspective*. Oxford University Press, Oxford, UK.

World Bank. 2001. *World Bank Development Report 2001. Attacking poverty: World Bank Development Report 2001*. WDR 2000/2001. http://www.worldbank.org/poverty/wdpoverty/

Author addresses

Aichi Kitalyi, Regional Land Management Unit (RELMA) at World Agroforestry Centre (ICRAF), P.O. Box 30677, Nairobi 00100, Kenya

John Morton, Natural Resources Institute (NRI), University of Greenwich at Medway, Chatham Maritime, Kent ME4 4TB, UK

Louis Mtenga, Department of Animal Science and Production, Sokoine University of Agriculture, P.O. Box 3004, Morogoro, Tanzania

Anni McLeod, AGAL, Food and Agriculture Organisation of the United Nations (FAO), Viale delle Terme di Caracalla, 00100 – Rome, Italy

Philip Thornton, International Livestock Research Institute (ILRI), P.O. Box 30709, Nairobi 00100, Kenya and Institute of Atmospheric and Environmental Sciences, University of Edinburgh, Edinburgh EH9 3JG, UK

Andrew Dorward, Department of Agricultural Sciences, Imperial College London, Wye, Ashford TN25 5AH, UK

M. Saadullah, Formerly of Department of Animal Science, Bangladesh Agricultural University, Mymensingh 2202, Bangladesh.

3

Livestock systems

Canagasaby Devendra[*], *John Morton*[†], *Barbara Rischkowsky*[†], *Derrick Thomas*[‡]

Cameo - Livestock in a pastoral system in Sudan

The Northern Beja of Halaib Province, Sudan, rely on livestock and a pastoral system to survive in an area of extreme aridity (<25-100 mm annual rainfall). Two overlapping rainfall systems, non-pastoral livelihood opportunities on the Red Sea Coast and a relatively high fertility in some wadi areas caused by run-off, give rise to complex 'systems' of pastoral migration. Some households migrate over 300 km twice a year in a quasi-regular pattern, whilst others are sedentary. The Beja have a strong ethic of home territory and permission to graze or browse elsewhere must be sought from the owners, though it is given freely. Accurate knowledge of recent rainfall and grazing conditions is at a premium and is passed on in the context of a 'greetings ceremony', but is liberally shared among adult men. The Northern Beja are famous for their camels, exported in large numbers to Egypt and the Gulf States both for slaughter and for riding. Camels become habituated early in life to very specialised diets varying by area: mangrove-like *Suaeda* species, the succulent shrub *Salvadora persica*, and *Acacia* browse. Sheep and goats probably make a greater contribution to livelihoods, being sold in Port Sudan or unofficially exported. Much of their diet comes from *Acacia* and similar species: flowers and pods are shaken from the trees by herders, and some species and sub-species are heavily, though sustainably, lopped for foliage. The staple diet of the Beja is sorghum-based, but large amounts of dairy products are also consumed. Some people practise a simple form of flood recessional cultivation, but most of the grain consumed is obtained with the proceeds of livestock sales (Morton, 1988; 1990).

---

[*] Principal author
[†] Co-author
[‡] Consulting author

Introduction

The livestock systems that have evolved in any particular location are a function of the agro-ecological conditions and farming systems (Duckham and Masefield, 1970; Spedding, 1975; Ruthenberg, 1980; Jahnke, 1982). Climatic, edaphic and biotic factors determine whether cropping is feasible and what crops can be grown. These, in turn, determine the link with animals through the quantity, quality and distribution of animal feed resources throughout the year.

Given the variations across regions in climate and farming systems, available species, production objectives and external factors, it is not surprising that there are many classifications of livestock systems. Nestel (1984) described types of livestock systems on a regional basis. Wilson (1995) used a farming system approach to classify crop-animal systems. Seré and Steinfeld (1996) used the agro-ecological zones (AEZ) described by Technical Advisory Committee (TAC, 1994) and provided a comprehensive description of global livestock production systems using quantitative statistical methods. Of the 11 systems reported, six are mixed farming systems. More recently, ILRI (2000) has broadly classified livestock systems into three groups: grassland-based systems, mixed crop-livestock systems, and industrial systems, which are implied in Seré and Steinfeld (1996), and used in the toolbox of the Livestock, Environment and Development [LEAD] Initiative (LEAD, 2003). In response to the Millennium Development Goals, DFID's Livestock Production Programme (LPP) re-classified production systems according to the specialisation of resource-poor keepers: landed milk producers, crop/livestock farmers and small stock keepers, landless urban livestock-keepers, and pastoralists and transhumants (LPP, 2000).

This chapter will focus principally on four main livestock systems - landless, mixed crop-based, agro-pastoralist, and range-based - which prevail in Asia, Sub-Saharan Africa (SSA), Central Asia (CA), West Asia and North Africa (WANA) and Latin America and the Caribbean (LAC). These four are the higher-order systems and therefore the most important, when considered from the standpoint of the diversity, distribution and concentration of animal species, the role of animal species within production systems, land use patterns, trends, extremes of poverty, environmental concerns and emerging issues. The chapter also identifies priorities for research and development (R and D) in each region and opportunities for improvement in the face of growing demand – production gaps.

Landless systems

Seré and Steinfeld (1996) defined landless systems as those where less than 10 per cent of the dry matter consumed is produced on the farm where the livestock are located, and where annual average stocking rates are above 10 livestock units (1 LU = 1 cattle or buffalo or 8 sheep or goats) per hectare of agricultural land. Furthermore, Seré and Steinfeld distinguished between landless mono-gastric and landless ruminant systems. The former are mainly industrial, intensive and vertically-integrated pig and poultry enterprises whose economic outputs are higher than those of ruminant enterprises. Examples of these are pig production in Asia and poultry production in Central and South America. In landless ruminant systems, the value of production of the ruminant enterprises is lower than that of the pig and poultry enterprises. Examples are sheep production systems in WANA, sheep-fattening operations in Nigeria, and nomadic systems in north India.

For mapping of industrial landless production systems Thornton *et al.* (2002) used the threshold of more than 450 persons/km^2 and further recognised landless metropolitan and landless systems according to the presence or absence of city lights in night-time satellite imagery. The authors (of this chapter) considered the threshold of population density (>450 persons/km^2) set by Thornton *et al.*, and that of livestock density (10 LU/ha) proposed by Seré and Steinfeld (1996) as too rigid, as these thresholds exclude landless rural livestock keepers, e.g. agricultural labourers in South Asia who often keep a few buffalo, cows or goats. A distinction is made between metropolitan and rural landless systems as this differentiates marketing and management practices which differ considerably between urban, peri-urban and rural farming.

Rural landless livestock production

Thornton *et al.* (2002) reported that rural landless systems were mainly found in South Asia (India and Pakistan), China and Indonesia (Java), but also in some countries in SSA and North Africa. With the exception of the densely populated Ethiopian Highlands and the coastal areas of Ivory Coast, most of the examples in Africa are peri-urban areas, where city lights were not prominent enough to appear on the night time satellite imagery that was used to differentiate between metropolitan and other landless systems. Geographically-stratified surveys are needed to identify the number of landless among the rural farmers.

The species kept depend mainly on traditional food preferences, availability of capital and (increasingly) market demand and opportunities. In India and Pakistan, landless agricultural labourers often keep buffalo or goats based on zero-grazing practices, grazing of roadsides, and hired land with forage or leguminous trees to harvest leaves and pods. In China, landless rural households often keep poultry and pigs.

Urban and peri-urban landless livestock production

In Asia, urban agriculture is well established, based on a long tradition of recycling waste for agricultural uses. There are modern and intensive production systems for poultry, pigs and fish. Grazing public land is officially accepted. Urban agriculture in Africa also has traditional roots, but its importance has become prominent only recently. The low density of built-up areas in many SSA cities even permits the raising of large livestock, although city by-laws often prohibit such activities. Landless livestock production systems involving pigs and poultry, dairy cattle and, to some extent, feedlot fattening, are mainly found in urban and peri-urban areas. The scale and intensity of production are often a response to increasing urbanisation, market opportunities and strong private sector participation.

Some case studies have reported the importance of urban livestock keeping. In Hong Kong, 10 per cent of the urban area is used to produce 45 per cent of the vegetables, 15 per cent of the pigs and 68 per cent of the chickens consumed in the city. Fish ponds occupy 31 per cent of all agricultural land. In the largest 18 cities of China, over half of the meat and poultry consumed is produced in the urban area. In Kathmandu, Nepal, 11 per cent of the animal products eaten were produced by urban farmers, and in Singapore 80 per cent of poultry products (UNDP, 1996).

Urban livestock production is a large industry, involving many small-scale farmers and some large agri-businesses. The systems are diverse and undergo a continuous process of change. A whole range of livestock is kept in cities; the choice will depend on traditional food preferences and capital availability, as in rural areas, but also on availability of space. Buffalo, cows, goats and sheep are raised under zero-grazing in backyards or grazed in parks, on road sides or open spaces. 'Mini' and 'micro livestock' (rabbits, guinea pigs, grasscutters, chickens and giant snails) are suited to the scarcity of space. Ducks and geese fit particularly well into aquaculture systems.

Monogastric production requires less space than ruminants and is more efficient in conversion of concentrates. These enterprises are knowledge- and management-intensive to satisfy an increasingly quality-conscious urban population. The use of land, labour and feed is maximised, backed by intensive husbandry, improved genotypes, veterinary care, and links to markets and feed mills, as is common in many parts of Asia. In WANA, subsidised industrial production units demonstrate state-of-the-art production, especially in poultry and dairy enterprises. These systems use many imported inputs, although locally-cultivated forage is also used for cows. In the Nile valley in Egypt, labour-intensive technology is applied in competitive dairy enterprises. Where agro-ecological conditions are more favourable in SSA, urban semi-intensive and intensive dairying systems use cultivated forage and agro-industrial by-products, whilst poultry production has started to become industrialised. In East Africa, in addition to the poor, a number of elite and civil servants are involved in urban livestock production. In Latin America, intensive poultry production and dairying are on the increase (Delgado *et al.*, 1999).

The demand-led expansion in landless livestock production has led to large concentrations of animals in urban environments, especially where regulations governing livestock production are weak, such as in Addis Ababa (Ethiopia), Beijing (China), Lima (Peru) and Mumbai (India) (Delgado *et al.*, 1999). Some associated problems are nuisance to neighbours (odour, noise, and manure on pavements), clogging of sewerage systems, traffic congestion and/or contamination of water sources (UNDP, 1996). Even more severe are the associated risks to human health, namely increases in zoonotic diseases such as tuberculosis (bovine tuberculosis from cattle and buffalo in South Asia), leptospirosis, anthrax, salmonellosis, brucellosis, trichinosis, swine flu from pigs and avian flu from poultry. Informal, unsupervised slaughtering of animals in the neighbourhood exacerbates the problems. Diseases are often ignored by resource-poor livestock-keepers, as treatment costs are high. Also there are often no adequate testing facilities, and farmers can easily evade the regulations of the public health systems. Environmental problems include overgrazing of native grasslands and the slashing of primary forests for agricultural use.

Future perspectives

Industrialised livestock production will continue to play an important role in meeting the expected increasing demands for meat and milk (the 'Livestock Revolution') in the cities and, to a lesser extent, in rural areas. In recent years such systems grew globally at twice the rate (4.3 per cent) of more traditional mixed farming systems (2.2 per cent) and more than six times the rate of grassland-based systems (0.7 per cent). Between 1982 and1994, higher annual increases in production were realised for poultry (7.8 per cent) and pork (6.1 per cent) than for beef (3.1 per cent) or milk (3.7 per cent). Pig and poultry production depend on imported grain in industrial production systems and are therefore vulnerable to availability, price changes, human requirements and alternative uses in the future (Delgado *et al.*, 1999).

Crop-based livestock systems

Crop-based farming systems, where crops and animals are integrated on the same farm, form the backbone of smallholder agriculture throughout the developing world. In global terms, these systems provide over 50 per cent of the output of meat and 90 per cent of milk (CAST, 1999). An estimated 678 million poor are livestock-keepers, of whom 57 per cent are on mixed rain-fed farms, 23 per cent are landless and 20 per cent practise extensive grazing (ILRI, 2000, based on Livestock in Development, 1999). These systems are especially dominant in both the irrigated and non-irrigated areas in humid and sub-humid environments. In Asia, more than 95 per cent of the total population of large and small ruminants, and a sizeable number of pigs and poultry, are reared on these small farms. The traditional small farm scenario is characterised by low capital input, limited access to resources, low levels of economic efficiency, diversified agriculture and resource use, and conservative farmers who are ill-informed of new technology, living on the threshold between subsistence and poverty.

Mixed farming systems provide farmers with an opportunity to diversify risk from a single commodity, to use labour more efficiently, to have several sources of cash for purchasing farm inputs and to add value to crops or their by-products. Combining crops and livestock also has many environmental benefits, including the maintenance of soil fertility by recycling nutrients, and providing entry-points for practices that promote sustainability, such as the introduction of improved forage legumes. Mixed farming systems maintain soil biodiversity, minimise soil erosion, conserve water, and make the best use of crop residues that might otherwise be burnt, leading to carbon dioxide emissions. The closed nature of mixed farming systems makes them less damaging or more beneficial to the natural resource base. It is in mixed farming systems that the best opportunities exist for exploiting the multi-purpose role of livestock (Devendra, 1995). Feed resources provide a direct link between crops and animals and the interactions between the two dictate, to a large extent, the development of the systems.

Types of crop/livestock systems

Two kinds of systems are identified: systems combining livestock with annual or with perennial crops. The potential of the latter is underestimated. Annual crops and perennial tree crops are grown in most parts of the developing countries, and both ruminants and non-ruminants are integrated into these systems. Examples of integrated annual crop/animal systems include rice/maize/cattle/sheep in West Africa; rice/wheat/cattle/sheep/goats in India; rice/goats/ducks/fish in Indonesia;

rice/buffalo/pigs/chickens/ducks/fish in the Philippines; rice/vegetables/pigs/ducks/ fish in Thailand; and vegetables/goats/pigs/ducks/fish in Vietnam. Examples of integrated perennial tree crop-animal systems include rubber/sheep in Indonesia; oil palm/cattle in Malaysia and Colombia; coconut/sheep/goats in the Philippines; and coconut/fruit/cattle/goats in Sri Lanka. In the WANA region, integration of sheep with wheat, barley, peas and lentils is common, together with olives and tree crops.

In rain-fed annual cropping systems, ruminants graze native grasses and weeds on roadside verges, on common property or on stubble after the crop harvest. There are few examples of improved pastures being utilised in these systems. Crop residues and by-products are fed throughout the year or seasonally, depending on the availability of grazing land. Animals are tethered, corralled or allowed free access to grazing. In areas of intensive cropping, stall-feeding is practised.

In the perennial tree crop-systems, ruminants graze the understorey of native vegetation or leguminous cover crops. Non-ruminants in these systems mainly scavenge in the villages on crop by-products and kitchen waste. However, village systems can evolve into more intensive production systems depending on the availability of feeds, markets, and the development of co-operative movements. This is evident in many parts of Central America, West Africa (e.g. Nigeria), South East Asia (e.g. Indonesia) and in South Asia (e.g. Bangladesh). In areas where root crops are produced, pig production is based on cassava and sweet potato.

Crop-animal interactions

The rationale for integrating crop and livestock production is associated with marked complementarities in resource use, with inputs for one sector being supplied by others, such as using draught animal power and manure for crop production and crop residues as feeds. The benefits of crop-livestock interactions are many (Table 3.1). Animal traction can improve the quality and timeliness of farming operations, thus raising crop yields and incomes. The transfer of nutrients from grazing lands to croplands through manure contributes considerably to the maintenance of soil fertility and the sustainability of the farming systems. Livestock provide a least-cost, labour-efficient route to intensification. The types of crop-animal interactions are similar in all mixed farm systems, but the extent and implications of the interactions will differ in different agro-ecological zones. In Asia, for example, the benefits of these interactions are considerable (Devendra and Thomas, 2002; Devendra and Chantalakhana, 2002).

Table 3.1 Main crop-animal interactions in crop-based livestock systems

Crop production	*Animal production*
Crops provide a range of residues and by-products that can be utilised by ruminants and non-ruminants.	Large ruminants provide power for operations such as land preparation and for soil conservation practices.
Native pastures, improved pastures and cover-crops growing under perennial tree crops can provide grazing for ruminants.	Both ruminants and non-ruminants provide manure for the maintenance and improvement of soil fertility. In many farming systems it is the only source of nutrients for cropping. Manure can be applied to the land or, as in South-East Asia, to the water which is applied to vegetables whose residues are used by non-ruminants.
Cropping systems such as alley-cropping can provide tree forage for ruminants.	The sale of animal products and the hiring out of draught animals can provide cash for the purchase of fertilisers and pesticides used in crop production.
	Animals grazing vegetation under the tree crops can control weeds and reduce the use of herbicides.
	Animals provide entry-points for the introduction of improved forages into cropping systems. Herbaceous forages can be undersown in annual and perennial crops and shrubs or trees established as hedgerows in agro-forestry-based cropping systems.

Source: Devendra and Thomas (2002).

The Three-Strata Forage System (TSFS) in Bali, Indonesia, is an example of how integrating crops with animals provides many benefits. The TSFS is a people-centred smallholder system for dry areas (eight months dry and 1000 mm annual rainfall), involving grasses and herbaceous legumes, shrub legumes, and forage trees in strata one, two, and three respectively. The TSFS aims to enhance year-round feeding and increase productivity in integrated systems involving food cropping (cassava, groundnuts and beans) and ruminants (cattle and goats). Nitis *et al.* (1990) reported that the additive effects of integration and crop-animal-soil interactions from applying TSFS resulted in many improvements: increased forage production, higher stocking rates and total annual weight gains (375 kg/ha compared to 122 kg/ha), 57 per cent less soil erosion, 64 per cent self-sufficiency in household fuel-wood requirements, 31 per cent more farm income, and economically-stable farm households. The concept and technology originally applied to 32 farmers were subsequently extended to another 144 farm households. The technology has now been institutionalised and officially promoted in Indonesia.

Under crop-based livestock systems, it is also pertinent to draw attention to the pasture-based systems that are common in many parts of Latin America, such as Brazil and Colombia, and also in Mongolia. In these systems, ranching is common and is a relatively low-cost way of producing beef. Uncontrolled grazing is the norm, but in recent years there has been an increasing shift to more intensive systems to produce both beef and milk through the use of dual-purpose cattle.

Photo 3.1 Forages being fed to a Bali cow in the Three-Strata Forage System (TSFS), in Bali, Indonesia. Crop-based mixed farming systems will continue to be dominant in Asia, LAC and some parts of SSA. (courtesy of C. Devendra)

Future perspectives

Although meat and eggs from non-ruminants in industrial systems continue to be the main source of proteins, it is unlikely that these systems will meet all the projected demands of the Livestock Revolution. Consumption of ruminant meats and milk is increasing, and systems of production will need to be made more efficient. It is suggested that crop-livestock systems will see important growth in the future and will remain the dominant systems in Asia (Devendra, 2002) and elsewhere in the humid tropics. It is further suggested that mixed farming systems will continue to be the main avenue for intensification and specialisation of food production. Some environmental concerns exist and include land quality in terms of erosion and soil fertility, but domestic animals will continue to enhance the natural resource base, the livelihoods of poor people who own them, and sustainable agriculture.

On the other hand, Thornton *et al.* (2002) envisage that current areas of mixed systems in close proximity to landless systems could change to the latter by 2050, due to increased crop-animal intensification through spatial integration, increased market integration and income diversification, and greater opportunities for off-farm income generation. This shift remains to be seen, and will be constrained by available feed resources, high capital investment needs and strong private sector participation.

Agro-pastoralism

Agro-pastoralism is a widely used but imprecise term. Wilson *et al.* (1983) produced the definition of pastoralism discussed below and Swift (1988) used it to describe production systems where more than 50 per cent of household gross revenue comes from farming, and 10-50 per cent from pastoralism. That definition might cover a great range of mixed farming systems, including those heavily reliant on cultivated forages, but in practice 'agro-pastoralism' is used to describe systems where livestock are dependent on natural forage, and where other indicators of crop-livestock integration, such as the management of manure for soil fertility and the use of draught animal power, are at a low level. In particular, it is used for systems where cropping is important, but where some individuals, or the entire household, leave the home, and the crop fields, at certain times of year to migrate with the livestock. Agro-pastoralism as a term is usually used to describe systems in Africa, where agro-pastoral systems (or mixed crop-livestock systems with extensive grazing and a low level of integration) are found over large areas. These are very diverse and can indicate different trajectories towards or away from specialised livestock production, as pastoral communities adopt agriculture because of changing dietary habits, or insecurity of livestock production, and farming communities invest in livestock and adopt different strategies, involving family members or hired herders, to manage it extensively. This dual movement into agro-pastoralism is a feature of the Sahel countries, where it is the dominant form of livestock production. Agro-pastoralism can be a relatively stable adaptation, as with the Nuer and Dinka peoples of the Sudd swamp in Southern Sudan, whose lives and cultures are dominated by cattle that are moved seasonally to take advantage of grazing and avoid floods, but who in fact rely heavily on small cultivated plots. In other cases, agro-pastoralism represents a diversification from settled agriculture. The Konso of Southern Ethiopia, for example, whose traditional terraced hillside agriculture is intensive and involves stall-feeding relatively small numbers of cattle, are increasingly using a system of mobile cattle camps in lower-lying rangeland areas, managed by men and boys on their own. In parts of Southern Africa, agro-pastoralism is the dominant form of livestock production for small farmers. In Botswana, farmers maintain 'cattle posts' on open rangeland, managed by family members or (increasingly) hired herders, and which can be moved to take advantage of new grazing, always at some distance from the villages and the cultivated lands. In South-West Zimbabwe, livestock-owners made long-distance migrations with their cattle at least up to the 1970s. Dry-land farming is now the norm, using draught oxen or donkeys and keeping livestock in kraals at night, but livestock mainly graze on communally-owned, open rangeland by day. The use of crop residues for feeding is of significance for only short periods of the year and even then is not particularly carefully managed. As can be seen, the point at which an agro-pastoral system can be regarded as a crop-livestock system is largely a matter of definition.

It is important to stress that livestock production in many of the higher rainfall areas of Africa, where integrated mixed farming might otherwise be possible, is severely limited by the presence of tsetse fly and trypanosomiasis, which increase the relative importance of agro-pastoralism (in semi-arid, tsetse-free areas) for the continent.

Future perspectives

According to one view, there is a trend shift from pastoralism to agro-pastoralism, and from agro-pastoralism into mixed integrated farming, due to population growth and stress on natural resources. This process is dynamic and its evolution will be influenced by various crop-animal interactions, as well as by population and resources (Scoones and Wolmer, 2002).

Range-based livestock systems

A number of different, overlapping and sometimes confusing terms are used to describe livestock-production systems that depend on the use of open rangelands.

This confusion has arisen because classifications are being used to answer several different questions:

- What proportion of animal feed comes from rangelands?
- Do livestock-producing households, or some individuals within them, traditionally migrate with their livestock?
- Who owns the rangelands: are they private property, common property or open-access?
- To what extent do households depend on livestock for their survival?

The first of these questions gives rise to a category of 'grassland-based systems' (Seré and Steinfeld, 1996), also known as 'extensive grazing systems' (LID, 1999) or simply 'grazing systems' (LEAD, 2003). These are defined as 'Livestock systems in which more than 90 per cent of dry matter fed to animals comes from rangelands, pastures, annual forages and purchased feeds and less than 10 per cent of the total value of production comes from non-livestock farming activities. Annual stocking rates are less than 10 livestock units (1 LU = 1 cattle or buffalo = 8 sheep or goats per hectare of agricultural land' (LEAD, 2003). A more precise definition of LU would be to include weight of animals, e.g. 250 kg for cattle. Grazing systems in turn can be subdivided according to the agro-ecological zone in which they are found: arid, semi-arid, sub-humid and humid temperate, and tropical highland.

The second question gives rise to terms like 'nomadism', 'semi-nomadism' and 'transhumance' depending on the regularity of migrations, and the extent to which the household maintains a home base (Johnson, 1969). However, such terms fail to capture the complexity of migration strategies, and the way in which they can vary over time and between households of the same community. These terms are now used less than they were.

Answering the third question involves terms like 'ranching', where large areas of rangeland are the private property of individuals or companies, and 'group ranching', where they are the formal private property of small groups. Private ranches are often historically the result of seizure of, or encroachment on, lands used by other, poorer livestock-producers, who may as a result be forced into less than optimal rangeland use. The reality in many parts of the developing world is that 'ranchers' can be surprisingly flexible about others (kin, neighbours, or simply 'poach grazers') grazing their animals on their land, but can also rely on grazing their own animals outside their ranches (e.g. the so-called 'dual-grazing' problem in Botswana). Group ranching was proposed as a means of development for traditional rangeland users, but has proved controversial, impeding flexibility of resource use and favouring the more affluent strata. 'Common property' is a term used in different ways by different writers, and is frequently confused with 'open-access'. Extensive grazing systems in the real world frequently rely on combinations of private, common (in the strict sense) and open-access grazing and browse resources (Behnke, 1994; Mendes, 1988).

Answering the fourth question has given rise to the system classifications most used in development literature, those of 'pastoralism' and 'agro-pastoralism'. Wilson *et al.* (1983) and Swift (1988) defined pastoral production systems as those in which 50 per cent or more of gross household revenue (i.e. the total value of marketed production plus the estimated value of subsistence production consumed within the household) comes from livestock or livestock-related activities. This in itself does not define a livestock production system, but it is generally assumed that "most of the feed that pastoralists' livestock eat is natural forage rather than cultivated forage or pasture" (Sandford, 1983). This helps distinguish pastoralists from specialised landless or urban livestock producers.

The definition of pastoralism emphasises that it is not just a livestock production system but also a form of livelihood. Sandford states "pastoralists devote the bulk of their own, and their families', working time and energy to looking after livestock rather than to other economic activities." Pastoralism is also a culture; livestock serve not only practical but also social ends, such as expressing status and forming social ties. Some authors therefore maintain that the definition needs to be expanded to include members of pastoralist ethnic groups who have been forced to withdraw from pastoralism, and agro-pastoral groups who hold to pastoral values (Baxter, 1994; Morton and Meadows, 2000). Since the definitions are hazy, and because pastoralists are often poorly covered by national censuses, it is difficult to produce reliable population figures. Global estimates vary from 20 million individuals to 20 million households (Blench, 2001).

Pastoralists make use of areas where cropping is difficult or impossible, i.e. areas of low and variable rainfall or of high altitude, although they may live and herd for some of the year in areas of higher potential. While mobility is not part of the definition of pastoralism, it is a characteristic of many pastoralist systems, and determines many aspects of the pastoral situation; the need for flexible systems of land tenure, the difficulty governments face in delivering appropriate human or animal services, and the general marginalisation and distrust that pastoralists often suffer.

Pastoralists are found in every continent. African pastoral systems, especially in the Sahara, Sahel and North-East Africa are the best known and best described. This is particularly because of the impacts of recurrent drought, and the common but problematic assumption that overgrazing by pastoralists is responsible for desertification. Many important studies of pastoralists have been produced by anthropologists; a more technically-based study is that of Coppock (1994) on the Boran of Ethiopia. This shows the Boran as in the grip of recurrent drought and progressive impoverishment, and points to developing market linkages and alternative (non-livestock) forms of investment as a way out. Pastoralists are also found in the Middle East and in parts of South Asia – the more arid areas of Western India and Pakistan, and in the Himalayas (Sharma *et al.,* 2003; Morton, 1994). Himalayan pastoralists winter in lower-lying areas, then ascend the great Himalayan valleys in spring to take advantage of high-altitude alpine pastures when the snows melt. They may maintain a variety of relationships with settled farmers near their winter quarters and along their route. The transformations in Central Asia, Mongolia and neighbouring parts of China have re-focused attention on the ancient but highly-adaptable pastoral cultures of those areas (Humphrey and Sneath, 1996). In Mongolia, after years of collectivisation, when herders were employed in specialist single-species production teams, livestock were redistributed to herding households, who generally tried to return to the tradition of mobile multi-species herding of cattle (or yak in higher altitudes), horses, sheep, goats and Bactrian camel in desert areas. New forms of communal management of rangeland are slowly emerging to suit greater mobility. Development has been hampered by a series of extreme winters for which old Soviet-era mitigation measures are no longer in place.

Future perspectives

There is no doubt that pastoralism worldwide is deeply threatened as a livestock production system (Blench, 2001) and as a livelihood, by interlinked pressures. These include inappropriate government policies on land tenure leading to encroachment on rangelands by cultivation, private ranching and protected areas, failure to develop appropriate livestock and human services, recurrent drought and (in Africa) armed conflict. Previous views that common-property grazing systems were inevitably associated with overgrazing have been subject to challenge in recent years (see papers in Behnke *et al.,* 1993) but with these policy pressures, and with human and livestock population increase, the capacity of rangelands to support livestock and livelihoods is diminishing. Yet pastoralism remains the only sustainable way to make use of large areas of the earth's surface too dry, cold or mountainous for cultivation, and contributes significantly to livestock production and export earnings in some countries. Ways need to be found to sustain and encourage pastoralism to adapt to new conditions; ways which will include better policies, better design of services, and diversification of pastoral livelihoods (Morton and Meadows, 2000).

Livestock systems and future research and development issues

It is appropriate to relate the earlier discussion on the four main types of livestock systems to different regions and to the main issues that impinge on future R and D in each region.

Asia

Asian livestock production systems will continue to evolve. Intensive pig and poultry systems will continue to be the dominant systems, driven by the private sector because of high capital and resource inputs and economies of scale, and will contribute the major share of animal protein production. On the other hand, the ruminant sector (buffalo, cattle, goats and sheep) presents major opportunities for expanding production through improvements to systems in priority agro-ecological zones. In this context, the less-favoured rain-fed areas, with a total of about 268 million ha in the arid, semi-arid, sub-humid and humid zones, have enormous potential.

Crop-animal systems involving annual crops and perennial tree crops will continue to be the main farming systems throughout the region. These systems will become increasingly intensive, diversified and specialised. There will be attendant varying levels of economic, social and environmental costs, which need to be taken into account in the process of prioritisation and resource allocation. A review of opportunities for improved livestock production in South-East Asia (Devendra *et al*., 1997) and South Asia (Devendra *et al.*, 2000) identified problems facing farmers and issues for R and D in the following priority production systems:

South-East Asia

The integration of animals with annual food crops (e.g. rice, maize, root crops).

The integration of animals with perennial tree crops (e.g. coconuts, oil palm and rubber).

South Asia

Dairy production systems in rain-fed and irrigated mixed farming systems.

Goat and sheep production systems linked to annual crops in semi-arid and arid areas.

Among ruminant species, dairy production is expanding rapidly in most parts of Asia, followed by beef. Goat and sheep production are more at the subsistence level, but offer great possibilities for intensive commercial production, especially in South Asia.

Problems facing farmers and issues for R and D

- Increasing efficiency in natural resource management involving integrated systems with annual and perennial crops. The integration of ruminants with perennial crops is underestimated.

- Improving understanding of the implications of crop-animal soil interactions to include socio-economic issues.

- Managing feed resources and identifying year-round feeding systems that can sustain higher productivity.

- Improving understanding of nutrient flows, and the use of animal manure for the maintenance of soil fertility for crop growth.

- In mixed farming systems, indigenous breeds are especially important for smallholders. There is a need for better understanding of genotype x nutrition x environment interactions and the effects of animal health interventions.

- From the standpoint of a poverty focus, small ruminants play a critical role in the livelihoods of marginal farmers and the landless, especially in the harsher arid and semi-arid areas. The landless are totally dependent on common property resources for grazing, with associated environmental degradation. Research is needed on socio-economic and policy issues that address these complexities and also on technological interventions.

Sub-Saharan Africa

Livestock production in SSA is characterised by systems which are intrinsically difficult to define and which merge into each other geographically and over time.

They cannot be attributed to entire communities or ethnic groups, as individuals with different resource endowments follow different strategies. In this situation it is very difficult to attach population figures to livestock systems. However, data from Seré and Steinfeld (1996) suggest that in SSA, of the seven systems identified, the greatest numbers of cattle, goat and sheep were found in the first three (grassland-based arid and semi-arid, mixed rain-fed humid and sub-humid, and mixed rain-fed arid and semi-arid systems). Whilst mixed systems in the higher potential (i.e. higher rainfall) areas of West and Central Africa account for the highest human populations, these systems have lower livestock : human ratios and account for relatively small livestock populations. This is due to the greater income per hectare from crops, and animal health constraints such as trypanosomiasis. Grazing systems, which include pastoralism and agro-pastoralism, account for large populations of livestock.

In the arid and semi-arid grassland system and all mixed systems, goats and sheep together outnumber cattle. This is important to note, because the traditional cultural importance and higher unit value of cattle often obscure the greater contribution of goats and sheep to the livelihoods of the poor. There are also important populations of camels (in North-East Africa and the most arid parts of West Africa), donkeys (in Ethiopia and Southern Africa) and pigs (in Madagascar). Poultry and pigs are important to the poor in all sedentary production systems.

Problems facing farmers and issues for R and D

- Clarifying the concepts of 'overgrazing' and livestock-related desertification, and relating them to socio-economic and policy contexts, to develop ways to mitigate environmental degradation in grazing systems, particularly in the Sahel and Southern Africa.

- Coping with drought by the best combinations of early-warning, long-term policies and short-term mitigation measures.

- Identifying appropriate feeding strategies for grazing and mixed systems to overcome the dry-season bottleneck.

- Enabling small livestock-producers in higher-potential and peri-urban systems to capitalise on rising demand for dairy and other livestock products.

- Assisting poor people in highland mixed farming systems to manage increasing shortage of land and feed, and declining soil fertility.

- Promoting improved understanding of available technologies, and the costs of production.

Central Asia

Central Asia comprises Kazakhstan, Kyrgyzstan, Tajiikistan, Turkmenistan and Uzbekistan, as well as Armenia, Azerbaijan and Georgia. Except for the productive crop farming in the moist sub-humid agro-ecological zone in the west, the larger part in the arid and semi-arid zone has only limited production potential.

Mixed farming systems

The mainly rain-fed mixed farming systems cover approximately 25 per cent of the total land area (Thornton *et al.*, 2002), sustaining 35 per cent of the total population. They are found especially in the more humid parts of the steppes in Northern and Southern Kazakhstan, Turkmenistan and Uzbekistan. Livestock, especially cattle and sheep, play an important role in the mixed systems in the drier areas and in pastoral systems. In the more arid areas, cereal cultivation and sheep raising are common. Large-scale farms, mostly collective, are still the dominant production structure. Since existing water resources are already over-exploited, reversion to some form of pastoralism in the most arid parts is unavoidable.

Pastoral livestock farming

The system is typical for much of South-East Central Asia. Thornton *et al.* (2002) categorised at least 51 per cent of the total land area as rangeland-based livestock production systems in arid and highland regions. Principal livestock species are sheep, goats or cattle and camels. Communal pastures close to the villages are grazed in spring and autumn; summer grazing is on distant mountain pastures, whilst zero-grazing predominates in winter. Since independence in 1991, all farming systems have undergone transformation. The collectivisation which began in the 1920s has replaced the traditional nomadic pastoral system. All range and arable lands were consolidated into large state-owned farms or collectives on which the pastoralists were forced to become sedentary and be labourers. Livestock production was oriented towards exporting livestock and livestock products to Russia (Miller, 2001; Kerven *et al.*, 2003). Ownership patterns are now in transition from collective state farms to cooperatives. Only a few small farms exist. The dramatic decline in the sheep and cattle populations (but not goats) between 1992 and 2002 (FAO, 2003), suggests that poverty is now widespread among pastoral livestock-keepers in this area (Dixon *et al.*, 2001).

Problems facing farmers and issues for R and D

- Promoting markets for livestock products, particularly animal fibres.
- Developing strategies for the sustainable use of rangelands with herders, using survey data and geographical information system techniques.

- Identifying appropriate winter-feeding strategies for emerging sedentary forms of sheep husbandry.
- Rehabiliting salinised and desertified areas by planting stress-tolerant halophytic plants.

West Asia and North Africa

The WANA region stretches from Morocco in the west to Pakistan and Afghanistan in the east, and from Turkey in the north to Yemen in the south. The principal types of production systems in WANA (irrigated farming, rain-fed mixed farming in drylands and highlands, pastoral and urban farming) do not differ from Central Asia, but smallholder agriculture is predominant (Dixon *et al.,* 2001; Thornton *et al.,* 2002).

Rain-fed mixed farming systems

Thornton *et al.* (2002) classified 12 per cent of the total land area as under rain-fed mixed farming systems, sustaining about 37 per cent of the population. Small ruminants are integrated with cropping to produce meat, yoghurt, cheese and ghee. Wool is of little economic importance. Dairy cows are kept in close proximity to cities. The drier arable areas (receiving an annual rainfall of between 200 and 350 mm) form the so-called barley belt (Thomson, 1987). Barley is the principal crop and is sometimes grown in rotation with wheat; the rotation also used to include a year-long fallow. The incidence of drought is high and food insecurity is an issue. Small ruminants are an important component of the system and are integrated closely with cropping, in which rain-fed barley is grown for grain and lightly grazed in spring. When there is insufficient rainfall, the whole crop will be grazed instead of harvesting it. Cereal stubble is grazed in summer, and cereal straws, barley, and wheat bran are used as winter feeds.

Pastoral livestock keeping

Pastoral livestock keeping of mainly sheep and goats, but also some cattle and camels, is found throughout the WANA region where annual rainfall is low (<150-200 mm). Where water is available, small areas are cropped to supplement the diets and income of pastoral families. However, the main income is from cheese and ghee and the sale of lambs at weaning. Pastoral livestock keeping is closely linked to the rain-fed and irrigated farming systems through the seasonal movements of sheep flocks to graze barley and wheat stubbles in summer and cotton residues in autumn, with peri-urban feedlots for fattening. In the highlands of Iran and Morocco, transhumant owners of sheep flocks migrate seasonally between lowland steppe in the more humid winter and upland areas in the dry season (Dixon *et al.*, 2001).

The highest proportion of poor people is represented by farmers or pastoralists. They are found in the drier areas and depend exclusively on agriculture, especially livestock, for food and their livelihoods. Drilling bore-holes and wells enable supplementary winter irrigation, and cash crops are grown in summer. Thus intensification has helped to compensate for the decreasing farm sizes due to rapid population growth (Rischkowsky *et al.*, 2003).

Problems facing farmers and issues for R and D

- Promoting more efficient use of supplementary irrigation for rain-fed farms.
- Introducing improved Awassi sheep and Shami goats to use the available forage legumes grown in rotation with cereal cropping.
- Developing common property resource management strategies to regulate pressure on the steppe rangelands.
- Creating off-farm income opportunities to reduce the dependence on natural resources in the drier cropping areas and to reduce the fragmentation of farms.

Latin America and the Caribbean

Given the heterogeneous nature of the region, several smallholder livestock systems are identifiable in the humid tropics and sub-tropics, sub-humid tropics and sub-tropics, arid and semi-arid tropics, and cool tropics. Crop-animal systems are common, especially on small to medium sized farms, as are also dual-purpose cattle in all agro-ecological zones. These mixed farms contribute 26 per cent of the beef and 50 per cent of the milk produced in Latin America (Janssen *et al.*, 1991). In the humid and sub-humid tropics, specialised dairy and beef cattle production is practised in the irrigated areas, such as northern Mexico and the Brazilian Cerrados. Goats and hair sheep are reared extensively by smallholders, especially in the arid and semi-arid regions, such as North Mexico, North-East Brazil, coastal Peru and North Chile. Wool sheep and camelids are found in the Altiplano (highlands). Cereals, root crops and bananas and plantains are common among the crops grown, and the residues from these constitute very important feed resources for livestock.

Among the livestock systems, pasture-based mixed systems will continue to be dominant, with cattle production for beef and dairy especially important.

Specialised ranching will also continue to be important, but there is already a shift from ranching to dual-purpose (beef and dairy) production. Especially on small farms, the development of more sustainable and specialised crop-animal systems, involving maize, rice, root crops, beans, bananas and dual-purpose cattle is also increasing. In addition, the development of silvi-pastoral systems is expanding in countries such as Costa Rica, Colombia and Venezuela.

The development of dual-purpose cattle will involve further shifts towards intensification and specialisation for both beef and dairy production in the future. Improved dairy production systems are especially important in view of the broader benefits of immediate cash income and impact on household nutrition. In the more arid and semi-arid areas, commercial goat production for meat and milk may become increasingly important through intensification and specialisation, if marketing can be more organised and target regional needs. Opportunities for improved livestock production systems include ranching, dual-purpose and specialised beef and dairy production, development of sustainable crop-animal systems based on cash crops and ruminants, and development of sustainable silvi-pastoral systems (Devendra and Pezo, 2004).

Problems facing farmers and issues for R and D

- Improving understanding of the shift from ranching to dual-purpose cattle (beef and milk) production systems and the link to socio-economic and policy issues.
- Reducing overgrazing and using the land better.
- Integrating cash crops and forage legumes in the development of food-feed systems.
- Developing silvi-pastoral systems that integrate livestock with agroforestry.

Major emerging issues

Table 3.2 summarises the type of livestock production systems, priority production systems, and the major issues relevant to the latter, across regions. The emerging issues are varied and numerous, and the major ones are highlighted. Research and development on these will obviously need to be addressed in interdisciplinary terms. Table 3.2 also highlights the role and potential importance of individual animal species in the different regions.

Table 3.2 Summary of livestock systems, priority production systems and major issues across regions

Livestock systems	*Priority production system*	*Regions*					*Major issues*
		Asia	*SSA*	*CA*	*WANA*	*LAC*	
Landless	• Peri-urban/urban dairy production	√	√		√	√	• Surface water contamination
	• Peri-urban/urban poultry and pig production	√	√	√	√	√	• Zoonoses • Waste disposal
	• Feedlot (cattle or small ruminants)	√	√	√	√	√	• Nutrient flows
	• Goat and sheep production	√	√	√	√	√	• Overgrazing
Crop-based mixed	• Integrated systems with annual crops (ruminants and non-ruminants plus fish)	√	√	√	√	√	• Food-feed systems
	• Integrated systems with perennial crops (ruminants)	√	√			√	• All year feeding systems • Nutrient flows/soil fertility
•	Beef and dairy production	√	√	√	√	√	• Productivity enhancement
•	Goat and sheep production	√	√	√	√	√	• Intensification and specialisation • Overgrazing
Agro-pastoralist	• Cattle		√	√	√		• Feed supplies/drought strategies • Property regimes
	• Goat and sheep production		√	√	√		• Overgrazing • Trypanosomiasis
Range-based	• Sheep and goat production	√	√	√	√	√	• Drought strategies • Overgrazing • Property regimes • Marketing

Note 1. SSA = Sub-Saharan Africa; CA = Central Asia; WANA = West Asia and North Africa; LAC = Latin America and the Caribbean.
Note 2. √ indicates that both the production systems and animal species are the most important within the region.
Note 3. Major issues inter alia are those that currently merit R and D attention. Across regions, the issues are broadly similar as is the case with dairying.
Note 4. Dairy production includes buffalo and cattle, especially in Asia.

Conclusions

The four main higher-order livestock systems are landless, crop-based, agro-pastoralism and range-based. These are the most important in Asia, Sub-Saharan Africa, Central Asia, West Asia and North Africa, and Latin America and the Caribbean, when considered from the standpoint of the diversity, distribution, role and contribution of livestock to poor people and their livelihoods. Some 57 per cent of the rural poor keep livestock in crop-based mixed farming systems, which provide over 50 per cent of the meat and 90 per cent of the milk. These systems, involving both annual and perennial crops, will continue to be dominant in Asia, LAC and some parts of SSA, within which there needs to be better understanding of the implications of crop-animal-soil interactions. Integrated systems involving ruminants and tree crops, and silvi-pastoral systems are underestimated. Agro-pastoralism and range-based livestock systems will be the most important in SSA, CA and WANA. All four systems provide major opportunities, and challenges, in all regions, for increasing productivity and alleviating poverty.

While industrialised non-ruminant systems will continue to produce the major share of meat and eggs, emphasis on R and D with ruminants (buffalo, cattle, goats and sheep) within prevailing livestock systems, coupled to integrated natural resource management, can make a substantially increased contribution in the future. Dairy development is expanding rapidly in all parts of the developing world, and good opportunities exist for poor farmers in high-potential and peri-urban areas to capitalise on the rising demand for milk and dairy products. Small ruminants play a critical role in the livelihoods of marginal farmers and the landless, especially in the arid and semi-arid areas in WANA, SSA, north India, north Brazil and north Mexico. Increasing their contribution and improving the livelihoods of the poor will need more R and D attention. The appropriate development of common property resources, and the threat of overgrazing, are problems in all systems and in all regions, problems in which socio-economic and policy issues are deeply involved.

Further reading

Dixon, J., Gulliver, A. and Gibbon, D. 2001. *Farming systems and poverty – improving farmers' livelihoods in a changing world* (ed. M. Hall), Food and Agriculture Organisation of the United Nations (FAO), Rome, Italy. http//www.fao-org./DOCRP/ Y1860E/y1860e00.htm

Duckham, A.N. and Masefield, G.B. 1970. *Farming systems of the world.* Chatto and Windus, London, UK.

Jahnke, H.E. 1982. *Livestock production systems and livestock development in*

Tropical Africa. Kieler Wissenschaftsverlag Vauk, Kiel, Germany.

Nestel, B. 1984. Animal production systems in different regions. In *Development of animal production systems*. World animal science, A2, (ed. B. Nestel), pp.131-140, Elsevier Scientific Publishing Company, Amsterdam, The Netherlands.

Ruthenberg, H. 1980 *Farming systems in the tropics* (3rd edition). Clarendon Press, Oxford, UK.

Seré, C. and Steinfeld, H. 1996. *World livestock production systems. Animal Production and Health Paper No. 127*. Food and Agriculture Organisation of the United Nations (FAO), Rome, Italy.

Spedding, C.R.W. 1975. *The biology of agricultural systems.* Academic Press, London, UK.

Wilson, R.T. 1995. *Livestock production systems*. The Tropical Agriculturalist, Series Editors R. Coste and A.J. Smith. Macmillan Education Ltd., London, UK, in cooperation with Technical Centre for Agricultural and Rural Cooperation (CTA), Wageningen, The Netherlands.

Author addresses

Canagasaby Devendra, Formerly of the International Livestock Research Institute (ILRI), P.O. Box 30709, Nairobi, Kenya. Current address: 130A Jalan Awan Jawa, 58200 Kuala Lumpur, Malaysia

John Morton, Natural Resources Institute (NRI), University of Greenwich at Medway, Chatham Maritime, Kent ME4 4TB, UK

Barbara Rischkowsky, Animal Genetic Resources Group, Food and Agriculture Organisation of the United Nations (FAO), Viale delle Terme di Caracalla, 00100 Rome, Italy

Derrick Thomas, Formerly of Natural Resources Institute (NRI), University of Greenwich at Medway, Chatham Maritime, Kent ME4 4TB, UK

4

The livestock and poverty assessment methodology: an overview

Claire Heffernan[*], *Federica Misturelli*[†], *Ruth Fuller*[†], *Sonali Patnaik*[†], *Nessa Ali*[‡], *Wijaya Jayatilaka*[‡], *Anni McLeod*[‡]

Photo: Participants creating a community resource map in Qaqachaka Community, Oruro Department, Bolivia (courtesy of Claire Heffernan)

Cameo - Participation and poverty

The origins of participation as a paradigm have been largely attributed to the limited success of 'top down' development theory and practice that failed to understand the needs of the people for whom development was intended (Escobar, 1985). In a reflective pause during the late 1970s,

---

[*] Principal author
[†] Co-author
[‡] Consulting author

development practitioners began to relate the low impact of projects and programmes directly with the alienation and exclusion of the beneficiaries from the process itself (Nelson and Wright, 1995). The full and unfettered involvement of the poor in development processes was considered the solution to the shortcomings of previous approaches (Cernea, 1991). Hence, during this time, the terms 'participation' and 'participatory' began to enter development discourse. However, what began as a hypothesis to explain the poor uptake of projects and programmes soon became a major shift in development thinking that came complete with a specific outlook, methodology and its own unique tools.

Introduction

The origins of participation as a paradigm have been largely attributed to the poor outcomes of 'top down' development programmes that failed to meet the needs of the people for whom it was intended (Escobar, 1985; Rahnema and Bawtree, 1997). Indeed, it was during the late 1970s that development practitioners began to relate the low impact of projects and programmes with the alienation and exclusion of the beneficiaries from the process itself (Nelson and Wright, 1995). The full and unfettered involvement of the resource-poor in development processes was considered to be the solution to the shortcomings of previous approaches (Cohen and Uphoff, 1977; Cernea, 1991). During this time, the terms 'participation' and 'participatory' began to enter development discourse. However, what began as a hypothesis to explain the poor uptake of projects and programmes soon became a major shift in development thinking that came complete with a specific outlook and methodology, and its own unique tools.

Today, participatory tools are commonly used in development research and practice. Although the methods are not without biases (Livestock Development Group, 2003a), the widespread application of the tools has increased our comprehension of the reality faced by the global poor. No longer is poverty viewed simply as a deprivation of income, but the wider implications of being poor have become much better known and understood. For example, a recent World Bank report (World Bank, 2001) described poverty as follows:

> "Poor people live without fundamental freedoms of action and choice...They often lack adequate food and shelter, education and health...And they are often exposed to ill-treatment by institutions of the state and society and are powerless to influence key decisions affecting their lives...Poor people's description of what living in

> poverty means bears eloquent testimony to their pain."

With the introduction of Participatory Poverty Assessments (PPAs) the voices and experiences of the poor were able to inform decision makers more directly. Thus PPAs have redefined and shaped our notions of what it means to be poor.

Although poverty is now viewed as multi-dimensional, our approach to poverty alleviation tends to remain one-sided. Projects and programmes often do not distinguish between the different groups of the poor and their varying needs. Poor livestock-keeping households are one such sub-group of the poor. However, the majority of poverty analysts still treat livestock as an asset to be counted rather than as a holistic tool to enable poor households to escape cycles of poverty and deprivation. Although it has been estimated that there are approximately one billion resource-poor livestock-keeping households, research has only recently begun in mapping their geographic location (cf. Chapter 2; Thornton *et al.,* 2002).

This chapter outlines core components of the livestock and poverty assessment methodology (Livestock Development Group, 2003b). The methodology is a collection of participatory tools to assess poverty and well-being among poor livestock-keepers. For those readers familiar with participation, many of the tools are not new, but rather have been adapted for use within the livestock sector. The overall intention is to provide a holistic toolbox to enhance the understanding of both the needs and the strengths of the poor within the livestock sector and thus to inform the debate on livestock and poverty. At the community level, the Livestock and Poverty Assessment (LPA) methodology was devised to assist practitioners in answering the following broad questions:

- How important are livestock to livelihoods and well-being, past and present?
- How many, and who, are the resource-poor livestock-keepers?
- What are the major issues in animal health and production?
- Are livestock interventions the most appropriate form of aid?

Like PPAs, the LPA methodology is comprised of a range of traditional participatory tools such as semi-structured interviews (SSIs), focus groups (FGs) and participatory exercises. The exercises, however, have been specifically adapted to gain information regarding the functioning of livestock-based livelihoods. As such, the LPA methodology offers a means both of identifying poor livestock-keepers and diagnosing the issues of relevance.

Table 4.1 offers an overview of the participatory exercises included in the methodology. The grouping of the methods is not intended to offer a blueprint to practitioners, but rather the aim is to provide an array of tools, which can be used individually or in tandem to obtain as much information as is required. In the remainder of this chapter, only the core tools from each section are described. For more detailed information on the methodology please see Livestock Development Group (2003b).

Table 4.1 The collection of methods (source: Livestock Development Group, 2003b)

I. Setting the Scene:

The Simplified SL (Sustainable Livelihood) Approach
Historical Trend Analysis
Community Resource Mapping
Livestock Production and Management Calendar
Livelihood Changes Diagramming
Livelihood Opportunities and Constraints Diagramming
Migration Mapping

II. Profiling the Livestock Keepers:

Livestock and Poverty Ranking
Compound Mapping
Household Resource Maps
Community Rangeland Mapping

III. Assessing Issues in Animal Health and Production

Livestock Health-Care Provider Maps
Consumer Preferences Regarding Animal Health-care
Livestock Disease Prioritisation
Livestock Problem Ranking
Assessing Knowledge Pathways for Livestock Production
Participatory Herd Assessment
Linkage Diagramming in Urban Production Systems

IV. Determining the Feasibility of Livestock Aid

Assessing Motivation
Community Values Diagramming

Prior to detailing some of the tools, a brief overview of practical issues in performing participatory exercises is offered. As each participatory process is unique, bad practice and poorly performed exercises can often go unnoticed. Hence, the following sections offer tips and hints on how to create more effective outcomes and enhance the experience for both practitioners and participants.

Introducing the practitioners, topics and methods

True participation is built upon trust and openness. Vital to facilitating an open environment is an effective introduction to the participants of the facilitator of the exercise, the topics to be discussed, and the methods to be performed. If the facilitator or practitioner is new to the community then obviously even greater time should be spent in developing a familiarity with both the geographic location and resident households. For participation to be a means and mechanism of empowerment, the communities themselves must agree to collaborate in the process without prompting, coercion or the expectation of future aid. Once a community has agreed to collaborate, participants may then be selected to perform the different exercises.

Identifying the participants

At the scoping stage of projects and programmes, participatory tools are commonly utilised to interact with key informants, focus groups and the community as a whole. As the process evolves, however, it is important to remember that the requirement for a multiplicity of viewpoints remains. Indeed, during the project cycle, there is a tendency for participatory processes to become more, rather than less, exclusive, i.e. dominated by community leaders and other gatekeepers. Practitioners must be aware that the poor often do not participate in large meetings and, further, being seen interacting with persons of perceived higher status may be viewed as socially inappropriate.

Indeed, research has demonstrated that in large groups the person with the most influence normally decides the tone and tenor of the responses (Livestock Development Group, 2003a). Hence, general opinions may be hidden in favour of the viewpoints of the more elite and powerful. Thus, smaller groups are recommended to ensure greater levels of participation because the danger of a single person dominating the discussion is minimised and it is also possible for the practitioner to elicit more personal responses. Therefore, after community-level introductions, participants should be split in smaller groups (7-8 people maximum), to facilitate discussion and the exchange of opinions.

To avoid selection bias, practitioners should consult as many stakeholder groups as possible. Gender, age and socio-economic standing are factors that can influence viewpoints. Therefore, it is advisable that the same exercises are held separately with men and women, the young and the aged, to gain as wide a perspective as possible. Livelihood criteria may also be utilised to filter participants, e.g. cattle owners vs. small-stock owners. In this manner, the viewpoints of the different socio-economic strata may be more effectively represented.

However, communities can also introduce biases. For some exercises, such as those utilised for targeting, practitioners must be sensitive to how expectations of future aid may either influence responses or even change community dynamics. Therefore, targeting exercises should be packaged together with other more neutral tools, to ensure that both clients and practitioners have a wider perspective.

Tips for successful focus groups

Focus groups (FGs) are targeted discussions with specific stakeholder groups or sub-groups, i.e. men, women, community members, government workers etc. In addition to generating detailed information regarding specific issues (e.g. access to veterinary services), FGs are also useful for illuminating the attitudes of the participants (notions regarding poverty etc.). However, practitioners should view themselves as facilitators, and should be careful not to influence the ensuing discussion by offering personal opinions. In addition, it is important to remember when analysing the subsequent discussion that the exercises are subject to social and group dynamics.

For example, in many communities, gender and economic standing greatly influence one's ability to express an opinion in a social forum. Therefore, when interpreting the results of an FG, practitioners need to assess not only the responses of the participants, but also *who i*ntervened in the debate, *how* often and the level of agreement generated by their input. In this manner, it is possible to assess whether the responses have been dominated by only a few individuals or if a wider consensus was reached.

The topic to be investigated may also impact the flow and reliability of FG discussions. Indeed, some subjects are not easily addressed, particularly in a group. If the topic is not socially acceptable, or if the participants find the questions too intrusive, the group may at best offer misleading information and at worst become suspicious or uneasy. Therefore, prior to commencing FGs, the facilitator needs to be aware of topics which may cause offence. During the discussion, practitioners must also be sensitive to the reluctance of group members to divulge personal and/or sensitive information.

Hints for semi-structured interviews

Semi-structured interviews (SSIs) are interviews with a single client on pre-determined issues which are not restricted to set questions and/or multiple choice answers. SSIs are generally used to generate in-depth information and as a means of obtaining comparable data across a sample group. Hence SSIs may be performed to obtain background information regarding household income and expenditure, livestock herd size, etc. Normally a practitioner will develop a list of questions or topics that he/she would like to investigate as a guideline for the interview. Depending upon the flow of the conversation, new issues may be raised or, conversely, less fruitful areas dropped.

Prior to initiating an LPA, it is very important that all individuals involved in SSIs are given comprehensive training regarding the aim and objectives of the exercise. The meaning and relevance of each topic has to be explained and clarified, in order to ensure that each issue is approached in the most appropriate manner.

As is further discussed below, to ensure the comparability of responses in large-scale exercises, it is advisable to provide practitioners, with a question guide, rather than simply a list of topics to be investigated. In this manner, the interviewer will be able to become familiar with the information required in a relatively short time, and will be able to develop his/her own interviewing style. Questions should be open-ended to enable clients to respond in the manner in which they feel comfortable. By creating a list of core questions, interviewers and clients have scope freely to explore, in an unstructured way, the different issues and topics that arise.

Finally, when conducting SSIs, developing an effective rapport between the practitioner and the client is vital. A properly administered SSI should resemble a conversation, rather than the question-answer schematic that characterises more formal surveys. Equally, information should flow between both parties. Local people are generally very interested in learning about the 'outsiders' that visit their community. For example, participants are often very curious to know personal details of the practitioner, such as whether or not he/she is married, has children and so forth. Equally, the practical issues regarding the management of livestock or livelihood activities conducted in far away places is often a topic of interest. Practitioners should share their experiences regarding the issues in question. A successful SSI is a lively discussion around specific issues in which both parties share and exchange information.

Problems and issues in visual exercises

Visual tools such as mapping and ranking are among the most frequently utilised methods in the PRA (Participatory Rural Appraisal) toolkit. Indeed, the exercises can be employed to gather information on a wide variety of topics, such as access to services, land-use patterns, social structures, livelihood activities and so on.

Nevertheless, visual exercises are not exempt from problems. As with focus groups, literate participants may take the lead and exclude others from the illustration process. Similarly, participants who enjoy a higher status within the community may exercise greater influence than those perceived to be of a lower standing. Indeed, many individuals (particularly those who are less educated) do not feel at ease performing the task, and may delegate the job to someone they perceive as more capable, such as a community leader.

Practitioners must therefore be aware that many mapping and diagramming exercises, rather than being more inclusive of the poor, may actually make participants feel uncomfortable. Consequently, due diligence must be taken to ensure that more marginalised community members are able to express themselves in a neutral and supportive forum.

The following sections explore some of the core tools that comprise the methodology.

Participatory tools for 'setting the scene'

The following are detailed examples of three of the tools in the LPA methodology that may be used to understand better the wider issues and constraints impacting livelihoods.

Livestock production and management calendars

The objective of a Livestock Production and Management (LP&M) Calendar is to identify the seasonality of livestock-related events and to detail inputs regarding the livestock production cycle.

However, when producing LP&M Calendars, it is useful to separate out the different groups of resource-poor livestock-keepers that may be present in the specific community involved. For example, small-stock keepers will depict a very different livestock production cycle from those households owning cattle, camels etc. Equally, among many livestock-keeping communities, there are significant gender differences with regard to the livestock production cycle. Further separating the groups by gender can lead to a more nuanced understanding of the different gender roles.

When producing a calendar, the first step is to discover how local people categorise seasons and time periods. Next, for the different periods of the year, major events in livestock production may be noted and discussed. Thus the seasonality of disease outbreaks and the reproductive cycle, i.e. calving, lambing and kidding seasons, are catalogued. Differences across the year in milk production and income from livestock may also be recorded at this time. Equally, access to labour and herding patterns may also be discussed and diagrammed. In addition, livestock markets and fluctuations in the price of livestock may also be documented. Figure 4.1 offers an example of a livestock production and management calendar.

Livelihood changes diagramming

The Livelihood Changes Diagram (LCD) utilises a Venn diagram format to document major changes in livelihood activities and gain a comparative notion of how communities believe the changes have impacted households. The technique can also be used to assess trends in community-level values regarding livestock keeping and other livelihood strategies (which will be described later).

To develop a LCD, group discussions are first held to determine any shifts in livelihood strategies and/or living standards over a time period of at least 10-20 years. It is generally easiest to begin by first discussing the specific livelihood issues of interest and, then exploring changes that have taken place over the time-period in question.

To keep the length of the exercise within a reasonable time, it is best to limit the discussions to five or six parameters. For example, the technique may be utilised to analyse changes in the number of households keeping cattle and small stock, or those involved in alternate income generating activities, the overall size of livestock herds, or changes in the socio-economic status of community members and the number of children attending school, etc. Each issue may be represented by a circle, and the level of change depicted by altering the size of the circle. Figure 4.2 offers an example of a LCD.

Month *Local season*	*J*	*F*	*M*	*A*	*M*	*J*	*J*	*A*	*S*	*O*	*N*	*D*
LIVESTOCK BIRTHS (seasonality of calving, lambing and kidding)												
# Calves born	2	1										1
SALE OF LIVESTOCK PRODUCTS (milk, cheese, meat, etc. and price changes)												
Daily average litres sold	5	12	12	11	8	6	3	3	3	0	0	4
Milk price/l (Bolivianos)	8	8	8	8	9	9	9	10	10	10	10	8
LIVESTOCK DISEASE PREVENTION (vaccination, de-worming, etc.)												
FMD vaccination campaign			x*									
De-wormed calves					X*							
LIVESTOCK DISEASE OCCURRENCE (seasonality of major livestock diseases)												
FMD				x	X	x						
INPUTS (seasonality of any feed, water or health inputs)												
Purchased feed	x	x										x
LIVESTOCK MANAGEMENT (livestock-related labour over the course of the year: herding, milking, etc.)												
W=Wife; C=Children												
Herding	C	C	C	C	C	W	W	W	W			
Milking	W	W	W	W	W	W	W	W	W	W	W	W

*x=denotes smaller amounts of inputs or animals infected; *X denotes larger amounts of inputs or animals infected.

Figure 4.1 Livestock production and management calendar

After identifying perceptions regarding the level or amount of change for each of the issues, the subsequent discussion should focus upon causality and how the shifts have influenced attitudes. For example, by asking community members to chart differences in school attendance over a 20-year period, the discussion should then focus on why those changes have occurred. Thus, perceptions regarding the benefits and drawbacks to educating children may be illuminated.

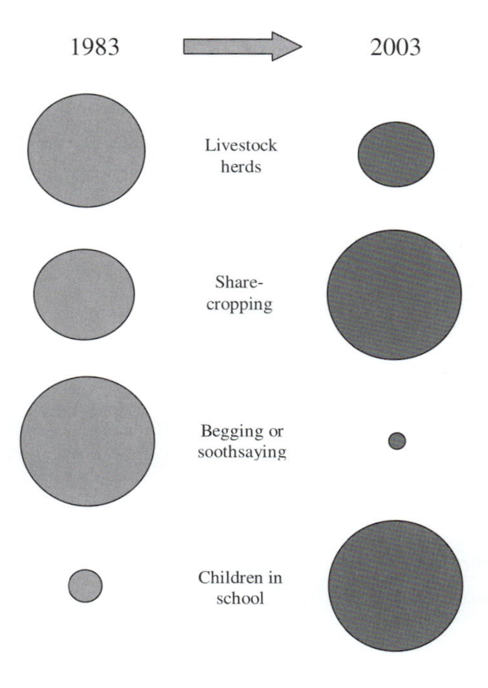

Figure 4.2 Livelihood changes diagram (source: Livestock Development Group, 2003b)

Historical trend analysis

The overall aim of a Historical Trend Analysis (HTA) is to explore how communities have reacted to, or recovered from, major change such as that relating to institutional, policy or environmental events. The analysis enables project planners and researchers to gauge the ability of households to cope with these stresses and, therefore, the robustness of current livelihood strategies. The technique is also useful for exploring different interventions that have taken place and their subsequent success or failure.

To perform a HTA, a group of elderly participants is generally required, although the technique may also be performed with an appropriate key informant. In general, exploring major events within the last twenty years is sufficient, although a greater time-span may be possible depending upon the age and interest of participants.

To initiate a HTA, practitioners may begin with the following questions:

- What are the major events that have impacted your community in the past 20 years?
- What happened to your livestock herds as a result of these events?

Once the major events have been detailed, how these events have impacted on household well-being in general and livestock keeping, more specifically, may be determined. Indeed, issues such as changes in the price of livestock or access to markets etc. may be explored using the same historical reference points. Figure 4.3 offers an example of a HTA performed among farmers in Orissa, India.

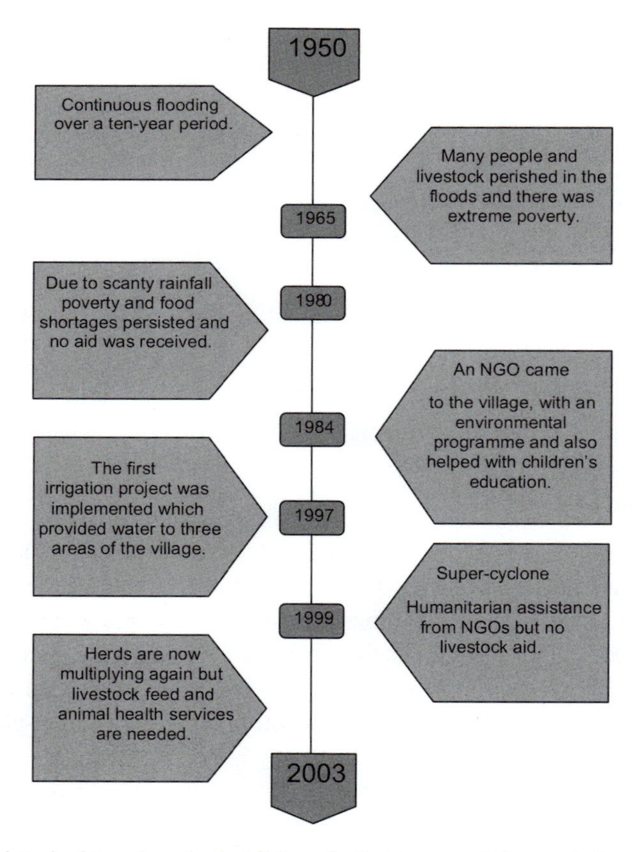

Figure 4.3 A historical trend analysis (Orissa, India) (source: Livestock Development Group, 2003b)

Participatory tools for profiling livestock-keepers

Correctly identifying the population of interest is imperative for better targeting livestock aid. While a range of participatory methods can be used to identify resource-poor livestock-keeping households, two such tools are presented below.

Compound mapping

Across the globe, resource-poor households generally do not live in isolation. However, in different countries and continents the configurations of living spaces differ dramatically. For example, resource-poor households in India tend to live in extended families whereas resource-poor pastoralists in Kenya live in multi-household compounds comprised of family members, friends and neighbours. Within this milieu, livestock keeping is often a shared activity. As such, compound-mapping exercises are useful tools to illuminate both the individuals responsible for livestock care-taking and the level of resource sharing between the households involved.

To produce a compound map, clients are asked to draw how and where their families live. Potential items of interest include the number of compound/household members, such as brothers, sisters, in-laws, friends, etc. The age and sex of children and school attendance may also be noted. Livestock herds can also be mapped at this time. By mapping the collective herd, discussions can ensue about an individual household's access to livestock products and the responsibility for specific livestock-related activities such as milking, herding, etc. In this manner, the exercise explores the social and financial connections both within and between households. Figure 4.4 offers an example of a compound map.

Household resource maps

Household Resource Mapping is a frequently utilised participatory technique, which yields information regarding the inputs and outputs of farm-level enterprises, such as crop and livestock production. With regard to resource-poor livestock-keepers, household resource maps may be utilised to delineate access to livestock markets (including where and how information regarding prices is obtained) and the availability of livestock-related services, e.g. the purchase of livestock drugs (km and time), nearest tick dip (if relevant), etc. Equally, the distance that livestock must travel daily for grazing and water can also be delineated. The inputs and outputs for other livelihood activities such as petty trade, e.g. firewood and charcoal selling, are also outlined. It is also important to note major household expenses and the income earned from the different tasks.

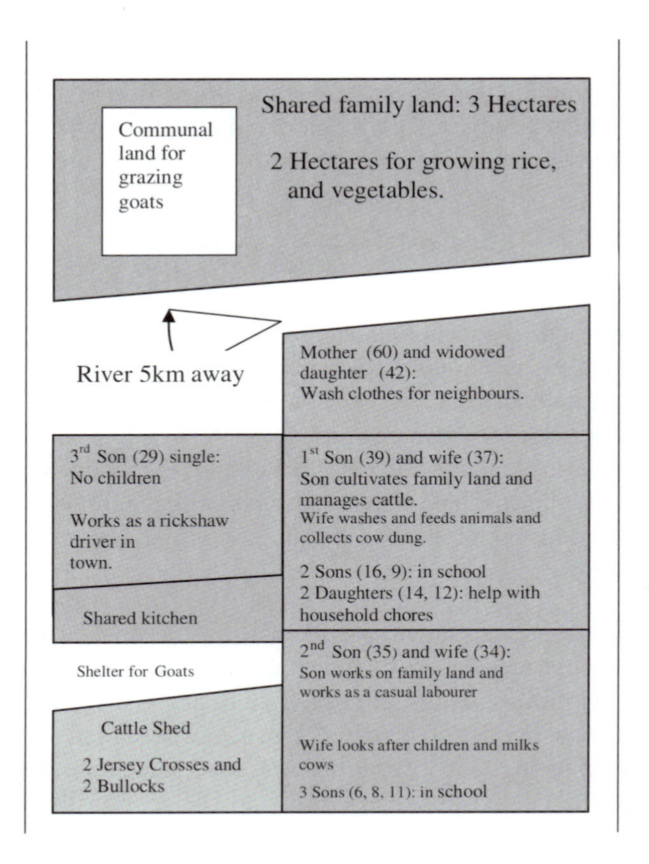

Figure 4.4 Compound map (Tamil Nadu, India) (source Livestock Development Group, 2003b)

The maps may also be utilised to explore the relationships between the different enterprises and any labour conflicts. Indeed, by mapping inputs and outputs, practitioners may obtain an idea of how households prioritise activities. Figure 4.5 offers an example of the type of information that may be derived from a household resource map in Bolivia.

Participatory tools for assessing issues in livestock production and health

A range of different tools are presented in the LPA methodology for assessing issues in livestock health and production. Selected examples are presented below.

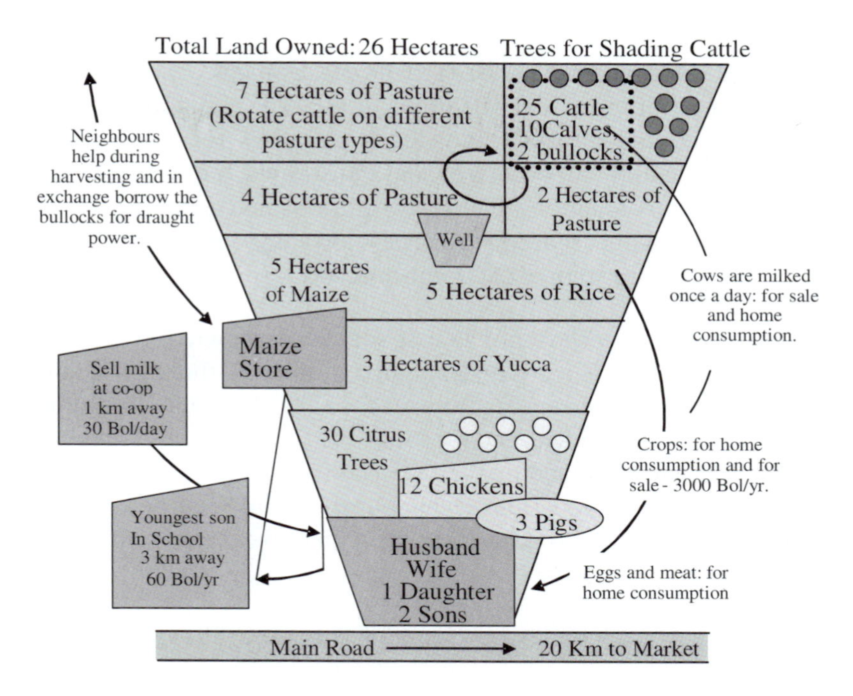

Figure 4.5 Household resource map (Bolivia) (source Livestock Development Group, 2003b)

Livestock health-care provider maps

To evaluate access to animal health-care, focus groups and key-informant interviews may be held to detail Livestock Health-Care Provider maps. Particular areas of interest include the distance to, and availability of, government, NGO and private animal health-care and livestock drug providers. In this manner, specific information regarding the distance, time required and frequency of use of both animal health-care providers and livestock drug stores may be documented. By periodically updating the maps, both project staff and community members can evaluate the influence of any project-related inputs. For example, the uptake of a Community-Based Animal Health-Care Worker project may be evaluated by analysing changes in the maps pre- and post-project implementation.

Therefore, the maps create base-line information regarding service delivery in addition to providing a means of assessing change.

Assessment of knowledge pathways for animal health and production

Knowledge and information have been identified as major constraints for poor livestock-keepers (Heffernan and Misturelli, 2000). The aim of this exercise is to identify the formal and informal knowledge pathways for animal health and production, and identify opportunities and constraints for strengthening such pathways. Perceptions regarding the trustworthiness and reliability of different knowledge and information sources influence knowledge-seeking behaviour. Such perceptions can also be explored using this method.

There are two stages in performing an assessment of knowledge pathways. The first stage of the exercise is a group discussion to identify the myriad sources of livestock-related information ranging from the formal to the traditional. Information points or sources should be identified for a range of subjects such as animal health, management, reproduction, and marketing. The frequency of use, and form and type of information acquired from each source should then be discussed and recorded. Once the different sources have been identified, each source should be assigned an agreed-upon symbol by the participants.

The information sources are then ranked for the following parameters: trustworthiness, proximity, ease of access, affordability and overall usefulness of information provided. The results of the ranking exercise should be recorded and the facilitator should explore the reasons for the different rankings of knowledge sources. Figure 4.6 shows an example of a matrix from ranking different sources of information for different parameters.

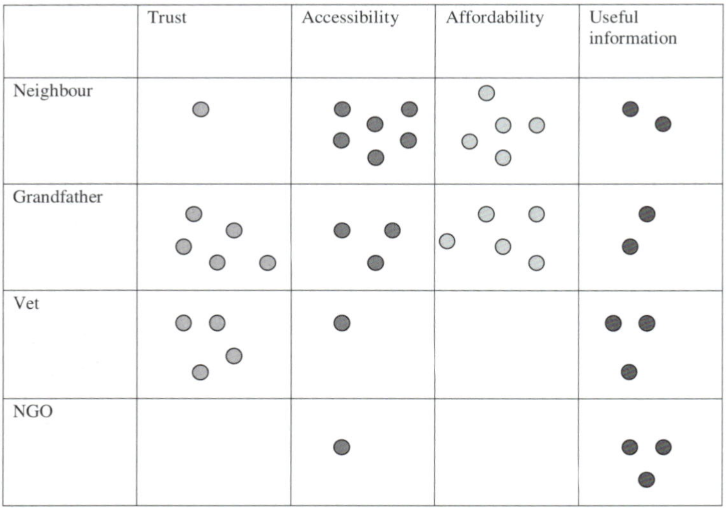

	Trust	Accessibility	Affordability	Useful information
Neighbour	○	●●●●●●	○○○○○	●●
Grandfather	○○○○○	●●●	○○○○○	●●
Vet	○○○○	●		●●●
NGO		●		●●●

Figure 4.6 Assessment of knowledge pathways regarding animal health (source: Livestock Development Group, 2003b)

Conclusion

The Livestock and Poverty Assessment Methodology provides practitioners with a set of tools to explore different aspects of livestock-based livelihoods. The intention is not to offer a blue print but rather to provide an iterative process that can be adapted to the wide variety of environments which resource-poor livestock producers inhabit. Although beneficial to project outcomes, it is recognised that it may be difficult for project planners and field staff to perform all of the exercises of the methodology. As such, use of the core tools described earlier will enable practitioners to gather at least an initial understanding of the complexities of livestock-based livelihoods in the communities involved. However, the above tools are not meant to be utilised in one-off exercises. Indeed, it must be counselled that developing a true understanding of livestock keeper communities and supporting their participation is a long process that is not amenable to rapid appraisals and assessments.

Further reading

Conroy, C. 2005. *Participatory livestock research: A guide*. ITDG Publishing, London, UK.

Holland, J. and Blackburn, J. 1998. *Whose voice: Participatory research and policy change*. ITDG Publishing, London, UK.

Krueger, R. and Casey, M. 2000. *Focus groups: A practical guide for applied research*. Sage Publications, London, UK.

Wengraf, T. 2001. *Qualitative research interviewing*. Sage Publications, London, UK.

World Bank. 1996. *The World Bank participation sourcebook*. The World Bank, Washington D.C., USA In http://www.worldbank.org

Author addresses

Claire Heffernan, Federica Misturelli, Ruth Fuller, and Sonali Patnaik, Livestock Development Group (LDG), School of Agriculture, Policy and Development, University of Reading, Earley Gate P.O. Box 237, Reading RG6 6AR, UK

Nessa Ali, Department of Rural Sociology, Bangladesh Agricultural University, Mymensingh 2202, Bangladesh

Wijaya Jayatilaka, Department of Agricultural Extension, University of Peradeniya, Sri Lanka

Anni McLeod, AGAL, Food and Agriculture Organisation of the United Nations (FAO), Viale delle Terme di Caracalla, 00100 – Rome, Italy

5

Livestock development and poverty

Claire Heffernan[*], *Louise Nielsen*[†], *Ahmed Sidahmed*[†] *and Tofazzal Miah*[‡]

Cameo - The Livestock Guru: An interactive, multi-media, learning platform for resource-poor livestock- keepers, Mannadipet, India
(www.livestockdevelopment.org/guruvideo.htm)

Lack of access to information, and the knowledge derived from it, is one of the largest barriers for the poor in the fight for livelihood security, self-respect and well-being. Nowhere is this more apparent than in the livestock sector. To address this constraint an interactive, multi-media, learning platform for poor livestock-keepers has been created. The software is tailored to individual users, particularly those livestock keepers with little formal education. In addition, the program is easily accessible by children. Children are an important channel for new information for their parents and grandparents. Benefits of the programme include increased household income and a lower incidence of zoonotic diseases as a consequence of improved livestock production and management strategies. Another unique element of this programme is its ability to record user preferences and priorities. In this manner, policy makers and planners may obtain specific and timely information regarding the knowledge needs of the poor. By creating new knowledge pathways between decision-makers and the poor, the *Livestock Guru* is an important tool in the fight for global poverty eradication.

A wide variety of in-country institutions have contributed to the development and delivery of the programme including non-government organisations, community-based organisations and government institutions. For example, in Mannadipet in Pondicherry India, a kiosk was set up in a local milk co-operative.

---

[*] Principal author
[†] Co-author
[‡] Consulting author

My name is Valliyammal. I am from Mannadipet and I have two cows. I come to the co-operative to sell my milk and I came to know about the computer programme. By using the program I learnt about mastitis and that will be very helpful for me. I never had any training and now I would like to go in order to learn more.

When I saw the computer here I realised that I could learn many things from this programme. This is the computer where I learned about mastitis and how to maintain my cows and the cowshed properly. If people discover this facility they will also explain what they have learnt to others.

Introduction

As the previous chapters demonstrate, livestock are vital to the lives and livelihoods of some of the most marginalised and vulnerable populations in the world. Therefore, the next question to be addressed is how can livestock development enhance the sustainability of livestock-based livelihoods? Over the past thirty years livestock development, in common with the development industry in general, has undergone a great transformation. Furthermore, in the coming years, with the prevailing global economic and environmental trends, further changes in the livestock sub-sector are predicted.

Indeed, the World Bank (de Haan *et al.*, 2001) identified the following broad 'driving forces' that will further influence changes within the livestock sector in future decades:

- Increasing demand and the consequent changes in global livestock production ('The Livestock Revolution')
- Altering macro-economic forces and institutional environments
- Changes in the role of livestock within 'Southern' countries.

Therefore any discussion on livestock development must allow for these changes at the macro-level. However, in order to fully understand the implications of these forces upon the poor, it is vital that the issues and constraints faced by resource-poor livestock-keepers are adequately understood. This chapter is divided into two parts. The first section explores the problems of the poor with regard to their livestock. The second examines the role of the different actors and agents involved in livestock development and the impact of projects and programmes.

The problems of the resource-poor

As mentioned in Chapter 4, the poor face a variety of constraints with regard to sustainable livestock production. Indeed, LID (1999) divides the problems of the poor into three basic types: herd acquisition, maintenance and the marketing of livestock products. Herd acquisition refers to the ability of households to access capital and credit facilities to purchase livestock. Maintenance denotes the ability of households to obtain animal health and production services. The final categorisation refers to the frequent inability of the poor to access reliable markets for their off-take.

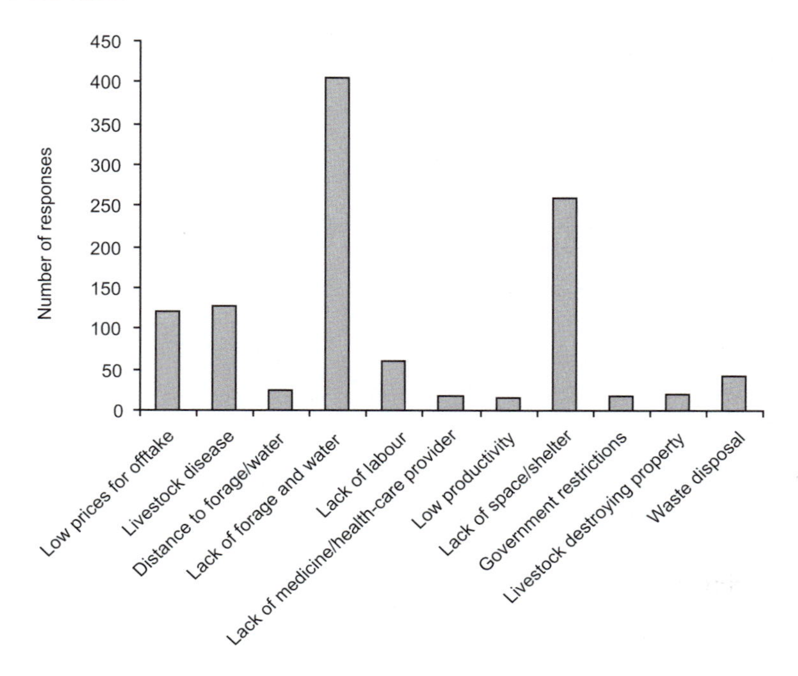

Figure 5.1 Ranking of primary problems with livestock in India (Livestock Development Group, 2002)

Nevertheless, the above description may ignore some of the more nuanced issues regarding livestock keeping and the poor. For example, for many resource-poor livestock-keepers, particularly in urban areas, welfare and hygiene constraints are paramount. The lack of sufficient space and adequate housing in addition to waste disposal are key issues for the urban poor. For example, Livestock Development Group (LDG) (2002) in an open-ended ranking exercise with over 1,300 households in India (Figure 5.1) found the top four problems for the poor included the lack of forage and water, space and/or shelter to keep livestock, animal diseases, followed by low prices for off-take. Overwhelmingly, obtaining sufficient feed resources proved to be the biggest worry for producers. Hence access to capital for inputs is one of the largest constraints faced by resource-poor farmers in India. There were various additional problems which, while localised to specific communities, were nonetheless overriding for the households involved. For example, urban pig producers in Delhi and Chennai encountered problems with low farrowing rates and high piglet mortality due to worry from street dogs.

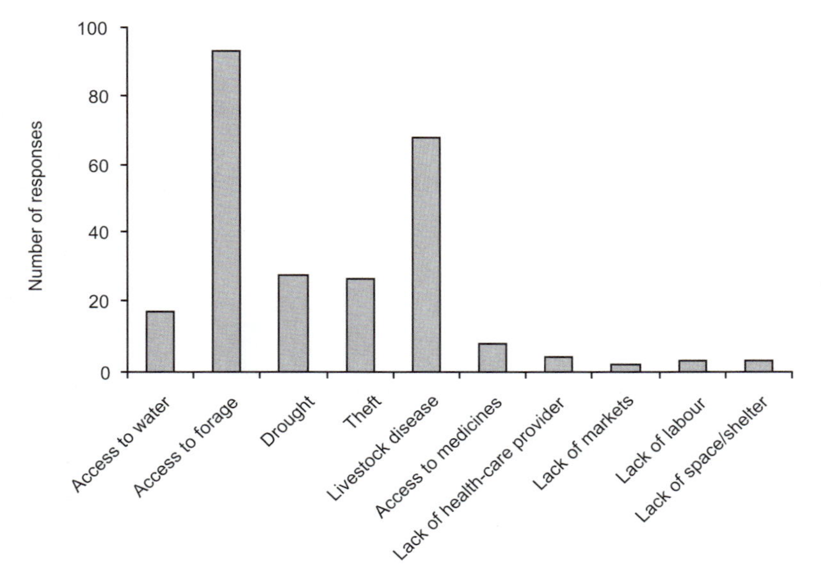

Figure 5.2 Ranking of primary problems with livestock in Kenya (Livestock Development Group, 2002)

In Kenya, the study also found that access to inputs, particularly forage ranked first in farmers' and pastoralists' perceptions regarding their livestock (Figure 5.2), with livestock disease ranking second. Not surprisingly, the lack of space/shelter for livestock was not deemed a large problem, with the exception of urban livestock-keepers in Nairobi. Theft was also considered to be problematic, particularly amongst pastoralists.

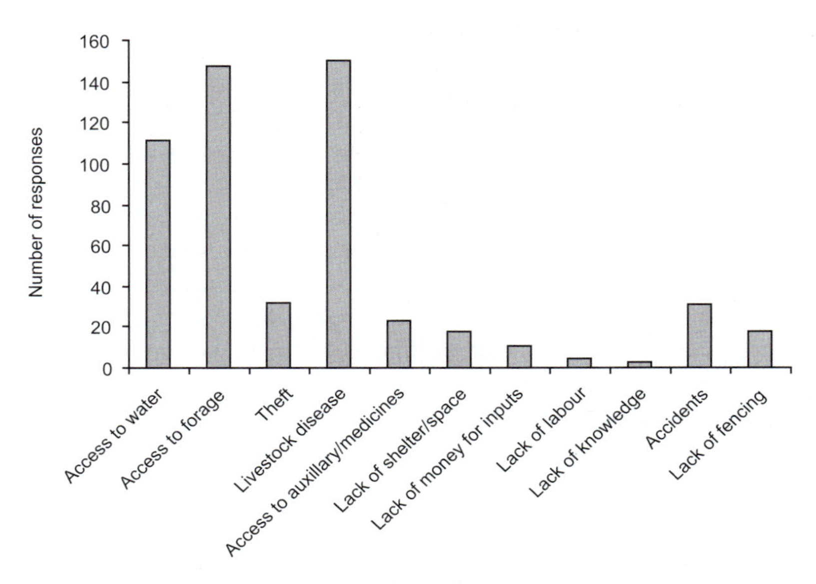

Figure 5.3 Ranking of primary problems in Bolivia

A similar study in Bolivia found that, although a number of different problems arose, livestock disease and the lack of access to forage and water were the primary concerns of farmers (Figure 5.3). Interestingly, livestock accidents (mainly cattle ensnared in barbed wire) ranked quite highly. Additionally, a few participants believed that their knowledge regarding animal husbandry and health was insufficient. Thus, as in India and Kenya, Bolivian farmers perceived access to inputs as the primary problem.

The above analysis of the problems of the poor demonstrates the following. Firstly, although there were regional and country variations, the problems faced by the poor regarding their livestock were surprisingly consistent. Across the three countries, the lack of basic inputs and poor animal health were considered the primary constraints to livestock-based livelihoods. Secondly, the findings demonstrate that even our most fundamental notions of the poor must be driven by actual evidence rather than informed perceptions. In contrast to the study by LID (1999), problems with livestock acquisition did not feature. Equally, the lack of markets for off-take was not considered an insurmountable problem. Thus, perceptions regarding the problems of the poor and their livestock are at variance with the problems as described by the poor themselves. However, understanding the problems of the poor is only the first step. In order to be successful, livestock development projects and programmes must ultimately understand the motivations of the clients toward livestock keeping.

Motivation for livestock keeping

Psychologists believe that integral to any assessment of motivation is an understanding of the values and attitudes of the individuals or communities involved. Therefore, in order to understand the motivating forces for successful livestock-based livelihoods, Nielsen (2004) explored the attitudes and values of livestock keepers across three countries: Kenya, India and Bolivia. Nielsen used socio-psychological methods to explore how resource-poor livestock-keepers perceived livestock, in relation to their perceptions regarding the future. The findings were illuminating from a number of perspectives. For example, in India, not surprisingly, livestock were largely associated with security and wealth accumulation. Indeed, animal ownership had a protective function in preventing households from participating in activities associated with extreme poverty such as begging (Table 5.1). Nevertheless there were also associations with debt and landlessness.

Table 5.1 Associations with livestock keeping in India

Associations	*Number of associations*
Source of income and wealth	8
An essential livelihood for survival	3
A form of security	2
Prevents begging	1
Inability to cultivate (landlessness)	1
Requires hard work/lots of attention	1
A source of debt	2
Offers no security/unable to make a living	2

Source: Nielsen (2004).

The notion of livestock as a source of debt is not surprising as many of the study participants were involved in micro-credit schemes. Overall, the findings appear to provide further confirmation of the importance of livestock to the livelihoods of the poor. However, when poor households were asked what aspirations they held for their children, a different picture emerged. Overwhelmingly households in India desired their children to be educated and seek full-time employment. As such, livestock keeping provided security for the present generation, but not for the children of those involved. The associations with education are displayed in Table 5.2.

As Table 5.2 demonstrates, the associations with education tended to be much stronger and more nuanced than those listed in Table 5.1 for livestock, revealing a deeper interest and hence potentially stronger motivational forces. While it is not surprising that the values and attitudes toward education were very positive, there are clear implications for development projects and programmes, particularly those related to livestock. Under these conditions, projects and programmes should support livestock keeping as a potential means to an end rather than as an end in itself. Further, practitioners need to recognise that the debt burdens that households accrue from livestock keeping may be counter-productive to a household's longer-term aspirations.

Table 5.2 Associations with education in India

Education	*Associations*
Ability to get a job, including government employment	71
Literacy/numeracy/communication	36
Enhancing knowledge and understanding	25
Better standard of life/improvement/prosperity	42
Source of family support	11
Ability to do business	7
Good citizenship/morality	5
Moving around/meeting influential people	4
Escape rural life/working in the fields	3
Marriage into better/wealthier families	5
Ability to maintain good health	2
Ability to gain respect/family prestige	2
Ability to improve community/village	2
Elitist/privileged/inaccessible/too expensive	6
Lack of jobs means it is futile	2

Source: Nielsen (2004).

The following section details the actors and agents currently involved in livestock development.

The actors, agents and projects

Today, there are five broad areas in which most livestock interventions fall:

- direct livestock aid
- service delivery
- technical responses to specific production or disease problems

- institutional and capacity-building projects
- product value-added or storage/processing and marketing.

Like other sectors, the formal stakeholders may be divided into three general groups:

- those that fund
- those that implement projects and programmes
- those involved in research.

Formal actors include:

- the government
- private sector
- donors
- non-government organisations (NGOs)
- community based organisations (CBOs)
- agricultural research institutes (ARIs)
- universities.

However, there are various stakeholders involved in service delivery that fall outside the above categorisation and form part of the informal sector, e.g. traditional healers and community groups. The role of informal service providers in the provision of livestock services to the poor is being increasingly recognised. Many recent projects have attempted to incorporate support for alternate providers.

The following discussion explores first, the service providers and their traditional roles (as described by Heffernan and Sidahmed, 1998). Second, a more detailed examination of the relationship between the actors and agents and the different types of livestock development projects and programmes is provided. Finally, an overview of current evidence regarding the impact of livestock development on the poor is offered.

Governments

Typically governments, often supported by funds from the donor community, handled the totality of livestock services from disease control to clinical and diagnostic services. Governments generally ran vaccination campaigns, implemented disease surveillance systems, supported clinical services and artificial insemination (AI), funded veterinary laboratories and extension services and often controlled livestock markets and marketing. However, in recent decades, government veterinary services have been forced to respond to a variety of changes. Indeed, at the farm level, production has shifted away from subsistence to a more commercial orientation. Consequently, services have had to change the focus away from the herd to the individual animal.

Equally, and perhaps more fundamentally, there has been an increasing trend towards the privatisation of veterinary services over the past two decades. Consequently most governments now favour the delivery of livestock services on a cost-recovery basis. Equally, with the increasing levels of debate over the role of governments in the provision of public services, many governments are limiting their roles to the following priority areas:

- Regulatory measures and quality control
- Trade, marketing and price policies
- National disease eradication/vaccination campaigns
- Production/importation of livestock drugs and vaccines
- Public health
- Extension and training
- Research.

Nevertheless, in many countries, the growth of the private sector is further limiting the role of governments.

The private sector: veterinarians, para-vets, auxiliaries, community-based animal health-care workers

A diverse group of actors, ranging from professionals to para-professionals to laymen and trained members of the public, operates within the private sector. The private sector has now taken over many aspects of clinical service delivery. Equally, in many countries the private sector has a much wider mandate and is involved in extension services, vaccine production and, with increasing frequency, animal health and production research. The majority of private sector activities have been biased towards peri-urban and accessible rural areas, although this has been changing in recent years. Private practitioners primarily operate in mixed farming systems and dairying zones. Reasons include the lack of incentives for travel and poor access to the herds of the more remote populations, e.g. pastoralists and agro-pastoralists. Although NGOs have been involved in the training of Community-Based Animal Health-Care Workers (CBAHW) to provide basic clinical services as well as to improve the supply chain of veterinary drugs to remote areas, many pastoralist areas are still under-served.

The full privatisation of clinical veterinary services faces a variety of hurdles. Firstly, rural financial systems to support private practice in remote areas are still under-developed so that private practitioners often face difficulty in accessing capital to start businesses. Secondly, the regulatory and policy environment of many nations is still hostile to private veterinary professionals and para-professionals. Quality issues and safety standards in clinical services and the distribution of veterinary pharmaceuticals remain key concerns for many national regulatory bodies.

Community organisations and co-operatives

At present, community organisations and co-operatives play only a minor role in the delivery of livestock services at both the community and the household level. Part of the problem is the lack of a suitable legislative framework to support the localised delivery of services by non-professionals. Equally problematic is the lack of knowledge in implementing agencies of community priorities, interests and leadership capabilities. However, some successes have been noted when there is an incentive for the community to work together due to common goals and interests (e.g. ownership of livestock watering facilities, communal management of grazing areas, tsetse fly eradication campaigns, etc.).

The benefits of such an approach are as follows. Firstly, farmers have greater ownership and control and hence, stake in the outcomes of services provided in this manner. Secondly, the demand-driven nature of the formation of such groups assures that even in remote areas the client base will be sufficient to support service delivery. Therefore, as the delivery landscape continues to evolve, community-based organisations may potentially play a much greater role in the future.

Traditional animal health-care providers

Although the wide-scale adoption of Western veterinary medicine has pushed traditional healers to the peripheries of animal health-care in many areas, pockets of activity remain. Often the poorest households make the greatest use of traditional healers, although evidence has demonstrated that this is now changing (Heffernan and Misturelli, 2000). However, with the deterioration of public services in many countries, increasing attention is being paid to alternative healers and therapies. It has recently been acknowledged that the creation of community-based and farmer-led delivery systems demands an understanding of local notions and perceptions of livestock disease and healing to enhance sustainability.

Donors

As outlined in the introduction, donor-driven livestock aid has undergone a wide variety of changes since the 1960s. In general, the donors have been involved in all aspects of the delivery of livestock services. Traditionally, however, donor support has been targeted at the delivery of both clinical and diagnostic services and epidemiological surveillance. The majority of donor programmes have been implemented in co-operation and partnership with national governments. Indeed, it has been the donors that have instigated large policy shifts in the delivery of livestock services and have generally changed the delivery landscape in a variety of countries. For example, donors initiated the trend towards the privatisation of livestock services within the wider context of support for structural adjustment and decentralisation. In recent years, however, donors have recognised that large-scale national-level programmes may not be appropriate for poverty alleviation objectives. Therefore, funding has been increasingly channelled toward NGO partners.

NGOs

Not surprisingly, NGO activity within the livestock sector has mainly been focused at the community level. Indeed, NGOs have supported a wide variety of projects and programmes. For example, NGOs have been involved in direct livestock aid and animal health delivery, community outreach, training and technology transfer. Traditionally, however, projects and programmes have focused upon capacity building and support for alternate delivery systems such as CBAHW programmes. International and national-level NGOs have also supported the start-up of Community-Based Organisations involved in animal health, particularly in remote areas. Nevertheless, NGO activities are currently evolving with many NGOs taking a more active role in setting the animal health policy frameworks at the national level, and many agencies are becoming increasingly involved in the legislative environment and regulatory frameworks (Catley, 1997). Equally, there has been greater involvement of NGOs at the institutional level, with support for reform of animal health institutions a priority activity.

The following section further explores the differing roles of the aforementioned actors in livestock development projects and programmes.

The relationship between the providers and interventions

Over the years, actors have developed varying interests and levels of expertise in specific areas of livestock development. Table 5.3 offers a rough approximation of the level of involvement of the different actors in the livestock projects listed. Although exceptions naturally exist, the aim is not to categorise strictly the implementing agencies but rather to offer a general overview of both the level of involvement and the different interests of the wide number of players involved in livestock development.

Table 5.3 The level of involvement of actors in projects

Type of project	*Government-sponsored*	*Donor-sponsored*	*NGO-sponsored*
Direct livestock aid			
Restocking	x*	xx	xx
Micro-credit	x	xx	xxx
Support for improved production/animal health			
Animal health-care	xx	xx	
Service delivery	xx	xx	xx
Disease surveillance	xxx	xxx	x
Public health	x	x	
Feed development	xx	xxx	
Extension services	xx	xx	x
Community-based training		x	xxx
Institutional development			
Advocacy			xx
Capacity building and institutional development		xx	xx
Policy development	xx	xxx	x
Product development			
Processing/cooling/ storage	xx	xxx	x
Marketing	xx	xxx	xx

*the number of x's denotes the relative level of involvement of the different actors
Source: Livestock Development Group (2002).

As Table 5.3 illustrates, direct forms of livestock aid, such as restocking and micro-credit initiatives, have been primarily the domain of NGOs and donors. Indeed, historically NGOs have dominated the livestock micro-credit sector with the exception of multi-laterals such as the International Fund for Agricultural Development (IFAD) and the World Bank. Donor interest in restocking and micro-credit is by and large a more recent phenomenon.

Conversely, donors and governments have primarily supported projects intended to increase livestock production via improved animal health-care. As Table 5.3 demonstrates, livestock extension, disease surveillance and public health activities have almost exclusively been the domain of the donors and governments. The NGO sector in this area has concentrated mainly on alternate delivery systems such as CBAHW projects and programmes. NGOs have also been active in training programmes at both the community and household level. Advocacy in such diverse areas as land tenure and conflict resolution is also the traditional domain of NGOs, with donor-driven projects and programmes in this area only recently being implemented. Conversely, capacity and institutional development, at the national level, have been funded primarily by the donors, with community-level programmes the domain of NGOs. Policy development, as mentioned above, has generally been driven by the multi- and bi-lateral donors in co-operation with national governments. Finally, numerous actors have been involved in the processing of livestock products and the size and scope of projects are generally directly correlated to the implementing agency. For example, donors have generally supplied large-scale milk cooling and processing inputs with smaller and more local initiatives funded by NGOs.

Thus, similarly to other sectors, donors and governments have responded to public goods issues in the livestock sector, with NGO programmes more focused upon addressing local needs and realities. Nevertheless, the roles are changing and the remit of NGOs and donors appears to be broadening as government service provision declines. However, even with the predicted rise of the private sector and the further decline of government inputs, it is anticipated that the influence of donors and NGOs in the delivery sector will increase rather than decrease in the coming decades. The following section explores the evidence for impact of livestock development on poverty alleviation.

The impact of livestock development on the poor

There are three potential ways of determining the impact of livestock development projects and programmes on the poor:

- At the global level, information and evidence may be gathered on a project-by-project basis for the nations involved

- Criteria can be devised to evaluate the overall impact of specific types of projects, e.g. animal health, technology transfer, etc.

- An individual agency approach may be undertaken with specific institutions offering an assessment of their projects and programmes.

However, numerous obstacles have been noted with all three approaches. At the global level, co-operation and partnership between the actors and agents involved are often weak and fragmented, so determining the impact of specific livestock projects is difficult. More success has been achieved in analysing different forms and types of livestock projects. For example, Oakley (1998) and Martin (2001) offer overviews of Community Animal Health-Care Projects, and Heffernan *et al.* (2001a; 2001b) performed a large-scale review of restocking projects. Nevertheless, obtaining sufficient information from the actors and agents involved is often difficult with the project-level approach. At the agency level, little information is available in the public domain regarding the impact of specific livestock development projects and programmes. Few critical analyses of livestock sector activities have been performed, with the notable exceptions of DFID, the World Bank and IFAD.

Indeed, in 1998, DFID undertook one of the most comprehensive reviews of livestock projects and programmes to date (LID, 1999). Over 800 livestock development projects were reviewed for their impact on the poor. Overall, the authors concluded that the majority of livestock projects and programmes had not had a significant impact on the poor for the following reasons:

- Technologies were developed, but not delivered to the poor

- The technologies that were delivered were inappropriate to the poor

- In cases where appropriate technologies were successfully delivered, wealthier farmers or herders tended to capture the benefits.

LID (1999) stated:

> "Our review of project documentation on technical and service-related projects revealed little evidence of widespread sustainable impact on the livelihoods of the poor. Although there are some islands of success, the overall tenor of the literature, donor assessments and evaluation

> reports that we reviewed is that technical and service projects were not successful at benefiting the poor on a sustainable basis."

This finding was corroborated by de Haan *et al.* (2001) who reported the following in regard to World Bank projects:

> "The livestock portfolio analysis shows that our current World Bank operations still lack a specific policy and environmental focus...This lack of focus is shown by the low level of investment in the poorest regions of the world (Central Asia, South Asia and Sub-Saharan Africa) in pastoral development and small stock, and to some extent, in the low share of investments to improve animal health and nutrition, which are critical constraints faced by the poor."

De Haan *et al.* (2001) further note that since the 1970s there has been a decline in support for livestock projects:

> "Currently, six active agricultural projects are livestock only, and about 50 projects (of a total agricultural portfolio of 270) have livestock components. The decrease in lending is partially in response to the poor performance of the projects during the 1970s and 1980s."

Thus it is apparent that at the donor level there is the perception that livestock projects and programmes have not had their intended impacts.

Nevertheless, part of the problem is that the impacts on poverty of many livestock projects are difficult to define. Lessons learned are generally confined to the agencies involved and are often lost with the closure of projects and programmes. There are few formal communication pathways for disseminating project-related information between institutions. Indeed, no consistent framework exists to aid practitioners in project design and delivery across the livestock sector and best practice recommendations are few and far between. For example Heffernan *et al.* (2001a; 2001b) found that the low sustainability of restocking projects was often due to confusion over the project purpose: was it intended as a means of relief, rehabilitation or development. Poor targeting has also been noted to be a problem by practitioners involved in both restocking and CBAHW programmes. With sufficient attention, many of the problems impacting on sustainability may be easily resolved.

Conclusion

Resource-poor livestock-keepers are an important subset of the global poor. However, as a group they have widely varying opportunities and constraints, aspirations and motivations with regard to livestock keeping. In order for livestock production to be a positive force for poverty alleviation, a much more nuanced understanding of the role of livestock in the livelihoods of the poor is required. The first step, however, is to outline the numbers and location of resource-poor livestock-keepers. While progress has been made, it is clear that much further work is required in defining the different subsets of resource-poor livestock-keepers (cf. Chapter 2). Equally important is a better understanding of both the economic and the social role of livestock for the communities and households involved.

As a form of financial capital, livestock are not neutral assets, which simply accrue value. The risks of production are often large, particularly for the poor. Nevertheless, livestock are perceived by both the poor and the better-off as being the best investment strategy available. Consequently, projects and programmes need to address the specific risks involved for the communities in question while supporting the aims of producers. However, projects and programmes also need to incorporate more effectively the actual, rather than the perceived, needs of the poor. The farmers themselves viewed the lack of inputs as the largest constraint to production. Consequently, support for livestock-based livelihoods need not always entail complex and expensive technical solutions. Equally important is accounting for the future aspirations of the individuals and communities involved. Motivation is a key factor that has been largely overlooked in the field of development, yet it is a fundamental influence on the outcome of projects and programmes.

Livestock can be a key asset in the fight to eradicate global poverty but the rhetoric regarding the role of livestock in poverty alleviation often exceeds the actual evidence. Without sufficient understanding of the role of livestock in the livelihoods of the poor and of the constraints to livestock production, it is unlikely that current projects and programmes will live up to expectations. As such, there is an urgent need for a critical analysis of best practices with regard to pro-poor impacts within the livestock sector.

Further reading

Bravo-Baumann, H. 2000. *Gender and livestock: Capitalisation of experiences on livestock projects and gender*. Working Document, Swiss Agency for Development and Cooperation, Bern, Switzerland.

Bruggeman, H. 1994. *Pastoral women and livestock management: Examples from Northern Uganda and Central Chad.* Issue Paper No. 50. International Institute for Environment and Development, London, UK.

Conroy, C. and Rangnekar, D. 1999. *Livestock and the poor in rural India with particular reference to goat keeping.* Paper presented at the DSA Annual Conference, September 12-13, 1999, University of Bath, UK.

De Lasson, A. and Dolberg, F. 1985. The causal effect of landholding on livestock. *Quarterly Journal of International Agriculture* 24 (no. 4): 339-354.

Fattah, K.A. 1999. *Poultry as a tool in poverty eradication and promotion of gender equality.* Proceedings of a Workshop. In http://www.husdyr.kvl.dk

FAO (Food and Agriculture Organisation of the United Nations). 1999. *Poverty alleviation and food security in Asia: Role of livestock.* FAO, Regional Office for Asia and the Pacific, Bangkok, Thailand In http://www.fao.org

Leonard, D.K. 1993. Structural reform of the veterinary profession in Africa and the new institutional economics. *Development and Change* 24: 227-267.

Lukefar, S. and Preston, T. 1999. Human development through livestock projects: Alternate global approaches for the next millennium. *World Animal Review* **93**: 24-25.

Perry, B.D., McDermott, J.J., Randolph, T.F., Sones, K.R. and Thornton, P.K. 2001. *Investing in animal health research to alleviate poverty.* International Livestock Research Institute (ILRI), Nairobi, Kenya.

Author addresses

Claire Heffernan, Livestock Development Group, School of Agriculture, Policy and Development, University of Reading, Earley Gate P.O. Box 237, Reading RG6 6AR, UK

Louise Nielsen, Livestock Development Group, School of Agriculture, Policy and Development, University of Reading, Earley Gate P.O. Box 237, Reading RG6 6AR, UK

Ahmed Sidahmed, International Centre for Agriculture Research in the Dry Areas (ICARDA), P.O. Box 5466, Aleppo, Syria

Tofazzal Miah, Department of Agricultural Economics, Bangladesh Agricultural University, Mymensingh 2202, Bangladesh

6

Knowledge – key to empowerment

Chris Garforth[*], *Ramkumar Sukumaran*[†], *Dan Kisauzi*[‡]

Cameo 1 - Information kiosk in India

Seeta is a landless dairy cattle owner in a rural area of India. Her only regular income is from selling milk each day to the local co-operative. Last year, she had to sell one of her two cows because it developed severe mastitis and could no longer be milked. She now has difficulty buying food for her family. Her remaining cow has not conceived after six inseminations at the village cattle insemination centre run by a Veterinary Field Assistant (VFA). At her most recent visit to the centre, she saw a new touch-screen information kiosk, which displayed information about animal husbandry stored on a computer inside the kiosk with an audio backup. After being shown how to use it, she browsed through the information on the causes and prevention of mastitis and on dairy cow fertility. She has now decided to do all her milking herself instead of leaving it to a professional milker, making sure she washes her hands and the cow's udder first. She has also taken her cow to the nearest veterinary surgeon to diagnose the reason for non-conception.

Cameo 2 - Knowledge through a farmers' association in Kenya

Elizabeth and John farm one hectare of high potential rain-fed land in central Kenya. They grow coffee, but the price they can get for their coffee beans no longer covers the cost of producing and transporting them to the factory. Three years ago, they joined a farmers' association that promotes rearing of cross-bred goats. Through the association's training, they learned how to house and manage the goats to give a good level of production. They have planted forage species on some of their coffee land. They have also learned the importance of controlling breeding to maintain the genetic quality and

---

[*] Principal author
[†] Co-author
[‡] Consulting author

composition: they take their female goats to one of the association's bucks and castrate the male offspring. The local market for goat's milk is picking up and they sell the offspring for a good price. Their new knowledge has helped them fill the gap left by the collapse of coffee prices and membership of the association has improved their access to other services.

Introduction

The two cameos above show how lack of knowledge can contribute to poverty while new knowledge can open new livelihood opportunities. For a small-scale rural or peri-urban cattle owner, loss of a single cow is a devastating event. Preventing that loss and increasing the cow's productivity can make the difference between a sudden slide into poverty and a slow increase in the family's ability to meet essential expenditure for food, children's education, housing and health care. Taking advantage of new enterprises and opportunities requires learning new knowledge and skills. Improved access to knowledge can therefore increase resource-poor livestock-keepers' control over their production systems and livelihoods. This access is influenced by the institutional context in which knowledge is created and made available.

This chapter first explores the meaning of 'knowledge' and discusses the relationship between poverty, power and knowledge. It then looks at how the way knowledge is generated and disseminated can either improve or worsen the situation of resource-poor livestock-keepers.

Knowledge and information

People often use these two words together and they sometimes seem to mean the same. However there is a difference: we can pass information from person to person while what we know – our knowledge – is personal to us (Röling, 1988). Livestock-keepers can gain new knowledge only by being exposed to information, which may come from their own observations, talking with other farmers, listening to the radio, attending seminars or training activities, or other sources.

However, we also talk about 'the body of knowledge' within a discipline as something which exists separately from the people who have that knowledge. The body of knowledge on poultry management, for example, is documented in textbooks, databases and scientific papers. It grows and changes as new research is published. Scientists cannot keep all this detailed knowledge in their heads, but they know how and where to find it. A challenge for resource-poor livestock-keepers, on the other hand, is knowing what is available within the body of knowledge relevant to their animals, and then how to access it.

'Local knowledge' is the body of knowledge among people in a particular area. Among livestock-keepers in a rural area, there is a body of local knowledge related to the management of animals. This does not mean that everyone in the area has all this knowledge. Some people are more knowledgeable, or have different knowledge, than others. These patterns are often related to social and economic differences (Chapters 2 and 5). Where women collect forage for animals while men carry out draught operations, women may have more knowledge about the nutritive values of different species while men know more about techniques of harnessing animals and maintaining equipment. Those who look after animals may develop more detailed knowledge than those who own them. People who travel outside their home area will have a richer set of knowledge than others. This uneven distribution of knowledge raises the question of how resource-poor livestock-keepers can gain access to knowledge within the bodies of both scientific and local knowledge. We return to this question later.

We can group the types of knowledge relevant to livestock production under five main headings (Table 6.1). We can also distinguish between different levels of knowledge, in particular between principles, facts and skills. There is a difference, for example, in knowing: (a) that restricting an animal's movement increases its rate of weight gain for a given level of food intake; (b) that stall-fed animals produce more; and (c) how to build a low-cost goat shed from local materials. Knowing 'how' (skills) is not much use unless we know 'why' (principles and facts). Any programme to increase resource-poor livestock-keepers' access to knowledge needs to address all three levels.

Table 6.1 Types of knowledge in livestock production

Knowledge of:	*Examples*
Biological processes	Life cycle of parasites
	Ruminant digestive system
Management practices	De-worming
	Castration
	Washing udder before milking
New technology	Planting of new forage species
	Artificial insemination
Markets	Current prices, and price trends, in local markets
Institutional processes and requirements	Regulations governing animal welfare
	How to obtain credit
	Rights to service and support from public institutions
	Livestock insurance

and livestock within mixed farming systems. Various international organisations play important roles in funding public research or making the findings available to livestock-keepers, either directly or through intermediaries such as advisory services and NGOs. Table 6.2 shows a selection of organisations which support the generation of and access to knowledge for resource-poor livestock-keepers.

When the CGIAR was first set up in 1971, IARCs were not particularly focused on poverty issues. Their task was to generate knowledge from basic scientific research that could be applied in farming systems around the world. Downstream applied and adaptive research was the responsibility of the national agricultural research systems (NARS) which typically consist of government research institutes and universities. Two things have happened since then. First, scientists and those who use their research findings have realised that much of the knowledge and technology they have generated does not fit into the production systems and the economic situation of resource-poor farmers. Second, governments around the world and international agencies have agreed a set of Millennium Development Goals for the early years of the twenty-first century, in which the alleviation of poverty has a high priority (MDG, 2003). The IARCs have, therefore, refocused their work to ensure that it contributes to the achievement of this goal (Perry *et al.,* 2002).

This refocusing has occurred also within the NARS, particularly in countries where the research system receives financial support from donors. Donors have encouraged a reorientation in national development policies and programmes towards alleviating poverty. One of the main vehicles for this is the Poverty Reduction Strategy Papers (PRSPs) that governments are expected to have in place as a framework within which donors provide development assistance. In this context, NARS are reforming themselves to ensure that their research is driven by the needs and demands of resource-poor farmers. Mechanisms for this include:

- decentralisation, so that more decisions about what research is done are made at regional and local levels, rather than at national headquarters
- giving farmers a direct voice in deciding research priorities
- competitive research funding, where research institutes submit proposals and the criteria for deciding between competing proposals include the extent to which they address farmers' problems.

Local and individual knowledge expand and develop through less formal processes: observing what happens to livestock during routine management; sharing and exchanging information between farmers and within households; trying out new ways of doing things and drawing conclusions from the results; interacting with

Table 6.2 International organisations which support the generation of livestock knowledge in developing countries

Organisation	*Acronym*	*Mandate*	*website (http://.....)*
International Livestock Research Institute	ILRI	Basic and applied research on a wide range of topics relevant to improving the livelihoods of resource-poor livestock-keepers	www.ilri.org
World Agroforestry Centre	ICRAF	Integration of livestock within agro-forestry systems, including selection and management of forage species	www.worldagroforestrycentre.org
International Institute for Tropical Agriculture	IITA	Research on crop-livestock systems, including dual purpose crops providing improved human and animal nutrition	www.iita.org
International Food Policy Research Institute	IFPRI	Research on institutional and policy issues affecting production, marketing and availability of food, including livestock products	www.ifpri.org
Food and Agriculture Organisation of the United Nations	FAO	Provides technical support and information resources in all agricultural sectors	www.fao.org
International Bank for Reconstruction and Development	World Bank	Provides funding for projects and programmes, including reform of national research and extension systems	www.worldbank.org
International Development and Research Centre (Canada)	IDRC	Provides funding for research that benefits people in developing countries, including knowledge systems, urban agriculture and management of natural resources	www.idrc.ca
Department for International Development (UK)	DFID	Funds bilateral development programmes including support to agricultural sectors in developing countries; funds livestock production and animal health research	www.dfid.gov.uk
Department of Research Co-operation (Sweden)	Sida-SAREC	Provides funding to national and regional research institutes in a wide range of subject areas, including animal husbandry	www.sida.se

people from other areas; and being exposed to information based on the body of scientific knowledge.

People disagree whether any of these local knowledge processes can be labelled 'research' and 'experiments'. Many scientists argue that these words can only be used to describe investigations designed according to established scientific protocols, with rigorous control of the conditions under which the investigation is conducted and systematic recording of observations for subsequent analysis. Others argue that farmers who try out new ideas and compare the results with existing practices, and exchange their experiences and observations with one another, are doing the same as scientists. One thing, however, is clear: collaboration between the two systems is essential in order to generate knowledge that can be applied by, and bring benefit to, resource-poor livestock-keepers.

Collaboration between formal and local knowledge generation processes generally comes under the label of 'farmer participatory research' (FPR). There are several forms of FPR, depending on the level of involvement of farmers in: the selection of themes and topics for research; the design of experiments; recording of observations; and interpretation of results. The rationale for FPR with resource-poor livestock-keepers is that the knowledge generated is likely to:

- be more *relevant* to the needs and opportunities of resource-poor livestock-keepers
- be better *adapted* to their situation and resources
- contribute directly to both *scientific* and *local bodies of knowledge*
- be *taken up more quickly,* at least in the area where the research is conducted (see Box 6.2).

Figure 6.1 shows three different ways of representing the interaction between scientific and local knowledge. In part (a), interaction is seen as a simple linear process, with the NARS taking the findings of basic research done by IARCs and developing practical innovations through applied and adaptive research, which farmers then integrate with their own knowledge. Part (b) is a more realistic representation of the relationship between IARCs and NARS, showing that both are involved in basic and applied research, though in different proportions. In part (c), three circles represent the knowledge generation activities of IARCs, NARS and farmers, with the shaded areas indicating activities in which farmers interact directly with scientists in FPR.

Box 6.2 Farmer Participatory Research

Rearing pigs on a small scale is common in towns in Vietnam. Farmers feed them sweet potato roots and vines with rice bran, corn and cassava roots. Labour and storage are major constraints to using sweet potato vines. Vines must be chopped into small pieces, a time-consuming task. Using vine silage overcomes both constraints: women can process vines during the off-season when there is more labour, and store the silage for use when feed is limited. Research trials by scientists and farmers in a village setting showed that fermenting vines with dried chicken manure, both readily available and cheap, improves weight gain and yields better profit. At a meeting after the feeding trial, forty women enthusiastically copied the suggested method. Along with profitability, the women considered the labour saving and storage potential very significant.

Source: Peters *et al.* (2000)

Knowledge transfer and communication

Knowledge is useful only if it is available to those who can benefit from it. The process by which knowledge is transformed into management and production practices by livestock-keepers is generally called 'knowledge transfer'. As suggested earlier this is a complicated and uncertain process, because we cannot directly transfer knowledge from one person to another. We can provide information, based on a body of knowledge, and share that with others through communication. We cannot predict, however, precisely what effect this will have on their knowledge. They may reject the information as incorrect or irrelevant, or add it to their existing knowledge, or modify their existing knowledge in the light of the new information. Among the resource-poor, who for obvious reasons try to minimise the risk that comes from changing the way they do things, it may take a long time for a new piece of information to be accepted so that it leads to a change in the individual's knowledge and practice. Some people may never accept it, which leads to a situation where there are big differences in knowledge between the individuals within a community.

We need to take two facts about communication into account when planning or analysing knowledge transfer:

(a) traditional view of the link between scientific and local knowledge generation

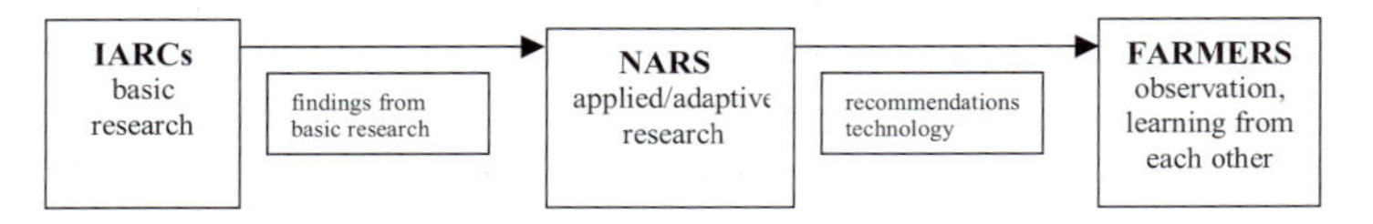

(b) overlapping roles of IARCs and NARS in basic and applied research

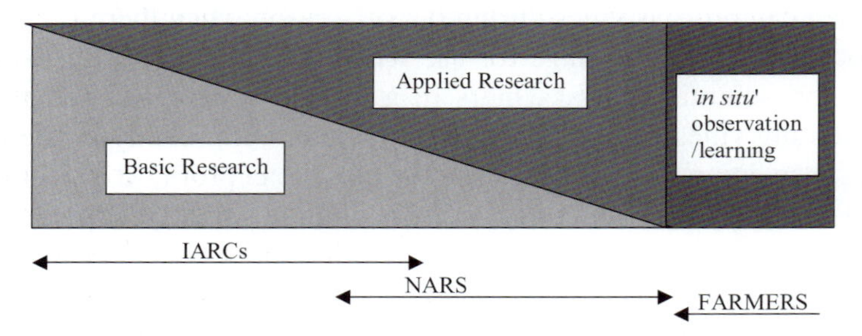

(c) interaction between scientific and local knowledge through farmer participatory research

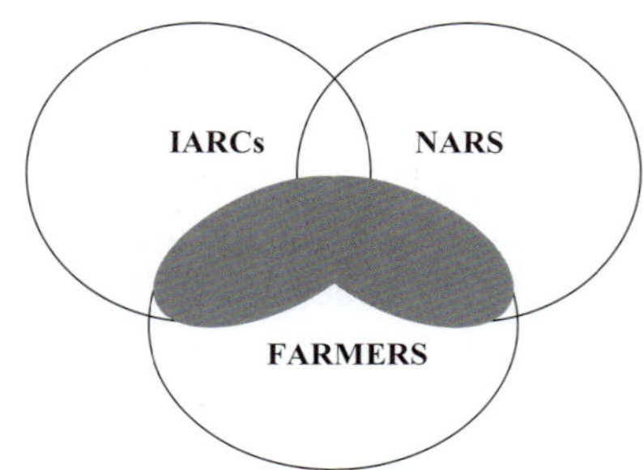

Figure 6.1 Changing views on interactions between formal and local knowledge systems

- Information is expressed in codes or language. The language of a scientific discipline is a code used by those who write papers for journals and books that will be read by others working within that discipline but will not necessarily be understood easily by scientists in other disciplines and certainly will not mean much to the general public. The language of newspaper journalism is another code, within which words have particular meanings and which people who regularly read newspapers can interpret easily. Photographs and diagrams represent two different kinds of visual language, which we learn to interpret through exposure to them in our particular culture. The language which people use in everyday conversation is very different from these other codes. The language in which people express local knowledge is very different from that of scientists. Resource-poor livestock-keepers are less likely than their better off neighbours to have had the opportunity to learn the languages or codes in which information coming from outside their local social and production system is expressed. This puts a responsibility on those who provide information for resource-poor livestock-keepers to make sure the code they use is appropriate.

- Information travels through channels which include face to face communication, broadcast mass media, printed material of various forms from posters to leaflets and newsletters, and telecommunications. Access to channels varies between individuals and households. Radio, for example, is widely available in many rural areas: research in Uganda showed that nearly all men and women in subsistence farming households listen to the radio while only one in five men and one in ten women had used a telephone over the previous twelve-month period. In the hills of Nepal, women have less interaction than men with sources of information from outside their village. Information will only reach resource-poor livestock-keepers if it is available through channels to which they have access (Box 6.3). Poorer farmers generally have less access to external sources of information and knowledge than farmers with more resources: while they may be rich in local knowledge, they are 'information poor' in terms of access to external bodies of knowledge.

Box 6.3 Wambui finds out

Researchers in Kenya produced a series of highly illustrated booklets on livestock production with content specifically designed for small-scale producers in central Kenya. Nearly all poor livestock-keeping households in the area have children at primary school. With the agreement of head teachers, the research team distributed the booklets through primary schools. Teachers used them as teaching material, and school students took them home to their parents. This ensured that nearly all households in the area received the booklets. Where the parents could not read, their children could explain the contents to them. Earlier testing showed that the intended recipients of the information appreciated and could interpret the code used – a highly illustrated, story-book format, with a conversational style of dialogue between the characters in the stories.

Source: LPP Project R7425

One way of exploring which channels are suitable for making information available to resource-poor livestock-keepers is by drawing up an information map (Box 6.4).

Box 6.4 Developing an information map with livestock-keepers

- Assemble a group of resource-poor livestock-keepers.
- Divide a large sheet of paper into three sections, representing the local area, the district or regional level, and the national and international level. Draw a circle representing the farmers in the middle of the 'local area' section.
- Beginning with the local area, ask participants who they get information from, and who they communicate with. Draw a circle for each one and draw a line between each circle and the circle representing the farmers.
- For each of these sources or contacts, ask participants to describe (and make notes against the lines on the paper, or on a separate piece of paper):
    - what kinds of information are exchanged between them and the contact?
    - how frequently are they in contact?
    - what are the contact's advantages and disadvantages, as a source of information on livestock production?
    - how easy it is to make contact (e.g. are they always available, or is it difficult to find them?).
- Repeat the process for sources at district/regional and national/international levels, including mass media (radio, television, newspapers).
- Once the 'information map' is complete, ask participants what they think are their main information gaps, and their main difficulties in getting access to useful information.
- Compare the information maps produced by different categories of farmer (for example, men and women; those with dairy cattle and those with poultry) and see if there are any differences which have implications for the design and implementation of knowledge transfer activities.

Changing models of communication

The way we think about communication affects the way we do it. Typically in agricultural college and university courses, students learn that communication is a process where a source sends a message through a channel to a receiver, with the intention of creating an effect. This is the conventional Source-Message-Channel-Receiver-Effect (SMCRE) model of communication (Figure 6.2), which leads us to think of channels of communication as ways of sending information to other people. But most human communication is a two way exchange, not a one way transfer.

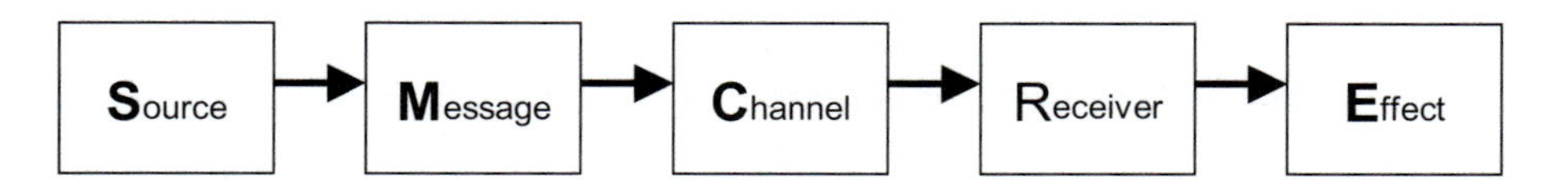

Figure 6.2 The SMCRE Model of Communication (Source: Berlo, 1964)

A more helpful way of thinking about communication is shown in Figure 6.3. This 'convergence' model shows how two parties in a communication process start from different sets of knowledge and different perspectives on the subject of their dialogue, but are seeking to find a common understanding. If successful, the communication process will result in convergence of those perspectives: both parties will have changed to some extent. If we think of communication in this way, we can see channels as opportunities to conduct a dialogue and achieve convergence. We develop this point further when we consider the interaction between livestock production professionals and livestock-keepers in the final section of the chapter.

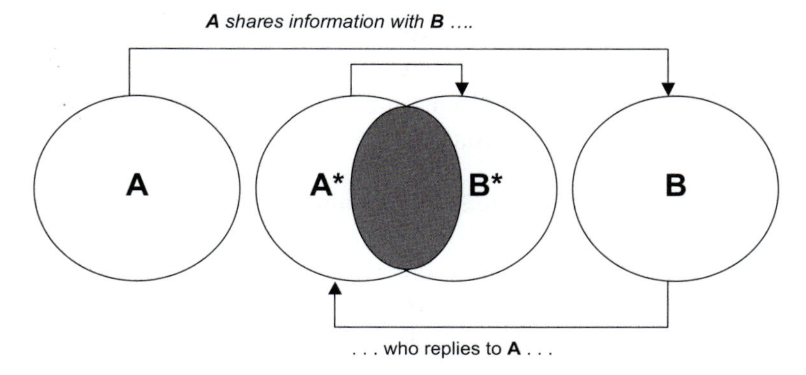

Figure 6.3 Convergence model of communication (Source: Rogers and Kincaid, 1981)

Supply driven and demand driven knowledge and information systems

Changes in models of agricultural research (Figure 6.1) and communication (Figures 6.2 and 6.3) indicate a shift from thinking mainly about the supply of knowledge and information, to thinking about how we can meet the demand for knowledge and information. This perspective recognises that people are not just passive receivers, but active seekers of information. Figures 6.1(a) and 6.2 represent supply-led processes. Figures 6.1(c) and 6.3 bring the voice of the user of knowledge directly into the picture. But demand itself requires a certain level of knowledge among resource-poor livestock-keepers, including knowledge of what kinds of information are available or could be generated, knowledge of how institutions and systems work and knowledge about their own rights. Expressing demand for knowledge and information also involves costs. A woman who keeps a few poultry in her backyard may not express demand for information on how to prevent her birds developing common diseases, either because she does not know that there is a poultry specialist in the district animal production office, or because she cannot afford the transport fare or the time to travel to the district town, or because within the social system it is not acceptable for a woman to travel on her own.

Changing a supply driven to a demand driven system may, therefore, require some action to stimulate demand. Figure 6.4 illustrates how this can be done through the mass media. Various organisations have an interest in helping resource-poor livestock-keepers improve their livelihoods, including government departments, politicians, NGOs and community based organisations (CBOs). Through radio or other mass media, they can tell livestock-keepers what services are available, and provide information on livestock management practices that will encourage them to seek further information and training from relevant professionals. Other ways of doing it include encouraging the formation of groups whose members can then share the cost of seeking out information and find a stronger voice and greater confidence in approaching organisations and professionals to demand services.

Rapid developments in information and communication technologies (ICTs) are making it easier for livestock-keepers to express their demands and for organisations to respond with appropriate information. Telephone infrastructure, both landlines and mobile phone networks, has expanded rapidly into rural areas. Countries where there used to be a single, government-owned radio station, now have multiple local stations, some run by governments, but others run by commercial companies or NGOs. Digital cameras and desktop computers and printers are making it cheaper and easier to produce quality print material for local use. However, as noted earlier, poorer people are less likely than others to have access to the means of communication. Governments and NGOs have a role here in bridging the gap between resource-poor livestock-keepers and the opportunities which ICTs offer.

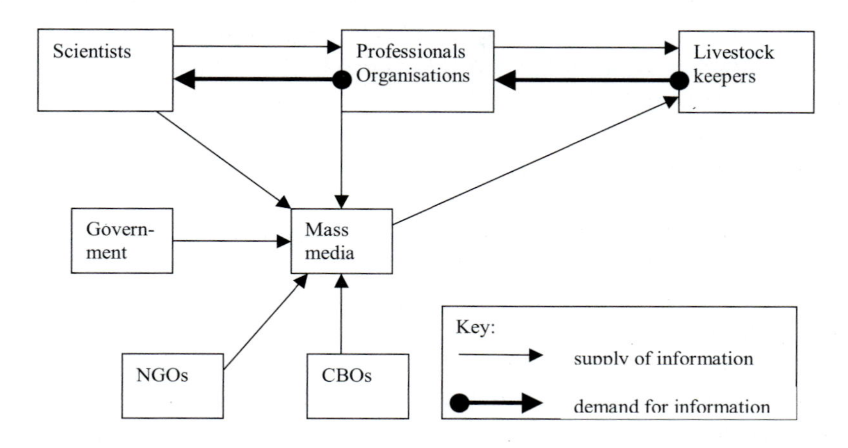

Figure 6.4 Stimulating the demand for information and knowledge transfer

Extension and advisory systems, too, are changing. Until the late twentieth century, most extension services were both funded and delivered by governments. Now systems are more pluralistic, with livestock-keepers able to access knowledge, information and other services from NGOs, private individuals, firms and farmers' organisations as well as from government departments. This greater variety of sources is able to meet the diverse needs within the livestock sector; but, again, the needs of resource-poor livestock-keepers will not be met adequately unless governments or other bodies take action. One way of doing so is to contract out the delivery of advisory services to private service providers, whose contract specifies the criteria for selecting clients and the nature and level of service they must provide (see Rivera and Zijp, 2002, for examples).

Livestock professionals as knowledge brokers

For many resource-poor livestock-keepers, a key figure in their own knowledge and information system is the local professional, working in either the private or the government sector, who has been trained in livestock production or animal health. He or she may be based in a veterinary facility, an animal production department of central or local government, a commercial or co-operative dairy, an NGO or a local agricultural college.

Knowledge gained from formal training forms the foundation of the professional's knowledge base. This will always help in the search for solutions to specific problems or questions of livestock-keepers, but will rarely be enough to provide a complete answer. Professionals need to blend their knowledge base with an understanding of the local context of livestock production and farmers' wisdom. This interaction between formal and local knowledge is important in arriving at practical options to address livestock-keepers' problems (Box 6.5).

Box 6.5 Blending local with professional knowledge

> A farmer who wants to improve her backyard pig production approaches a lecturer in the local agricultural college for advice on what to feed the animals. Instead of giving a scientific feeding schedule based on what the standard textbooks say, the lecturer should first of all find out from the farmer what is available locally, in the form of agricultural and industrial by-products, and household and kitchen wastes. He or she can use that information, together with their own knowledge of pig nutrition, to come up with a recommended schedule that provides reasonable quality and quantity of nutrients at low cost.

Livestock owners know their situation better than the professionals. The inputs or resources they have and their objectives from rearing livestock (see Chapter 2) need to be recognised before venturing into professional knowledge dissemination. Listening to farmers is an important dimension of knowledge transfer. Listening initiates a process of thinking and dialogue, which improves the scope for incorporating farmers' ideas with the professional's thoughts. This helps to develop an integrated and useful version of knowledge.

An important factor in the relationship between livestock professionals and livestock-keepers is credibility. Resource-poor livestock-keepers will not accept any information or advice offered unless they feel that it can be trusted: they cannot afford the risk of following bad advice. In assessing the trustworthiness of the information and of the knowledge that lies behind it, they will ask themselves two questions: does this information make sense, does it seem reasonable? and is this professional someone I can trust to give me good, reliable information? In other words, the credibility of the information is linked closely to the credibility of the person from or through whom it comes. Information that is rejected as untrustworthy will not be absorbed into the livestock-keeper's knowledge.

Credibility works in both directions. If livestock-keepers do not trust a professional, they are unlikely to be open in what they say about their situation, their problems and their aspirations. Knowledge-generating institutions, who have greater interaction with livestock professionals than with livestock-keepers, may then end up with an inadequate or incomplete understanding of the context for which they are conducting research. Establishing and maintaining the trust of the livestock-keepers one works among is a vital for livestock professionals.

One way in which local and scientific knowledge can interact to create local solutions to resource-poor livestock-keepers' problems is the Farmer Field School (FFS: Box 6.6).

Box 6.6 Livestock Farmer Field Schools (FFS) in Kenya

Ten FFS groups of 30 to 35 farmers with similar interests were established in five different agro-ecological zones in Kenya. Facilitators worked with groups of farmers to identify the main constraints to improved milk production. Groups prepared grant proposals with detailed workplans and budgets. A maximum grant of US$ 600 was deposited in an account controlled by elected members of the group to cover the cost of field activities and facilitation. Management of this budget empowered the farmers to demand and control activities covered by the FFS and ensured that the extension services offered responded to farmers' priority problems and needs. All FFS groups meet weekly, from 9 to 12 am. In each session farmers work with the extension facilitator in a structured manner, organised by a different set of farmers each week nominated from the group. The main activity is participatory technology development (PTD) in which farmers focus on solving local problems through a process of collective and collaborative inquiry. The PTDs empower participants (both farmers and facilitators) with analytical skills to investigate cause and effect relationships of problems in farming practices.

Source: Minjauw *et al.* (2004)

Conclusion

Poverty and powerlessness are closely related. Lack of economic and political power makes it difficult for the poor to access resources that would help them move out of poverty. Improving their access to knowledge and greater participation in the creation of knowledge can make an important contribution to their empowerment. For this to happen, research systems, extension and advisory services and individual professionals must continue to change the way in which they interact with resource-poor livestock-keepers.

Questions / exercises for class discussion

1. In the field, with different groups of resource-poor livestock-keepers: draw up an information map with each group and discuss the implications of the maps for how you might design a knowledge transfer strategy.

2. In class: look back at the three scenarios in Box 1. Discuss what could be done to reduce the power that one person in each scenario is exercising over the other.

Further reading

Garforth, C. 2004. Knowledge management and dissemination for livestock production: global opportunities and local constraints. In: *Responding to the Livestock Revolution – The role of globalisation and implications for poverty alleviation* (ed. E. Owen, T. Smith, M. A. Steele, S. Anderson, A.J. Duncan, M. Herrero, J.D. Leaver, C.K. Reynolds, J.I. Richards and J.C. Ku-Vera), pp. 287-298. British Society of Animal Science Publication No. 33, Nottingham University Press, Nottingham, UK.

Leeuwis, C. (with contributions from van den Ban, A.). 2004. *Communication for rural innovation: Rethinking agricultural extension*. 3rd Edition. Blackwell Publishing, Oxford, UK.

Rivera, W.M. and Zijp, W. 2002. *Contracting for agricultural extension. International case studies and emerging practices*. CABI Publishing, Wallingford, UK, and New York, USA.

Websites with useful information on knowledge issues relevant to resource-poor livestock-keepers include:

FAO Sustainable Development Department: http://www.fao.org/sd

The Communication Initiative: http://www.comminit.com

World Bank's Agricultural Knowledge and Information Systems network: www.worldbank.org and follow the links Topics in Development, Agriculture and Rural Development, Topics, Agricultural Knowledge and Information Systems (http://lnweb18.worldbank.org/ESSD/ardext.nsf/26ByDocName/AKISResearchExtension)

Author addresses

Chris Garforth, Department of International and Rural Development, School of Agriculture, Policy and Development, University of Reading, Earley Gate, P.O. Box 237, Reading RG6 6AR, UK

Ramkumar Sukumaran, Department of Veterinary and Animal Husbandry Extension, Rajiv Gandhi College of Veterinary and Animal Sciences, Kurumbapet, Pondicherry 605009, India

Dan Kisauzi, Nkoola Institutional Development Associates, P.O. Box 22130, Kampala, Uganda

7

Livestock products – valuable and more valuable

Trevor Wilson[*], *Anne Pearson*[†], *Nicola Bradbear*[‡], *Ashini Jayasuriya*[‡], *Henry Laswai*[‡], *Louis Mtenga*[‡], *Sarah Richards*[‡], *Rose Smith*[‡]

Introduction

Livestock are not raised simply for their own sake but to fulfil interacting and often conflicting human nutritional, economic and environmental needs. They contribute to human welfare and people's livelihoods in many ways (cf. Chapters 2 and 5). In particular, however, they serve in four ways:

- In poverty alleviation, they are often the only assets of many of the landless poor, their products (milk, meat, eggs, wool, hides and skins) provide a direct or indirect source of income throughout the year, and they are a means of capital accumulation and provide a cash buffer in times of need.
- In food security, the milk and eggs they produce are the only agricultural products that can be harvested every day throughout the year. Livestock can be productive throughout the year where crop agriculture is difficult or impossible. Animals provide draught power, without which crop production in many areas would be severely compromised, and they make use of crop and agro-industrial by-products and waste and convert them to high quality human food.
- With regard to the environment and its conservation, they produce manure that contributes to sustainable nutrient cycling and maintenance of soil fertility and structure, as well as to bush and weed control in many areas.
- Small animals are often owned by women. Furthermore, women and children have priority access to livestock products for consumption or sale. Draught animals reduce much of the drudgery of women's work. Therefore livestock keeping increases gender equity.

---

[*] Principal author
[†] Co-author
[‡] Consulting author

The outputs from the keeping and husbandry of livestock can be considered under the three families of immediate, intermediate and indeterminate products (Table 7.1). In each of the three product families it is possible to provide added value through increased quantities and improved qualities of product and, in many cases, through better presentation of processed items and more aggressive marketing. There are also opportunities, particularly in the immediate family of products, for greater use of non-conventional species and promotion of niche and speciality articles.

Table 7.1 Animal products by type of 'family'

Immediate	*Intermediate*	*Indeterminate*
Meat	Farm draught power	Reduction and spread of risk from crop operations
Milk	Off-farm transport	Generation and accumulation of capital
Eggs	Industrial applications (oil mills, etc)	Generation of income and smoothing out cash flow
Fibre	Manure as fertiliser	Fulfilling social, cultural and religious needs and obligations
Hides and skins	Manure as fuel and for biogas production	Providing status or 'prestige' in the immediate community
Feathers	Weed control	Sport, culture and recreation

Setting the scene

Protein-energy malnutrition, vitamin A deficiency, iodine deficiency disorders and nutritional anaemia are the most common nutritional problems in the developing world (Latham, 1997). Livestock food products and other products of domestic animals that assist in generating income can contribute greatly to alleviating these problems. Two examples help to set the scene and highlight the possibilities of adding value to livestock and livestock enterprises:

- "Honey production was not a big traditional economic activity in Somaliland. Some pastoral people harvest honey and would either consume, give out or sell the product away. Today, honey has a big market in urban areas and pastoralists are aware of the fact. There is a trend now for some pastoralists catching bee queens and selling them to urban people who produce honey. Many pastoralists are becoming very much aware of the domestication of the beehives and its commercial use as an alternative income. Some of the favourite plants species known for honey making are in danger of getting extinct like the Dibow, which gives a distinct taste and logging down forests for charcoal and fencing are also endangering the survival of bee colonies. In agro-pastoral areas, the case is different. Some families have started to harvest honey and sell it in urban areas. However, the amount is not sufficient for consumption and the price is very high. At the time of writing this report, a kilogram of honey cost around 40,000 Somaliland shillings, equivalent to 8 US dollars. Honey is often in high demand because people value it for medicinal purposes." (Sadia M Ahmed *et al.*, 2001).

- "From 1970 to 1995 poor countries increased their consumption of milk and meat by 175 million tonnes. That is more than twice the increase in developed countries and over half as large as the increased consumption of cereals made possible by the Green Revolution. The market value of the increase in milk and meat production was 153 billion US dollars or more than twice the value of the increased consumption of wheat, rice and maize." (ILRI, 2003).

Food products

Overview

Animal products provide the best quality protein in the human diet as they are able to provide essential amino acids, such as arginine, that the human body cannot manufacture for itself. Meat, milk and eggs are the major products but many other outputs are important in some areas and among some societies. The British statesman Winston Churchill may have been guilty of a terminological inexactitude (as he himself would have said) when he stated that "there is no better investment than putting milk into babies" but he was certainly on the right track. Low animal protein intake can result in a high incidence of kwashiorkor in children and high infant mortality; and in adults protein malnutrition can lead to weakness and impairment of the immune system, which can predispose to disease. A restricted protein intake can also lead to disturbances in growth and development in children that not only lead to growth retardation but also to irreversible reduced mental development. The recommended total minimum protein intake for an adult is 85.9 g daily of which 34 g (40 per cent) should be of animal origin (Bender, 1992).

Animal products supply about 17 per cent of the energy and 32 per cent of the protein eaten by people (Bender, 1992). There are of course considerable regional differences, but the main demand in the future can be expected to come from the developing countries as incomes increase and the demand for a more varied and higher quality diet intensifies (cf. the Livestock Revolution). There is thus an almost tailor-made market for higher output of food products from domestic animals in areas where the demand is greatest, the cost of production is lowest and the potential for adding value is most favourable. There is also increasing evidence that high-protein high-energy diets may have a role in reducing susceptibility to HIV and reducing the impacts of AIDS (Beaugerie *et al.*, 1998; Kadiyala and Gillespie, 2004; Stack *et al.*, 1996). Presumably this suggests diets rich in animal protein.

The above notwithstanding, it should be noted that there is currently a vogue in developed-country medical teaching for advocating vegetarianism as the answer for a long and healthy life. This is based on the perception that vegetarian diets lead to better bowels, whereas red meat consumption is linked to colorectal cancer.

Meat

Meat and its products are sources of high quality protein. The composition of amino acids in meat usually compensates for those that are deficient in staple diets that rely largely on cereals as food. Meat also supplies iron that is easily absorbed and assists the absorption of iron from other foods, in addition to assisting absorption of zinc. These products are also rich sources of some B group vitamins. The consumption of meat and meat products can clearly help to alleviate some common nutritional deficiencies (Bender, 1992).

The over-consumption of meat has been associated with some health risks. The ingestion of large amounts of saturated fatty acids can raise blood cholesterol levels that may predispose to blood vessel narrowing, resulting in coronary heart disease or stroke. In many areas up to a quarter of the saturated fatty acids in a diet derive from meat and this has been used by some organisations to discourage the consumption of meat. However, these health issues are primarily seen in the westernised world where there is a much higher intake of foods rich in saturated fatty acids and processed food; it is of far less relevance in the developing world. There are also many other risk factors associated with heart disease and stroke including family history, smoking of tobacco, diabetes, lack of exercise, high blood pressure and other disease states. Almost all dietary cholesterol is supplied by meat, milk and eggs but reduced intake of these foods gives rise to other nutritional problems. They are therefore crucial dietary components but should not be consumed to excess, particularly by susceptible individuals. In short, most national dietary guidelines recommend a reduction in cholesterol intake, with a maximum ingestion of 300 mg per day (see Pearson and Dutson, 1990, for broad-ranging views and recommendations). However, it is now thought (in medical circles) to be the amount of dietary fat rather than the amount of dietary cholesterol per se that is important in controlling cholesterol levels in terms of heart disease (because fat becomes cholesterol). Dietary guidelines usually include some recommendation to substitute red meat (including especially pig meat) with white or poultry meat, although some such meats are also high in saturated fatty acids. In addition lean red meat, such as that from tropical fat-tailed sheep and goats, contains no more fat than white meat.

A great deal of meat is eaten fresh but this may mean that its value is reduced as in many hot countries it cannot be transported far and its 'shelf life' is short. The shelf life of fresh meat can, however, be lengthened, and thus the value of the primary product increased, by paying attention to hygiene at all stages from slaughter through to the shop or the consumer. Value can be added to the basic fresh meat - and has indeed been added for many thousands of years - by traditional methods of preservation. These include salting, drying and smoking or combinations of these processes. Dried meat is known by a number of traditional names in various parts of the world, including 'pastrami' in the Near East, 'charqui' in South America, 'kwanta' in Ethiopia and 'biltong' in southern Africa. Under suitable climatic conditions of hot sun and a dry atmosphere, very good quality dried meat can be produced. Damp conditions or poor storage lead, however, to spoilage due to bacterial infection. Simple drying is suitable for small-scale production and avoids the high capital, operating and maintenance costs of sophisticated equipment. For slightly larger lots, simple solar driers can be constructed and used. Flavour can be added to dry meat as required by the use of salt and various spices. Curing by the use of salt is another traditional way of preserving meat and adding to its value. Concentrations of greater than 4 per cent of salt (sodium chloride) in curing inhibit the growth of spoilage organisms. Smoking is less satisfactory as a preservative method as light smoking delays the onset of spoilage by a relatively short period. Heavy smoking is more satisfactory from the preservation point of view but can have severely negative effects on the flavour of the final product. Smoking therefore is often regarded as an emergency measure, with other traditional methods of preservation generally being preferred. Modern methods of meat preservation include the use of refrigeration for chilling and freezing, and canning. Frozen meat is not regarded very favourably by some groups of consumers and may attract a lower price.

FAO statistics suggest that total annual meat production at the end of the 20th century was about 200 million tonnes from poultry, 78 million tonnes from pigs, 50 million tonnes from cattle, 3 million tonnes from buffalo, 7 million tonnes from sheep and 3 million tonnes from goats. Most developing-world countries have large deficits in supply in relation to demand. In many of these countries, however, meat production is relatively efficient with slaughtering and processing making use of many modern techniques. There remains, nonetheless, very considerable scope for improvement in the final product in terms of taste, presentation, food quality and food safety. This applies from the producer level during the growth of the animal right through to sale of products on local or international markets. Improvements in the nutrition and welfare of live animals and in hygiene through the slaughtering and processing stages would add value at all stages and provide incentives for increased output.

Further value can be added to meat-producing animals by making better use of by-products, including the offal for food, intestines for sausage skins and various organs in cosmetics or as traditional or modern medical products.

Milk

Cattle are the most important milk producing species on a world scale. There are, however, regional differences. In much of India and Pakistan the buffalo is the major producer of milk. In arid areas the camel is the dominant milk producer. In many nomadic societies and in many small-scale mixed crop-livestock systems goats and sheep are the major source of milk for the family, even though output per individual animal is small.

Milk is a fundamental product in human nutrition (Tuszyñski *et al.*, 1983). It is the neonatal or 'baby' food of all mammal species. People are especially fortunate, however, in that the milk of many species of domesticated animals is generally suitable for human consumption, in spite of minor differences in composition. Fat and nutrient content, especially immunoglobulins, do, however, vary between species and for infant or neonatal nutrition same species milk is *always* preferable. Some people are intolerant of lactose (due to a deficiency of the enzyme lactase – see Swallow, 2003) and especially the lactose of bovine milk, but many such people are able to make use of the milk of other species. Goat milk is often a substitute for cow milk in these cases and is well tolerated by almost everyone. In addition to its value as a food, milk has (and is often considered to have) medicinal properties and, in some societies and for some species of animal, is also believed to have magical properties.

In general milk consists of about 85 per cent or more water. Fat and 'solids not fat' (SNF) make up the rest. Of the main domestic species it is the milk of bovines that contains the least fat and that of buffalo and yak the most (Table 7.2), but there are also differences due to breed, and, of course, to feeding and management, as well as stage of lactation. Milk is an important source of dietary protein and calcium, which are important for growth and bone formation. Potassium, phosphorus and trace elements are also present in milk. Milk is also usually a good source of vitamin C (camel milk is particularly rich in this vitamin), vitamin B_{12} (cyanocobalamin) and some other B complex vitamins (riboflavin and thiamine), vitamin D and of carotene, which is the precursor of vitamin A. Drinking 0.5 litre of cow milk at 4.5 per cent fat will provide 2 per cent of the energy, 40 per cent of the protein, 70 per cent of both the calcium and riboflavin and 30 per cent of the vitamin A and thiamine of the daily requirements of a 5-year old child (Kon, 1972).

Table 7.2 Main composition of milk of various species and breeds of domestic animal

Species and breed	*Type of solid (per cent weight/weight)*				*Total solids (per cent)*
	Fat	*Protein*	*Lactose*	*Ash*	
Bovine					
India	4.6	3.3	4.9	0.67	13.47
East African zebu	5.5	3.3	4.7	0.76	14.30
Holstein	3.4	3.3	4.9	0.68	12.28
Jersey	5.2	4.0	4.8	0.75	14.50
Sheep					
Average	5-8	5-7	3.9-4.4		16-20
Goat					
Average	3.5-8.0	2.8-3.0	3.9-4.4		11.5-13.5
Anglo-Nubian	4.1		4.2		12.2
Buffalo					
India	6.6	3.9	5.2	0.8	15.8
Yak					
Nepal	7-9				17-18

Source: Crawford and Coppe (1990).

Milk is a highly perishable product because it is an excellent medium for the growth of many micro-organisms that can cause spoilage. This defect can be countered by many traditional and modern conservation processes. Milk is generally conserved by traditional methods as one of four groups of products: fermented milks; butter and butter oils; cheese and curds; and other milk products (see Crawford and Coppe, 1990, for a discussion of how complicated milk processing can be).

Fermented products are often the result of natural souring and indeed this is such a common product that many classes of people who rely on milk for much of their nutrition actually prefer soured to fresh milk. Unfortunately sour milk is itself susceptible to spoiling if kept for longer periods and souring does not kill many potential pathogenic organisms, such as the bacteria of tuberculosis. Fermented milks that are essentially the same product are known by different names in different areas. Thus in Spanish speaking countries it is known as 'leche agria', in South Asia as 'dahi', in Arabic speaking countries as 'laban' and in Central Asia as 'yoghurt' or 'kumiss' (with the latter often being the product of horse milk).

Fat has always been one of the most important components of milk, as it is often the major or the only source of fat available to most people in the tropics, especially Asia. It is easy to separate from the other components and does not deteriorate as quickly as some of them. Cream is used by some people in the developing world and notably that of sheep (known as 'kasha') by Bedu arabs. Butterfat is churned in several traditional ways by simply shaking a bag (often a goat skin) or gourd, or by turning paddles inside a container. Butter made from soured milk is usually used for cooking or further transformed to 'ghee' (India) or 'semn' (Arabic). This liquefied product is the best way of conserving milk in tropical climates. In South Asia more than 40 per cent of all milk is converted to 'ghee' and even a poor quality product will keep for several weeks. 'Ghee' is made by evaporating off most of the moisture from the fat or butter and then settling out the SNF to leave pure butterfat or butter oil.

Cheese is solid curd with most or all of the whey drained off. It can be used fresh or dried and matured in storage. Curd is formed by the action of lactic acid bacteria, some organic acids or, most commonly, by the use of rennet (traditionally obtained from the stomach membranes of a calf). Cheese is preserved by acidity, addition of salt or reducing the moisture content to a minimum. Some form of cheese is made in most tropical areas and the milk of most domestic species can be used. The nutritional value of cheese depends on the type of milk used and the treatment it receives. However, cheese is generally a high quality protein concentrate that is rich in calcium, riboflavin and vitamin A and has great value as a supplement to cereal-based diets.

Many other products are made from milk but they are generally related to those already mentioned. The market for flavoured milk drinks, milk-based sweets and ice creams is expanding rapidly.

In the late 1990s world milk production was about 600 million tonnes annually, of which 85 per cent was from cattle, 10 per cent from buffalo and most of the remainder from goats and sheep. Some 15 million tonnes of cheese and 7 million tonnes of butter were produced. Most of the raw milk and its manufactured derivative output came from the developed world. Demand for milk and milk products has been outstripping the supply for many years and especially so in the developing countries. Where local supply is insufficient to meet the demand, attempts to reduce the shortfall have often depended on imports (frequently at subsidised prices) of powdered milk for reconstitution as liquid milk, sweetened condensed and evaporated milks, ice cream and milk drinks (Brumby and Gryseels, 1985). The potential for fulfilling the demand for such speciality products from local sources is immense and must be seen as an opportunity to add value to increased and improved local production. Further value would be added to the whole system through the additional employment opportunities created in production, processing and marketing of milk.

Eggs

Most eggs consumed by people are from the domestic fowl or 'chicken'. Other species that produce eggs that are eaten on a regular basis include ducks, geese, turkeys, pigeons, Guinea fowl and some wild or semi-domesticated game birds such as pheasants, partridges and quail. Ostrich eggs are also eaten in some areas and are either gathered in the wild or collected as surplus from the specialist breeding industry.

Eggs are an important food, relatively inexpensive and a valuable source of nourishment. Weight-for-weight, eggs contain the same amount of high quality protein as pork or poultry meat, about 75 per cent that of beef and 67 per cent of whole milk cheese (FAO, 2003). As a nutritional source of vitamin D, eggs are second only to fish liver oils and they are also good sources of iron, phosphorus and some other vitamins. Eggs are, however, low in calcium (which is mainly locked up in the shell) and vitamin C. Low energy value, ease of digestibility and high nutrient content mean that eggs are valuable for therapeutic diets and for older people. The nutrient make-up of eggs also makes them suitable in the diets of rapidly growing humans in childhood and early youth. Eggs are generally acceptable to a very wide range of consumers, but there are beliefs in some areas, such as Central Africa, that women should not eat eggs at some physiological stages, such as pregnancy. There have been occasional (but usually highly publicised) food safety scares, most often associated with contamination by *Salmonella* organisms. Attention to management, feeding and general bird and environmental hygiene coupled to appropriate marketing will, however, greatly reduce the risk of any problems of food safety.

Egg sizes vary from about 12 g for pigeons up to 1900 g for ostriches. Chicken eggs from most birds in the tropics are small and in the 30-60 g range, with most being at the lower end of the scale.

We should note that poultry species are found in almost all resource-poor households and are a major source of animal protein for immediate family use, for visitors and for income generation.

Honey

Pollination is the main service obtained from bees, but honey is usually the main product in quantitative and financial terms of traditional bee keeping activities. Most of the world's honey is produced by the honey-bee *Apis mellifera* but other species of bees also produce honey. The general definition of honey is that it "is the natural sweet substance produced by honey-bees from the nectar of blossoms or from the secretion of living parts of plants or excretions of plant sucking insects on the living parts of plants, which honey-bees collect, transform and combine with specific substances of their own, store and leave in the honey comb to mature" (Codex Alimentarius, 1989). Sugars account for 95-99 per cent of honey dry matter and 85-95 per cent of these are the simple sugars fructose and glucose (Krell, 1996). Water is the second most important constituent but must be less than 18 per cent of the whole if honey is to be stored without risk of fermentation. Minerals are present in very small quantities, as are nitrogenous compounds amongst which are the enzymes that originate from the saliva of the worker bees. Honey was the only source of concentrated sugar available to people for thousands of years. Honey is considered to facilitate better physical performance, reduce fatigue, promote higher mental efficiency, improve food assimilation, be useful for chronic and infective intestinal problems and act as a remedy for colds and mouth, throat and bronchial irritations. Honey is used in moisturising and nourishing cosmetic creams and in pharmaceutical preparations that are applied directly to open wounds, sores, ulcers and burns as well as being useful for assisting tissue generation and reduction of scars due to various types of wounds.

Honey is most commonly consumed in its unprocessed or natural state in the developed countries. In some countries of Africa its main use is for making beer or wine, as indeed it was in the past in Europe (honey wine is known as 'mead' in English), or for sale as 'comb honey' (that is not separated from beeswax but eaten together with it). Value can be added to honey in a variety of ways but the best opportunities for further added value from the keeping of bees arise from the use of other hive products. Wax is a valuable product in itself and other (and often highly valuable) possible outputs have rarely been considered. Wax can be used for making candles, in cosmetics, in food technology, in varnishes and polishes and in medicine. Other marketable products include propolis (a plant material used by the bees themselves as a natural antibiotic and preservative but used by people in cosmetics, medicine and food technology), royal jelly (fed by bees to very young larvae and to larvae that will develop into queens and used by people as a dietary supplement), venom and queens themselves for the formation of new colonies.

World production of honey during the 1990s was in excess of 1.2 million tonnes per year. The world demand for honey is substantially in excess of this and could be increased even further. Value could be added through packaging and by the use of a range of different presentations. Most added value will arise in the future from supplying niche markets for honey and by marketing the other high value products mentioned in this section.

Other immediate products

Conventional fibres

Wool is the coarse or fine fibre produced by sheep and is by far the most important of all natural fibres produced by animals (Table 7.3). World wool production was estimated at about 2.6 million tonnes in 1996 (FAO, 1996), with just over a quarter of this being produced in developing countries. Mohair, whose fibres are smoother than wool, is the non-medullated fibre produced by Angora goats (non-medullated means the fibre is solid with no hollow core as occurs in the coarse wool of sheep). The finest quality mohair is produced by kids and young goats and has a mean diameter of 25-32 μm and is used for fashion fabrics and fancy yarns, whereas that of older goats has a variety of uses including blending with wool (Van der Westhuysen, 1982). World mohair production is small compared with wool and fluctuates rather widely, as for example it did between 1975 and 1996, in the range of about 13,000 to 30,000 tonnes per year. Cashmere, also known as 'pashmina', is the fine downy non-medullated undercoat produced by some types of goat. The very finest cashmere has a mean diameter 14-15 μm and is used for speciality ladies' clothing, sweaters, scarves, stoles and other luxury items. Goat hair is rarely used in commercial trade but a considerable amount from longer haired breeds is used in Asia and Africa for traditional production of tents, ropes and bags. The alpaca is the most important species of South American camelid for fibre, with an annual production of about 4,000 tonnes. The Suri type produces a longer, more desirable fibre than the more numerous Huacaya type (Novoa, 1981). The llama is the largest of the South American camelids and is primarily a pack animal but does produce a coarse fibre of little commercial value that is used nonetheless in the manufacture of local garments. The vicuña is a small wild species of South American camelid that produces extremely fine fibre. Fibre production by the one-humped camel is generally of minor importance (Schwartz, 1992) but the hair from camels in Saudi Arabia is used for weaving blankets, rugs and wall hangings (Sohai, 1983) and in India there is some interest in blending camel hair with wool or polyester for cloth manufacture (Gupta *et al.*, 1989).

Table 7.3 Production, price and some characteristics of the major animal fibres

				Parameter		
Animal species and type	*Fibre type*	*World production (tonnes)[a]*	*Price[a] (US$/kg)*	*Mean diam. (μm)*	*Mean length (mm)*	*Fleece weight (kg)[b]*
Tropical wool sheep	Wool	2,600,000	1-15	26-66	40-190	0.9-3.5
Merino sheep (tropical Australia)	Wool			20-28	80-120	1.4-2.3
Hair sheep	Hair			40-100	10-50	-
Angora goat	Mohair	11,000	10-30	25-41	150-250	2.5-5.6
Down goat	Cashmere	6,000	50-100	11-19	60-80	0.2-0.7
Goat	Hair			40-100	10-100	-
Alpaca	Fibre	4,000	20-25	22-40	80-150	1.5
Llama	Fibre			28-50	250-300	-
Vicuña	Fibre	50	200-300	10-20	30-50	-
Bactrian camel	Hair	3,000	15-20	15-55	230-320	2.8-6.8
One-humped camel	Hair			20-140	20-400	0.7-1.4
Buffalo	Hair			80-107	10-70	-
'Angora' rabbit	Angora 'wool'	10,000	30-40	11-15	40-100	0.2-0.4

Source: Petrie (1995); [a]1996 data; [b]Adult female greasy-fleece weight

Value can be added to animal fibres at two main stages: on the animal and at or after harvesting (Petrie, 1995). Attention needs to be given to management and husbandry during the growth of animals. Nutritional checks can result in 'breaks' in the fibre which reduce its harvested value because there is an introduced weakness. Wherever possible, animals should be kept away from areas that could introduce thorns or burrs into the fleece as these are difficult to remove after harvest. Dried mud and faeces in the fleece reduce its value. At harvesting ('shearing') care should be taken to remove the fleece as clean and as whole as possible and consideration must be given to correct handling and packing. If further processing, such as degreasing or spinning, is undertaken at the point of or near to the harvesting site, high standards of quality control must be maintained. Only then can producers and people in the early part of the chain to the final consumer product reap adequate rewards in relation to the final added value.

Hides and skins

Hides and skins are valuable products at both the farm and national level. For a variety of reasons they have rarely contributed their full value, either to household income or to national accounts (Wilson, 1992). They do, nonetheless, contribute substantial amounts to national incomes in some countries. One example is Botswana, where revenue to the Meat Commission from goatskins was 20 per cent that of meat and of sheepskins 25 per cent that of meat (BMC, 1986). In Ethiopia in the early 1980s, some 1.8 million goatskins and 3.3 million sheepskins earned export revenue equivalent to 9.2 per cent of all agricultural exports and 29.5 per cent of exports excluding coffee (CSO, 1984). Value can be added to these outputs quite easily at primary and secondary levels. At the primary level, branding should be avoided on the major areas of the body skin and damage by biting and burrowing insects and acarids should be avoided as far as is possible. Skins should be removed from animals with care not to cut them, fat and flesh should be carefully scraped off and skins should be dried while stretched vertically on frames or partially cured. Many countries have now passed legislation that prohibits the export of raw skins. Partial processing, such as to the 'wet blue' stage before export, can multiply national income several fold, as it has done in Ethiopia, and also helps to even out the income which otherwise may fluctuate wildly due to large variations in the demand for and prices offered for raw products.

Ostrich skin is a luxury leather that can compete with the skins of crocodiles and snakes for the most demanding markets. The finished leather is thick and durable but very supple and has a pleasant pitted appearance which is due to the follicles from which the feathers grow. A small breeding group of ostrich can produce about 50 m^2 of finished leather, which is very favourable compared with the less than 3 m^2 from a single large steer (Shanawany, 1996).

Pelts

Pelts are a specialised product used mainly in the fur-fashion trade. Pelts from Karakul sheep and rabbits are the most important ones from domesticated animals, although they are also obtained from many wild animals and perhaps most notably from beavers in North America. The best pelts of Karakul sheep are obtained from lambs that are killed immediately after birth before the skin has dried (Wilson 1992). Paradoxically, better pelts are obtained from lambs whose mothers have suffered nutritional stress during pregnancy. Prices of pelts are related to colour, size, curl type, hair length and hair quality and pattern.

The market for Karakul pelts has traditionally been dominated by the Southwest African Karakul Association, who sell mainly by auction on the London market. There are probably limited but potentially highly lucrative opportunities for pelt production from animals in arid and semi-arid areas.

Feathers and down

Duck and goose feathers and down are used in bedding and for other applications. They have been replaced to some extent by synthetics but the very best quality pillows and padded clothing still make use of natural products. Down is the most valuable product and provides the best insulation, the softest texture and the lightest weight. Most commercial products contain a mixture of feathers and down but the higher the proportion of the latter the more valuable the product (Buckland and Guy, 2002). Feathers and down can be obtained from birds at slaughter or during their life cycle. If harvested from live birds, down can be an important supplementary income to meat (and *foie gras* in the case of geese). In geese the first down harvest can be taken at 9-10 weeks when the birds first moult and then at successive 6-week intervals. First harvest yield is about 80 g and that of subsequent pluckings 100-120 g, with the down being 15-20 per cent of the total yield. In 1994 the international trade from all species of waterfowl was in excess of 67,000 tonnes of raw feathers valued at US$ 650 million, of which 30 per cent by weight and 40 per cent by value was from geese.

There was very high demand for ostrich feathers, mainly for the fashion industry but also for cleaning, in the second half of the 19th century with the first commercial ostrich farm being set up in South Africa in 1860 (Shanawany, 1996). Ostrich farms quickly spread to other countries, including several in Africa (Wilson, 1976) and by the start of the First World War (1914-1918) more than one million birds were being raised on commercial farms. The feather market crashed during the middle of the 20th century but built up again in later years. Feathers - of which the large black and white primary wing feathers of the male are the most valuable - provide a useful additional income to the meat and skin and can of course be harvested throughout the life of a bird. A male and two breeding females producing about 40 young per year will produce about 36 kg of prime feathers.

Feathers and down are generally products harvested in conjunction with other bird products. There are good if rather demanding markets for these items and there is a ready sale for top quality output.

Silk

Silk is an animal fibre that is produced by caterpillars of the butterfly *Bombyx mori*. 'Silk' is a continuous thread of a single filament that can be as much as 1500 metres in length. For production purposes several filaments are twisted into one strand by a process known as reeling or filature (Lee, 1999). World production of silk as a natural fibre is less than 1 per cent of the world market with a total of 91,000 tonnes in 1995, compared to 1.6 million tonnes wool, 18.2 million tonnes cotton and 17.8 million tonnes of synthetic fibre. Traditional producers have been Japan and Korea but in the 1990s China and India were the major producers, with Brazil and the former Soviet Union being the main sources of silk outside Asia. Silk is a much sought-after luxury commodity for which the international demand is likely to increase. Production is not too difficult and there are clear opportunities for expansion of silk production to other areas of the developing world.

Intermediate products

Animal power

Despite increased mechanisation and use of motorised forms of power over much of the world in the 20th century, many people at the beginning of the 21st century still rely on animal power to complement machines and human labour in agriculture and transport (Figure 7.1). Draught animals and humans provide an estimated 80 per cent of the power input on farms in developing countries where enterprise size and scale rule out mechanical power. Cattle are the species most commonly used for work. Water buffalo are used in the humid tropics and donkeys, horses and camels mainly in the semi-arid areas. Draught animals are maintained over a wide range of agro-ecological zones and production systems but are particularly

Further reading

Bender, A. 1992. *Meat and meat products in human nutrition in developing countries. Food and Nutrition Paper No 53.* Food and Agriculture Organisation of the United Nations (FAO), Rome, Italy.

Crawford, R.J.M. and Coppe, P. 1990. *The technology of traditional milk products in developing countries. Animal Production and Health Paper No 85*. Food and Agriculture Organisation of the United Nations (FAO), Rome, Italy.

Latham, M.C. 1997. *Human nutrition in the developing world. Food and Nutrition Series No 29.* Food and Agriculture Organisation of the United Nations (FAO), Rome, Italy.

Author addresses

Trevor Wilson, Bartridge House, Umberleigh, Devon EX37 9AS, UK

Anne Pearson, Centre for Tropical Veterinary Medicine, Division of Animal Health and Welfare, University of Edinburgh, Easter Bush Veterinary Centre, Roslin, Midlothian EH25 9RG, UK

Nicola Bradbear, Bees for Development, Troy, Monmouth NP25 4AB, UK

Ashini Jayasuriya, Department of Medicine, Nottingham City Hospital, Nottingham, UK.

Henry Laswai, Department of Food Science, Sokoine University of Agriculture, P.O. Box 3004, Morogoro, Tanzania

Louis Mtenga, Department of Animal Science and Production, Sokoine University of Agriculture, P.O. Box 3004, Morogoro, Tanzania

Sarah Richards, Department of Surgery, Southmead Hospital, Westbury-on-Trym, Bristol, Avon BS10 5NB, UK

Rose Smith, Wales College of Medicine, Cardiff University, Heath Park Campus, Cardiff CF14 4XN, UK

8

Marketing to promote trade and development

Martin Upton[*], *Stephen Mbogoh*[†], *Jonathan Rushton*[†], *Serajul Islam*[‡], *Louis Mtenga*[‡]

Cameo - Local wool growers' association in South Africa

The Republic of South Africa is a major producer of wool on world markets. However, the industry is dominated by commercial producers, brokers, exporters and spinners. On commercial farms, sheep are sheared on the farm and the wool is traded through brokers on the auction markets.

Until recently smallholder sheep farmers in the Transkei region, a former homeland, had limited access to a profitable market for their wool. They operate on too small a scale with inadequate resources to hire brokers on an individual basis. Their only market outlet is to sell to wool traders who pay very low prices. Traders face the uncertainty and the physical costs of grading, sorting, packaging and transport. These costs, together with the associated transaction costs, are high because supplies from individual households are small and scattered. The residual payment to farmers is necessarily small. In these circumstances some smallholder sheep producers do not harvest wool, as the economic returns do not justify the cost.

As part of a development programme in the deprived areas under the Land Care Project of the National Department of Agriculture, funds were provided, mostly by the National Woolgrowers' Association, for building and equipping shearing sheds for use by local wool growers' associations. Group collaboration in shearing captures economies of scale and justifies the costs of grading, sorting and packaging, and the hire of brokers to negotiate sales. However farmers must wait for payment until after the wool is sold on the auction market, whereas traders make payment on the spot. This explains why some farmers choose not to join the local wool growers' association.

---

[*] Principal author
[†] Co-author
[‡] Consulting author

A recent study in three Transkei villages, comparing gross margins from sheep production between members and non-members of wool growers' associations, has shown a significant increase in income, of about R 6.5 (about US$ 1.00) per sheep, resulting from membership and use of a shearing shed. It is concluded that use of the shearing shed increases the wool price received, reduces uncertainty, increases efficiency of shearing, grading and packaging and increases equity and bargaining power of the farmers. (D'Haese *et al.*, 2003)

The importance of markets

Earlier chapters in this book have emphasised the contribution of livestock to agricultural production, household subsistence and rural welfare. However, the additional benefits of providing finance for the purchase of other goods and services, increased investment and growth, and cash reserves against risk, depend upon sales of surplus production over that used within the household. The importance of markets was briefly outlined in Chapter 2 (See also Abbott, 1993, and Scott, 1995, on agricultural marketing in developing countries, or Kohls and Uhl, 1998, for a more general review). Limited access to markets is a major cause of rural poverty. Despite the increasing opportunities offered by the rapid growth of demand for livestock products in developing countries, small-scale producers in rural areas face large constraints in reaching the major urban and global markets. The constraints faced are both physical and institutional.

Physical constraints on marketing include the low population densities in rural areas, remoteness of livestock producers from the main urban market centres and poor communication systems which result in high transport costs. For producers with better communications and easier access to densely populated urban centres, the potential market for livestock products, as for other agricultural produce, is larger than that in remote areas. This is the case not only because there is a much greater concentration of potential purchasers and consumers in the towns and cities, but also because average incomes are generally higher in urban areas than in rural regions. Peri-urban livestock producers benefit from their ready access to markets, but may face problems of forage supply and waste disposal.

Institutional constraints are linked with the weaknesses of formal institutions, including a legal framework, needed to make market information more widely available, to protect property rights and to provide effective mechanisms for enforcing contractual agreements. It has been argued that "the inability of societies to develop effective, low-cost enforcement of contracts is the most important source of both historical stagnation and contemporary underdevelopment in the Third World" (North, 1990, p. 54). This issue will be discussed in more detail later (see Section `Failure of input delivery systems').

Markets link producers and consumers. Within rural communities, links are generally direct and informal; for instance where a producer sells milk or eggs to another household for final consumption. In larger and more complex market systems, there is usually a chain of intermediaries between the producer and the consumer. Traders include wholesalers, who aggregate consignments from various sources before selling on to other traders or retailers, the latter group selling directly to consumers. At some stage in the chain, products are transported from one location to another, and may be processed from one form into another, for instance where animals are slaughtered and butchered or milk is processed into more durable products by pasteurisation, fermentation and cheese-making. Most livestock products are perishable, so storage generally requires suitable cold stores.

All these functions require funding, and risks of losses exist at all stages of the market chain. They may be subject to formal rules and possibly legal constraints on methods of sale, processing, grading, hygiene and health inspection. The costs of all these marketing functions have to be met, as well as the 'transaction costs' of negotiating and enforcing transactions. The institutional framework for market transactions has an important influence on how income and costs are shared between livestock producers and traders.

Opportunities offered by the rapid growth in developing country markets, known as the 'Livestock Revolution', were discussed in Chapter 1. This rapid growth of demand for livestock products, which has occurred over the last quarter century, is largely driven by increases in per capita incomes, population growth and urbanisation of the developing countries. In these countries average per capita consumption of milk, eggs and meat has grown, over the last decade, by 20.5 per cent, 68.3 per cent and 48.0 per cent respectively. The so-called 'landless production systems' are largely responsible for the rapid growth in average meat consumption in the developing countries. For instance the per capita intake of pig meat has increased by nearly 50 per cent over the last decade (by 4 per cent annually) while that of poultry meat has more than doubled (over 7 per cent annual growth).

Two key points should be noted in this context. First, much of the growth in consumption has been linked with increasing imports, by the developing countries as a group, of all livestock products and of key food crop cereals. Clearly, developing-country producers must now compete with exports from the developed countries in terms of both product price and quality. It appears that, for the developing countries as a group, the greatest opportunities for import substitution and expanded sales of livestock products lie in dairy and poultry, followed by pig-meat production.

However, these enterprises are suited to intensive, large-scale industrial-type production and/or processing and marketing, which are subject to economies of scale. Such systems are largely dependent on purchased grain and exotic breeds of animals and birds. They are concentrated in peri-urban areas and contribute to local environmental pollution. It is argued that these enterprises are of little benefit to the resource-poor, and may impose unfair competition on smallholder producers (LID, 1999, p. 24). Yet, given that current trends in consumer demand look likely to continue, that expansion of production creates employment opportunities and that resource-poor consumers may benefit from these relatively cheap sources of animal protein, a case can be made for encouraging this type of production system. That might involve drawing on the capital assets and technical, managerial and marketing expertise of the large commercial processors, whilst promoting smallholder production through out-grower schemes and producer co-operation.

The second key point is that there is much variation between individual developing countries. The Latin American and South Asian groups of countries are net exporters of livestock products, mainly poultry, pig-meat, cattle and honey from Latin America and buffalo and their meat from South Asia. Most African countries and many countries in Asia fall into the category of 'Low Income Countries'. These countries, as a group, are net exporters of ruminant meat, hides and skins and live sheep and goats, despite being net importers of livestock products in total and of dairy products and pig and poultry meat in particular. Some of the 'Least Developed Countries' of the African Sahel and South-East Asia are net exporters of live cattle, sheep and goats.

Markets for specific livestock products

Livestock

Live animals may be sold locally, for breeding, fattening or to be slaughtered and consumed. One of the advantages of small stock, such as poultry, sheep and goats, is that they are of an appropriate size for consumption within a single household, so direct sales may take place, without the need for intermediary slaughter and cutting-up operations. However, local markets may exist for all types of livestock, with the involvement of intermediaries, both butchers and local traders.

Larger markets exist in the urban centres while livestock production, particularly that of grassland-based ruminants, is spread out in remote areas. The higher prices obtainable in urban markets may be sufficient to justify the high costs of transport. Large animals may be moved large distances on the hoof, but are likely to lose condition as a result. Where motorised transport is available, it may well prove a cheaper alternative. Small animals and poultry require transport, but are bulky and therefore costly to move over large distances. Traders are involved in buying from producers and moving livestock to the larger urban markets, where they are probably sold to wholesalers or directly for slaughter. Commission agents may also be involved in the buying chain.

Peri-urban producers benefit from higher product prices and lower costs of inputs than those of producers in more remote areas, where the majority of resource-poor livestock-keepers live. The limited availability of rural transport routes and the poor condition of many roads greatly increase the costs of transport. Small-scale producers are at a particular disadvantage due to the higher unit costs of moving small consignments. High transport costs also result in big inter-regional differences in the net price received for produce, and the cost of inputs. Outside the peri-urban zone, the sparse road network and poor postal and telecommunication systems limit the spread of market information and result in high transaction costs for both producers and traders. The public dissemination of market information would reduce these costs and improve market efficiency. Where sufficient numbers of traders are competing to buy stock, auction markets provide an effective means of transmitting price information and limiting transaction costs.

Slaughter facilities – meat marketing

The optimal location of slaughter facilities depends upon the relative transport costs of live animals and frozen or chilled meat. Refrigerated vehicles are needed for meat to avoid deterioration of quality during transport. In industrialised countries, refrigerated transport is readily available, and transport of the less bulky and higher valued product is cheaper than that of live animals. Thus abattoirs are generally located in producing areas and meat is transported to the towns. In addition, the animal welfare problems of long distance transport are limited or avoided. However, in developing countries, the least costly alternative is likely to be location of slaughter facilities near urban centres, with the trekking of live animals on the hoof or their transport by trucks to these centres. Economies of scale can be derived from the larger, possibly mechanised, slaughter and processing units that are justified near major market centres. Equipment for the processing of by-products may be justified, while the need for duplication of costly refrigerated storage in both slaughter and terminal markets is avoided, together with the costs of refrigerated transport.

Economies of scale in slaughter facilities arise from the high costs of establishing hygienic working conditions, cold storage and by-product processing facilities, together with the employment of skilled staff. However, some large abattoirs established in developing countries, operate at well below full capacity, so the fixed costs are spread over a less than optimal throughput with a consequent increase in average cost per animal slaughtered. Smaller, less sophisticated slaughter facilities may be more appropriate where a low throughput is expected, for instance where animals are slaughtered in production areas.

A single slaughter facility serving a large area is in a 'monopsony', or buyer's monopoly, position. This means that by restricting throughput, the operator can hold down the price paid, per animal, below that which would obtain in a competitive market. Although such price fixing is possible under a public sector meat corporation, it is more likely to occur under private enterprise. Regulation is needed to promote competitive pricing, and to improve quality, public health and hygiene. Parastatal meat marketing corporations have been established in Botswana, Kenya and other African countries, with the main aim of promoting beef exports. These agencies were never in a monopsony position, since informal slaughter establishments existed elsewhere. Apart from the Botswana Meat Commission, most of them have ceased operation following major financial losses and national policies of structural adjustment. In Botswana, and other exporting countries, costly control of animal disease is necessary to comply with developed country import regulations.

Apart from the sanitary requirements of export markets, meat inspection is important for the protection of human health, and should be publicly funded. The cost of employing skilled meat inspectors is more easily justified at a central abattoir, with a large throughput, than when distributed over many small scattered slaughter points.

The slaughter and processing of small stock, including poultry, small ruminants, pigs, and South American camelids, raise fewer problems, since these operations can be carried out within individual households. However, for poultry, there are major economies of scale in the mass production and marketing of a standardised, plucked and dressed product. This, in turn, requires a regular and reliable supply of a standard type of chicken, usually a hybrid, reared rapidly on a concentrated feed diet. Poultry production companies in developing countries are often subsidiaries of trans-national corporations and operate mainly in peri-urban locations. Suitable birds can be raised on smallholdings, provided that day-old chicks, medication, concentrate feeds and expert advice can be made available. Vertical integration of producers with processing and marketing companies has made this possible in some developing countries.

Milk and dairy products

Milk marketing differs from that of other products in that milk is produced every day over the lactation period and that it is very bulky, consisting of over 85 per cent water. Raw milk is a highly perishable product, which deteriorates in quality within two to three hours from milking. Hence markets for raw milk are limited to potential consumers that can be reached within this short period. Milk may be delivered from cans by producers, by local milk traders or by peddlers using bicycle or donkey transport. There is rarely any means of controlling quality, and adulteration is difficult to detect. Risks of the spread of human disease, such as tuberculosis, are limited by the normal practice, in some countries, of boiling milk before use. Similarly milk produced in peri-urban areas may be marketed directly to townspeople in a raw state. Most of the milk produced in developing countries is consumed as raw milk, without further processing, other than in simple small-scale cottage industries.

The usable life of milk may be extended by cooling, normally using electrical power, or by low-technology fermentation. These technologies are relatively unsophisticated and, although control of hygiene is essential to avoid microbial and other contamination, can be managed by farmer representatives and justified economically for relatively small-scale operations. Thus in producing areas where there is sufficient density of milk producers, a network of small dairies based on these technologies may be justified. The products may be marketed locally but may also be transported, if all-weather roads exist, in tanker-lorries to urban centres for sale or further processing.

High levels of capacity utilisation are necessary, in treatment, storage and transport, to keep unit costs low. Pasteurisation is desirable, to further extend the usable life of milk and to reduce or eliminate risks of disease transmission to humans. However, the process is capital-intensive since the necessary equipment is expensive and subject to economies of scale. Pasteurisation plants are usually located near the large markets in urban centres. Although there is no reduction in bulk, there are advantages in treatment close to the point of sale, linked with packaging and retail sales. One reason is that even after pasteurisation in the production regions, milk quality may deteriorate as a result of transport and handling problems.

As stated earlier, much of the milk in developing countries is sold raw, in an unpasteurised state, particularly in rural areas. Some governments have established pasteurising plants in rural areas and attempted to ban the sale of raw milk. Despite the health risks associated with the consumption of untreated milk, it may be difficult to justify such policies in economic terms, particularly because of problems of enforcement and/or the constraint placed on the development of rural milk markets. Capital-intensive co-operative dairy processing plants established in countries such as India, Kenya and Bolivia produce pasteurised, brucellosis- and tuberculosis-free milk, but at high cost.

It is difficult to ensure that plants are operated near to full capacity throughout the year, not least because of the seasonality of milk production. The establishment of adequate capacity for the peak production season (normally the tropical rainy season or the temperate summer season) necessarily results in excess capacity in the dry or winter season. Where facilities are available, seasonal shortages of milk may be met by reconstitution from imported skimmed milk powder (SMP) and butter oil. In India, and less-successfully in other countries such as Tanzania, reconstituted milk based on food-aid in the form of SMP and butter oil, was used on a longer term basis to promote growth in domestic demand for milk and dairy products. This growth in demand may then provide the necessary incentive for increased domestic milk production and sales.

Further processing is involved in the production of a range of nutritious dairy products, such as butter or ghee, cheeses, fruit-flavoured yoghurts and ice cream. Ghee and cheese have lower water content, less bulk and are less perishable than milk. Hence processing reduces transport costs, and is therefore best located near where the milk is produced in rural areas. This is not the case for flavoured yoghurts and ice cream, which are more likely to be produced in an urban or peri-urban environment. Many processing plants are small-scale 'cottage' activities, but there are probably economies of scale in maintaining quality and hygiene, packaging, transport and selling.

Developing-country processors may find it difficult to compete with low priced imports from the developed countries, in some of which dairy products are still subsidised. Processing plants in the developing countries must keep costs low in order to compete. Large-scale processors may take advantage of their monopsony position to hold down the price paid to milk producers. Such price distortions discourage further growth of both milk production and processing. Government subsidies might be justified in some cases, in order to promote infant industries.

Other food products

Other food products, such as honey, after extraction from the comb, and eggs, require no further processing before sale in local markets. However transport to more distant urban or even international markets necessitates packing and is relatively costly since the products are still perishable. In seeking larger markets, there may be economies of scale in packing, labelling and marketing these products. As noted earlier, egg consumption in developing countries has increased much more than milk and meat during the past decade. Most of the egg marketing appears to be direct from producer to consumer, particularly in peri-urban areas.

Wool

Wool production in the developing countries is restricted to temperate regions, or to high-altitude tropical regions, such as in Bolivia, Peru and Nepal, where wool-sheep thrive. Washing, carding and spinning or felt-making may be carried out locally, with transactions occurring between near neighbours and acquaintances. However, where manufacture of woollen goods or international markets are important, industrial-scale processing is likely to take place in urban centres. Traders and other intermediaries are likely to be involved. Contractual agreements with agents may be used to limit transaction costs. A similar situation exists where silk is produced.

Hides and skin

Hides and skins are by-products of meat production. There is very little processing of hides and skins from home-slaughtered animals, other than karakul pelts. The necessary salt is not always available and the returns may be insufficient to justify the cost. Tanneries are usually associated with large-scale abattoirs, and livestock producers are not directly involved in the associated transactions.

Marketing of manure and hire of draught animals

Local markets may exist for manure and the hire of animal draught. Manure may be sold as an organic fertiliser, as fuel or as building material. However, its financial value is relatively low, while it is bulky and costly to transport. Despite this low value to cost ratio, there are instances of goat manure being transported over quite large distances in Bolivia, for use in potato and onion growing areas (J. Rushton, personal communication). Market transactions are most likely to be direct, between seller and buyer in local markets. Arrangements for the post-harvest grazing of crop residues, and manuring of the soil, by pastoralist-owned livestock, are of this nature. Similarly, the hire of draught animals is likely to be a direct transaction, possibly repeated under relational contracting between neighbours or kinsmen. The extent of the market is limited by competition for the use of draught teams during the rather short cultivation period, and the costs of moving draught animals from one site to another.

Market access for resource-poor livestock producers

With rapidly growing urban populations and increasing per capita incomes the main markets for livestock products exist in the towns and cities. The remoteness of many livestock producers from these markets creates physical problems of access. As a result, rural poverty is often worst in remote regions of developing countries. Problems of access are exacerbated in all developing countries, but particularly in sparsely-populated African countries, by the limited road network. Not only is the length of metalled roads per square kilometre and per head of population, much lower than in the developed countries, but much of the existing road network suffers from lack of maintenance and is in poor condition. Many roads are impassable in the rainy season, which is often the peak season for the production of livestock products such as milk. Associated with the poor road network, transport facilities are often limited and costly.

Telecommunications and other means of disseminating information on markets, products and prices are also limited. As a result livestock producers in remote areas are at a serious disadvantage in seeking markets and negotiating sales with traders and commercial firms offering contracts. The livestock producers have little knowledge and information on market opportunities and the prices prevailing in the larger urban markets. The traders and commercial buyers are generally much better informed about market conditions.

The above notwithstanding, the increasing availability of inexpensive mobile telephones with facilities to send short text messages is proving very useful. For example, Kenya has recently started a community telecommunication system which allows producers to access information on prices of produce in different parts of the country using the Kenya Agricultural Commodity Exchange Institution.

While there are large numbers of livestock producers, and consumers of the products, there are often very few traders or market intermediaries. This situation results from the small-scale and scattered distribution of producers, and inadequate transport and communications. The costs of setting up a trading agency are high and there may not be enough business to justify many traders becoming involved. Hence there is a lack of market competition and an inequality of bargaining power between the few traders, who largely control the market, and the many small producers. In addition the producers lack the experience and the necessary skills for negotiating contracts to ensure that the terms are equitable. They may be forced to sell to meet an emergency, such as a drought, or to clear a debt. Hence the producers have to accept whatever terms are offered; they are 'price takers'.

In the past many developing-country Governments intervened in marketing, especially through parastatal Marketing Boards, of meat and of milk. Marketing Boards were supposed to serve the livestock producers by providing assured markets for meat and milk at pre-set prices, quality control and food safety in processing, and efforts to seek out and expand product markets. However, Marketing Boards in many countries have experienced financial difficulties, while their services have deteriorated. Following structural adjustment, and identified weaknesses in their operation, many Marketing Boards have ceased operations. It was hoped that private enterprise or Non-Government Organisations (NGOs) would fill the gaps and provide marketing services.

However, the experience has been mixed. There has been a tendency in many cases to revert to traditional methods of more direct marketing from producer to consumer, often associated with declining food hygiene and quality standards in processing. Where producers previously knew the prices they would be paid, often at the start of the season, they now face a situation which is open ended and in which prices may vary from day to day. Producers have little or no information on market conditions, on prices and quality, and little experience of how the market works, of negotiating contracts and of their ability to influence the terms and conditions of sale. Agricultural extension advice generally concerns technical production issues and little guidance is given on marketing. Hence small-scale producers are poorly equipped for negotiating transactions with specialised traders and processors.

Marketing co-operatives have long been recommended as a means of allowing small-scale producers to benefit from economies of scale and exercise greater control over the marketing process than is possible in dealing with private traders alone. This form of organisation is widely adopted, for milk marketing, in the industrialised countries of Europe, North America, Australia and New Zealand. Among the developing countries, India's 'Operation Flood', for the development of the dairy industry, is of particular note. Local supply of milk is organised through co-operatives, modelled on that founded in Kaira District, Gujarat State, known as the 'Anand Model'. All the milk co-operatives in a district form a union that, ideally, has its own processing facilities. Unions are federated in each state and in the National Co-operative Dairy Federation. The programme appears to have been successful in matching rapid increases in milk production and consumption per capita, since the mid 1970s, so that today India is self-sufficient in milk and produces more than any other country in the world.

Dairy Marketing Co-operatives have been established in other developing countries assisted, as in India, by international food aid and Government supported prices and investment in processing equipment. The outcomes have been mixed. In many cases governments sought to influence or directly control the co-operative movement. Problems of mismanagement arose through lack of business skills and training, lack of finance, corruption and state interference. Thus many co-operatives have failed at the local level. In addition, the high costs of processing and competition from other sources of dairy products have reduced the importance of co-operative processing and marketing in many countries. As mentioned earlier local milk processors face competition from cheap imports of dairy products and from local traders who are not members of the co-operative. These problems have been exacerbated, since the 1980s, by market liberalisation, and the ending of price-fixing and state support for the co-operatives. With the ending of price control, producers may be offered a higher milk price by local traders than by the co-operative processor. Nonetheless, despite failures in the past, co-operative group action remains an important means of strengthening producers' bargaining power and deriving economies of scale in marketing.

Failure of input delivery systems

The problems of poor communications, inadequate transport systems and lack of competition among traders applies equally, or with more force, to the supply of inputs. This applies particularly to the supply of animal health services and advice on livestock production. It may also apply to genetic material where exotic breeds or cross-breeding are introduced to raise production. In the past, basic services in many developing countries were provided by staff of the Ministry of Agriculture and Livestock Development. The quality and standard of the services provided has always been open to criticism and subject to budgetary constraints. However, with liberalisation these services have been cut back and the private sector is expected to play a greater role in service provision.

Initially, cost recovery from livestock producers and privatisation of the services were seen as the appropriate policy instruments. However, it is clear that public sector involvement is still necessary, since some of these services have the characteristics of 'public goods', for which cost recovery is difficult, if not impossible. This applies for instance to epidemic or zoonotic disease control, and the dissemination of extension advice from which everyone may benefit. In addition some central organisation and quality control is needed for the services provided. Hence an appropriate sub-division of responsibility between the private and public sectors must still be found.

Progress in privatisation has been slow. For instance, private veterinary practices have only been established and survived in areas of intensive livestock production (see Otieno-Oruko *et al.*, 2000). The costs of establishing a practice are high and must be recovered, along with travel and other operating costs, from livestock producers. In remote areas, with relatively sparse livestock populations and high transport costs, there may not be enough business to justify the private provision of animal health services. Where private practices have been established, each will probably be a monopoly service provider and will need to charge high prices. Similar problems are associated with the delivery of artificial insemination (AI) services. Meanwhile public sector delivery of extension and other 'public goods' is generally under-funded and weak.

Credit is another important input, needed particularly for consumption smoothing, coping with disasters, or for investment in new livestock enterprises. Costs and risks associated with provision of credit to small and scattered livestock-producers in rural areas, are high. Resource-poor producers lack suitable assets for use as collateral security for loans. Informal local moneylenders have the advantage of personal knowledge and ability to screen potential borrowers. However, they generally charge high interest rates to cover costs and restrict lending to short-term loans only. Formal lending institutions, such as banks, have often needed subsidies to support rural lending at low interest rates, and have tended to concentrate their lending to large-scale commercial producers.

Many commentators believe that micro-finance institutions offer the best prospects for improving the access of rural people to credit, saving and, possibly, insurance facilities. These are generally privately- or NGO-funded agencies aimed at providing small loans, often focused on women entrepreneurs. Their objectives are to generate productive income streams, to reduce poverty and to support sustainable rural financial institutions. Many have achieved sustainability by combining high repayment rates with interest charges at the market rate. The risks and costs of individual loans are often reduced by group lending. However, despite the poverty-reduction orientation of these institutions, there is a natural tendency to concentrate on the more credit-worthy and to by-pass many of the poor. The alternative of credit in kind is adopted in development projects, such as the 'heifer-in-trust' schemes (Afifi-Affat, 1998).

The economic environment in which livestock producers operate, is made up of many inter-locking markets for livestock and their products, and for the necessary inputs. Sometimes the inter-linkage is direct, for instance where a landlord also supplies credit to tenants. With increasing commercialisation, linkages may develop between the processing and marketing agency's purchase of the product and the provision of necessary inputs on credit. For instance, inputs of day-old chicks and pre-mixed concentrate feeds may be provided, together with veterinary services and advice, to broiler producers, by the commercial processor. However, for the reasons given above, smallholder producers may be at a disadvantage in negotiating contracts.

Opportunities offered by international trade and globalisation

The expansion of international trade opens up further opportunities for livestock producers to specialise in products for which they have a comparative advantage over other countries. Patterns of trade have evolved, and are still evolving, as a result of market forces in response to differences between countries in the relative costs of production of alternative commodities and consumer preferences. However, patterns of trade are also influenced by the imposition of trade barriers by individual countries or groups of countries. Barriers to free trade result from government policies, including agricultural support and protection in developed countries, as well as taxation and industrial protection in developing countries.

Among the developed countries, the European Union, the USA, Japan and others such as Norway and Switzerland, assist their domestic livestock producers by means of price-support, import tariffs and export subsidies. Whilst raising the incomes of domestic producers and encouraging home production, the increased supplies on world markets tend to depress prices. As such, these policies to support livestock producers represent barriers to free trade, and unfair competition for producers in other countries.

A widely quoted example is that of cheap sales of beef from the European Community to coastal West African Countries in the early 1990s. These imports met a large proportion of the demand in these countries (60 per cent of the beef supply for Ghana and 40 per cent of that for Côte d' Ivoire) and caused a serious drop in exports from the Sahelian Countries to the Coast (Van Ufford and Bos, 1996). However, the experience of 'Operation Flood' demonstrates that cheap imports of dairy products from the developed countries can be used effectively, to develop domestic markets and the same may apply to meat.

Some African, Caribbean and Pacific Ocean (ACP) States have benefited from concessionary reductions in tariffs by the EU under the Lomé Convention, now replaced by the Cotonou Agreement. Tariffs are reduced on beef and veal, and other commodities, from Botswana, Zimbabwe, Madagascar, Swaziland and Kenya. These trade advantages over other potential exporters to Europe would not exist if the trade barriers were abolished.

The World Trade Organisation (WTO), which derived from the General Agreement on Tariffs and Trade (GATT), is aimed at global trade liberalisation. The Agreement on Agriculture, signed in 1994, called for reduction of export subsidies, reduction in financial support for agricultural producers and improved access through replacement of non-tariff barriers by tariffs which would be gradually reduced. Global economic benefits were predicted to result from the growth in trade following liberalisation. However, the main beneficiaries would be found in the developed countries, either the consumers in countries where the producers had previously been supported, or the producers in countries, such as Australia and New Zealand, where they had not.

Some developing countries, in Latin America, South and East Asia might also be able to expand their export markets following trade liberalisation. In those which are net importers of livestock products, local producers would benefit from the predicted modest increases in prices. Developing country consumers would face higher prices for livestock products but the greatest impact will be on urban middle classes rather than on the rural poor.

Hitherto there has been limited progress in reducing protection of developed-country producers. Although policies in the European Union and North America have shifted away from trade distorting price supports, the overall level of support remains high. The Doha Round of WTO negotiations, launched in November 2001, together with the collapse of the meeting in Cancun, served to demonstrate the wide diversity of levels of commitment to trade liberalisation, as might be expected given the differing effects of barriers on producers and consumers in different countries. Globally, however, the benefits of liberalisation are expected to exceed the costs, so there is a case for maintaining pressure on the developed countries to reduce their protective trade barriers. Nevertheless it is predicted that the impacts on developing countries would be limited and that local producers would benefit more from domestic policy reforms.

Biosecurity measures

Most countries attempt to prevent the import of animal and human diseases by appropriate checks and controls (either at the border or by monitoring the production processes in the country of origin). High-income countries often impose tighter, more rigorous rules than those currently in force in developing countries. Hence the high-income country rules serve as non-tariff barriers to exports from developing countries.

The Sanitary and Phyto-sanitary (SPS) Agreement of the WTO is aimed at rationalising the rules and adjudicating on inter-country disputes, in particular to ensure that they are not used as disguised measures to protect developed-country producers from competing imports. However, the costs of meeting the legitimate SPS rules of the developed countries, and of negotiating dispute settlement, are high for potential developing-country exporters. Apart from the impact on trade, improved biosecurity is beneficial for domestic livestock populations and for human health. Hence improvements in developing country SPS measures are desirable. International support and assistance are justified to promote these improvements.

Policies for improving markets

Direct government intervention and price control in agricultural markets, through the operation of parastatal Marketing Boards, are no longer favoured options. In many cases they failed to meet development objectives, while many commentators believe that liberalised competitive markets can provide a more efficient and effective service. Nonetheless, governments have an important role in improving the market environment, lowering transformation and transaction costs and helping markets to function better. International trade negotiations, with the WTO and other agencies, may well affect domestic prices and are clearly the responsibility of national governments. Governments have a clear role in providing the physical infrastructure of roads, stock routes, watering facilities, market places and slaughterhouses. Most of these facilities are unlikely to be provided by private enterprise, other than where the operator has monopoly control of the service and can recover the costs from users. In addition postal and telegraphic communication systems are important for the flow of information as, perhaps more fundamentally, is the general level of education and literacy of the people.

This links with the important public role of stimulating the flow of market information. Rural livestock producers are at a severe disadvantage in lacking knowledge of demand and prices in accessible markets. Timely and clearly-analysed price data should be made available through mass media, notices, leaflets or radio programmes. Including such data in village computer-information systems (cf. Chapter 5 Cameo) should be considered. Agricultural extension staff require training to provide advice on marketing issues as well as on technical matters. Supervision and guidance are required for small-scale producers entering contracts with large-scale processing and marketing enterprises.

Governments have a role in improving and strengthening the institutional environment. This involves putting in place an enforceable legal framework for defining property rights and establishing market contracts. It may include the setting of standards for weights and measures and for quality grades. Such rules and norms should reduce transaction costs for all market activities. Appropriate organisational structures for equitable co-operative group activity by livestock producers, should be promoted. Rather than imposing new institutions on producers, the aim should be to encourage and support traditional or innovatory indigenous organisations.

It is generally accepted that governments have a role in correcting for market failure and promoting competitive conditions. In many developing countries, the dispersed nature of agricultural production and the high market-transaction costs result in markets being dominated by a small number of produce buyers and input sellers. These traders may then have local monopoly power. Although the promotion of competitive conditions may be impossible without direct market intervention, governments still have a role in price and quality control of produce marketing and input delivery.

An important case of market failure, where private enterprise cannot be relied upon, is in the provision of public goods. Governments need to be involved in providing public goods, such as those occurring in the field of human and animal disease control and health protection, together with innovations resulting from research and development. Publicly funded action is needed for control of major epidemic (Class A) and zoonotic diseases, with monitoring, surveillance and meat inspection. These measures not only protect the human and animal populations from disease hazards, but also may also widen export markets by meeting SPS requirements.

Further reading

Abbott, J. 1993. *Agricultural and food marketing in developing countries: selected readings*. CABI Publishing, Wallingford, UK.

There is no piped water in Giaki, so everyone depends on local rivers, streams, springs or wells for their supply. Animals share many of the same sources. As temporary streams and pools dry up during the dry season, animals concentrate around permanent water sources, where surroundings are eroded and water is contaminated by urine and faeces. Pollution is particularly severe towards the end of the dry season, when water sources become stagnant.

Cameo 2 - Livestock rearing in the highlands of Lesotho

Lesotho is a land-locked, mountainous country with a temperate climate and relatively high rainfall, which provides Johannesburg and Gauteng, South Africa's industrial hub, with much of their water through the Lesotho Highlands Water Project. Most of the highlands above 2,000 m are unsuitable for cultivation and are used for grazing by sheep, goats, cattle, donkeys and horses, in descending abundance. Livestock are kept mainly by smallholders.

Merino sheep and Angora goats are raised for wool and mohair, slaughter and ceremonial purposes. Cattle are raised for milk, meat, fuel (dung) and draught power, as well as for investment and socio-cultural reasons, such as bohali (bride-wealth) and other ceremonies. Donkeys and horses are widely used as pack animals and for personal transport. A system of livestock borrowing and lending, known as mafisa, in which the holder has rights to wool, mohair, milk and draught power, whilst the owner retains title to the animal and any progeny, is widely practised (Marake *et al.*, 1998).

Lesotho's extensive rangelands are used under two management regimes: maboella that govern the use of village common lands in the lowlands and foothills; and a 'cattle-post' regime for the control of highland grazing. Maboella were instituted in the mid-1800s by Lesotho's first Paramount Chief to protect communal lands at particular times during the year. Maboella rules were enforced at village and local levels by the Paramount Chief's designates, who were compensated for their efforts. The cattle-post regime was established in the 1920s and evolved as a seasonal transhumance of lowland livestock to mountain cattle-posts for the summer months so as to co-ordinate the activities of increasing numbers of livestock grazing mountain pastures and protect those pastures from over-exploitation. As with maboella, the cattle-post regime was enforced by the Paramount and Principal Chiefs, who granted permission to individual livestock owners to use specific cattle-posts (Swallow and Bromley, 1998).

Land degradation, overstocking and grazing control have been major concerns of Government and development planners for decades, resulting in various laws and regulations that have led to considerable controversy and calls for the reform of Lesotho's land tenure system (Phororo and Letuka, 1993). Community-based land use planning and identification of specific uses for specific areas are essential for the future sustainable use of natural resources, whilst local and national authorities also have vital roles to play in co-ordinating activities and ensuring that wider environmental concerns are addressed.

Photo 9.1 Lesotho cattle with reflectors. Use of reflective neck and/or girth bands is a recent innovation to aid night movements and reduce road accidents. Note the encroachment of cultivation on former rangeland and erosion gullies in foreground and on distant hillside (courtesy of B. Motsamai).

Introduction

This chapter outlines the main interactions between livestock and the environment, and identifies interventions and strategies to enhance the positive effects and/or mitigate negative impacts of livestock production, focusing on smallholder animal husbandry. Resource-poor livestock farmers are at the forefront of livestock interactions with the environment, as highlighted in the cameos above.

The cameos also illustrate the variety of roles and complexity of livestock interactions with the environment, and the need for careful consideration of specific local circumstances and concerns in the prioritisation and implementation of interventions.

Interactions between livestock and the environment

Livestock interactions with the environment have been the subject of much discussion and heated debate over the past decade, and have spawned numerous studies and a substantial literature, encompassing:

- Destruction of the Amazonian forest to produce beef for the hamburger society (Nations and Komer, 1987)
- Livestock as sources of greenhouse gases (IPCC, 1995)
- Advocacy for a more harmonious balance between livestock and the environment (de Haan *et al.*, 1997; Steinfeld *et al.*, 1997; Nell, 1998)
- The environmental risks and recuperative effects of animal agriculture in the developing world and the unwarranted blanket condemnation of livestock production (Nicholson *et al.*, 2001; Blake and Nicholson, 2004).

A major component of the International Livestock Research Institute's (ILRI) programme is directed at people, livestock and the environment. The World Bank, the European Union and many bilateral donors have sponsored the Livestock, Environment And Development (LEAD) initiative and a compilation of 130 publications on CD-ROM concerning livestock, environment and development interactions (FAO, 2003).

The natural environment can be conceptualised as a complex and dynamic entity, composed of a variety of interlinked ecological systems and cycles that are buffered against disturbance by their relative size and feedback loops, and are reasonably stable over time, within certain limits of tolerance. Livestock interact with and impact on their environment through various activities, inputs, outputs, and management practices. Direct interactions include respiration, drinking, feeding, gaseous emissions from digestion, excretion and movement. More general interactions and indirect impacts relate to animal husbandry, land management and mode of production. The relative importance of these interactions and impacts depends on their magnitude, extent and duration, and specific local circumstances, as well as the standpoint and perception of the observer. The main interactions between livestock and the environment are outlined below, together with a summary assessment of key monitoring indicators.

Atmosphere

Agriculture, including livestock production, is a major source of greenhouse gases and other atmospheric emissions, which contribute to climate change, acid rain and the eutrophication (nutrient enrichment) of water bodies, as summarised in Table 9.1. Crude global estimates, such as in Table 9.1, are subject to considerable uncertainty but provide an indication of relative importance.

Carbon dioxide

Land use changes, including the clearing and burning of forests, woodlands and grasslands for cultivation and the grazing of livestock, account for 15 per cent of total anthropogenic releases of carbon dioxide, the balance coming largely from burning fossil fuels and industrial processes. Livestock also contribute to carbon dioxide emissions through their basic metabolism and respiration (Bruinsma, 2003).

Methane

Livestock account for 30 per cent of all anthropogenic emissions of methane. Although methane is less persistent in the atmosphere, it has 20-25 times the global warming potential of carbon dioxide. Livestock emissions of methane come from enteric fermentation of ingested plant material and releases from animal excreta. Other important anthropogenic sources include rice paddy fields, biomass burning, landfills, coalmines and the exploitation of oil and gas fields (Moss, 1993; Bruinsma, 2003).

The amount of methane produced by livestock depends on their size, age, digestive system and the quantity and quality of feed intake. Ruminants (buffalo, cattle, camels, goats and sheep) emit the greatest quantities of methane: 25-118 kg per head per annum for cattle, and 5-8 kg per head per annum for small ruminants (IPCC, 1995). Pseudo-ruminants (horses, donkeys and mules) and mono-gastrics (pigs and poultry) produce less methane, because their digestion is not so dependent on enteric fermentation.

Livestock manure consists mainly of organic matter, which decomposes under anaerobic conditions to produce methane. The amount of methane produced depends on the quantity of manure and the proportion that decomposes anaerobically. When stored or treated in liquid form, as in the slurry lagoons, ponds, tanks or pits common to more intensive systems, anaerobic decomposition is favoured, and greater quantities of methane are produced. When manure is managed as solid in heaps or stacks, or when deposited on pastures and rangelands, it tends to decompose aerobically and little or no methane is produced (IPCC, 1995).

Table 9.1 Agriculture's estimated contribution to global greenhouse gas and other emissions

Gas	*Carbon dioxide*	*Methane*	*Nitrous oxide*	*Nitric oxides*	*Ammonia*
Main effects	Climate change	Climate change	Climate change	Acidification	Acidification & eutrophication
Agricultural source (% contribution to total global emissions)	Land use change, especially deforestation	Ruminants (15)	Livestock (incl. manure on farmland) (17)	Biomass burning (13)	Livestock (incl. manure on farmland (44)
		Rice production (11)	Mineral fertilisers (8)	Manure and mineral fertilisers (2)	Mineral fertilisers (17)
		Biomass burning (7)	Biomass burning (3)		Biomass burning (11)
Agricultural emissions as % of total anthropogenic sources	15	49	66	27	93
Expected changes in agricultural emissions to 2030	Stable or declining	From rice: stable or declining From livestock: rising by 60%	35-60% increase		From livestock: rising by 60%

Source: Bruinsma (2003).

As livestock populations increase and are managed more intensively, as projected for most regions, the quantities of methane and manure produced are projected to rise by up to 60 per cent by 2030 (Bruinsma, 2003).

Nitrous oxide

Nitrous oxide is the most potent greenhouse gas, some 320 times more powerful than carbon dioxide. Agricultural emissions account for most of the anthropogenic sources of nitrous oxide and come primarily from microbial nitrification and denitrification in soil, and biomass burning. Increasing application of nitrogen to the soil, from whatever source including animal manure, produces more nitrous oxide (Bruinsma, 2003).

Nitric oxides

Just over a quarter of all anthropogenic sources of nitric oxides are agricultural, emanating from biomass burning and microbial nitrification and denitrification of animal manure and organic fertilisers.

Ammonia

Agriculture accounts for nearly all anthropogenic emissions of ammonia, with livestock production, including the application of manure to farmland, accounting for nearly half of the global total. Ammonia emissions are potentially even more acidifying than sulphur and nitrogen oxides, and releases from intensive livestock systems contribute to both local and long distance deposition of nitrogen, with damage to trees and the acidification and eutrophication of aquatic systems (Table 9.1).

Odours

Livestock may also emit unpleasant odours, especially from the accumulation of wastes from intensive production units, which may be a nuisance and cause offence in populated areas. The most prominent gas is hydrogen sulphide, which is very toxic at high concentrations.

Vegetation

As herbivores, ruminant and pseudo-ruminant livestock are dependent entirely on vegetation for their nutrition. Monogastric livestock, although technically omnivorous, are often fed cereal-based diets and/or vegetable wastes, and are thus also largely dependent on plants for their survival. However, the interactions of livestock and impacts of livestock production on vegetation extend far beyond the consumption of plant material, to the shaping of agricultural landscapes around the world. Some 34 million square kilometres of land (26 per cent of the total) are used for grazing livestock and an additional 3 million square kilometres (21 per cent of all arable land) are used for producing cereals for livestock feed (Steinfeld *et al.*, 1997).

Cattle ranching has been accused of contributing to the deforestation of Central and South America through the clearance and conversion of land to pasture (Nations and Komer, 1987; Kaimowitz, 1995), and pastoralists have been blamed for the degradation and desertification of rangelands through 'overstocking' and reduction in vegetation (Hardin, 1968; Sinclair and Fryxell, 1985). There are many detailed discussions and rebuttals of these complex and controversial issues (Homewood and Rodgers, 1991; Behnke *et al.*, 1993; Brockington and Homewood, 1996; Nicholson *et al.*, 2001; Blake and Nicholson, 2004).

Animal husbandry in mixed-farming systems has a strong influence on vegetation and land-use patterns, through individual and/or collective choices of:

- What areas should be set aside for livestock use as fallow, or rough grazing
- Where cereal and forage crops are grown
- Use of live fencing for field boundaries and stock routes.

Livestock also have important roles in arable agriculture through manuring, ploughing, seeding, weeding and consuming crop residues, whilst in cut-and-carry and more intensive systems, vegetation is collected, or crops are grown, to be fed to animals kept in confinement. In East Africa, nutrient hot spots from seasonal kraaling of pastoral livestock can remain visible for decades and facilitate tree regeneration (Reid and Ellis, 1995).

Soil

The physical impacts of livestock on soil include:

- Breaking and penetrating surface crusts
- Compaction
- Creation of denuded pathways along which rainwater run-off may flow

- Disturbance through digging and rooting
- Consumption of mineral and salt deposits
- Tillage for crop production.

All forms of soil disturbance have the potential to cause erosion, but may also facilitate rainfall infiltration, organic matter incorporation and seed germination, depending on circumstances and management.

Livestock influence soil chemistry through the supply of nitrogen, minerals and organic matter. Livestock excreta contain a high proportion (60-90 per cent) of the nitrogen, minerals (phosphorus, potassium, magnesium etc.) and heavy metals obtained from their feed (de Haan *et al.*, 1997). Only for high-yielding dairy cows, do the excreta contain less than 80 per cent of nutrients consumed. In low-intensity systems, 85-100 per cent of these nutrients are excreted (Sundstøl *et al.*, 1995). Accumulations of solid and liquid manure from feedlots, piggeries and poultry houses are a potential source of various environmental pollutants. Whilst the judicious application of animal excreta on fields and recycling of plant nutrients are beneficial to soil composition and fertility, the indiscriminate disposal of excessive quantities on arable land can result in accumulations of nutrients and heavy metals that threaten soil fertility. Some diseases and pathogens harmful to human and animal health may be spread in a similar manner (Menzi, 2001).

Water

For their survival livestock depend on regular supplies of water which may be obtained from a variety of naturally occurring springs, streams, rivers and lakes, or man-made sources. Problems arise from the concentration of animals and the contamination of water sources by dung, urine and run-off, especially where those sources are shared by other users. Whilst drinking, or crossing streams and rivers, livestock may also reduce water quality by disturbing sediments and increasing downstream turbidity.

Run-off from heavily-manured fields and discharges from intensive production units, abattoirs and processing plants into streams and rivers can have severe impacts on aquatic systems, in particular eutrophication of water bodies and consequential algal blooms, composition of fish populations, ecological balance and water quality (de Haan *et al.*, 1997). Nitrates may also leach out to contaminate ground water and threaten drinking water supplies.

Table 9.2 Monitoring indicators for grazing and mixed-farming systems

Category	*Natural resource base*	*Livestock*	*Socio-economic*
		Grazing systems	
Soil/land	Erosion: universal soil loss equation	Arid: herd mobility	Arid: human carrying capacity of the land
	Presence and use of legumes	Sub-humid and humid: stocking rates and productivity trends	Land tenure and recent trends in fencing and encroachment of cultivation in key areas
	Manure collection and application practices		Vulnerability to drought (reliance on food aid)
Vegetation	Proportion of ground cover	Forage demand	Infrastructure
	Proportion of land cultivated	Diet preferences	Cohesion of user groups
	Plant species composition	Animal productivity and species composition	Diversity of land use
	Rate of firewood extraction		
	Presence and use of leguminous plants		
	Utilisation of crop residues, tame pastures and native rangelands		
	Rate of deforestation		
Water	Quality: turbidity, oxygen, nitrogen, phosphorus, pesticides content etc.	Use requirements	
	Number of boreholes		
	Number of new surface watering points		
Air	Greenhouse gas balance	Greenhouse gas balance	

Table 9.2 (Contd.)

Category	*Natural resource base*	*Livestock*	*Socio-economic*
	Mixed-farming systems		
Soil	Erosion: universal soil loss equation	Access to animal traction	Rate of integration and degree of reliance on outside inputs
	Nutrient levels: nitrogen, phosphorus, copper, zinc	Manure storage and utilization Quantity of concentrates brought into system	Community cohesion in watershed and regional landscape
	Organic matter: cation exchange capacity Farm and regional nutrient balances Presence and use of legumes		
Vegetation	Proportion of ground cover Proportion of land cultivated Plant species composition Presence and use of legumes	Forage demand Diet preferences	Farm income
Water	Surface water quality: turbidity; oxygen, nitrogen, phosphorus, pesticides content etc. Ground water quality, nitrogen, phosphorus content	Use requirements	
Air	Manure application techniques	Quality of animal diets Number of animals	

Source: Derived from de Haan *et al.* (1997).

Monitoring indicators

From the foregoing, livestock clearly have a multitude of interactions with, and potential impacts on, the environment. Whether or not these interactions are considered to be beneficial or harmful depends on the specifics of local circumstances, and the observer's point of view. With continuing human population growth and agricultural expansion, and the predicted further intensification of livestock production in years to come (Delgado *et al.*, 1999), ever-widening land-use changes and impacts on the environment are inevitable. Key indicators for monitoring the environmental interactions of different farming systems and processing plants are summarised in Tables 9.2, 9.3 and 9.4.

Table 9.3 Monitoring indicators for industrial livestock production systems

Input-related	*Production-related*	*Output-related*
Land use changes and land requirements for feed production	Conversion efficiencies for nitrogen and phosphorus by animal species	Manure discharge nutrient balances
Percentage of grain in concentrates and diet	Farm nitrogen and phosphorus balance	Fertilising value of manure
Rangeland requirements for young stock	Ammonia emissions	Methane emissions
Livestock breeds used	Methane emissions	Quantity of live-weight slaughtered
Inputs to feed production (fuel, fertiliser)	Fossil energy consumption	Quantity of raw milk produced
Animal welfare index	Weight of raw hides processed	
Chemical use	Manure storage	

Source: Derived from de Haan *et al.* (1997).

Table 9.4 Monitoring indicators for processing plants

Direct indicators	*Indirect indicators*
Amount of solid waste	Proportion of industrial and traditional processing
Total Biological Oxygen Demand (BOD) of wastewater N.B. not percentage of effluent, as this would provide incentives just to increase water use)	Proportion of by-product utilisation
Carbon dioxide, carbon monoxide and nitrous oxide emissions	

Source: Derived from de Haan *et al.* (1997).

Nutrient cycles, flows and accumulations

Nutrients are a collection of chemical compounds, minerals and elements essential to the survival of living organisms. Plants obtain these nutrients from their surroundings, and animals obtain theirs from what they eat. The passage, or flow, of a nutrient through the environment, via various plants and animals and alternative pathways, is a nutrient cycle. Nutrient cycles are a sub-set of a broader class of global biogeochemical cycles, including water, carbon, oxygen, nitrogen and mineral cycles.

Agriculture has a significant influence on several biogeochemical cycles, especially those relating to nitrogen, phosphorus and potassium (the three primary ingredients of manufactured fertiliser) and sulphur (Bruinsma, 2003). Livestock have an important role in the recycling of nutrients, both by defecation and urination whilst grazing, and farmers' application of manure to fields as organic fertiliser. Environmental problems may arise, however, from excessive applications, and from the uncontrolled disposal and discharge of accumulated animal wastes from 'industrial' livestock production units (de Haan *et al.*, 1997; Steinfeld *et al.*, 1997). In such circumstance, where animals are kept at high density in piggeries, poultry farms and feedlots, they are dependent entirely on cereal-based diets containing supplementary nitrogen, potassium, phosphorus and various heavy metals, including copper and zinc.

Relatively small proportions of these supplementary nutrients are absorbed, with most passing through the gut to be excreted with other animal wastes. Net imports of animal nutrients lead to local accumulations and excess soil nutrients, with adverse effects on soil fertility, increased run-off, eutrophication of aquatic systems and contamination of ground water and drinking water supplies. Hot-spots of nutrient surplus are found in North-Western Europe, Eastern and Mid-Western regions of the USA, Eastern China, Japan, Korea, Malaysia, Thailand and parts of Java and Sumatra in Indonesia (de Haan *et al.*, 1997; Steinfeld *et al.*, 1997), and are a potential problem for intensifying peri-urban livestock production in general.

With the anticipated further intensification of livestock production to feed increasingly urbanised populations in years to come (Delgado *et al.*, 1999), the eutrophication of terrestrial and aquatic systems is likely to become more widespread around the world, unless preventative measures are taken. Such interventions might include zoning land use and agricultural practices, regulating the distribution of intensive livestock production units, taxing excess nutrients, reducing nitrogen and phosphate excretions by improved feed formulation and utilisation, and minimising emissions from manure storage and spreading.

Interventions to enhance positive effects and/or mitigate negative impacts

Given the range of environmental impacts outlined above and the limited degree of control over many traditional livestock production systems (Chapter 3), there are no panaceas, or 'quick-fixes' that will solve all problems; but there are various measures that can be taken to enhance positive effects and mitigate negative impacts. In the first instance, there is a general need at all levels, from central government to farmer, and from university to primary school, to promote greater awareness of livestock related environmental issues and the options available for dealing with them.

In reviewing the range of interactions and impacts that livestock may have in a specific location, a useful distinction can be made between:

- Intensive (high input/high output) production systems, including dairying, feedlots, piggeries and poultry farms
- Extensive (low input/low output) production systems, including pastoralism and ranching and
- Intermediate mixed-farming systems.

Whether or not such interactions will result in serious adverse impacts on the environment depends on their magnitude and extent, specific local circumstances and individual perspective. It is worth noting, however, that with increasing production per animal, livestock emissions and excretions decline per unit of product.

Clearly, much depends on specific circumstances and technical capacity, and it is always essential to build on indigenous knowledge and experience of local conditions, but the following smallholder and community-based measures should be considered:

- Application of manure to fields to recycle nutrients, improve fertility and increase organic matter content
- Use of crop residues by livestock to improve nutrition and reduce biomass burning
- Use of animal traction for tillage and carting to increase efficiency and reduce engine emissions

- Use of urea-molasses blocks by ruminants to enhance microbial fermentation, increase digestive efficiency and reduce methane production (see Chapter 11 on feeds and feeding)

- Protection of water sources and segregation of use to reduce contamination and ensure access and adequate supplies to all users

- Anaerobic digesters to produce biogas (methane and carbon dioxide) as an energy source, rather than release into atmosphere, for when burnt, the methane in biogas produces a mixture of carbon dioxide and water vapour, which are less powerful warming agents than methane

- Measures to maintain vegetative ground cover and use of live fencing and planting of forage tree legumes (Gutteridge and Shelton, 1998) and various plant species, such as Vetiver grass (Anon, 1993), to reduce erosion

- Construction of drainage ditches and biologically active settling ponds to recycle nutrients and minimise downstream contamination

- Use of leguminous crops, both to increase nitrogen fixation and as forage crops.

At policy setting, strategic planning and regional levels, priority should be given to promoting:

- Mixed farming systems and the closer integration of arable and livestock production

- Zero-, or reduced-grazing, cut-and-carry systems in areas of high human-population density, with safeguards for animal welfare

- Community-based, catchment-oriented land use and environmental management planning

- Monitoring key environmental indicators, especially water quality and discharges from livestock markets, intensive production units and processing plants, including: dairies, feedlots, piggeries, poultry houses, abattoirs and tanneries.

Given current global environmental trends, particular attention needs to be given to the zoning, monitoring and control of livestock production activities in urban and peri-urban areas.

In general evolutionary terms, even after 10,000 years of development, the future course of arable and livestock production is by no means certain, but it would seem reasonable to assume that agricultural production systems are most likely to be sustainable, if their adverse environmental impacts are minimised.

Strategies to sustain/improve farm animal genetic diversity

Chapter 12 considers breeding strategies to achieve sustainable improvement in livestock productivity. It is considered relevant to address the question of genetic diversity here in the context of 'livestock and the environment'.

More than 40 species of mammals and birds have been domesticated, and some 5,000 livestock breeds are recognised, 20 per cent of which are endangered because their populations have declined rapidly in recent years, or their breeding stocks are low (Scherf, 2000). This loss of genetic diversity limits the options for improved production and sustainable agriculture, and reduces the gene pool available for new breeds. The need to conserve livestock diversity and the means by which this might be achieved have been reviewed by Blench (2001).

The maintenance of farm animal genetic diversity is of crucial importance to food production, food security and the sustainability of farming systems. Indigenous livestock breeds are important because, through many generations of natural selection and selective breeding, they are adapted to local conditions and farming systems, and are more likely than exotic breeds to tolerate seasonal variations in climate and forage supply, and resist local diseases and parasites. Despite the superficial attractions of modern, 'high-tech' methods, traditional farming systems still account for most of the food produced in the developing world. Just as some local crop varieties are resistant to drought, or tolerant of poor soils, indigenous livestock breeds and landraces provide an invaluable reservoir of genetic variation and breeding potential for adaptation to changing climatic and farming conditions.

Conservation strategies include:

- *In situ* maintenance of breeding populations in their regions of origin
- *Ex situ* maintenance of breeding populations away from their ancestral homes
- 'Cryo-conservation' of genetic material, such as semen, embryos, DNA, cells or ova, by freezing (FAO, 1998).

The choice of conservation strategy obviously depends on the breed and specific local circumstances, but wherever possible *in situ* conservation is the preferred option, through the promotion of livestock breed associations and stud books to maintain and enhance breed characteristics (see Chapter 12).

Strategies to accommodate global warming

Global warming refers to the progressive increase in mean surface temperatures over the past century. Despite general agreement amongst scientists that global warming is a reality (IPCC, 2001), great uncertainty remains about its consequences in different parts of the world. Predictions from climate models are subject to considerable doubt and different climate models, based on different assumptions, produce different results (Hadley Centre, 2004). Nevertheless, some general predictions can be made about the consequences of global warming. Sea levels will continue to rise, with increased risk of flooding in estuarine and other low-lying coastal areas. Climatic conditions are also likely to become more variable, with more frequent extreme events. Some regions will become wetter, others will become drier. The poorest and weakest members of society and the least-developed regions and countries of the world will be most vulnerable to adverse impacts because of their limited assets, weak infrastructure, and restricted access and ability to invest in technological solutions.

A wide variety of impacts on global agriculture and food security has been predicted for the 21st century (UNEP and UNFCCC, 2002; Bruinsma, 2003), but these remain largely speculative because of the uncertainties inherent in their derivation. One of the few predictions about the future that can be made with virtual certainty, however, is that farming systems and agricultural environments will be subject to increasing pressure to produce more food to feed an ever-increasing number of people. Whilst specific changes at precise locations are impossible to predict with accuracy, they are likely to be progressive and cumulative, which means that communities and societies should have some time to adapt to them as they occur.

Given the highly variable and uncertain impacts of global warming on agriculture and renewable natural resources, six sets of activity are considered to be of critical importance to the establishment of effective strategies to meet the challenges of environmental change at national and local levels:

- Assess what changes may occur and identify representative areas for retrospective and future monitoring

- Conduct periodic, standardised surveys of agricultural resources, including characterisation of livestock breed attributes, population size and distribution

- Review agricultural and economic significance of potential and actual change, and act accordingly

- Maintain flexibility in agricultural development scenarios at national level

- Ensure diversity of choice in livelihood, farming and livestock production options at local level

- Increase awareness of livestock-environment interactions at all levels from producer to policy makers.

The long-term consequences of global warming on farmers and farming systems depend very much on their response to change and ability to take advantage of potential benefits and mitigate adverse impacts. Sustainable agriculture can be promoted and food security can be enhanced through the adoption of coherent policies and the implementation of appropriate strategies at national, local and farmer levels.

Options available to promote sustainable agriculture, include:

- Characterisation and conservation of existing genetic resources

- Selective breeding programmes to enhance local crop varieties and livestock breeds

- Improved water-management and irrigation systems

- Modification of planting schedules and tillage practices

- Closer integration and intensification of arable and livestock production through mixed farming

- Promotion of more effective participatory land-use planning and watershed management.

National and local strategies to accommodate and adapt to global warming must obviously be developed to meet specific national and local circumstances. Key principles for such strategies should be to promote adaptability and sustainability by maintaining collective diversity of livelihoods and livestock resources, i.e. by not putting 'all your eggs in one basket'.

Access to water and wetlands is already a source of conflict between arable and livestock farmers in many regions, not just in arid and semi-arid zones, and is likely to become an increasingly contentious issue in decades to come, as will the provision of water to satisfy growing demand in expanding urban and peri-urban areas. Local communities and natural resource user groups, including arable farmers, settled livestock producers and transhumant pastoralists, must be encouraged to participate actively in the preparation, implementation and monitoring of adaptive natural resources/land-use management plans, to ensure equitable access to, and sustainable use of, water and wetland resources. Implementation of such plans will be fraught with difficulties and conflicting demands, but these must be resolved by stakeholders themselves, through negotiations, compromises and trade-offs, if sustainable use is to be achieved.

Conclusions

- Livestock and livestock production have a wide range of interactions with, and impacts on, the environment, which can have both adverse and beneficial consequences.

- Individuals, communities, governments and international bodies need to be aware of both the positive and the negative impacts of animal husbandry and livestock production, so that appropriate measures can be taken to maximise benefits and minimise, or mitigate, adverse consequences.

- Hasty decisions, blanket judgements and all-purpose, 'quick-fix' solutions, based on inadequate information and limited understanding of local conditions, are unlikely to be sustainable in the long term.

- Each situation should be examined and evaluated separately in relation to its own specific environmental and socio-economic circumstances.

Further reading

FAO/ILRI. 1999. *Farmers, their animals and the environment.* CD-ROM. Food and Agriculture Organisation of the United Nations (FAO), Rome, Italy, and International Livestock Research Institute (ILRI), Nairobi, Kenya. http://www.virtualcentre.org/en/enl/vol1n2/read.htm

FAO. 2003. *Livestock, Environment and Development (LEAD) Digital Library.*

CD-ROM. Livestock, Environment and Development (LEAD) Initiative, Animal Production and Health Division, Food and Agriculture Organisation of the United Nations (FAO), Rome, Italy.

Hall, S.J.G. and Clutton-Brock, J. 1989. *Two hundred years of British farm livestock*. British Museum, London, UK.

Mason, I.L. 1996. *A world dictionary of livestock breeds, types and varieties*. CABI Publishing, Wallingford, UK.

Scherf, B.D. (ed.). 2000. *World watch list for domestic animal diversity*. Food and Agriculture Organisation of the United Nations (FAO), Rome, Italy.

Websites

Animal Production and Health Division of the United Nations Food and Agriculture Organisation (FAO AGA):
http://www.fao.org/ag/againfo/home/en/home.html

Domestic Animal Diversity Information System (DAD-IS):
http://dad.fao.org/en/Home.htm

Ecological Society of America (ESA) Issues in Ecology:
http://www.esa.org/sbi/sbi_issues/

Intergovernmental Panel on Climate Change (IPCC):
http://www.ipcc.ch/

International Livestock Research Institute (ILRI)
http://www.ilri.cgiar.org/

Livestock and the Environment/Livestock-Environment Interactions:
http://www.fao.org/ag/aga/LSPA/LXEHTML/Default.htm

Livestock, Environment And Development (LEAD) Initiative:
http://www.virtualcentre.org/en/frame.htm

United Kingdom Meteorological Office, Hadley Centre Climate Change Projections:
http://www.met-office.gov.uk/research/hadleycentre/models/modeldata.html

United Nations Framework Convention on Climate Change (UNFCCC):
http://unfccc.int/

Author addresses

David Bourn, Environmental Research Group Oxford Limited, Department of Zoology, South Parks Road, P.O. Box 346, Oxford OX1 3QE, UK

Joseph Maitima, International Livestock Research Institute (ILRI), P.O. Box 30709, Nairobi, Kenya

Bore Motsamai, Formerly at Ministry of Environment, Gender and Youth Affairs and National Environment Secretariat, Maseru, Lesotho

Robert Blake, Department of Animal Science, Cornell University, Ithaca, NY 14853, USA

Charles Nicholson, Department of Applied Economics, Cornell University, Ithaca, NY 14853, USA

Frik Sundstøl, Centre for International Environment and Development Studies (Noragric), Agricultural University of Norway, N-1432 Ãs, Norway

10

Animal response to nutrient supply

Peter Buttery[*], *Robert Max*[†], *Abiliza Kimambo*[†], *Juan Ku-Vera*[‡], *Ali Akbar*[‡]

Cameo 1 - Feeding high quality browse to Pelibuey hair sheep in South Mexico

Sheep production has been expanding in South Mexico during the last few years mainly due to the increased demand for mutton from tourist resorts in the Caribbean, but also from the traditional 'barbacoa' markets in Central Mexico. *Señora* Adda Ramos-Be, a Maya Indian living in Hotzuc, a small village near Merida, the capital of Yucatan State, owns 60 Pelibuey sheep which browse on the natural vegetation, where many different species such as *Leucaena leucocephala* (known as huaxín in the Maya language), *Brosimum alicastrum* (ox), *Piscidia piscipula* (jabín) and others, are available. With funds from an International Atomic Energy Agency (IAEA) Technical Cooperation project, researchers at the University of Yucatan have characterised the natural vegetation browsed by *Señora* Adda Ramos-Be's sheep. This involved determining the chemical composition, the presence of anti-nutritional compounds, such as tannins, and the nutritional value by *in situ* and *in vitro* techniques. Trials carried out on-station have demonstrated that dry matter intake of sheep on a basal diet of low-quality grasses can be doubled by incorporating between 30 and 40 per cent of a high-quality browse in the ration. Selected varieties are gradually being planted by farmers in field plots adjacent to the sheep houses. Silages from these species are being prepared to conserve forage for the dry season, when browse is scarce. With this technology, it is expected that efficiency of sheep production will be increased. Efforts by university staff are being made to disseminate the technology to other sheep farms around Hotzuc. With the new technology, resource-poor sheep farmers do not have to spend their scarce cash buying expensive 'balanced feeds' which are usually imported. Sheep keeping is becoming a more productive activity for Mayan families in Yucatan, and is helping to improve their livelihoods.

---

[*] Principal author
[†] Co-author
[‡] Consulting author

Cameo 2 - Increasing milk production in the Kilimanjaro region of Tanzania through improved feeding

The Kilimanjaro region has the highest population of improved dairy cattle in Tanzania. However, milk yield per animal is lower than the genetic potential. This low milk production has been associated with the low quality of feeds and the inadequate amounts of feed offered to the animals. The FAO Smallholder Dairy Development Project in Kilimanjaro and Arusha districts was aimed at increasing the quantity of milk produced per animal. This was achieved by improving the quality of the feeds offered to the animals, especially during the dry season, by advocating urea-treatment of straw and stover, and appropriate supplementation with mixture of urea and molasses, and some concentrate.

This is what a secondary schoolteacher, who is also a peri-urban dairy farmer, had to say about her benefits from the project "I used to think that cows could produce milk from straw and grass alone and that is what I used to feed to my cows. To increase intake of straw I used to sprinkle molasses and sometimes magadi (a naturally-occurring form of sodium sesquicarbonate) on the straw. I used to supplement lactating cows with about one kilogram of maize bran or wheat bran and a very small amount of oilseed cake during milking. Water was offered only during the afternoon. Daily milk yield was low, about 6 litres per cow, and the animals used to have problems of retained placenta and long calving intervals. One day I was visited by researchers from Sokoine University of Agriculture who were involved in the FAO project on improvement of feeds and feeding for dairy cattle in the region. They advised me on how to feed my animals for higher milk production - a technology that I adopted. I started by mixing a concentrate of maize bran, wheat bran, cottonseed cake, sunflower seed cake, chickpea hulls and minerals. I also increased the amount I offered to the animals from 1 kg to 3 kg/animal per milking. Instead of sprinkling molasses alone I started using a urea-molasses mixture. I also provided water to my animals all the time. By adopting this technology I managed to triple the milk yield, obtaining an extra 3000 Tanzanian shillings (US$ 3) per day from the sale of milk, and to reduce reproductive problems. I can say boldly to my fellow smallholder farmers that it is profitable to feed your dairy animals well. Increasing the amount of well-formulated concentrate offered to lactating cows during the dry season is profitable since the price of milk is good and the forage of low quality".

Introduction

In order to improve the productivity of animals, a prerequisite is a practical and effective strategy to utilise the available feed resources with the aim of maximising the animals' response to nutrients. It is generally agreed that the more feed an animal consumes each day, the greater will be the opportunity for increasing its daily production. However, nutritive value, including palatability, digestibility, and the presence of anti-nutritive factors, and animal factors, e.g. genotype, digestive efficiency and health, need to be taken into account. The management of manure should also be considered in view of its importance for resource-poor livestock-keepers.

One of the major aims of studying nutritional responses in domestic livestock is to generate information that can be used to predict the response of an animal to a particular diet in a variety of productive situations. To do this the feedstuff and the animal need to be adequately characterised. We also need to be able to predict the response of the animal to a particular feed or diet. Various feeding standards are available for developed countries, which are relevant to high-producing animals (see Further Reading at end of chapter), but schemes which are relevant to low-producing animals in the tropics are scarce. It is possible to use the schemes designed for high-producing animals in the temperate zones for low-producing animals in the developing countries, but the results need to be interpreted with caution because of the differences in intake, genotype, feed quality and availability, and the environment. Developing countries should develop feeding schemes based on their own situation. This is discussed further in Chapter 11.

It is not possible to cover adequately the basic principles of animal responses to nutrients in one chapter. The reader is therefore directed to the general textbooks listed in 'Further Reading'. The purpose of this chapter is not to duplicate the information in these standard texts, but to draw the attention of the reader to aspects of the subject which are of particular relevance to feeding livestock in the developing world, and also to provide background information to many of the nutritional topics dealt with elsewhere in this volume. Many of the feeding schemes given at the end of the chapter also serve as good sources of information on the factors that influence the response of animals to nutrients.

Palatability and feed intake

The term palatability generally describes the degree of readiness with which a particular feed is selected and eaten and it involves the senses of smell, touch and taste. It appears that animals can recognise both unpleasant and pleasant sensations associated with feed prior to or during eating. Whilst they often readily consume some feeds (e.g. green pastures, ammoniated straws, silage, cereal grains and molasses) they tend to have low appetite for certain feeds, particularly those containing anti-nutritional substances such as tannins and other phenolic compounds. Ruminants also differ in their ability to select feed: sheep and goats graze more selectively than large ruminants and therefore mixed species of grazing animals will often use pastures more efficiently. Feed processing, such as chopping, grinding or pelleting of straw or elephant grass, usually tends to increase intake. Dusty feed materials, such as finely ground feeds, can cause irritation of the nasal cavity and eyes, hence decreasing feed intake; sprinkling of these dusty materials with water (dampening) has been used by Indian farmers to increase the intake of straw. Animals also have complex neurological control mechanisms, which relate nutrient supply to the brain with the process of food intake. Feed intake is discussed further in Chapter 11 and by Forbes (1995).

Partition of and responses to nutrients

Once ingested, nutrients have several possible fates: they can remain undigested and passed out in the faeces, they can be excreted, used for maintenance of the animal, used for a productive process, for example, growth, milk production, reproduction, wool growth, egg production, or work. The partition of nutrients is an exceptionally complex process controlled by the genotype of the animal, the stage of development of the animal, the quantity and quality of feed available, and the environment.

Maintenance

Energy requirements for maintenance are related to the metabolic weight of the animal (i.e. live weight to the power of 0.75) and this energy leaves the animal as heat, and is approximately 0.3 MJ/kg $W^{0.75}$ /day. This does vary from species to species, but the value gives an indication of the amount of digested feed that is required to maintain the animal. The partition of energy by an animal is illustrated in Figure 10.1, and is discussed at length in most textbooks on nutrition. Maintenance energy is used for a variety of processes, for example ion transport across cell membranes, repair of tissues, resting muscle activity and the obligatory turnover of tissue protein, etc. The environment also influences maintenance energy requirements. For example it is increased by a cold environment when the animal will need to generate more heat to maintain body temperature. It is also increased during exposure to heat for evaporative cooling. Changes in physiological status, such as pregnancy and lactation, also increase maintenance energy requirement.

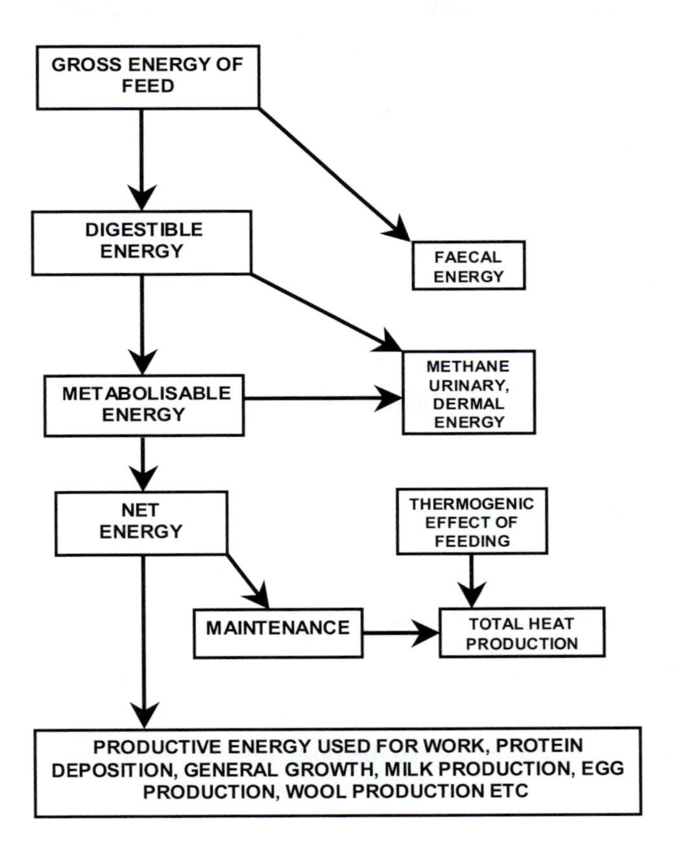

Figure 10.1 The partition of dietary energy.

Animals fed a nitrogen-free diet continue to excrete nitrogen. Substantial amounts of protein are lost in the faeces as a result of the secretion of digestive enzymes and the sloughing of the cells from the gut lining into the intestinal tract. Hair loss and sloughing of skin also contribute to endogenous losses. Efficient utilisation of amino acids by animal tissues requires that the supply of individual amino acids is balanced with the demand of the tissue. Excesses of an individual amino acid are catabolised. Excess nitrogen originating from the catabolism of amino acids is largely converted into urea (uric acid in the case of poultry) and excreted. Nitrogen is also lost as a result of the obligatory turnover of protein in the tissues.

In a 400 kg steer, for example, approximately 1.5 kg of protein can be synthesised in the tissues per day and almost an equivalent amount degraded. Especially during periods of feed shortage, there may be no increase or even a reduction in total protein mass (changes in protein mass are the result of the difference between protein synthesis and protein degradation). Amino acids released as part of the degradation of protein are not reused with 100 per cent efficiency, partly contributing to the maintenance requirements for nitrogen. This obligatory turnover of protein also accounts for a significant amount of the energy requirements for maintenance: estimates of around 20 to 30 per cent of the fasting metabolic energy expenditure have been made. Ruminants have the ability to recycle urea-nitrogen back to the rumen and convert it into microbial protein. The extent of this recycling is influenced by the concentration of ammonia in the rumen (see below).

Nutrient use for growth and productivity

Having satisfied the requirements for maintenance, the animal uses additional available nutrients for productive processes. The efficiency with which these nutrients are used varies between the nutrient and the particular productive process; consideration of this is a component of the majority of feeding schemes. John Hammond, one of the founders of animal science, introduced the concept of metabolic priorities operating during times of nutrient shortage. It is now generally accepted that neural tissue has the first call on nutrients, followed by the lymphatic system and the immune response, the viscera, bone, muscle and, finally, fat. Foetal development is maintained at the expense of lactation, although during periods of nutrient shortage many animals use the body reserves to support milk synthesis. Nutrient shortages normally result in a failure or a delay in the animal becoming pregnant.

Nutrient use for muscular work

Energy is required for general activity, for example shivering in cold climates, reducing body temperature in hot climates and even the physical process of eating. Allowances for, or consideration of, these energy-consuming processes are also built into most feeding standards. Work can be a major component of energy expenditure in draught livestock and, as a consequence, work can have effects on milk yield and on increasing post-partum anoestrus. It should be noted, however, that undernutrition has a greater effect upon milk yield than work. The partition of energy in a dairy cow being used for draught power is illustrated in Table 10.1.

Lawrence and Pearson (1998), from LPP Projects R6609 and R6166, describe a system for ruminants used for draught power which enables the nutrient requirements for work, and /or the effect of work on live-weight gain and milk production, to be calculated.

Table 10.1 Calculated daily energy requirements of dairy cows used for draught. An example of 4 hours work a day, a body weight of 450 kg, speed of 0.5 m/sec and a milk yield of 5 kg/day

Function	*Energy requirement (MJ)*
Work output	3.6
Net energy for work	12.0
Net energy for walking	9.0
Net energy for work plus walking	21.0
Metabolisable energy for work + walking	31.8
Metabolisable energy for maintenance	45.1
Net energy for milk (/kg)	3.1
Metabolisable energy for milk (5 kg/d)	26.5
Metabolisable energy for gestation	8.2
Total metabolisable energy requirement	111.6

Source: Zerbini and Alemu Gebre Wold (1999).

Kinetics of responses to nutrients and limiting nutrients

Phrases such as 'the requirement of an animal for a nutrient' are often used. Such statements and concepts can be misleading. What is of more interest is the response of an animal to its nutrient supply. Figure 10.2 illustrates the theoretical response of an organism to the supply of a nutrient. It should be noted that the supply of individual nutrients needs to be balanced. For example, if all nutrients are adequate with the exception of say, one amino acid, then the animal's productivity will be dictated by the lack of the limiting nutrient. The same concept applies to the supply of essential vitamins and minerals. With some nutrients there is an interaction, for example, an excess of the amino acid leucine, in the diet, can result in an apparent deficiency of isoleucine and valine, although the valine and leucine content of the diet might appear to be adequate. Such interactions have greater effects on productivity at low planes of nutrition. Interaction between minerals is also seen, for example, the interaction of copper, molybdenum and sulphur.

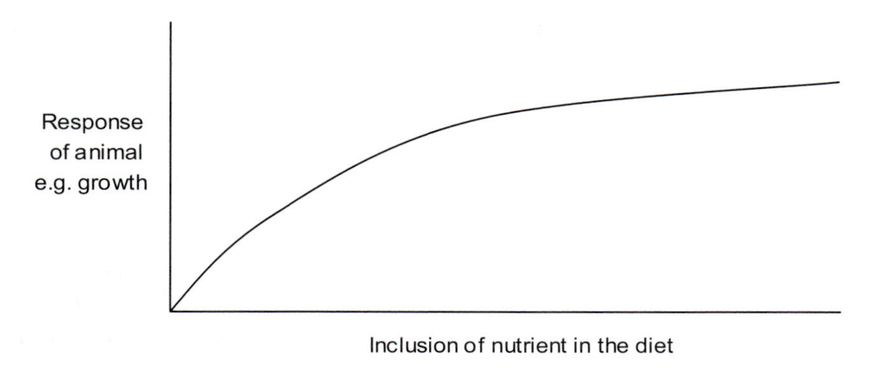

Figure 10.2 The theoretical response of a productive process to increasing nutrient supply

Anti-nutrients

Many plant materials contain toxic substances. Tropical legumes have considerable potential for enhancing productivity due to their prevalence and high crude protein content, but the use of these legumes can be limited by the presence of a diverse array of natural compounds which may act as anti-nutritional factors, causing undesirable effects such as reduced feed intake, low digestibility and acute toxicity. Tannins are by far the most frequently encountered anti-nutritional factors. Their impact can be considerable, for example, the digestible organic matter content of the tannin-containing leaves of carob (*Ceratonia siliqua*), was doubled when the animals were supplemented with 25 g/day of polyethylene glycol (PEG) (a substance known to complex tannins) in their concentrate ration. Tannins may not always be deleterious; there is some evidence that tannins also protect ruminants against the effects of intestinal parasites. Further discussion of the nutritional implications of tannins is presented in Chapter 11. Other anti-nutrients include amino acid analogues, such as mimosine that is found in *Leucaena leucocephala*, oestrogenic compounds found in some pasture plants and the protease inhibitors that are found in soyabeans. These anti-nutrients reduce the nutritional value of a feed and in severe cases can be extremely toxic. The extent of the toxicity is often dependent upon the nutritional status of the animal as well as the animal species; for example, ruminants tend to be more tolerant than non-ruminants. Again, undernourished animals tend to be more susceptible than well-fed animals.

Seasonality and variation in nutrient supply

In many tropical developing countries there is a marked variation in climatic conditions, for example a dry season and a wet monsoon season. This disparity, coupled with financial hardship of the livestock-keeper, can result in great fluctuations in the nutrient supply to domesticated animals.

Compensatory growth

Apparent accelerated weight-gain, following an increased supply of nutrients after a period of nutrient shortage, has been shown to occur in most animals. If the apparent rate of growth during the period of re-adjustment exceeds that which would normally be expected from the new diet, then the animal is said to be undergoing compensatory growth. Realimentation is associated with an increased feed intake, and this is often associated with an increased gut-fill, which gives an apparent increase in body weight. Great care must be taken not to confuse increases in weight due to gut-fill with genuine increases in the rate of tissue deposition. There is some evidence that the efficiency of nutrient utilisation is increased during compensatory growth. The long-term effects of undernutrition depend on the stage of development. There is increasing evidence that restriction of nutrient supply *in utero* can induce changes in the developing embryo which cannot be reversed in later life.

Effects of foetal undernutrition

Many domesticated animals, especially in the developing world, are subjected to wide variation in the supply of nutrients. Extensively-grazed animals are often a good example of this. As discussed above, undernutrition can have a detrimental influence upon conception rates in livestock, but there is also evidence that nutrient restriction at specific times during pregnancy can imprint changes in the embryo which manifest themselves during subsequent growth of the offspring: a phenomenon often referred to as ‘foetal programming’. This phenomenon is illustrated by the effects of nutrient restriction on muscle fibre development. The number of muscle cells in the adult is set *in utero* when the myogenic cells proliferate and then differentiate to form muscle fibres. In the sheep this occurs at around 50 to 80 days of gestation, while in cattle it is around 100 days. The subsequent growth of muscle is by hypertrophy of existing fibres rather than by the production of additional fibres. The extent of cell proliferation is nutritionally sensitive and therefore reduction of nutrient supply during this phase of development reduces the number of muscle fibres produced, and hence subsequent lean development. It is suggested from studies with humans and laboratory animals, that this reduced capability to deposit lean can result in an increased deposition of fat. Other aspects of development in later life (e.g. kidney function) have been shown to be influenced by undernutrition during embryo development. The been reported not to be seen until rumen ammonia nitrogen reaches 150 mg ammonia N per litre (Mehrez *et al.*, 1977). Other values were discussed in Pisulewski *et al.* (1981) and illustrate the difficulty in coming to a definitive answer; a value of 50 to 80 mg ammonia N per litre could be taken as a working figure for the minimum concentration for effective rumen function. Critical levels of ammonia can be achieved through supplementation of the diet with proteins degradable in the rumen or other soluble-N materials such as urea or poultry manure. Nitrogen absorbed from the GI tract as ammonia, or as a metabolite from tissue protein metabolism, is converted into urea by the liver. There is considerable evidence that urea is recycled to the rumen via passage through the rumen wall and via the saliva. This helps the ruminant to conserve nitrogen, and also to balance the availability of energy-yielding metabolites with nitrogen supply to promote microbial protein synthesis. Synchronisation of energy and protein supplies to the rumen is desirable. This can be achieved by careful consideration of the diet, for example, by supplementation of a rapidly-degradable source of protein with a rapidly-degradable source of carbohydrate. Experimental studies have shown that in practice the ruminant has adapted to variation in the synchronisation of nitrogen and energy supply to the rumen microbes, principally by the recycling of absorbed urea back to the rumen, and also by the storage of carbohydrate polymers by the rumen bacteria. Excess ammonia in the rumen can be very toxic to the animal and can be fatal, for example, when urea supplements are fed without a ready supply of rumen-fermentable carbohydrate. For efficient microbial growth, an adequate supply of sulphur, principally for synthesis of sulphur-containing amino acids, and of phosphorus, principally for the synthesis of microbial nucleic acid, are also required. The rumen also requires a supply of other minerals. An adequate supply of water is required to maintain active rumen fermentation.

A variable amount of the dietary protein escapes rumen fermentation and passes out of the rumen along with the microbial protein and endogenously-produced protein, e.g. sloughed cells. The extent of the degradation (fermentation of the dietary protein) largely depends on the nature of the protein, the rate of passage of the digesta through the rumen and the nature of the fermentation.

Protection of nutrients from rumen fermentation

Usually most of the protein leaving the rumen for absorption in the lower GI tract is of microbial origin. However a variable amount of the dietary protein will escape ruminal degradation and increase the quantity of protein reaching the duodenum. Tannins are known to interact with dietary protein and form complexes, which are partially resistant to microbial fermentation but dissociate in the lower tract to release the protein for absorption (i.e. post-ruminal supply). Many forages contain moderate levels of condensed tannins (i.e. 4-6 per cent of DM) which under certain circumstances have been shown to confer ruminants with several advantages, including improved protein digestibility and general performance. This may be due to partially protecting the protein from rumen degradation. As indicated above, higher concentrations of tannins can have deleterious effects. Protection of dietary protein from microbial fermentation can also be achieved by heat and chemical treatments.

Protozoa defaunation

Protozoa have high maintenance-energy requirements and they also consume bacteria in large numbers. The bacterial protein is degraded to amino acids before being incorporated into the protozoal protein. This process may introduce inefficiency into the overall production of protein by the rumen. Some studies have shown that defaunation of protozoa can increase the quantity of microbial protein reaching the duodenum by about 25 per cent. In general, defaunation increases productivity of animals receiving low-protein diets based on low-quality forage; with concentrate-based diets well-supplied with proteins, the presence of protozoa has been reported to be beneficial. However, methods for reducing protozoa have not been practicable in routine animal production. More recently, there has been some interest in the use of plants with anti-protozoal properties (such as those rich in saponins) as a possible means of suppressing or eliminating rumen protozoa.

Methane production

Another approach in ruminants has been to modify the rumen ecosystem to suppress undesirable processes such as methane production. During the microbial fermentation of carbohydrates up to 10 per cent of the gross energy is lost in the form of methane. Increasing the rate of passage of digesta and lowering the rumen pH can reduce methane production. Methane production can also be reduced by using long-chain, polyunsaturated fatty acids. Manipulation of the diet to promote propionate production will reduce methane production. The use of ionophores (such as monensin) to suppress the methane-producing bacteria is used in some countries with intensive agriculture to improve the efficiency of rumen fermentation.

Animal genotype

Where nutritional and environmental constraints are absent, the response to nutrients is determined by the animal's genetic potential for production. In principle, selection for high-yielding dairy cattle results in animals of large body size, with a capacity to consume a large amount of feed. Such animals are usually inappropriate in the developing countries, since the resources available will support only moderate levels of production. Breeding strategies in the developing world should focus on genotypes which can withstand heat stress and low-quality feed, and ability to tolerate important diseases that are endemic in a particular agro-ecological zone. For example under the conditions that prevail in the developing countries cattle with *Bos indicus* genes may have greater tolerance to diseases and a better ability to disperse body heat than their *Bos taurus* counterparts. As a result, the former are usually more productive in the long-term, although their lactation yields may appear to be relatively low. There is also evidence that rates of passage of nutrients through the digestive tract can be higher in *Bos indicus* than in *Bos taurus*. However, selection of local animals and subsequent cross-breeding with high-yielding *Bos taurus* up to certain limits has been found in some circumstances to be useful, in terms of both production and management. Breeding strategies are discussed in Chapter 12.

General health

There is no doubt that the general health of an animal is the single most important factor as far as nutritional response and productivity is concerned. The effect of stresses and diseases on the nutrition of any animal is dependent on the type of tissue affected, severity and duration. Feed intake, digestion, absorption and tissue metabolism are all negatively affected by disease and stress. The influence of diseases, particularly those caused by parasites, on nutritional status and hence productivity of animals, is extensively documented. Effective and sustainable control of diseases must therefore be an important strategy if maximum response to the available nutrients is to be achieved. Strategies for maintaining health are discussed in Chapter 13.

Reproductive status

Pregnant ruminant animals are constrained by the decrease in rumen volume and an increased nutrient demand by the developing foetuses.

Water supply

Water is a nutrient and is probably the most important substance in nutrition; it is so commonplace that it is often not discussed with other nutrients. Water is vital to the life of an organism and it is the largest single component of the animal body at every stage of its development. The animal obtains its water from three major sources, i.e. free drinking water, water contained in feed (bound water) and from oxidation of hydrogen in organic nutrients (metabolic water). Voluntary ingestion of clean and palatable drinking water is widely accepted as the best means of meeting the water needs of animals. Water contained in or on the feed is extremely variable and may range from a low of 5 per cent in dry grains to about 90 per cent in young lush pastures; it fluctuates widely with climate and weather conditions. Metabolic water meets a small but a significant part of the need; this source is particularly important in periods of negative energy balance, i.e. when depot fat and protein are being utilised. Water produced during metabolism depends upon the nutrients being metabolised; the catabolism of 1 kg of fat, carbohydrate or protein produces approximately 1190, 560 or 450 g of water, respectively.

Function

As a universal solvent, water facilitates cell reactions and performs the vital functions of transporting nutrients and metabolites throughout the body to supply the needs of cells and to excrete waste products. Essentially, all chemical reactions in the body take place in the presence of water; reactions in digestion and intermediary metabolism involve the chemical addition or release of water. The high specific-heat and high heat-of-vaporisation of water make it an ideal agent for regulation of body temperature. Water makes it possible for the large amounts of heat produced from metabolism to be dissipated with very little change in body temperature. Other functions of water include lubrication (of food bolus during mastication, various mucus membranes and joints) and cushioning of the nervous system. Most minerals, which are essential as dietary nutrients for livestock, are present in water and therefore water is a medium through which animals can acquire these nutrients. Water forms 87 per cent of milk and 66.7 per cent of eggs and therefore water needs are increased to facilitate milk and egg production. In ruminants, pregnancy and milk production increase both energy and water consumption rates by 40-60 per cent and, given the opportunity, lactating cows may drink as often as eight times a day while dry beef cows may drink only every other day.

fertilisers is limited because of being expensive or not available. The situation in South Asia is slightly different because chemical fertiliser is often used, as it is both available and cheap. Moreover, some farmers consider artificial fertilisers are for the plants, while farmyard manure is important for improving soil fertility.

Poultry manure

Poultry manure includes only the droppings if it is from cages, or droppings and bedding materials if from a deep litter system. In some improperly-managed farms large quantities of feed can also be found in the manure, increasing the nutritive value (particularly energy content) of the waste. The nutritional quality of poultry manure is therefore potentially higher than that from other farm animals, due to relatively high nutrient densities. Both plants and ruminants can utilise poultry manure as a source of nitrogen. In ruminant animals the apparent digestibility of nitrogen in poultry excreta is relatively high at 74 to 78 per cent. Studies undertaken in the highlands of Kenya have revealed that 22 per cent of farms use purchased poultry waste as cattle feed throughout the year, while 18 per cent of farms use their own poultry manure as cattle feed throughout the year. Poultry manure when used as fertiliser is often considered to have a rapid effect on soil fertility with lasting effects. It is preferred for use in vegetable gardens rather than in field and perennial crops. In some South Asian countries (e.g. Bangladesh) the disposal of poultry manure produced by the expanding poultry industries has become a problem.

It should be noted that the practice of feeding poultry manure to animals is banned in many countries, largely due to the possibility of spreading animal diseases.

Cattle manure

Cattle manure includes the dung, urine, bedding and feed refusals, which are often composted before being used in the field. The quantity and quality of cattle manure are influenced by the level of feed intake, the apparent digestibility of the feed, the type of livestock production and the storage systems applied to the manure. The ability to collect and utilise the urine also influences the quality. Supplementation of the basal diet with concentrate and minerals improves the quality of manure by increasing the levels of nitrogen, phosphorus and carbon in fresh manure. Generally, feeds rich in nitrogen and minerals, result in more plant nutrients in the excreta.

The quantity of manure produced depends on the quantity and quality of feed offered and the size of the animal. Theoretically cattle will produce about 0.8 per cent of live weight as faecal DM, daily, but this will depend on the quality and quantity of feed offered. Studies conducted in Kenya revealed that most smallholder farmers believe that they could increase the quantity of manure by increasing the use of crop residues, particularly maize stover, as forage and bedding. They could also improve the quality of the manure by mixing it with urine, poultry manure and by covering the compost to minimise volatilisation of soluble nitrogen and nutrient leaching. Nutrients returned to the soil through the excreta of farm animals represent a potential pathway in the recycling of nitrogen and other nutrients. On some farms cattle manure is used as a source of energy in the form of biogas production or dried and used as fuel for cooking. A similar scenario exists in South Asian countries. Using manure as a fuel represents a serious loss to nutrient recycling and is considered (e.g. by UNESCO) to be the first step to desertification in dry areas.

Other farm animals such as pigs, goats and sheep, also produce useful manure, which varies in quantity and quality.

Manure as a source of income

The sale of both cattle and poultry manure can provide direct cash to farmers. The price set on livestock-derived manure can be high; the value of manure was found to be about 30 per cent of the annual milk production income in the highlands of Kenya.

Concluding remarks

The basic principles of the quantitative responses of animals to nutrients are relatively well understood, particularly for intensively-fed livestock. Few quantitative data are available for animals on low planes of nutrition. In deciding on a feeding regime for animals it is essential to undertake a systems approach; striving for maximum production is not always the best objective. Will the food given to the animal compete with the needs of the human population or other animals? Is there enough labour for the feeding regime to be adopted? What will the value of the manure be? What economic return will there be for the feed and veterinary costs? Consideration of the systems approach is discussed in many of the other chapters in this volume.

Further reading

Agricultural Research Council. 1980. *Nutrient requirements of ruminant livestock.* Commonwealth Agricultural Bureau, Farnham Royal, UK.

Agricultural and Food Research Council. 1993. *Energy and protein requirements of ruminants.* CABI Publishing, Wallingford, UK.

Agricultural and Food Research Council. 1998. *The nutrition of goats. Technical Committee on Responses to Nutrients Report No. 10.* CABI Publications, Wallingford, UK.

Ayantunde, A.A., Fernandez-Rivera, S. and McCrabb, G. (ed). 2005. *Coping with feed scarcity in smallholder livestock systems in developing countries.* Animal Sciences Group, Wageningen UR, The Netherlands; Univeristy of Reading, Reading, UK; ETH (Swiss Federal Institute of Technology), Zurich, Switzerland, and ILRI (International Livestock Research Institute), Nairobi, Kenya.

Cronje, P. 2000. *Ruminant physiology: Digestion, metabolism, growth and reproduction. Proceedings of the Ninth International Symposium on Ruminant Physiology.* CABI Publishing, Wallingford, UK.

Feeding standards for Australian livestock. Ruminants. 1990. CSIRO Publications, East Melbourne, Victoria, Australia.

Forbes, J.M. 1995. *Voluntary food intake and diet selection in farm animals.* CABI Publishing, Wallingford, UK.

Lawrence, P.R. and Pearson, R.A. 1998. *Feeding standards for cattle used for work.* Centre for Tropical Veterinary Medicine, University of Edinburgh, Edinburgh, UK. See also http://www.vet.ed.ac.uk/ctvm

McDonald, P., Edwards, R.A., Greenhalgh, J.F.D. and Morgan, C.A. 2002. *Animal nutrition. Sixth edition.* Prentice Hall (Pearson Education), London, UK.

**Nutrient requirements for swine. Tenth revised edition.* 1998 National Academy Press, Washington D.C., USA.

**Nutrient requirements of poultry. Ninth revised edition.* 1994. National Academy Press, Washington D.C., USA.

**Nutrient requirements of goats; Angora, dairy, and meat goats in temperate and tropical countries.* 1981. National Academy Press, Washington, D.C., USA.

Ørskov, E.R. 1998 *The feeding of ruminants. Second edition.* Chalcombe Publications, Lincoln, UK.

Theodorou, M.K. and France, J. (ed). 2000. *Feeding systems and feed evaluation models.* CABI Publishing, Wallingford, UK.

Thomas, C. (ed). 2004. *Feed into milk.* Nottingham University Press, Nottingham, UK.

Van Soest, P.J. 1994. *Nutritional ecology of the ruminant. Second edition.* Cornell University Press, Ithaca, New York, USA.

* These publications and many similar ones are available to read free on line at www.nap.edu/

Author addresses

Peter Buttery, School of Biosciences, University of Nottingham, Sutton Bonington Campus, Loughborough, Leicestershire LE12 5RD, UK

Robert Max, Department of Veterinary Physiology, Biochemistry, Pharmacology and Toxicology, Faculty of Veterinary Medicine, Sokoine University of Agriculture, P.O. Box 3017, Morogoro, Tanzania

Abiliza Kimambo, Department of Animal Science and Production, Sokoine University of Agriculture, P.O. Box 3004, Morogoro, Tanzania

Juan Ku-Vera, Facultad de Medicina Veterinaria y Zootecnia, Universidad Autonoma de Yucatán, C.P. 97100 Merida, Yucatan, Mexico

Ali Akbar, Department of Animal Nutrition, Bangladesh Agricultural University, Mymensingh 2002, Bangladesh

11

Feeds and feeding to improve productivity and survival

Tim Smith[*], *Noble Jayasuriya*[†], *Victor Mlambo*[†], *Faustin Lekule*[‡], *Derrick Thomas*[‡], *Emyr Owen*[‡], *Anne Pearson*[‡], *Marion Titterton*[‡]

Cameo 1 - Crop/livestock farmer in Zimbabwe

Mrs. Ngwabi, of Gulathi Communal Land, Matabeleland South Province, near Bulawayo, in Southern Zimbabwe, is a smallholder farmer. The area in which she lives is semi-arid, characterised by a long dry season (6-9 months) and occasional droughts. She is a mixed crop/livestock farmer, with cattle, who decided to volunteer to grow Hybrid *Pennisetum* forage on some of her arable land, including the contours. Mrs Ngwabi decided to make silage, storing it in plastic bags, to help overcome the dry-season feeding constraint to milk production. Herd numbers have increased, with milk being sold daily. Mrs Ngwabi's success resulted in her being chosen as national smallholder dairy farmer of 2003. About 40 other farmers in Gulathi are now growing intercropped forages and legumes for ensiling.

Cameo 2 - Survey of smallholder dairy farmers in Sri Lanka

A survey in the Central Province of Sri Lanka, amongst 110 resource-poor smallholder dairy farmers rearing low- to medium-producing cross-bred cattle, showed that supplementation of low quality forage-based diets with 'urea molasses multi-nutrient blocks' (UMMB), manufactured using locally-available feed resources, increased milk production (by 1-2 litres/cow/day) and improved milk composition, and reduced age at puberty, length of the calving interval and the number of inseminations per conception. With UMMB, farmers stopped feeding traditional concentrates to cows producing less than 6 litres of milk daily, and replaced 40-60 per cent of the traditional concentrates in cows producing more than 6 litres/day. The average cost : benefit ratio was 1:3-5.

---

[*] Principal author
[†] Co-author
[‡] Consulting author

farmers for consumption by the household and the livestock is demonstrated (see also Table 13.1 in Chapter 13, which lists reasons for keeping livestock). Although there is no direct dietary link between grass/pasture and crop residues and humans, their conversion into animal protein plays a major role in human nutrition. There is clearly complementarity between ruminants and humans within the mixed-farming system.

The major options open to resource-poor farmers for feeding ruminants are based on forages, either from natural and sown pastures, including the contribution of trees and shrubs, or crop residues A major constraint to smallholder ruminant production is the lack of good quality feed throughout the year. Most smallholders do not plant forages but depend on the natural grazing available in communal and waste land, forests, waterways, etc.. During the rains natural grazing can sustain limited production. During the dry season forage is scarce and of low quality, being high in fibre and low in available nutrients. In extensive systems there is an annual cycle of weight gain and loss. Losses of up to about 15 per cent of body weight can be acceptable (O'Donovan, 1984), although subsequent recovery of weight in grazing animals is often incomplete (Manyuchi *et al.*, 1991). How can this loss be minimised and cessation of production averted?

Pigs and poultry owned by resource-poor farmers are often regarded as scavengers. In the growing season they often satisfy their needs for protein, minerals and vitamins but in the dry season, despite the substantial amounts of crop residues available, protein, minerals and vitamins are likely to be in short supply.

Intake

The value derived from a feed by an animal depends on the nutritive value of the feed, coupled with intake, which can be constrained by:

- the amount offered (affected by amount available and cost)
- palatability of the feed offered
- imbalance of a specific dietary nutrient (e.g. cereal stovers fed without a source of degradable nitrogen)
- the physiological state of the animal (e.g. maintenance, lactation, pregnancy).

Low intake results in lost production, reproductive failure and, especially in very young and old animals, death. Coping strategies are dealt with elsewhere.

Intake and its manipulation are important topics and are comprehensively discussed by Forbes (1995).

Appropriate methods of evaluating feeds

Much of the following discussion centres on ruminants, because methods of evaluation are more complex than in monogastrics (because digestion is mainly in the rumen) and ruminant feeds vary widely in quality, both within and between feeds. Acceptable feed evaluation techniques must closely approximate *in vivo* responses, address the objectives of farmers and researchers, and be affordable. Feed evaluations must meet the requirements of the end users. For resource-poor farmers, feed evaluation must include availability, affordability and minimal risk. The quantities of the major nutrients (i.e. energy and nitrogen [N]) need to be known, together with any anti-nutrients which may adversely affect nutrient availability and animal health. Chemical analysis needs the support of *in vitro* techniques which mimic the breakdown and digestion of feeds in the animal. Where available, *in sacco* methods are often the closest to what happens *in vivo*. Analysis of individual feeds is desirable, especially for novel and local resources, but there are data available in the literature for a wide range of feedstuffs (Gohl, 1981) and these can be used as a starting point for diet formulation. Mineral analysis is expensive and requirements are difficult to estimate. Leng (1995) suggests feeding mineral supplements, especially of sulphur and phosphorus, which are necessary for N utilisation in ruminants. Most feeds can be regarded as either basal diets or supplements. Basal diets are usually coarse forages or crop residues. Supplements are energy and/or protein-rich and the starting point for characterisation is to determine the amounts and availability of these major nutrients. With protein supplements, availability is affected by the presence of protein-binding tannins as well as other anti-nutritional factors that impact negatively on rumen microbial activity. As well as chemical analysis, techniques are available to determine rate of degradation in the rumen. These include:

- *in sacco* bag techniques (Ørskov and McDonald, 1979; Huntington and Givens, 1995)
- *in vitro* techniques (Tilley and Terry, 1963, Minson and McLeod, 1972)

- rumen simulation (Czerkawski and Breckenridge, 1977)
- *in vitro* gas production techniques (Menke *et al.,* 1979; Mauricio *et al.,* 1999).

Visual assessment of forages can be done by pictorial comparisons (LPP Project, R7855). Of course, visual assessment is routinely done by farmers to provide information on leaf : stem ratio and mouldiness.

Choice of technique depends on the researcher, but *in sacco* methods do not necessarily require electricity or chemical inputs. However, the use of fistulated animals to provide rumen fluid raises ethical and welfare issues; their future use cannot be guaranteed. Research on the use of faeces as an alternative to rumen fluid from fistulated animals, is on-going. Forage evaluation techniques have been reviewed by Givens *et al.* (2000).

Feeding trials

Feeding trials are probably the most accurate way to determine nutritive value of forages, but the required resources are rarely available. However, in feeding trials the animal has the opportunity to select, whereas the tests described above are usually based on very small samples. *In vivo* nitrogen balance data are useful for on-farm application, although energy requirements of on-farm animals are likely to be greater than those of animals confined in crates. Whole-animal trials are also the only way in which intake can be assessed. Measurements of preference and indicators of intake can easily be made using the short-term intake rate (STIR) method (Romney and Gill, 2000). Responses obtained in controlled experiments are rarely repeatable on-farm, which can be frustrating for the farmer. Priority setting within research centres has not always reflected the priorities of farmers or led to adequate dissemination of information.

Improving indigenous pastures

Areas of natural grazing are multi-functional. Woody species supply shelter, browse, tree fruits, fuel wood, building materials, raw material for wood carvings and a habitat for honey-bees (Chapter 14). There is a wide diversity of plants, insects and other fauna which depend on well-managed natural grazing. Natural grazing is usually a community asset. While this guarantees the right to graze livestock, there are disadvantages:

- Without proper controls there is no effective management (short-term aspirations of individual livestock-keepers may not favour sustainable grazing management)

- Usually little control of the number of animals grazed

- Where there is control, poorer households can be disadvantaged by wealthier households

- No incentive for improvement through introduction of sown grasses and legumes or fertiliser application

- Replanting of grasses or trees rarely succeeds because of the difficulties of establishment, not helped by the lack of fences and inadequate numbers and inappropriate location of water points. Changes in plant cover often reflect degradation rather than improvement

- Many communities control the removal of firewood. Where dung is removed for fuel it is a step towards rangeland degradation (indication of excessive removal of trees).

Mixed-species grazing, especially with grazers and browsers, is more efficient than single species (in the smallholder system this is normal; cattle, goats and other domestic livestock graze together). The two systems most used are continuous set-stocking or rotational grazing. Management tools such as fire are unlikely to be relevant in arid and semi-arid smallholder systems. Bush encroachment, at the expense of grass growth, could be a problem, although most seedlings fail to survive, especially where grazing pressure is high. However, control of bush encroachment is particularly important in tsetse areas.

In the humid tropics there is likely to be a build up of rank grasses during the growing season. This material will be unpalatable and also presents a risk of unplanned fires. Rotational grazing becomes an option, but there is the expense of fencing and a need for management skills if it is to work.

Communities and institutions should be encouraged to develop by-laws and regulations to manage grazing and related activities (LPP Project, R7432). Controls may be necessary to regulate removal of saleable resources (e.g. timber; thatching grass). Application of nutrients is usually restricted to manuring by grazing stock, a practice which does not ensure even distribution of nutrients. Encouraging legumes, including trees, will contribute nutrients. However, young trees will need protection from livestock and wildlife. Perennial grasses provide more nutrients than annuals and should be encouraged. The use of fencing or close herding (labour intensive) allows the creation of exclusion zones, allowing the setting of seeds, reduction of bare patches or the bulking-up of forage in a particular area.

A major problem in overgrazed rangeland during the rains is 'run-off' of top soil, containing nutrients and seeds. This degrades the slopes and the benefits of 'run-on' do not equal the losses. Formation of gullies is encouraged. Run-off can be reduced by encouraging penetration of the rain, e.g. by maintaining adequate ground cover. Regrowth on bare patches can be encouraged by spreading a layer of brushwood over them.

Improving sown forages and their utilisation

Techniques for developing and managing sown pastures are documented (Humphreys, 1987). However, despite considerable research and development in the second half of the twentieth century, there has been little adoption by smallholder farmers. When adopted, planted pastures have largely been unsustainable. For the resource-poor smallholder the likely constraints are:

- shortage of land (arable crops are increasingly grown on marginal land), forcing a choice between grain for household consumption (with residues for livestock), a cash crop or a forage crop
- shortage of cash or credit to purchase seeds and fertiliser, if they are available
- lack of a guaranteed market for the extra produce
- low farm gate returns for milk, making dairy farming a poor profit-making enterprise, especially in the Asian context
- unpredictable rainfall during the growing season (loss of inputs)
- labour availability
- lack of knowledge of conservation techniques.

In some situations dairy farmers are most likely to sow pastures because of the immediate returns from sales of milk and dairy products. For example, in the highlands of Kenya high milk prices are driving the planting of Napier grass

(Pennisetum purpureum) by smallholders. Where water (rain-fed or irrigation) is available, the choices are to grow forage for grazing, cut-and-carry or conservation. Grazing and cut-and-carry systems require forage species which regrow through the growing season. Biennials and perennials reduce the labour requirements and replanting costs usually associated with annuals. Exceptions include self-generating annual species, such as *Stylosanthes hamata*, a legume successfully used in mixed farming systems in parts of Asia and West Africa. The species selected should complement the other, usually low quality, feeds, although the major dietary constraints will be lack of bulk and nutrients.

Examples of successful species for conservation include forage maize, forage sorghums, *Pennisetum* hybrids and varieties of Napier grass. The last two are also suitable for intercropping with legumes and use in cut-and-carry systems (LPP Project, R7010). In the highlands of Tanzania planted pasture had a cost advantage over crop residues, because the latter had to be transported from the lowlands (Mdoe *et al.*, 1992). Intensive forage gardens of grass/legume intercrops in the Tanzanian highlands have produced yields of 48-254 t/ha of fresh material (Mtengeti *et al.*, 1992). In a semi-arid area of Zimbabwe irrigation increased forage yields by 50 per cent (Mhere *et al.*, 1995).

Appropriate methods of forage conservation

There is usually adequate biomass in the tropics during the wet season. However, this is also the time for planning the dry-season feeding strategy which, to allow for sustainability of production needed in dairying, will depend on conservation of forage as hay or silage.

Hay

Making high-quality hay depends on good drying conditions when the crop has reached a suitable stage (around flowering) for harvesting. This usually occurs before the end of the rains, making drying difficult. Tropical grasses are thicker-stemmed than temperate grasses and usually smallholders lack the equipment to facilitate the passage of air through the cut crop (dry air rather than hot sun is required for the drying process). Although large-scale hay making is usually mechanised, hand tools can be used but they are labour intensive. Although the use of racks and tripods has succeeded in research, they are not widely used in practice. Transporting the crop some distance will need draught animal power

(DAP) or mechanised transport, with cost being related to distance. With loose forage, use of the manual box-baler is possible (LPP Project, R6619)). The hay crop must be adequately dried (85-88 per cent dry matter) before storage. Movement of air within the stack is necessary to minimise the risks of mould formation and spontaneous combustion (small quantities are not normally at risk from the latter). A covered store will probably be necessary to prevent mould formation (LPP Project, R6993). Long storage periods increase the risk of damage from termites. Storage above ground will reduce damage from rodents.

Standing hay is where the residual grass crop is not harvested. Although use of this resource entails little more than the labour of herding, problems include uncontrolled grazing (domestic and wild animals), fire, termite damage and loss of leaf due to weathering. As the dry season progresses crude protein content and digestibility of standing hay fall rapidly and fibre content increases.

Hay made from legume residues, often where the haulm is collected as a by-product of oilseed or vegetable production (e.g. groundnut, lablab), is a valuable protein supplement for diets based on crop residues.

Silage

Silage quality depends on the material ensiled and the efficiency of the ensiling process. Once the crop has reached a suitable stage for harvesting, the decision on when to harvest depends on whether the need is for a high quality (high protein, low fibre) feed, or a bulk feed of lower quality but greater quantity. The crop must be stored in anaerobic conditions, obtained by compaction and excluding air, in which plant sugars are converted to lactic acid (McDonald *et al.*, 1991). Plants low in fermentable carbohydrates may need an additive (e.g. molasses) to guarantee this.

Small quantities of forage can be ensiled using relatively simple techniques. The key factors are:

- choice of crop for ensiling
- method of ensiling
- class of animal to be fed will influence whether quality or bulk of silage is most important
- cost of ensiling.

Most arable land is used for household food and cash cropping. However, where farmers are unable to change their cropping programme, fallows and contours are possible areas for planting forage. Climate and the likelihood of successfully growing a crop must be considered. In semi-arid Zimbabwe, yields of 6-10 t DM/ha/year from Napier grass hybrids and 12-14 t DM/ha/year from forage sorghum were recorded. Clay soils gave higher yields than sandy soils. Silage was successfully made from these forages in plastic bags (LPP Project, R7010).

In some areas forage maize is replacing grain maize, the method of ensilage (bag, clamp or pit) being dependent on farmer choice and cost. Forage maize is starch-rich and rarely needs an additive for good fermentation, but it is an annual crop needing adequate heat, nutrients and water for optimum yield. In some systems the weeds removed from the growing crop are regarded as valuable forage for green feeding (e.g. LPP Project R7955).

Multi-purpose trees

Leaves and fruits from multi-purpose trees (MPTs), shrubs and woody perennials contain medium to high levels of crude protein and are useful additives to low quality diets. Exotic species (for example *Leucaena leucocephala, Gliricidia sepium, Cassia, Calliandra calothyrus, Sesbania sesban*) grow quicker than indigenous species and are often more reliable as a forage source (Bennison and Paterson, 1993; Paterson and Clinch, 1993; Paterson, 1994)). Intercropping with field crops, or planting as fences, provides mulch, controls soil erosion and enhances soil fertility (Topps, 1992). Alley cropping, where crops are interspersed with other crops at regular intervals, to the mutual benefit of both and to the soil, is a development of intercropping and fence line planting. The presence of mimosine in several species of *Leucaena*, including *L. leucocephala,* can cause problems in unadapted animals, especially where the forage is offered in large quantities. However, exotic species are costly, thus their use is restricted to income generating systems, such as smallholder dairying in Kenya. Farmers often prefer local species, which meet needs beyond those met by introduced species. Preparations from many indigenous tree species are valued for their medicinal properties for preventing and curing disease in both humans and animals (LPP Project, R6953). Economically important exotic and indigenous species need assessing to determine which species can be cultivated and managed. Exotic species should produce good yields of palatable and nutritious leaves and fruits, be fast growing, withstand browsing, pruning, lopping and coppicing and be disease resistant (Preston, 1991). Making silage from exotic multi-purpose trees is discussed by Titterton and Bareeba (2000).

MPTs as feed for livestock

Tree leaves and fruits fed as supplements supply N to the rumen micro-organisms, thus aiding digestion and increasing intake of N-deficient feeds (Chapter 10). Leaf and fruit yield will partly depend on rainfall and in drier areas indigenous species may be the most appropriate. In Africa, *Acacia* spp., comprising trees and shrubs (thorn trees), are widespread. Animals browse the green leaves, the fruits being a useful dry-season protein source. The fruits of different *Acacia* spp. are available at different stages of the dry season (LPP Project R7351). The fruits are usually easy to collect and store, and so can be used as a strategic protein resource.

Nutritional characteristics of MPTs

The amount and quality of forage from MPTs depends on the species and the growth environment, which varies between seasons. Anti-nutritional factors, such as tannins, increase as stress in the growth environment increases, but the interrelationships are not fully understood. Total yield (forage plus fruits) may not be the sole determinant of value as perceived by the farmer. Crude protein content of leaves (140-179 g/kg DM, Barnes *et al.*, 1996) may be higher than fruits (127-183 g/kg DM, Tanner *et al.*, 1990; Ncube and Mpofu, 1994). Young leaves contain more phenolics than mature leaves and fruits (Ernst *et al.*, 1991), but are often available before the end of the dry season (browse flush), thus providing a valuable protein feed.

Indehiscent fruits are most valued for collection and storage because the protein-rich seed is retained. However, many seeds survive the digestion process and appear in faeces (Table 11.3). This is essential for the sustainability of some tree species. Grinding increases digestibility of seeds. The total number of seeds collected is small, compared to the number produced, and is unlikely to affect rangeland regeneration.

Many small-scale and commercial farmers in semi-arid areas of India (Joshi *et al.*, 2004) and Zimbabwe (Kindness *et al.*, 1999) collect and store pods for livestock feed in the dry season. Pods are also a marketable commodity. For pods containing cyanogenic glycosides, storage is necessary to avoid nutritional problems.

Utilisation of tannin-rich forages

Most beneficial effects of feeding tannin-rich forages result from work with ruminants rather than monogastrics. Laboratory studies tend to overestimate the

Table 11.3 Seed to hull ratio (w/w) and undigested seeds, from four *acacia* tree species, voided in the faeces of sheep

	A. albida[1]	*A. nilotica*	*A. sieberiana*	*A. tortilis*
Seed : hull ratio	23:77	26:74	27:73	36:64
No. of seeds eaten/day	397±12	500±20	370±10	1335±114
No. of seeds voided in faeces/day	64±16	45±21	15±1	610±14
Seeds voided in faeces (%)	16	9	4	46
Seeds apparently digested (%)	84	91	96	54

[1]Synonymous with *Faydherbia albida*
Source: Tanner *et al.* (1990).

negative effects of anti-nutrients, possibly because in practice tannin-rich feeds are a dietary component, usually offered together with low-grade roughage or residues, rather than the sole diet. Generally the strength of association between tannins and protein determines whether the tannin-protein complex will dissociate in the small intestine, thus providing 'by-pass' (i.e. that passes through the rumen undigested) protein as a nutritional benefit (other benefits include nutrient partitioning and control of bloat). However, there is potentially a risk that free tannins, post-rumen, could bind with endogenous protein, thereby 'cancelling out' the benefits from increased by-pass protein. It is recommended that tannin-rich supplements are fed sparingly when the tannin levels of other dietary components are unknown. Grazing animals rarely show nutritional stress from over-consumption of tannin-rich forage, probably because of the opportunity to select. There is a need to investigate tannin inactivating treatments, such as the use of wood ash (LPP Project, R7351).

Fibrous crop residues

Crop residues are an important dry-season feed in all crop/livestock systems. With supplementation and/or treatment they can support maintenance or limited production. Crop residues are mainly cereal straws characterised by their low contents of fermentable energy, protein and minerals and high contents of fibre, resulting in low intakes and digestibility. When legumes are intercropped with cereals, the soil and the residues are enriched (LPP Project, R6610). The use of crop residues as feed was reviewed by Sundstøl and Owen (1984). Cereal by-products represent about two-thirds of the crop residues produced in Africa and Asia (Kossila, 1984). Quantities have increased in recent years because of increased crop production.

Residues can be grazed *in situ* or removed from the field after harvest and stored. Grazing is wasteful because nutrients lost through weathering and trampling result in a falling plane of nutrition over time, as the best parts of the plant are selected first. However, labour costs are low and the soil benefits from hoof action and direct application of manure.

Care is needed to minimise nutrient loss in the storage process. Leaf material is the most nutritious part of the plant but the longer the residue is left in the field, the greater the danger of senescence from weathering and shattering (Manyuchi *et al.*, 1990). The benefits from storing correctly are similar to those for hay (LPP Project R6993). Depending on the location of cropping land in relation to livestock, use of the manual box-baler (LPP Project, R6619) may be justified.

Improvement in the nutritive value of residues can be achieved by:

- Grinding: this exposes internal cell material to immediate microbial attack in the rumen, thus speeding up digestion and increasing intake. Scope for selection by the animal is reduced, ensuring greater use of the residue. Fine grinding is expensive and needs sophisticated machinery. Grinding in a hammer mill with the screen removed is suggested

- Chopping, e.g. through a chaff cutter, minimises selection and facilitates mixing with other feeds, but has little effect on intake and rate of digestion

- Soaking in water (e.g. overnight): results are not consistent but dustiness, a problem with fine chopped or ground residues, is reduced. It affords an easy way to add urea in solution, with reduced risks of urea poisoning

- Removal of the nutritious leaves and top part of the stem for feeding, with the lower part of the stem being discarded, possibly for use as fuel or left in the field for soil conservation

- Treatment with industrial alkalis: sodium hydroxide is the strongest alkali but is difficult to use on-farm; it does not improve the nitrogen status of the crop. Urea in solution is safer and adds nitrogen. Urease is usually present in the tropical residues, to act as a catalyst to break the urea down to ammonia. Treatment procedures suitable for adoption under a range of small farm conditions have been developed (e.g. Smith, 2002), However, various constraints have restricted adoption (Owen and Jayasuriya, 1989a)

- Use of locally-occurring salts (Owen and Jayasuriya, 1989b): salts suggested include calcium hydroxide, calcium oxide, potassium hydroxide and sodium carbonate and local materials, for example Magadi salt (sodium sesquicarbonate) in Kenya

- Use of urine (Sundstøl and Owen, 1993): while this is effective and is probably the cheapest source of urea, it can be used only if acceptable to the community. Collection from animals is not easy

- If the amount of residue is not a constraint, increasing the amount offered will increase intake by increasing the scope for selection (LPP Project R5188), this is known as 'excess feeding' or 'self selection' (Osafo *et al.*, 1997).

Future methods of treatment are likely to involve bio-degradable materials such as enzymes (Colombatto *et al.*, 2003). Materials for use on-farm must be cheap, available, safe (both to handle and to feed) and easily transportable.

Supplements

*C*hoice is limited by the high cost of conventional supplements such as grain-based concentrate feeds, oilseed cakes, urea and minerals. However, supplements are necessary to optimise the utilisation of the low-quality feeds available (see Chapter 10), particularly during the dry season. Farmers are forced to use what is available locally. Young animals, especially those recently weaned, and lactating stock are particularly vulnerable to underfeeding at this time. In both Africa and Asia, the use of tree forage, either fresh, dry or processed as leaf meal, for milk production is increasing, because of the high costs of dairy meal. Legume residues from vegetable and oilseed crops, while normally only available in small amounts, are a useful source of protein. Legumes such as cowpea (*Vigna unguiculata*) and lablab (*Lablab purpureus*) are highly palatable and nutritious supplements (Smith *et al*., 1990; Singh and Tarawali, 1997). Cowpea and lablab are also often used as intercrops with cereals, thus producing enriched stover. In sub-humid coastal Kenya cross-bred dairy cows, fed Napier grass as the major forage, benefited from a legume supplement (Muinga *et al.*, 1995). Other supplements include molasses/ urea blocks, oilseed residues and tree fruits. Responses from true proteins are usually greater than from non-protein nitrogen (NPN), but a supplement of NPN is usually better than no supplement.

In Zimbabwe and Tanzania, the development of small ram-presses for sunflower oil production has greatly increased the amount of sunflower grown. The residue, after oil extraction, is a protein-rich by-product, the energy value depending on the degree of oil extraction. The fine residues are suitable for poultry (LPP Project, R7524), the coarser material for ruminants. When possible, young and breeding stock should receive supplements, including minerals. Non-ruminants, including herbivores such as equines, must not be offered feeds containing non-protein-nitrogen (NPN) as it is poisonous to them.

In dry areas there is increasing interest in spineless cactus (*Opuntia ficus* f. *inermis*) as a supplement to straw-based diets (Ben Salem *et al.,* 2004), because of its high water content. Spiny cactus is still grown, often on field boundaries; however, the spines must be singed off before feeding (see Photo 11.1).

Photo 11.1 Singeing of cactus spines before feeding to sheep in Tunisia (courtesy of T. Smith)

Multi-nutrient feed blocks

Feed blocks are usually based on molasses and urea, thus providing energy and protein together, which reduces the danger of urea poisoning (Leng *et al.*, 1992). The amount of block to be fed will depend on the size and nutritional requirement of the animal. While the main aim is to provide degradable nitrogen in the rumen, with a compatible energy source, minerals as well as rumen non-degradable protein can also be added, as can medication (Aarts *et al.,* 1990). Blocks are normally introduced post-weaning. Supplementation of low quality feed material with urea/ molasses blocks has had some impact in smallholder dairying in Africa and Asia.

For example, Thailand and Indonesia introduced this technology to small farmers and reaped benefits in terms of increased milk production and improved reproductive performance (Perera *et al.*, 2001). Constituents of blocks can be varied to include locally-occurring surpluses, for example the inclusion of olive cake in Tunisia (Ben Salem *et al.*, 2003). The percentage composition, by weight, of a typical block (Garcia and Restreppo, 1995) would be sugar-cane molasses 50, fibre 20, urea 10, cement 5, lime 5, sodium chloride (salt) 5 and bone meal 5.

Novel feeds and use of locally-occurring feed resources

Many of these result from the industrial processing of agricultural products. Compared to crop residues they are less fibrous and contain more fermentable energy. They are often rich in protein and minerals and include oilseed cakes and by-products of the sugar, brewing, distilling and catering industries. Some, like oilseed cakes, are exported or unaffordable to resource-poor farmers. Others (e.g. brewers' grains) are only available locally because of bulk or a short shelf-life and availability of transport may determine whether they are purchased or not. They are often excellent supplements for pigs, poultry and dairy cattle.

Novel feeds can be classified into four main types, according to source of origin:

- feeds from field and plantation crops
- feeds from tree crops
- feeds from fruit processing industries
- feeds from miscellaneous sources (plants, animals).

Novel feeds often have a scattered (e.g. off-farm) and seasonal production pattern, and frequently are not available in large quantities.

- Often they need collecting and physical processing, adding to the cost of production. An example is rubber seed meal, which can replace coconut meal in poultry rations (Rajaguru, 1973), but has to be collected and decorticated before processing into a meal
- Many novel feeds contain toxic components or anti-nutritional factors which limit their use as animal feed.

Examples of the feeds in different classes, together with an indication of their nutritive value are shown in Table 11.4. Many of the feeds are likely to be of local interest, either because they result from a micro-industry or are a locally-occurring forage source. However, similar products probably occur elsewhere.

Table 11. 4 Proximate composition of some novel feedstuffs found in developing countries in Africa and Asia

Feed resource	*Ash*	*Crude protein*	*Crude fibre*	*Digestibility*
Feeds from field and plantation crops				
Cereal straws/stovers	low/moderate	low	high	low
Sugarcane bagasse	low	low	high	low
Sugarcane tops	low	low	high	moderate
Corn cobs	low	low	high	low
Banana leaves/waste/stem	low	low	moderate	low
Banana – whole plant	high	low	high	moderate
Cassava pulp/waste	moderate	low	low	moderate
Cassava leaf meal	low/moderate	high	high	moderate
Coffee hull/pulp	moderate	moderate	high	low
Cotton straw	low	moderate	high	low
Sago pith/refuse	low/ moderate	low/ moderate	low/ moderate	moderate
Haulms/tops – ground nut, soyabean, cowpea	moderate	high	high	moderate
Cowpea seed meal	moderate	high	low	high
Rice grain (broken)	low	moderate	low	high
Rice husk	high	low	high	low
Cowpea vines/haulms	moderate	high	high	moderate
Sunflower straw/heads	moderate/high	low	high	moderate
Groundnut haulm/leaves/ tops	moderate	high	high	moderate
Brewers grain/waste	low	moderate	low	high
Feeds from tree crops				
Cocoa pod husk	high	low	high	low
Cocoa bean meal	low	high	moderate	moderate
Palm press fibre	moderate	low	high	low
Palm kernel cake	low	high	high	high
Oil mill effluent	high	moderate	high	low
Coconut frond	low	low	high	low
Tree fruits	low	moderate	moderate/ high	low/ moderate

Table 11. 4 (Contd).

Feed resource	*Ash*	*Crude protein*	*Crude fibre*	*Digestibility*
Feeds from fruit processing				
Mango seed kernel cake	low	moderate	low	moderate
Fruit processing/cannery waste	low/ moderate	high	moderate/ high	moderate
Citrus pulp	low/moderate	low	high	moderate
Miscellaneous feeds				
Neem seed cake	high	high	high	moderate
Guar meal	moderate	high	low	high
Castor seed cake	low	high	low	moderate
Rubber seed meal	low	high	low	moderate
Tamarind seed hulls	low	moderate	moderate	moderate
Legume tree leaves (*Gliricidea, Leucaena*)	low	high	high	moderate
Water hyacinth	moderate/ high	moderate/ high	high	moderate
Jack fruit leaves	moderate	high	high	moderate
Spent tea leaves/tea waste	low	high	high	moderate
Azolla	high	moderate	low	moderate
Spineless cactus	low	moderate	low	moderate
Silk worm pupae	low/moderate	high	low	high
Sawdust	moderate	low	high	low
Wood pulp	high	low	high	low
Mango leaves	high	low	high	low
Feather meal	low	high	low	low
Sal seed meal	low	moderate	low	moderate
Poultry litter	moderate/high	high	high	low
Paper waste	moderate/high	low	high	low/high

Key to Table 11.4
Ash (g/kg DM): low = <60, moderate = 60-120, high = >120;
Crude protein (g/kg DM): low = <60, moderate = 60-110, high = >110;
Crude fibre (g/kg DM): low = <60, moderate = 60-120, high = >120;
Digestibility (% DM): low = <40, moderate = 40-60, high = >60

General characteristics of novel feeds and their likely availability in the Asia and Pacific regions have been documented (FAO/APHCA, 1988; Devendra, 1988). Estimates of quantities available were based on the yield of by-product as a percentage of the total yield of the crop. For example, for cassava leaf as the non-conventional feed, the estimated yield was 6-8 per cent of the total, whereas for sweet potato vines the value was 24-35 per cent.

Of the total by-products available from field, plantation and tree crops, around 45 per cent (over 200 million tones) are novel feeds. This does not include many varieties of shrubs and tree forages, residues and wastes from animal sources and food processed for human consumption (FAO/APHCA, 1988). In many Asian countries poultry litter and animal by-products, such as blood meal, are considered novel, as they are rarely used in animal rations. Use of animal by-products is often controlled by legislation, which must be adhered to, especially where exporting of product is a possibility. Although over 80 per cent of novel feeds from field crops and over 90 per cent of feeds from tree crops are theoretically suited for feeding ruminants, information on their use is still lacking.

Waste paper is a largely unused resource, although there are several reports of ruminants eating paper, especially when the forage supply is restricted. Paper waste varies greatly according to its source (Gohl, 1981), particularly the proportions of chemical pulp (high digestibility) and mechanical pulp (low digestibility).

Water

Provision of water is raised again, deliberately, because it is often forgotten that water is the primary limiting factor for animal production. Lack of water will very quickly limit DM intake. Animals obtain water from three sources: drinking water, water contained in feed and metabolic water. Requirements will depend on:

- DM intake and dietary salt content
- Species (goats are more tolerant of water shortage than cattle, which are in turn more tolerant than buffalo [Devendra, 1987])
- Breed (exotics usually have a greater demand than indigenous breeds, especially at high ambient temperatures)
- Age and physiological state (young and lactating animals have a higher water requirement); work, encouraging sweating
- The presence or absence of shade (temperatures under metal roofs can also be very high).

Feeding of ruminant livestock

Smallholders' resources for feeding livestock are limited. Feeding standards have mostly been derived for temperate breeds of livestock kept in temperate conditions (e.g. AFRC, 1993) and may not be directly applicable to smallholder tropical-livestock production. Intake, especially in the dry season but also possibly in cut-and-carry systems, will be subject to availability and also to constraints imposed by forage quality, access and labour. Optimisation of resources through planning is essential. At present, there is little information regarding feeding standards for tropical breeds of livestock, reliance being placed on guidelines for temperate breeds in temperate conditions. However there is information regarding responses to supplementation of local basal diets, although care is needed when extrapolating on-station research results into a farm situation. Where grazing is the source of the basal feed, its contribution is rarely adequately described, making comparisons across trials difficult. A computerised feeding system for smallholder dairy cattle (DRASTIC) is available on diskette (Thorne, 1998). The data base used for diet formulation can be added to, and adjusted to suit local feedstuffs and conditions. An extension worker with a computer can advise groups or individual farmers. The colour indicators used for protein content of forages and use of the leaf : stem ratio as a further indicator of forage quality are simple tools of practical relevance (see also Chapter 10).

Methods of feeding

The system of management will dictate how an animal receives its feed, and will vary according to the availability of grazing land and labour. Systems imposing restrictions, such as stall feeding and tethering, may be applied all the year or only when it is necessary to keep livestock away from arable areas. Supplements are normally fed on an individual basis; with animals which graze this is usually before being released from the night pens, or at penning in the evening.

- Grazing requires a large area of land, with herding or fencing being available. Direct application of manure is likely to be beneficial. The advantages and disadvantages of hoof action are not clear. However, selective grazing may adversely alter species composition, especially of grasses in pastures

- Tethering controls the animal's movements and allows full use to be made of roadside verges, areas around cropping land etc., but the tether, either via a collar or a 'hobble' (i.e. by the leg), must allow adequate movement and not cause injury. With goats, where the herd has no buck, tethering can restrict breeding opportunities. It is important that animals tethered to graze are allowed grazing durations which maximise intake (LPP Project R5194)

- Stall or pen-feeding must allow adequate trough space and access for all animals. Labour will be required for cutting-and-carrying, and usually also for chopping of forage. Access to DAP and a manual forage chopper are desirable. Adequate water must be provided. Where crop residues are offered in large quantities to promote selection (cf. 'excess feeding'), the rejected residues can be re-fed after treatment with urea.

Conclusions

The amount and quality of feed is a major constraint to livestock production by resource-poor livestock-keepers. The feed resource is predominately forage from grazing, from cutting-and-carrying of green forage, or from crop residues. The opportunities to purchase feed are limited, unless there is a market for livestock products. Although the problems of dry-season feeding are generally recognised, the wet season can also be problematic in some situations: supply of forage to stall-fed cattle; in arable areas, closing of land in the growing season, resulting in reliance on road side grasses and the restrictions of tethering. Livestock-keepers need to be aware of opportunities to use locally-occurring resources, including 'novel' feeds. Upgrading of low quality forages (e.g. by urea treatment) and strategic supplementation can be achieved at low cost. For dairying, conservation of forage produced in the wet season, often as silage, can extend the period over which milk can be produced and sold. A key factor in reducing seasonal feeding stress is planning via an annual feed calendar, through which critical periods are identified.

Further reading

Bayer, W. and Waters-Bayer, A. 1998. *Forage husbandry*. The Tropical Agriculturalist, Series Editors R. Coste and A.J. Smith. Macmillan Education Ltd., London, UK, in cooperation with Technical Centre for Agricultural and Rural Cooperation (CTA), Wageningen, The Netherlands.

Cook, B.G., Pengelly, B.C., Brown, S.D., Donnelly, J.L., Eagles, D.A., Franco, M.A., Hanson, J., Mullen, B.F., Partridge, I.J., Peters, M. and Schultze-Kraft, R. 2005. *Tropical forages: an interactive selection tool.* CD-ROM.

CSIRO, DPI&F (Qld), CIAT and ILRI, Brisbane, Australia.

Cullison, A.E. 1979. *Feeds and feeding. Second edition*. Prentice-Hall Company, Reston Publishing Company Inc., Reston, USA.

Forbes, J.M. 1995. *Voluntary food intake and diet selection in farm animals*. CABI Publishing, Wallingford, UK.

Gohl, B. 1981. *Tropical feeds. Feed information summaries and nutritive values. FAO Animal Production and Health Series No. 12*, Food and Agriculture Organisation of the United Nations (FAO), Rome, Italy.

McDonald, P., Edwards, R.A., Greenhalgh, J.F.D. and Morgan, C.A. 2002. *Animal nutrition. Sixth edition.* Prentice Hall (Pearson Education), London, UK.

Minson, D.J. 1990. *Forage in ruminant nutrition*, Academic Press Inc., San Diego, New York, Boston, London, Sydney, Tokyo and Toronto.

Preston, T.R. and Leng, R.A. 1987. *Matching ruminant production systems with available resources in the tropics and sub-tropics*. Penambul Books, Armidale, Australia.

Sundstøl, F. and Owen, E. (ed.) 1984. *Straw and other fibrous by-products as feed*. Elsevier, Amsterdam, The Netherlands.

Author addresses

Tim Smith, Formerly of Department of Agriculture, School of Agriculture, Policy and Development, University of Reading, Earley Gate P.O. Box 237, Reading RG6 6AR, UK

Noble Jayasuriya, National Science Foundation, 47/5 Maitland Place, Colombo 07, Sri Lanka

Victor Mlambo, Faculty of Agriculture, University of Swaziland, P.O. Box Luengo, Swaziland

Faustin Lekule, Department of Animal Science and Production, Sokoine University of Agriculture, P.O. Box 3004, Morogoro, Tanzania

Derrick Thomas, Formerly of Natural Resources Institute (NRI), University of Greenwich at Medway, Chatham Maritime, Kent ME4 4TB, UK

Emyr Owen, Formerly of Department of Agriculture, School of Agriculture, Policy and Development, University of Reading, Earley Gate P.O. Box 237, Reading RG6 6AR, UK

Anne Pearson, Centre for Tropical Veterinary Medicine, Division of Animal Health and Welfare, University of Edinburgh, Easter Bush,Veterinary Centre, Roslin, Midlothian EH25 9RG, UK

Marion Titterton, Telford Rural Polytechnic, Private Box 6, Balclutha, New Zealand

12

Breeding strategies for sustainable improvement

Michael Bryant[*], *Carlos Galina*[†], *Alan Carles*[‡]

Cameo - A goat cross-breeding programme in Meru, Kenya

Goats are appropriate animals for resource-poor farmers with limited land holdings. The indigenous goats of Kenya are kept mainly for meat, milk yields being as low as 100 ml per day in the predominant East African breed. A cross-breeding programme was devised and supported by FARM-Africa with the objectives of improving both the meat and the milk production traits of the goat flocks while maintaining high levels of adaptation to local feed resources, resistance to disease, and heat tolerance. Previous experience suggested that the Toggenburg, a European breed, would be a suitable animal to use in the crossing programme. Breeding units producing pure-bred Toggenburgs were set up to multiply the exotic breed and produce males that were supplied to buck stations where the Toggenburgs were mated to local goats to produce the F1 generation. The F1 males were castrated while the F1 females were back-crossed to unrelated Toggenburg bucks to produce ¾ Toggenburg + ¼ local progeny. These animals will be multiplied by subsequent *inter se* matings and, by progressive selection, a synthetic breed will be formed.

Over 6 years, 20,000 cross-bred kids (F1 and back-cross) were born. The cross-breds have proved very popular with farmers because of their enhanced growth and dairying characteristics compared to the indigenous breeds. The average yearling weight of the cross-breds is 40 and 30 kg for males and females, respectively. The average daily milk yield of F1 does in their 1st and 2nd lactations is 2.6 litres. A family with three young children owning three cross-bred does, two of which are in milk at any one time, is provided

[*] Principal author
[†] Co-author
[‡] Consulting author

with an adequate protein intake, even when half the milk is sold. The cash from the surplus milk provides a daily income of US$ 0.66.

Ahuya and Okeyo (2002)

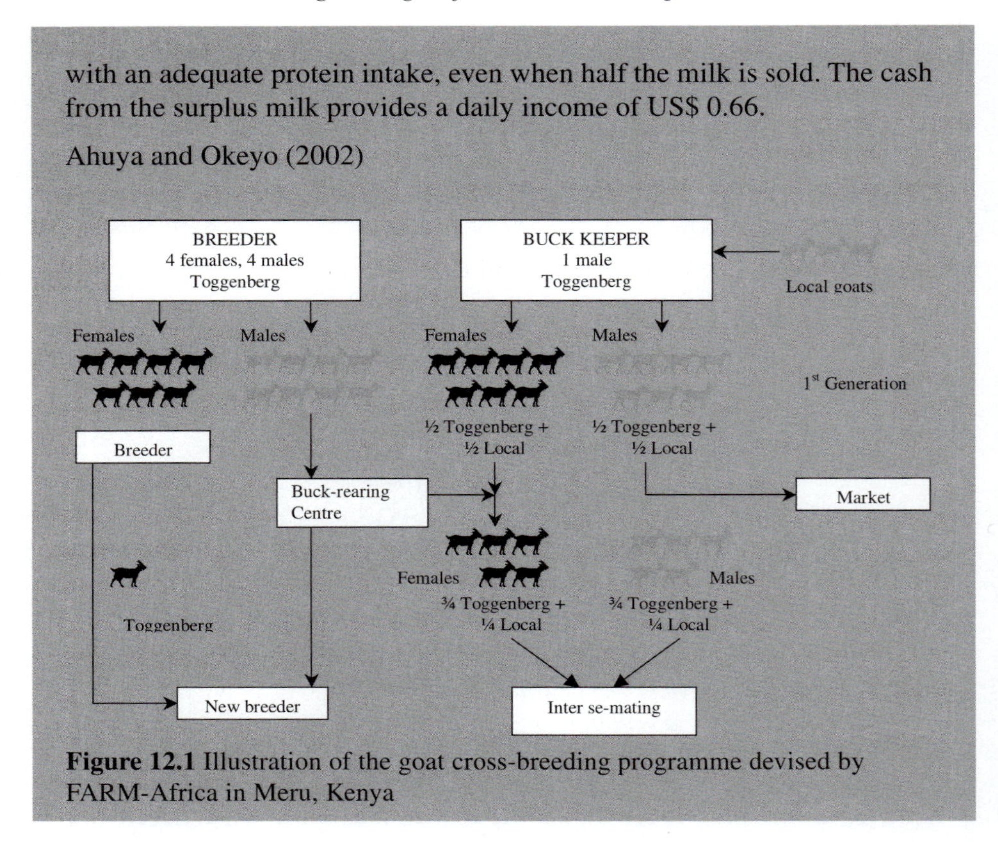

Figure 12.1 Illustration of the goat cross-breeding programme devised by FARM-Africa in Meru, Kenya

Introduction

The productivity of any livestock system is ultimately determined by the responses of a given genotype of animal to its environment. The animal's genetic make-up will establish the animal's ability to cope with stressors imposed by the environment, be those stressors the result of climate, either directly or indirectly, or the result of endemic disease and parasitism. Where the environment allows the animal to express its productivity, the animal's genetic make-up will determine the extent that it may respond to inputs, such as feed, in terms of enhanced fertility, growth or milk production.

The first archaeological evidence of the domestication of animals by man dates from about 10,000 years ago. While genetic change in the wild ancestors of domestic species was dependent on the selective forces of evolution, domestication resulted in the pressures of natural selection being tempered by human control over random matings and the selection of traits within the domesticated population favoured by livestock-keepers. Diversity is far greater in domestic than in wild species of animals and this diversity is a result of controlled breeding, either intentional or 'unintentional'. It seems likely that man's efforts in selective breeding had resulted in the appearance, in Egypt, of distinctive types or breeds of dogs, cattle and sheep about 4,000 years ago. A breed is a group of animals of similar appearance and characteristics that, when bred together, produce offspring with the same traits as their parents. The concept of 'breed' is widespread, although not all domestic animals can be categorised in this way and in some regions of the world a large proportion of animals may be labelled as 'nondescript'. There are approximately 6,000-7,000 breeds of mammals and birds in the world, although the survival of many is under threat. About 50 per cent of genetic variation in domestic species is between-breed variation, with an equal percentage of genetic variation represented within breed. The loss of breeds, therefore, represents a major threat to the genetic diversity of domestic animals.

Characteristics of breeds

The characteristics of breeds tend to reflect the environments in which they were bred. This applies to the physical characteristics of the environment (e.g. climate) as well as the needs and aspirations of their human attendants. The characteristics of animals of interest to humans are frequently called *traits*. Traits may be classified as follows:

- *Fitness traits* – associated with viability and reproduction
- *Production traits* – associated with the products derived from animals and, therefore, with rates of production, e.g. milk yield, growth, wool production
- *Quality traits* – associated with the quality of the product, e.g. milk composition, carcass fatness, the length and fineness of wool fibres
- *Type traits* – associated with appearance, e.g. coat colour, size and shape of the body or parts of the body
- *Behavioural traits* – associated with the animal's behaviour, e.g. docility in relation to ease of handling, skills in relation to herding ability (in dogs).

The animals kept in subsistence and smallholder farming systems are generally multi-purpose (cattle, for example, providing draught, manure, milk and meat) and they survive and produce with minimum inputs and management, often under difficult environmental conditions. Fitness traits rather than traits associated with production are likely to be of the greatest relevance under such circumstances although, given the social significance of animals, attention may also have been afforded to type traits, at least historically.

Subsistence and small-scale farming systems are found in many diverse climatic and agro-ecological zones. It is, therefore, difficult to generalise when describing forms of adaptation found in indigenous livestock. However, examples of adaptation might include the following:

- Adaptation to the thermal environment
- Adaptation to prevailing diseases and parasites
- Adaptation to feed and water resources.

The possession of such adaptations is of considerable benefit, providing survival advantages to animals within the environments to which they are adapted. Animals imported into the same environment may lack the adaptations to local conditions, and hence their survival and success is compromised. This 'fit' between a specific breed and its environment is known as a *genotype-environment interaction*. An example of the phenomenon is illustrated in Table 12.1. This shows the results of a cross-breeding experiment in Tanzania (latitude 6° S). European breeds of bull were crossed with Boran (a local East-African breed) cows and the survival, growth and carcass traits of the cross-bred steers evaluated and compared to pure-bred Boran steers. The experiment was carried out on two ranches, one almost at sea level, the other at 1,087 m above sea level and therefore providing a more equable climate. Furthermore, trypanosomiasis was endemic in the first ranch but not in the second. When productivity of the genotypes was expressed in terms of lean meat yield per 100 calves born alive, the pure-bred Boran competed favourably with the European crosses in the harsh environment, but was the least productive in the equable environment. The advantage of the Boran steers in the harsh environment was their high survivability (78/100 calves born alive on both ranches) compared to the cross-breds. However, in the more equable environment the Borans were unable to compete against the cross-breds because of their low genetic potential for growth. While carcass weights of Boran steers improved by approximately 12 per cent with the more favourable conditions, those of many of the European crosses increased by around 30 per cent. Thus the local breed excelled in survivability, an advantage in the harsh environment, but suffered from low productivity, a disadvantage in the equable environment.

Table 12.1 The ranking of sire breeds according to carcass lean weight productivity of steers born in two environments in Tanzania

Sire breed	*Calves surviving per 100 born alive*	*LSM[1] carcass weight (kg)*	*Percentage lean in the carcass*	*Lean productivity index (kg)[2]*
'Harsh' environment[3]				
Chianina	75	222	69.7	11598
Boran	78	197	66.1	10208
Charolais	64	216	68.6	9498
South Devon	59	216	69.0	8767
Limousin	57	202	71.3	8215
Friesian	53	218	68.6	7992
Hereford	47	227	65.6	6991
Angus	48	199	67.9	6422
'Good' environment[4]				
Chianina	70	293	68.1	13974
South Devon	72	282	67.7	13704
Charolais	70	261	67.4	12369
Hereford	68	275	65.8	12300
Limousin	66	257	69.7	11739
Friesian	68	264	64.9	11624
Angus	64	273	65.5	11523
Boran	78	222	66.2	11509

[1] LSM = least squares mean.
[2]Lean productivity index = number of calves surviving per 100 calves born alive x LSM carcass weight x % lean in the carcass.
[3]'Harsh' environment = ranch at sea level; trypanosomiasis endemic.
[4]'Good' environment = ranch at 1,087 m above sea level; trypanosomiasis not endemic.
Adapted from Said *et al.* (2003).

Selection

The expression of many of the traits possessed by a population of animals can be modified by the genetic characteristics of the population and the environment in which the population lives. What we see or can measure of a trait is called the *phenotype* and is a consequence of the interaction of genetic make-up and the environment. The genes carried by the animal are described as the *genotype*. While such characteristics as coat colour are a result of genotype alone, most characteristics expressed as fitness traits or production traits are the result of both genotype and environment. Moreover, while a characteristic like coat colour may be the result of the animal possessing one or a few specific genes, fitness and production traits may be the consequence of the interplay of very many genes. Animals differ in their ability to express traits (as illustrated in the cross-breeding experiment described above) and some of the variation will be as a result of the genes carried by the animals. When attempting to change the characteristics of the population, the breeder is attempting to change the genes carried by the animals, either increasing the number of 'desirable' genes or reducing the number of 'undesirable' genes, desirability being a flexible concept and determined by the natural and domestic environment.

Changing gene frequencies in a population can be achieved by:

- Breed substitution
- Cross-breeding
- Selection within breeds.

Breed substitution

Breed substitution is to replace one breed with another. This does not necessarily imply the disposal of *Breed A* on day one to be replaced by *Breed B* on day two. In most circumstances where breed substitution has occurred, the change has been achieved by 'grading up', a process of continuous cross-breeding (see below) where sires of *Breed B* are mated to *Breed A* cows over successive generations, such that there is progressive change from one genotype to another.

Cross-breeding

By crossing two, often distinctly different, breeds it is possible to combine the merits of the two breeds in the offspring, sometimes called *complementarity*. It is a breeding system that is frequently employed in environments where the genetic potential of production traits in the indigenous breeds is low and where crossing with a second, usually exotic, breed is seen as a means of achieving a quick increase in production. The example used in Table 12.1 illustrates many of the features of such schemes. Here the aim was to increase the rates of gain and carcass weights of beef animals, traits that were considered to be less than optimal in the Boran. European beef breeds were identified as providing these traits. By crossing European bulls with Boran cows the intention was to combine the favourable growth traits of the European breeds with the adaptive traits of the Boran in the resulting progeny (known as Filial Generation 1 or F1). As described above, this appeared to be successful so long as the environment was capable of supporting the cross-breds.

Cross-breeding also allows the breeder to exploit *heterosis* or *hybrid vigour.* It might be anticipated that when two breeds are crossed the offsprings' expression of a particular trait will be midway between that of the two parents. Thus if the Boran has a growth rate of 400 g/day and the Chianina 800 g/day then the gain of the offspring will be 600 g/day. This is known as *additive inheritance*. In practice, cross-bred offspring may express non-additive gene action, such that their performance may be greater than the midpoint of the two parents and in some cases greater than that of the best parent. Our Chianina x Boran calf may gain 700 g/day, 300 g/day more than the slowest gaining parent: of the 300 g, 200 g is as a result of additive inheritance and the additional 100 g as a result of heterosis. Heterosis is generally associated not with growth traits but with reproductive, survival and fitness traits. Such heterosis is usually expressed to the greatest extent when genetically diverse breeds are crossed. Thus crossing European breeds with indigenous breeds may be associated with substantial gains as a result of heterosis.

Cross-breeding has been used extensively in livestock development and there are abundant examples available from cattle, sheep, goats, pigs and chickens. However, there are few examples where cross-breeding can be shown to have sustainable benefits or, more accurately, where the full benefits are sustained. The classic examples of cross-breeding that have been sustained for centuries frequently require the cross-bred F1 to be the terminal cross. In other words the F1 generation is slaughtered and replacement F1 animals are produced from the two parent breeds. This clearly requires large numbers of the parent generation, particularly the dam line, to be maintained. Alternative strategies to this involve mating F1 animals (*inter se* matings) to produce the F2 generation, or back-crossing to one of the parent breeds. Both of these options result in a reduction in the benefits gained from heterosis by 50 per cent, as shown in Table 12.2. Further crossings of filial generations lead to no additional loss of heterosis but crossing back-crosses reduces further the gains from heterosis.

Table 12.2 Fractions of heterosis maintained in different crosses

Cross	Proportion of heterosis relative to that in the pure-bred
Pure-bred	0
P_X x P_Y = F1	1
F1 x F1 = F2	0.5
F2 x F2 = F3	0.5
F1 x $P_{(X \text{ or } Y)}$ = B	0.5
B x B	0.375

P = parent generation X and Y
B= back-cross

In small-scale farming systems, cross-breeding has been most extensively studied when it has been employed to provide cattle with worthwhile milk yields for dairying purposes.

Indigenous breeds of cattle, with few exceptions, have relatively short lactations, low daily yields, and will let down milk only in the presence of their calves. Bulls of European dairy breeds have been crossed with indigenous cows and the resulting progeny have enhanced yields while retaining some of the adaptive features of their dams. Syrstad (1990) summarised 54 data sets from published cross-breeding experiments that involved a number of both indigenous and exotic breeds. His findings on the effects of the inclusion of various percentages of *Bos taurus* blood in the crosses on milk yield are shown in Figure 12.2. As can be seen, increasing the percentage of *Bos taurus* blood is associated with increasing milk yield up to the 50 per cent cross (F1) when it plateaus. The implications of *inter se* matings can be clearly seen: the F2 loses approximately 500 kg of milk compared to the F1 counterpart. Back-crossing to the indigenous parent is an even less attractive proposition.

Back-crossing to increase the percentage of *Bos taurus* blood may maintain milk yields in the daughters but at the cost of reduced fitness. For example, a recent study in Tanzania has shown that animals with 75 per cent *Bos taurus* blood are three times more likely to die without realising their potential, than F1 animals (Swai, 2002).

An alternative way forward from the F1 is to introduce a third breed. For example, farmers practising small-scale dairying in Kenya might produce the F1 by crossing Holstein bulls to the local East-African zebu. This F1 is then crossed to the Sahiwal, a *Bos indicus* dairy breed. Such a three-way cross allows the maintenance of heterosis (i.e. the proportion of heterosis relative to the pure-bred would be one). However, the absolute amount would vary according to the breeds used.

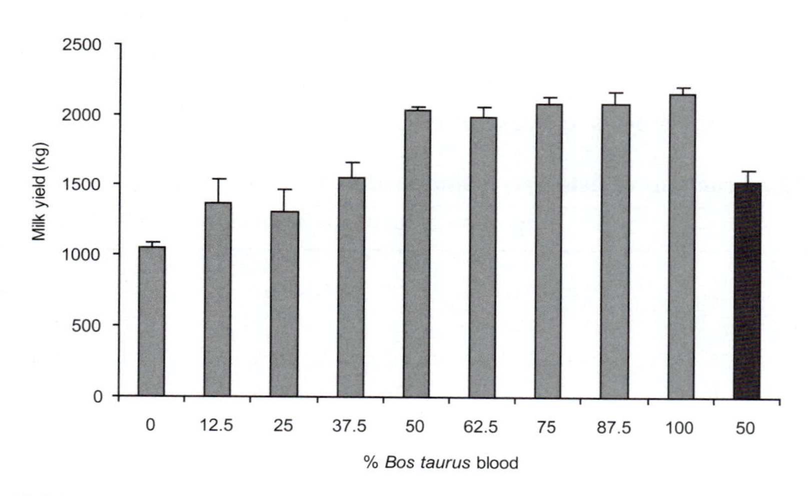

Figure 12.2 Lactation milk yield (LSM and SEs) of dairy cattle with varying percentages of *Bos taurus* blood. The light column labelled 50% represents the F1 while the dark column represents the F2. Source: Syrstad (1990).

A further possible strategy that may be employed is rotational crossing. This involves the use of two or more breeds in rotation. In the case of the cross-bred dairy cow, experience suggests that farmers are reluctant to use indigenous bulls on cross-bred cows because of the possibility of milk yield being lower in the daughters than in their dams.

It follows, therefore, that sustaining the benefits of cross-breeding is difficult, even just from the point of view of the principles of inheritance. However, within the context of smallholder and subsistence farming, the difficulties are compounded by the slim resources available to support such activities. Firstly, the animal resource available to the farmer is, by definition, the indigenous breed. The exotic breed, usually the sire breed, must be provided by government or some other agency. Attempting to inject exotic genes into the local population using live sires is fraught with difficulties, not least because of the health and fertility risks to the animals imposed by their new environment. The use of artificial insemination (AI) would seem a more practical means of disseminating new genetic material, but such technologies also have their disadvantages when employed in low-input livestock systems (see below). The difficulties in maintaining sustainable, planned cross-breeding schemes are frequently found to revolve around the difficulties of making the appropriate sires or semen generally available to livestock-keepers. As a result of these frustrations, the farmer often has to resort to the use of the village bull, regardless of its genetic mix and its relationship to the farmers' cows. Such default arrangements not only confound breeding plans but may also lead to inbreeding.

A further serious consequence of cross-breeding is the loss of pure breeds. Farmers anxious to breed more productive stock will use exotic sires indiscriminately, such that minor breeds decline in numbers and eventually disappear. This results in the loss of genetic diversity and is a major issue in the modern world. Most countries are committed to the Convention of Biological Diversity (Secretariat on the Convention on Biological Diversity, 2001-2004). This formally identifies domestic animal diversity as an important component of global diversity and the United Nations has a global strategy for the management of farm animal genetic resources. Farmers have an understandable desire to increase animal productivity that, unfortunately, often appears to be in conflict with the preservation of pure breeds. Ideally, the wants of both farmers and conservationists could be met through breed improvement by selective breeding.

Selection within breeds

There can be little doubt that selection within breeds represents the most sustainable way to change gene frequencies. However, gene frequencies only change as a result of reproduction and therefore rate of genetic change is ultimately dependent on the succession from one generation to another. As a consequence progress is slow, particularly in large animals, and considerable commitment is required to initiate and maintain programmes of genetic selection.

The steps involved in a within-breed improvement programme are as follows:

- Deciding upon the required outcome
- Deciding upon the traits to be measured and selected to achieve the required outcome
- Recording the desired traits in the population of animals
- Evaluating and selecting animals to be used to produce the next generation.

Simply deciding upon the required outcome is not necessarily straightforward. The livestock kept by smallholder farmers are frequently multi-purpose, so attempting to improve milk yield in a cattle breed is not necessarily an appropriate breeding goal where the animals are also expected to yield meat and provide power and traction. Selecting for reproductive rate and live-weight gain in sheep may represent sufficient challenge, but resistance to internal parasites can also be considered to be an essential objective.

The rate of genetic change for a given trait in a population is determined by the following:

- The heritability of the trait
- The selection differential
- The generation interval.

Heritability

The variation we see between individuals in the phenotypic expression of a particular trait is a function of the individual's genotype and the effects of the environment. Only the contribution of genotype to any superiority that the individual may show will be passed on to the offspring. 'Heritability', therefore, expresses that part of superiority that will be passed on to the offspring. Since each parent contributes 50 per cent of the genes of the offspring, only 50 per cent of the superiority that is of genetic origin will be passed on by each individual parent.

Heritability is usually expressed as a coefficient (h^2), either 0-1 or 0-100 per cent. The heritability of a particular trait represents a ratio between the contributions of genotype and environment to the variation in the trait. It follows that in attempting to measure heritability of a specific trait, reducing the influence of environment on the expression of the trait will increase the precision of identifying true genetic differences between individuals and hence increases the heritability coefficient. This explains why heritabilities cannot be regarded as absolute values but vary from one estimate to another. Nevertheless, it is known that heritabilities differ between traits, some traits being regarded as highly heritable while with other traits genetic differences between individuals seem to be minimal. Some estimates of heritability of common traits are shown in Table 12.3.

Table 12.3 Estimates of heritability (h^2)

Traits	*Range of h^2*
Cattle	
Milk yield (single lactation, mainly 1st), indigenous breeds	0.11-0.48
Fat percentage	0.09-0.41
Lactation length	0.06-0.51
Age at 1st calving	0.01-0.69
Calving interval	0-0.40
Service period	0-0.18
Services per conception	0.03-0.08
Weight gain (birth to weaning)	0.02-0.34
Weight gain (post-weaning)	0.13-0.38
Rectal temperature	0.17-0.31
Tick count	0.09-0.44
Gastrointestinal nematode count	0.28-0.44
Buffalo	
Milk yield (lactation)	0.19-0.67
Fat percentage	0.22-0.37
Adult weight	0.35-0.88
Sheep and goats	
Milk yield	
Lactation	0.20-0.53
Test day	0.14-0.31
Birth weight	0.03-0.43
Weaning weight	0.08-0.62
Adult weight	0.11-0.72

Table 12.3 (Contd.)

Traits	*Range of h^2*
Fleece weight	0.17-0.57
Fleece quality traits	0.13-0.72
Number of lambs at birth	0-0.49
Litter weight at birth	0-0.12
Gastrointestinal nematode count	0.06-0.37
Resistance/resilience to intestinal nematodes	0-0.33
Pigs	
Number piglets at birth	0-0.39
Number of piglets at weaning	0.02-0.34
Litter weight at birth	0-0.31
Litter weight at weaning	0.03-0.20
Post-weaning weight gain	0.40-0.88
Back-fat thickness	0.51-0.71

Adapted from Wiener (1994), with some additions

Selection differential

If we measure a particular trait in a population of animals we will find that the expression of the trait varies between individuals. For example, if we measure post-weaning gains in a population of goats we may find that the average daily gain is 120 g while the slowest gain is 20 g and the fastest 240 g. If we select the fastest growing male (240 g) and the fastest growing female (180 g) for breeding then the average gain of our selected pair is 210 g, which is 90 g superior to the population average. The 'selection differential', the superiority of the animals selected over the mean of the population from which they came, is, therefore, 90 g. Of course, to breed an adequate number of replacements for our goat herd we would need to breed from more than one female, although one male may be sufficient to impregnate the number of breeding females we select. If we select the 10 fastest growing females and if these animals have an average daily gain of 150 g compared to the average gain of 90 g for all females in the population, this gives a selection differential of 60 g for the females. Our selected sire has a gain of 240 g compared to the average gain of 150 g for all males in the population, a selection differential of 90 g. Because we can afford to select fewer males than females from our population, the selection differential imposed on males is usually greater than that imposed on females. Breeding our selected male to our 10 selected females gives a selection differential of 60+90 = 150/2 = 75 g.

Generation interval

Since gene frequencies can change only as a result of reproduction, the rate of genetic change is dependent on the 'generation interval', that is the average age of the parents when their progeny are born. This will of course vary considerably between species. Moreover, the environment and management within smallholder systems may not be conducive to rapid generation turnover, particularly in larger livestock. Therefore, the generation interval of small ruminants may be in the region of 3-4 years while for cattle it may be 9-12 years or more.

In summary, the rate of genetic change from one generation to the next is, therefore, a function of heritability (h^2) x selection differential. We can increase the rate of genetic change by the pragmatic choice of selection goals. Thus we are likely to make more progress with changing the genetic potential of our goats if we choose to select for growth traits ($h^2 = 0.3$-0.4) rather than number of kids at birth ($h^2 =$ 0.1-0.2). Alternatively we can increase rate of genetic change by increasing the selection differential. This could be done by recording a large population of animals, and therefore increasing the chances of identifying exceptional animals, and then selecting very few animals, for example selecting only exceptional males and using them in the population through the application of AI. Ideally, of course, we would do both. The rate of genetic change per year is a function of (h^2 x selection differential)/generation interval. Increasing genetic change per year involves reducing the generation interval, for example by replacing males (and possibly females) in the population after the first breeding cycle.

The use of artificial insemination, embryo transfer and other biotechnologies

As indicated above, AI and other technologies have a major role to play, not only in procedures of genetic selection but also in the dissemination of genetic progress. The use of AI in the tropics has been a technique routinely used for many years by commercial farmers dedicated to producing sires to be used in natural mating in other enterprises with less infrastructure or economical resources. In some countries, this well established symbiosis between farmers producing sires and users of these animals for natural mating has facilitated the implementation of AI. Conversely, the shortcomings of applying AI in the tropics have been known for some time. At an organisational level, constraints to the efficient use of AI include:

- Poor infrastructure (for example, roads and transport)

- Poor communications
- Difficulties in obtaining liquid nitrogen for semen preservation
- Poor training of technicians
- Difficulties in recouping the costs from farmers.

At the farm level, constraints to the efficient use of AI might include:

- Poor potential fertility, because of poor nutrition and management
- Difficulties with oestrus detection
- Poor timing of services
- Farmers unable to afford the costs.

Among these, oestrus detection has been the major constraint for the rapid development of this technique, as only 30 per cent of the incidents of oestrus manifestations are readily detectable in practice. However, a 50 per cent fertility rate has been achieved, particularly by commercial farmers who may use AI only on the elite cows in their herds.

A different scenario is the application of AI by smallholder farmers dedicated to the production of milk and meat, where little progress has been made up to the present in spite of the growing interest in producing F1 crosses between *Bos taurus* and *Bos indicus* breeds. The use of AI is generally associated with a decline in fertility when compared to natural service. For example, studies among smallholder dairy farmers in Tanzania found calving intervals of 444 ± 12.2 days where natural service was employed, compared to 553 ± 15.6 days for AI (Msangi, 2001). The limitations of this technique are apparent despite numerous local and international efforts to establish AI routinely.

The same principle can be applied to the other biotechnological approaches, such as semen preservation or embryo transfer (ET). In relation to the former, stud farmers in general have used professionals to preserve semen on the premises or to collect the biological material and transport it to the laboratory in a diversity of diluents. The shortcomings of this methodology have been discussed elsewhere (Galina *et al.*, 1996). Embryo transfer, on the contrary, has gained popularity among stud farmers for reasons mostly related to economics and prestige. The technique has yet to prove the claim of being a method to gain genetic progress in a herd.

Embryo transfer has obvious advantages, particularly on smallholder farms, as it overcomes the need for keeping sires. It also avoids the transmission of diseases if the farmer needs to borrow the sire from a neighbour, a very common practice within the smallholder system. However, the main advantage is the facility offered to farmers of raising the replacement genotypes of their choice – an important benefit in relation to a breeding programme. Embryo transfer also avoids the indiscriminate systems of cross-breeding referred to earlier. In spite of these obvious advantages, the use of ET among smallholder farmers is very limited and, surprisingly, its lack of impact is related not only to the expense incurred.

In two recent projects involving many farmers (Diaz *et al.*, 2002) it became apparent that implanting embryos in the cattle of smallholder farmers poses difficulties. If one selects the system of observing cows for spontaneous oestrus, the results are disappointing, as the detection rate by these farmers can be as low as 20 per cent, particularly in lactating cows. If the option is to synchronise cows to promote oestrus, for instance at 70 days post-partum, then results can be encouraging if the cows are cycling prior to treatment. In general, if this is the case, 64 per cent of the cows will ovulate following synchronisation. However, only 35 per cent of all cows are cycling prior to this date. More problematic is the fact that as many as 60 per cent of the cows not cycling prior to 70 days post-partum will express oestrous behaviour but will not ovulate. If one decides to extend the voluntary period to 100 days post-partum then most of the cows cycling early become pregnant, either because the farmer decides not to wait any longer, or because of the intervention of the neighbour's bull. Therefore, the outcome is similar to attempting transfer at 70 days, as the cows remaining that are available for ET are incapable of restarting ovarian activity. In two surveys carried out in dual purpose herds (Galina and Arthur, 1989; Anta *et al.*, 1989), the average interval from parturition to the onset of ovarian activity was 100 days. Hence the decision to implant at 70 days will increase the costs per pregnancy as there will be a need to stimulate ovulation and oestrus; however, a later intervention does not seem to decrease the costs substantially or increase the number of pregnancies obtained. More research is obviously needed before a sound decision can be taken with respect to the technical problems and also the economic implications. In a survey undertaken via a participatory rural appraisal (PRA) with almost 100 farmers (Molina, 2003), the cost involved was an important consideration, but not the only concern. Obviously, embryos produced *in vitro,* cheaper methods to synchronise cattle, easier transport of the embryos to the farms and other related activities will contribute to lower the costs of each pregnancy, allowing more farmers to participate in these programmes.

Of the other new technologies, gene mapping offers exciting possibilities. Gene mapping allows identification of quantitative gene loci and the discovery of markers to allow the direct selection of animals on the basis of genotype rather than phenotype. This would facilitate developing strains of animals with inherent resistance to the threats and challenges posed by their environment while retaining high levels of productivity. Unfortunately, the means of delivering new genotypes to smallholder farmers through AI and ET remain problematical. Impact can only be achieved when farmers appreciate the need and the value of change, as will be shown in the following section.

Breeding strategies for sustainable improvement – the importance of public ownership

Resource-poor farmers are unable to employ breeding strategies without outside help. Yet the commitment of the farmers to any proposed breeding strategy is of vital importance if it is to have impact and to be sustainable. Imposing and maintaining recording schemes, essential if on-farm testing is to be employed, are not always easy among smallholder farmers, where a proportion may be illiterate and where months and years of recording may be required before any pay-off becomes evident. Taneja (2000), describing cattle breeding programmes in India, claims that between 75 and 80 per cent of female calves identified and registered in progeny testing programmes are lost before completing their first lactation. He recommends that 320-400 cows should be allocated to each bull in the testing programme if complete records are to be available from 40 daughters. This adds considerably to the effort and expense of the programme.

In the case study described in the cameo, the farmers were involved from the beginning of the project. Farmer 'Self-Help' groups form the basis of the organisational structure. These are made up of resource-poor farmers with a common interest in goat farming. The groups are represented by a single organisation, the Meru Goat Breeders Association, which is a fully registered, local breed association with its own by-laws, by which members must abide. The objectives of the breeding programme were drawn up through participatory workshops and discussions. In some instances farmers wanted characteristics included in the programme that might not have been predicted; for example, they wanted certain coat colours to be retained. Each Self-Help group selects two members, one of whom will become the breeder of the pure-bred Toggenburgs and the other the buck keeper. Each group has a committee that oversees activities related to the breeding programme, including monitoring activities, and organising shows and auctions.

Record keeping, by the farmers, and the use of such records is considered essential in the breeding programme. Such record keeping includes individual animal records (reproduction, growth, milk production and health), herd events such as vaccinations, and pedigree records. Manually-collected data from the farms are stored electronically at the project office and are available to group members through the Meru Goat Breeders Association. To help overcome problems with illiteracy, school children are taught how to keep records on the farms. Records are checked regularly by a trained group member or by the local extension staff.

A further issue that has led to success has been the emphasis placed on animal health and improved management (especially nutrition), i.e. the necessity of matching genetic change with environmental change. The vulnerability of the pure-bred Toggenburgs and the loss of adaptation in the cross-breds are shielded by an improved farm environment, allowing the genetic potential of the cross-breds to be exploited. Here genetic selection goes hand-in-hand with better management and this possibly explains why this project has succeeded when so many similar initiatives have failed.

Conclusion

The livestock associated with subsistence and smallholder systems often have advantageous fitness traits, but have limited genetic potential for the production traits of meat, milk, eggs and fibre. Modifying the animals' genetic potential is hampered by the challenges to health and survival posed by the farm environment, and the limited inputs to such livestock systems. Improving existing breeds through selection is a long-term undertaking that is probably not appropriate given the physical and socio-economic climate of such systems. Introducing exotic genes through judicious cross-breeding seems more appropriate. However, the success and sustainability of such an enterprise is likely to be very sensitive to the involvement of participating farmers and farmer organisations. Success is most likely to be achieved where the breeding strategy is just one of the components of a programme aimed also at increasing livestock productivity through the general improvement of animal health, nutrition and management.

The problem is particularly acute for the resource-poor. Achieving genetic improvement of livestock is always a long-term business, requiring large investments of resources, including physical, financial, human and time; although once achieved it may cost much less to maintain. Yet the resource-poor are in great immediate need. While farmers with a moderate level of resources will not suffer greatly if significant improvement of their livestock's performance takes a few years, the poor will. This requires a very delicate balance in the apportioning of scarce resources, involving the cost effectiveness, over time, of all components affecting performance in the production system (genetic and environmental) and the minimum acceptable rate of improved productivity for each succeeding year. In the early years of a genetic improvement programme, this may mean greater investment in environmental improvement, as this usually yields quicker results, although it may be costly in the long-term due to the continuous need for maintenance; however, it would also provide the environment that would be required for the improved genotype to come.

Further reading

Galal, S., Boyazoglu, J. and Hammond, K. (ed.). 2000. *Workshop on developing breeding strategies for lower input animal production environments*. Bella, Italy, 22-25 September 1999. ICAR Technical Series, No.3. (available online at www.icar.org)

Wiener, G. 1994. *Animal breeding*. The Tropical Agriculturalist, Series Editors R. Coste and A.J. Smith. Macmillan Education Ltd., London, UK, in cooperation with Technical Centre for Agricultural and Rural Cooperation (CTA), Wageningen, The Netherlands.

Willis, M.B. 1998. *Dalton's introduction to practical animal breeding. Fourth edition*. Blackwell Science, London, UK.

Author addresses

Michael Bryant, Department of Agriculture, School of Agriculture, Policy and Development, University of Reading, Earley Gate P.O. Box 237, Reading RG6 6AR, UK

Carlos Galina, Departmento de Reproducción, Facultad de Medicina Veterinaria, Universidad Nacional Autónoma de México 04510, México

Alan Carles, Formerly of Department of Animal Production, The University of Nairobi, Faculty of Veterinary Medicine, P.O. Box 29053, Kabete, Nairobi, Kenya

13

Improving the health of livestock kept by the resource-poor in developing countries

Brian Perry[*], *Thomas Randolph*[†], *Amos Omore*[†], *Oswin Perera*[‡], *Adriano Vatta*[‡]

Cameo - Reliable veterinary services are crucial in livestock improvement programmes

Mr. George Kapere and his wife Paskazia of Buganda sub-county in Mbarara district (Uganda) acquired a cross-bred cow for milk production in 1999 through a donor support project. The animal suffered two abortions and died from what was suspected to be tick-borne disease after incurring high treatment expenses. The causes of the abortions were not detected, but may have been brucellosis, which is a common zoonotic disease. This farmer suffered heavy losses in terms of labour, money and even social trust, because this was a heifer-in-trust programme. These losses could be attributed mainly to poor veterinary diagnostic and delivery services.

Introduction

It has been estimated that livestock form a component of the livelihoods of 70 per cent of the world's resource-poor people (LID, 1999), something that many people living in the developed world have great difficulty understanding. Many are not aware that livestock are important in supporting the livelihoods not only of resource-poor farmers, but also of consumers of meat, milk and eggs, of traders in different products, and of labourers working with livestock or the processing of products derived from them. Animal diseases are an everyday occurrence to all of these people, as animals of the resource-poor are particularly vulnerable to disease (due to many reasons, including the presence of a wide range of disease-causing

---

[*] Principal author
[†] Co-author
[‡] Consulting author

organisms, lack of knowledge about their management and control, and lack of access to, and resources for, animal health and production inputs and services), and the resource-poor often handle and consume products that do not pass through normal inspection and quality assurance processes. Furthermore, resource-poor farmers usually have few animals, so the loss of an individual animal has proportionally greater significance.

But identifying the animal health constraints to resource-poor livestock-keepers, consumers of products, traders and labourers, is not a straightforward process, let alone that of providing services directed at the resource-poor to alleviate these constraints. There are some diseases that affect all regions of the world and all sectors of the community, and there are some that are of particular importance, individually and collectively, to the very poor. There are diseases that affect the particular species of animals that have special importance to the resource-poor as security, financial capital and social capital, as machines for cultivation, as producers of fertiliser, and, of course, as nourishment. They include diseases that affect the human populations of these resource-poor societies themselves, causing death, disability and suffering, and representing a barrier to escape from poverty.

Although in general terms most of the infectious diseases are shared by the animals of the rich and the poor, not all of them have the same importance to these different extremes of society. For example, foot and mouth disease has devastating effects on the milk yield and weight gain of many breeds exotic to developing regions, while the lower-producing indigenous genotypes of cattle are often less susceptible to the effects of the disease in terms of productivity parameters. But that is not the end of the story. The disease causes lameness, so if oxen, often used by the resource-poor for a variety of traction functions, are affected, it can have severely disruptive impacts on ploughing and moving of grain to markets, particularly if these animals succumb to secondary bacterial infection as a consequence of poor management and unhygienic conditions. And further, the movement restrictions put in place to contain foot and mouth disease outbreaks can limit significantly the flexibility of the resource-poor to sell animals and raise cash at the time they need to.

Not only are different diseases important to the poor, or the same diseases manifest their impacts in different ways, but we also have very little empirical information on the occurrence, dynamics and impact of disease in smallholder systems in developing countries, when compared with the mass of knowledge from more commercially orientated livestock-keeping enterprises. Animal health information systems that exist in the developing world are often available only to the more commercial livestock producers to whom they serve as a valuable aid to enhancing production efficiency. Where they do exist in the public sector, they are often targeted at the detection and monitoring of national priority diseases undergoing programmes of control or eradication. The public sector invariably focuses on gathering information on the infectious diseases of interest to the international community, and so information on the endemic conditions causing morbidity, mortality and production losses among the livestock of the resource-poor is often rudimentary at best. In many such countries, this unsatisfactory situation has deteriorated still further over the last decade as public sector support to veterinary services has declined dramatically.

What are the ways in which poor animal health affects resource-poor people?

Key to understanding the way in which animal health affects the resource-poor is an understanding of the role that different animals play in the livelihoods of resource-poor people, as was discussed in Chapter 2. Table 13.1 illustrates the contributions of livestock to the household assets of resource-poor livestock-keepers. Two key points characterise the role of livestock for the poor:

- *The resource-poor usually keep more than one species.* Resource-poor households rarely specialise in one species of animal, preferring to diversify and take advantage of the different types of role each species can play, as well as to spread risk, including the risk of disease. Obviously, the ability of the resource-poor to acquire livestock is constrained by the market value of each species, which increases as one moves up the 'livestock ladder', as roughly approximated by the order of the species from bottom to top in Table 13.1.

- *Each species serves multiple roles for the household.* Each species contributes in various ways to the different types of household assets. Resource-poor households often have multiple objectives in keeping livestock, and some of the most common have been highlighted in Table 13.1.

In a recent study of the ability of the perceived priority species to the poor to fulfil these varied functions in the different production systems of Sub-Saharan Africa, South Asia and South-East Asia (Perry *et al.*, 2002), the following conclusions were reached. In pastoral/range-based systems, several livestock species play an important role, but within these sheep and goats generally are the most important, usually playing a more important role than cattle. In the agro-pastoral (mixed) systems, cattle predominate. In the peri-urban landless systems, poultry, sheep and goats, and pigs play the most important roles.

Table 13.1 Animal species kept by the resource-poor, and their contribution to household assets

Species	*Contribution to household assets*				
	Financial	*Social*	*Physical*	*Natural*	*Human*
Cattle, Buffalo	• Sales of milk, meat, hides, animals, draught power services, transport • Savings instrument	• Networking mechanism • Social status indicator	• Draught power for crop cultivation • Draught power for transport	• Manure for maintaining soil fertility	• Household consumption of milk, meat
Camel	• Sales of milk, meat, hides, animal, transport services • Savings instrument	• Networking mechanism • Social status indicator	• Draught power for transport		• Household consumption of milk, meat
Donkeys, Horses	• Sales of animals, draught services, transport (esp. water)		• Draught power for crop cultivation • Draught power for transport (esp. water)	• Manure for maintaining soil fertility	• Provision of household water supplies
Goats, Sheep	• Sales of milk, meat, hides, animals • Savings instrument	• Networking mechanism • Social status indicator		• Manure for maintaining soil fertility	• Household consumption of milk, meat
Pigs	• Sales of meat, animals • Savings instrument			• Manure for maintaining soil fertility	• Household consumption of meat
Poultry	• Sales of eggs, meat, fowl	• Networking mechanism		• Manure for maintaining soil fertility	• Household consumption of eggs, meat

Source: Perry *et al.* (2002).

Within these production system groupings, each geographical region has a slightly different pattern to the priority species of the poor. For example, in South-East Asia, pigs and poultry were considered the most important species in both mixed rain-fed and irrigated systems. Moving further west to South Asia, buffalo ranked second after cattle, and yak are important in the grassland humid systems. In eastern and southern Africa, cattle ranked first in the mixed agro-pastoral systems, replaced in West Africa by sheep and goats, followed by poultry.

Different diseases, different impacts

To facilitate the discussion on the impacts of animal diseases, it has been found useful to distinguish four general groups of diseases: the endemic, the epidemic (or transboundary), the zoonotic and the food-borne (Perry *et al.*, 2001). However, as we shall argue, this classification is insufficient when considering how diseases constrain development processes, including poverty reduction.

Endemic diseases include the vector-borne haemoparasitic diseases, the multitude of helminth diseases, the enteric bacterial diseases of the neonate, bacterial and viral causes of reproductive inefficiencies, specific diseases such as haemorrhagic septicaemia (particularly in South Asia), anthrax and blackleg, among many others. They can be further divided for the purpose of priority setting for the world's poor into those that are 'tropical' and those that are 'tropical and temperate'. Many of the endemic diseases that still occur today in the temperate regions of the world represent some of the last hurdles to improving production efficiency there, and as a result many effective control technologies are available or under development through support from the public and private sectors of the developed world. Most of these technologies are not widely applied in much of the developing world, and certainly not in resource-poor communities.

Much less attention has been invested in the 'tropical' group of endemic diseases, as to the livestock industries of the developed world they are often considered 'somebody else's problem'. These include the vector-borne haemoprotozoan infections, for which effective control technologies appropriate for the majority of resource-poor livestock-keepers in the developing world are still lacking. Endemic diseases tend to be those that exert their greatest effect at the farm, village and community level, even though the aggregation of all the farm level effects can of course be translated into national-level losses.

Epidemic diseases (sometimes termed transboundary diseases) are those that characteristically occur at a frequency above the expected, are highly infectious, and exert their influence at both farm and national level on local marketing and international trade. This group includes the virus infections of foot and mouth disease, rinderpest, hog cholera, Newcastle disease (ND), and the influenzas, among others. Some epidemic diseases can result in devastating shocks to the resource-poor, by wiping out their entire livestock holding. Because of their potentially explosive nature, their tendency to cross international borders, and the need to protect valuable commercial livestock production systems and/or markets, public sector involvement in their control is common. This is particularly the case where lucrative export markets exist, and a country is trying to protect an existing or potential market by maintaining disease freedom or a certain level of disease control. It is important to note that these diseases are endemic in some countries and production systems, particularly in the developing world. Thus they may also have considerable impact at the farmer level, in cases where they are widely distributed and occur frequently.

Zoonotic diseases may cause significant productivity losses in livestock (or in other domestic or wild animal species), but their major impact is usually in causing human disease and suffering. Some can be characterised as endemic, such as many of the meat-borne helminth zoonoses, brucellosis and tuberculosis, and some are epidemic in nature, such as rabies and Rift Valley fever. Food-borne diseases, such as cysticercosis and trichinellosis, can be particular problems to the resource-poor due to inadequate hygiene and sanitation, and insufficient attention paid to cooking animal products. In addition, *Escherichia coli* O157 and salmonellosis are particular problems in more industrialised systems of the world, and thus their incidence is likely to increase in developing countries as livestock production and processing systems become more intensive. Food-borne diseases affect consumers, food-processing workers and livestock producers.

There are many features of the zoonotic diseases that render them particularly important to the poor, be they livestock-keepers, labourers working with livestock, livestock owners consuming products from their animals, or livestock non-owners consuming livestock products of their neighbours or of other resource-poor communities.

The first of these features is that many of these diseases produce fatal or disabling diseases in humans, the prevention of which is often through their control in animals. This requires the availability of appropriate animal health technologies to be in place for these diseases, and their delivery to and accessibility by the poor. Human sleeping sickness caused by *Trypanosoma brucei rhodesiense* is an important example, in which the mass treatment of cattle significantly reduces the risk of disease in humans. Another important example is human epilepsy, of which neurocysticercosis (caused by the intermediate stage of the pork tapeworm *Taenia solium*) is considered the main cause (Anon., 1994). The World Health Organisation (WHO) estimates that at least 50 million people are infected with the parasite, with more than 50,000 deaths annually (Schantz *et al.*, 1993).

The second of these features is that while there are some zoonotic diseases to which a large section of any given human population is equally susceptible, the resource-poor are particularly at risk to many of them. Examples include cysticercosis in pigs, in which poor sanitation is the underlying cause, and knowledge and resources in poor communities to adopt relatively straightforward preventive measures are limited or absent. Another example is leptospirosis, in which rats play an important role in the maintenance of infection, and rats often thrive in resource-poor communities and in rice paddies.

The third, and possibly the most important of these features, is that the lower down the income scale, the more likely is a high risk of multiple zoonotic infections. Consider the landless peri-urban setting found in many regions, in which a cow, some pigs and goats, and the household dogs all co-exist with the family struggling for survival. There is a potential risk of human infection from *Brucella abortus* in the cow and *B. melitensis* in the goats, as well as tuberculosis and leptospirosis in the cow, cryptosporidia in a calf, cysticercosis and trichinellosis in the pigs, and rabies in the dogs. The risk of multiple zoonoses is a factor of poor hygiene, the purchase of cheap animals that may be the culls of others having failed disease-screening tests, the purchase of cheap meat that has not undergone inspection, or has – but failed, and the lack of resources or knowledge to protect their dogs against rabies.

What are the impacts of animal disease on resource-poor people?

Animal diseases generate a wide range of biological and socio-economic impacts that can be both direct and indirect, having their effect locally or globally. A particularly useful way of understanding their impacts is by considering those impacts associated with overt disease, and those associated with disease risk (Figure 13.1).

The impacts of overt disease

When animal disease occurs, there are several different types of commonly-recognised impacts:

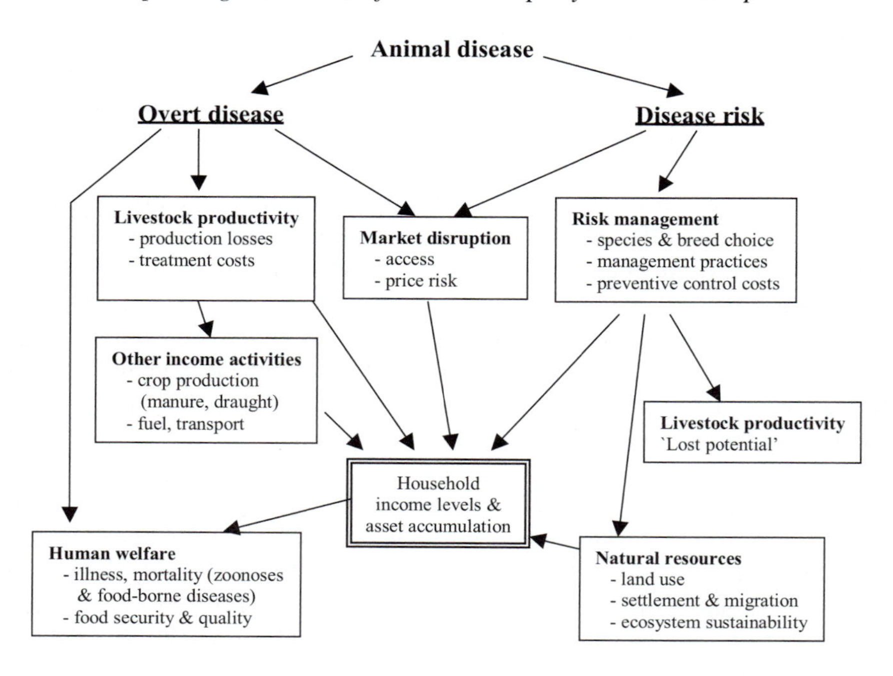

Figure 13.1 The impacts of animal diseases on resource-poor households and communities

- Loss of livestock productivity: The most important and readily measurable direct effects of diseases are manifested by losses in productivity. These include the effects of death, illness leading to condemnation, low weight gain, low milk yield, inefficient feed conversion, low reproductive capacity and low work capacity for ploughing or transport. Achieving effective control of them is long-term and difficult in the smallholder sector, with its limited resources and poor infrastructures, and extremely difficult with the poorest of the poor. Some of the causes of low productivity in livestock of the resource-poor have yet to be fully elucidated. Many of the constraints may require education, information and extension rather than technology.

- Treatment costs: Assuming that an appropriate veterinary technology is available, livestock-keepers, communities, and public services may incur direct financial and time costs responding to animal disease by seeking or providing treatment.

- Loss of farm productivity: Through their effects on performance, diseases of livestock have additional indirect impacts on other agricultural enterprises, in particular crops. This is through lowered traction capacity for ploughing, the effect of reductions in manure output on soil fertility and nutrient cycling, decreased capacity for harvesting and marketing of crops by animal traction, as well as the reduction in general transport, including transport of essential water supplies. All of these changes can severely affect livelihoods of smallholder farmers and the communities they support. This impact is often highly under-estimated, and has generally been inadequately quantified.

- Reduction or elimination of market opportunities: Outbreaks of infectious diseases in a community or a region may result in local market disruptions as movement restrictions are imposed, with farmers unable to market livestock and livestock products with optimal timing (such as moving fattener pigs to market) or at all (such as restricting milk collection, for example), or they may face dramatically depressed prices. The mere occurrence of certain diseases can also severely constrain cross-border and other international trade, and is most commonly associated with the highly infectious diseases such as foot and mouth disease, rinderpest, hog cholera, Newcastle disease, and the epidemic zoonoses such as Rift Valley fever. Restrictions on international trade typically affect primarily the large-scale commercial sector, with potential multiplier effects on employment and other auxiliary sectors.

- Disturbance of human health: Illness in people associated with zoonotic and food-borne diseases leads to losses in their productivity and quality of life, as well as costs incurred for treatment. Productivity losses in people are more difficult to quantify than in livestock, where there are more readily measurable indicators such as production of meat and milk. Currently the unit of the disability-adjusted life year (DALY) has been adopted as the standard measure of impact on humans used by WHO (1996). In the evaluation of human health research priorities conducted by WHO, many of the zoonotic diseases were ranked individually as relatively low on the scale of DALYs.

- Impairment of human welfare: Diseases of livestock have many additional direct and indirect impacts on human nutrition, community development and socio-cultural values (e.g. Curry *et al.*, 1996). Animal disease can significantly reduce farm income, contributing to food insecurity and poor nutrition.

The impacts of disease risk

Even if no disease occurs on a given farm or in a particular community, the mere threat of the disease occurring may induce significant impacts. The most obvious are economic losses from higher production costs or public expenditures incurred in attempting to prevent disease. These are typically related to prophylactic control strategies (vaccination, chemo-prophylaxis) and monitoring and surveillance programmes.

Less obvious, though, are the changes in behaviour or management in the face of disease risk that lead to sub-optimal production systems. At the extreme, disease risk may limit the use of susceptible species or high-productivity breeds. The low density of cattle in general, but especially of improved cattle, across the Sub-Saharan Africa tsetse belt, attributed primarily to the ever-present risk of trypanosomiasis, illustrates this impact. In the parts of its distribution in which infection challenge is particularly high, the disease actually prohibits the keeping of most livestock species, including the indigenous breeds well known for their hardiness. This represents economic losses from what is often referred to as 'lost potential' since farmers are discouraged from keeping cattle or trying dairy production, which otherwise might offer substantial financial rewards.

The impacts of disease control

Disease control efforts are undertaken to minimise the various impacts of diseases described above. In doing so, however, disease control may spawn yet further unintended impacts. The example often cited is that of potential environmental impact of effective control of trypanosomiasis. It has been estimated that herds in areas under trypanosomiasis risk are only 50 to 70 per cent the size of herds of similar areas with no risk (Swallow, 2000), and so cattle numbers would be expected to increase substantially with better disease control. If not properly managed, such growth in cattle populations may contribute to degradation of the natural resource base.

Poor implementation of disease control may also contribute to localised negative impacts. Improper use of chemicals and drugs, in particular, can expose animals, humans, and the immediate environment to possible toxic effects, either directly or through residues in livestock products. It can also lead to the emergence of resistance by parasites to control drugs, as has occurred for example with trypanocides, anthelmintics and acaricides.

The importance of disease risk to resource-poor people

The various types of impacts of animal disease outlined above are all likely to be proportionally greater for the poor. Focusing on *risk* is the key to understanding why this is true. The resource-poor are exposed to more animal disease risk and have less capacity to cope with that risk than the better-off, and this combination reduces yet further their chances of escaping poverty.

Their much higher exposure to risk

The resource-poor in the developing world face particularly high risk from animal disease. Firstly, there is more disease present. Much of the developing world lies within the tropical and subtropical regions of the world, where climates and ecosystems favour a wide range of parasitic infections and infestations, many of which do not occur in the temperate regions of the world. Unrestricted movement of animals for marketing, social and other reasons is widespread in and between many regions, and while it promotes market orientation for many resource-poor livestock-centred communities (such as in the Horn of Africa), it also can enhance the spread of certain diseases. Livestock production systems of the resource-poor further enhance the risk of disease through confounding factors such as inferior housing, multiple species, and inadequate nutrition.

Secondly, there is less disease control. Even if the appropriate technology exists, animal health services in developing countries, through financial, infrastructural, logistic and educational restrictions, often do not permit the optimal delivery and adoption of known disease control measures. Poor delivery is exacerbated by the fact that markets for animal health inputs in the developing world, such as vaccines and pharmaceuticals, are relatively small, given the low incomes of the majority of the populations, so financial incentives for technology development and application by international pharmaceutical industries are severely limited.

Production systems are evolving rapidly, with increasing human population growth and changes in the demands for livestock products, and many traditional disease-control strategies and policies are outdated and inappropriate. Examples include the need to consider how vaccines against endemic livestock diseases can best be delivered to the evolving peri-urban smallholder dairy sectors in many developing countries, and how rabies vaccines can be effectively delivered to an adequately high-proportion of stray dogs in high-risk urban and peri-urban communities.

Their much lower capacity to bear risk

While exposed to a wide array of risks related to animal disease, the resource-poor have yet less capacity to cope. Existing close to the survival threshold, the resource-poor tend to be more risk averse, and so less likely to 'take a chance' on preventive disease technologies. More importantly, low income and few assets mean that the poor have few options available for managing crises, are less resilient to shocks and are slower to recover. Livestock disease is particularly damaging since it threatens one of the few assets that the resource-poor keep for dealing with other shocks.

As the preceding discussion suggests, the impact of animal disease on the poor is complex, involving direct and indirect effects and multiple pathways operating at a variety of levels, depending on the particular disease or syndrome. The livelihoods approach (DFID, 2000; Figure 13.2) offers a ready framework for handling these various dimensions. Consider first the framework applied to the resource-poor livestock-keeping household.

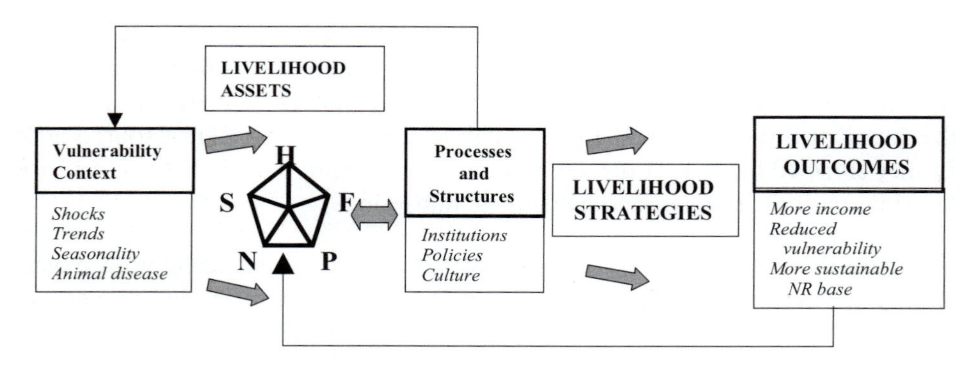

Figure 13.2 Sustainable livelihoods framework (H=Human captial; F=Financial capital; N=Natural capital; S=Social capital; P=Physical capital). Source: DFID (2000).

The vulnerability context (left box in Figure 13.2) represents the environment in which the resource-poor live, particularly as it translates into the various types of risk they face. The resource-poor face direct risks from livestock disease, and also indirect risks through a number of other different sources.

Livelihood assets

Within the livelihoods framework, the impact of animal disease can be described by the various ways it affects the poor household's asset base represented by the pentagon. Animal disease can threaten each of the five types of household assets.

- Financial capital: Livestock mortality and morbidity can directly reduce both income flows from livestock activities by cutting output and, more importantly, the financial investment value of the livestock assets themselves. It is also likely to raise production costs if control costs are incurred or production efficiency is lowered. Income from crop production or transport activities dependent on animal traction may also be affected. Even a small reduction in income flow is likely to have immediate impact on consumption for basic needs.

- Human capital: Zoonoses and food-borne diseases can temporarily or permanently impair an individual's ability to work, and thus deprive the resource-poor household of its principal income-generating asset. Other animal diseases may also indirectly affect the health of household members by reducing the supply and consumption of livestock products produced by the household (milk, eggs, meat), or through their effect on available income for food purchases and medical care. Chronic animal diseases have potential longer-term nutritional impacts, which for young children may jeopardise the quality of their future human capital.

- Social capital: In many societies, livestock serve as a mechanism for establishing relationships of trust within social networks. This includes gifts and loans of animals and their products for a variety of social functions, including dowry. These functions may be particularly crucial for the resource-poor to ensure an informal safety net in times of crisis, through the development of trusting relationships with others in the community. Disease lowers the number and quality of animals available for this purpose.

- Natural capital: In mixed crop-livestock systems, manure often plays a critical role in maintaining soil fertility, especially for the resource-poor who are less likely to be able to invest in chemical fertilisers. Disease may reduce the availability of manure.

- Physical capital: Livestock can be considered production assets as farm 'tools', and disease lowers their productive quality or even wipes them out. The use of larger stock for ploughing or transport is particularly important to the poor. Disease in the draught animal at critical periods during the crop year can reduce the area the household can successfully cultivate. Often forgotten, many resource-poor households depend on animal transport for their water supplies, and animal disease may, therefore, have indirect sanitary implications in terms of the quantity and quality of water supplies used by the household.

Transforming structures and processes

Poor households devise their livelihood strategies depending on their asset base and the risks they face, but conditioned by the structures (public, civil and private sector) and processes (policies, legislation, institutions, culture) they operate under.

For animal diseases, these structures and processes refer to the delivery of animal health services. The general failure of animal health services to reach resource-poor livestock-keepers is commonly recognised as perpetuating the particular vulnerability of the resource-poor in the developing world to animal disease and its impacts.

Livelihood strategies and outcomes

Animal diseases reduce the already limited asset base of the resource-poor livestock-keeping household, and currently existing structures and processes offer little assistance in helping the household to respond effectively and contain the often multiple impacts of disease. The result is a livelihood strategy that must accommodate lower than expected productivity from the household's livestock, and often rules out, due to the risk-averse nature of resource-poor households, adopting better management or more productive livestock activities. The outcome is continued low levels of income, asset accumulation, and investment, and thus poverty is perpetuated.

The resource-poor who do not keep livestock

The same type of analysis can be applied to resource-poor households that do not necessarily keep livestock. First, for the resource-poor who earn wages from working as hired labour or traders in livestock production or marketing enterprises, animal disease can put at risk one of their important sources of income (financial capital). Second, most poor, rural and urban, are consumers of animal products, and often can only afford low-quality products sold in informal, uncontrolled markets. They therefore face a higher chance of contracting zoonotic and food-borne diseases, putting at risk their key human capital (illness) and financial capital (wage losses and medical expenditures) assets. Poor consumers may also be affected by epidemic animal diseases when outbreaks disrupt markets, create product shortages and raise prices.

A novel typology of disease impacts on the resource-poor

By looking at the impacts of animal disease through a poverty lens with the help of the livelihoods approach, a new way of grouping the impacts of diseases begins to emerge that is much more functional in terms of supporting poverty-reducing policies (Perry *et al.,* 2002; 2003). Three general categories are proposed. The boundaries between the categories are certainly not distinct, and there is an inevitable degree of overlap. Nonetheless, the three categories provide a useful framework for organising appropriate research and development efforts.

Diseases that exacerbate asset insecurity

The first set of diseases includes those that threaten and degrade the asset base of the resource-poor household under current conditions of use of livestock within the household. Whether the household keeps livestock for consumption or to market, earns wages from livestock activities, or simply consumes livestock products, the focus here is on the impact of animal diseases in eroding the household's assets through the various pathways discussed above. These include many of the endemic diseases and production syndromes, as well as the common zoonoses. Through the continued high exposure to the wide array of risks associated with animal disease, and the lack of access to appropriate and effective means to manage those risks, resource-poor households are forced to adopt risk-averse livelihood strategies that do not allow them to accumulate assets or invest in better technologies. These types of animal diseases help to keep the poor trapped in poverty.

Diseases that limit market opportunities

The second set of diseases refers to those that restrict the resource-poor from exploiting market opportunities for their livestock and livestock products. Resource-poor livestock-keepers generally have open access to local markets. Their access to markets has in part been due to the lack of, or lax application of, animal sanitary controls. Where sanitary controls are applied, a parallel informal market usually exists in which the resource-poor can sell their livestock goods that do not meet standards, but at lower prices.

Market opportunities are changing rapidly for the resource-poor. First, local demand for livestock products is expected to increase dramatically in developing countries as income levels improve in what has been termed the Livestock Revolution (Delgado *et al.*, 1999). Most of these large increases in demand are expected to be satisfied to a large extent through expansion of intensive commercial production systems. With appropriate policies, the response to increased demand could also be harnessed as a mechanism for reducing poverty. This will require paying particular attention to enhancing the role of the resource-poor, including smallholder livestock-keepers, casual labourers, and petty traders.

Globalisation is a reality that will eventually revolutionise livestock markets in the developing world. For now, its impacts are being felt mainly in the large-scale export-orientated commercial sector in those countries that satisfy the particular health requirements of their trade partners. As globalisation gathers momentum, developing countries will be under increasing pressure to adopt a certain minimum standard of sanitary control even within local markets, if they are to continue participating in regional and international trade. If the resource-poor are to avoid being further marginalised, they will need access to better monitoring and control of the diseases that restrict trade. This may require adapting monitoring and trace-back systems from the developed world, making them appropriate to the context of rural markets in the developing world, and will probably require an innovative mix of private and public action.

From this perspective, many of the zoonoses, the epidemic and the food-borne diseases can be seen to limit, now and increasingly so in the future, access to markets for livestock products from the poor. This works both to reduce their ability to reap full income value from their livestock activities by restricting them to informal markets and their lower prices, and to exclude them from participating in new market opportunities which develop under globalisation.

Diseases that limit livestock-based intensification of farming systems

The first two categories have concentrated on the current livestock production activities undertaken by the poor, regardless of their relative importance within the household economy, even if it refers to the scavenging chicken kept in the backyard. The third category of diseases and their impacts turns the focus on to those livestock activities that would require a specific effort and investment by the resource-poor, because they involve upgrading an existing activity through a more productive management technique or adopting a wholly new, more productive, livestock activity. Increasing productivity is the classic pathway for intensification of farming systems by which households increase the value of output for their inputs, and is thus key to escaping poverty. Moving up the 'livestock ladder' is a common form of intensification. But, as emphasised above, the resource-poor tend to be risk-averse, and so are reluctant to invest in a new activity that may increase their vulnerability and threaten their already-constrained asset-base. The possibility of livestock disease would obviously be an important consideration. Some diseases have had a major impact by discouraging certain livestock activities. A well-known example is that of trypanosomiasis, which has been responsible for the under-utilisation of livestock across the tsetse belt of Sub-Saharan Africa. Similarly, the low adoption of improved dairy-grade cattle in the Great Lakes area of Central Africa has been largely attributed to the continuing threat of East Coast fever.

What can be done to improve the health of livestock kept by resource-poor people?

A clear understanding of how diseases impact the resource-poor is critical in designing programmes to reduce these effects and promote pathways out of poverty through livestock. From this understanding, a series of issues emerges that can help in guiding policies and strategies targeted at the poor, and which will contribute to a better understanding of the overall delivery systems tailored for the needs of the poor. Figure 13.3 illustrates a generic delivery system and its components.

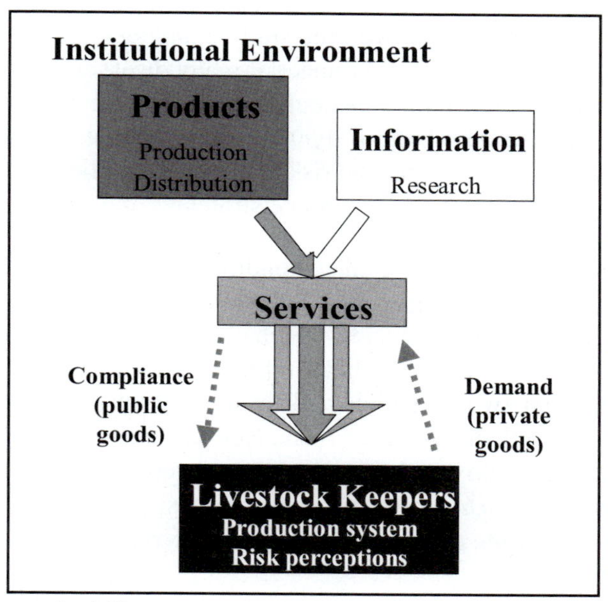

Figure 13.3 A generic delivery system

With respect to the poor, the issues that need to be considered within this generic delivery system can be summarised in the following three headings:

- Better understanding and disease information of the priority diseases of greatest direct importance to the resource-poor
- Better technologies that are more appropriate for use by the resource-poor and their service providers
- Better services, institutions and policies that consider the resource-poor as a specific entity

We will review each of these.

Better understanding and information of diseases of greatest importance to resource-poor people

There have been few studies to determine the specific disease entities that impact especially on the poor, so we draw heavily on the recent study focussing on priority animal health research opportunities for poverty reduction (Perry *et al.*, 2002). The results of this study identified a total of 76 syndromes, general diseases and specific disease entities as having impact on the poor. These included all the categories discussed in the previous section (endemic, epidemic, zoonotic and food-borne). Whereas some diseases were reported from all regions, others had more limited distributions. However, the study concluded that there is a lack of basic data on the epidemiology and impact of many diseases and syndromes that are important to resource-poor livestock-keepers. There is a need for better information on disease distribution, dynamics and impact, which would allow more effective disease forecasting and risk factor identification, and which in turn can support the development of appropriate strategies and policies.

On a global basis, the 20 highest ranked conditions with impact on the resource-poor comprised three syndromes (neo-natal mortality, reproductive disorders and nutritional/micronutrient deficiencies, which all rank in the top ten), four general disease categories (gastro-intestinal helminths, ectoparasites, respiratory complex and mastitis, the first two of which rank in the top ten), and thirteen specific diseases (foot and mouth disease, liver fluke, Newcastle disease, anthrax, *Toxocara vitulorum* infection, followed by haemorrhagic septicaemia, peste des petits ruminants (PPR), *B. abortus* infection, haemonchosis, African trypanosomiasis, coccidiosis, *Trypanosoma evansi* infection and rinderpest); see Table 13.2.

On a global basis, gastro-intestinal parasitism emerges with the highest global index as an animal health constraint to the poor. This is a reflection of the wide geographical distribution of gastro-intestinal parasitism, the wide host species range, and the importance given to its high economic impact at the resource-poor farmer level in all production systems, and particularly in camels, sheep and goats, and poultry.

The other general disease category ranked highly is ectoparasites, and this includes a range of parasites affecting cattle, pigs, sheep, goats and poultry, reported from all regions of the study. As with gastro-intestinal parasitism, there is extremely little documented evidence and quantification of their impacts on these species, but they are visible to resource-poor livestock-keepers, they are considered vermin by both the resource-poor and their veterinary advisors, they are in abundance both in terms of species and absolute numbers in resource-poor households with livestock, and as such are considered of significant impact.

The presence of the three syndromes of neonatal mortality, reproductive disorders and nutritional/micronutrient deficiencies in the top ten reflects the general recognition of production inefficiencies compounded by nutritional inadequacy across all of the species as being among the most important health impacts on the livestock of the poor.

Of the five specific diseases in the top ten, some were predictable, and some were less so. Among the more predictable is Newcastle disease, prevalent in all regions, and always identified as the major disease of village poultry. The high ranking of poultry to the resource-poor appeared to outweigh the poultry-specific characteristics of the disease.

In the zoonotic diseases, infection caused by *B. abortus* in cattle and buffalo ranked highest overall by a considerable margin, followed by the related *B. melitensis* in sheep and goats. With the exception of bovine tuberculosis, infection caused by *B. abortus*, and anthrax, there appeared to be significant regional differences in the priority rankings. Thus sleeping sickness ranked as third globally, but is only present in Africa, and even there only in limited areas, reflecting distribution of the tsetse fly.

Better technologies for use by the resource-poor and their service providers

Certain characteristics of a control technology can constrain its delivery to and adoption by, the very poor. The classic example is that of veterinary products that require a cold chain, requiring refrigeration at low temperatures to remain viable. Lack of infrastructure, especially in areas where livestock-keepers are widely dispersed or mobile, has made delivery of such products problematic in much of Africa. One way to overcome these constraints is to adapt veterinary products accordingly. An important theme has been the development of thermostable products, particularly vaccines such as those against rinderpest and Newcastle disease. Another has been simplifying techniques of administration to make control more assessable or improve coverage, such as the use of a simple eyedropper to administer thermostable Newcastle disease vaccines to village poultry.

There are many other issues regarding the appropriateness of technologies targeted at the poor. This includes the availability of therapeutic products in pack sizes

Table 13.2 Top 20 ranked diseases globally or regionally, according to impact on the resource-poor

Rank group	*Disease (Alphabetical within each rank group)*	*Global Index*	*Contribution to Global Index*[b]		*Species role in Global Index*[c]		*Regional*[d] *ranking*				*Production system*[e] *ranking*			*Species ranking*				
			Econ	*Natl*	*Import to Poor*	*Species*	*WA*	*ECSA*	*SEA*	*SA*	*Past-oral*	*Agro Past*	*Peri-urban*	*Cattle*	*Sheep Goats*	*Buff-alo*	*Pig*	*Poul-try*
		index[a]	*percentage*		*index*	*no.*	*Ranking Group*[f]											
	Anthrax	46	49.5	50.5	78	4	A	D	B	B	E	A	B	A	A	A	-	-
	Ecto-parasites	63	99.8	0.2	53	4	A	A	A	C	A	A	A	C	A	-	A	A
	Foot and mouth disease	64	68.9	31.1	64	5	B	C	A	A	C	A	A	A	B	A	A	-
	Gastro-intestinal helminths	100	99.6	0.4	49	6	A	A	A	B	A	A	A	B	A	-	A	A
A	Liver fluke (fascioliasis)	51	99.1	0.9	73	4	A	B	B	A	D	A	B	A	A	A	-	-
	Neonatal - mortality	79	100.0	0.0	32	6	-	A	-	A	A	A	A	B	A	-	A	A
	Newcastle disease	58	93.7	6.3	51	1	B	A	A	B	B	A	C	-		-	-	A
	Nutrition micronutrients	46	100.0	0.0	67	4	-	A	B	A	A	A	B	A		B	-	A
	Reproductive disorders	50	100.0	0.0	82	2	-	B	-	A	B	A	A	A		A	-	-
	Toxocora vitulorum	45	100.0	0.0	65	2	-	-	A	A	B	A	A	A		A	-	-
	Brucella abortus	38	66.8	33.2	79	3	C	D	-	A	A	B	A	A		A	-	-
	Coccidiosis	30	100.0	0.0	20	1	D	B	B	A	E	B	C	-		-	-	A
	Haemonchosis	42	100.0	0.0	57	2	C	A	-	B	C	B	B	-	A	-	-	-
B	Haemorrhagic septicaemia	40	78.5	21.5	70	3	B	D	A	A	B	B	A	A		A	-	-
	Mastitis	23	79.2	20.8	84	2	-	C	-	B	E	B	A	B		B	-	-

Table 13.2 (Contd.)

Rank group	*Disease (Alphabetical within each rank group)*	*Global Index*	*Contribution to Global Index*[b]		*Species role in Global Index*[c]		*Regional*[d] *ranking*				*Production system*[e] *ranking*			*Species ranking*				
			Econ	*Natl*	*Import to Poor*	*Species*	*WA*	*ECSA*	*SEA*	*SA*	*Past -oral*	*Agro Past*	*Peri-urban*	*Cattle*	*Sheep Goats*	*Buff -alo*	*Pig*	*Poul-try*
	Peste des petits ruminants (PPR)	40	90.0	10.0	47	1	B	D	-	A	C	B	C	-	A	-	-	-
	Respiratory complexes	35	99.0	1.0	82	4	A	A	-	C	A	B	B	-	A	A	-	-
B	Rinderpest	21	6.4	93.6	79	2	C	D	-	B	C	B	D	B		A	-	-
	Trypanosoma evansi	21	99.7	0.3	77	3	C	C	B	B	A	B	D	C		A	-	-
	Trypanosomiasis (tsetse)	32	96.8	3.2	69	4	A	B	-	-	A	B	B	A	B	-	A	-
	Babesiosis	14	91.8	8.2	80	1	B	B	-	C	C	C	E	B		-	-	-
	Contagious bovine pleuro-pneumonia (CBPP)	19	77.3	22.7	72	1	A	B	-	-	A	C	D	A		-	-	-
C	Contagious caprine pleuro pneumonia (CCPP)	12	100.0	0.0	70	1	B	C	-	D	C	C	D	-	B	-	-	-
	Diarrhoeal diseases	19	100.0	0.0	85	3	-	-	-	B	B	C	C	B		B	-	-
	Foot problems	12	91.5	8.5	88	1	B	B	-	-	C	C	D	-	B	-	-	-
	Heartwater	19	95.6	4.4	78	2	A	B	-	-	-	C	D	C	A	-	-	-
	Infectious coryza	12	100.0	0.0	56	1	-	A	-	-	B	C	C	-		-	-	-
	Rift Valley fever	17	2.2	97.8	82	3	B	A	-	-	A	C	C	C	B	-	-	-
	Sheep & goat pox	17	85.7	14.3	58	1	B	-	-	C	C	C	C	-	A	-	-	-
	Theileria annulata	13	100.0	0.0	91	1		-	-	B	-	C	E	B		-	-	-

Table 13.2 (Contd.)

Rank group	Disease (Alphabetical within each rank group)	Global Index	Contribution to Global Index[b]		Species role in Global Index[c]		Regional[d] ranking				Production system[e] ranking			Species ranking				
			Econ	Natl	Import to Poor	Species	WA	ECSA	SEA	SA	Past-oral	Agro Past	Peri-urban	Cattle	Sheep Goats	Buff-alo	Pig	Poul-try
	Anaplasmosis	10	98.0	2.0	73	1	B	E	-	C	C	D	B	C	-	-	-	A
	Black leg	11	87.1	12.9	61	2	A	E	B	D	-	D	E	B	-	B	-	-
	Dermatophilosis	11	88.7	11.3	62	1	A	C	-	-	D	D	E	B	-	-	-	-
	Duck virus enteritis (DVE)	8	89.5	10.5	73	1	-	-	A	E	B	E	-	-	-	-	-	A
D	East Coast fever	9	88.5	11.5	97	1	-	A	-	-	-	D	A	C	-	-	-	-
	Fowl cholera	11	98.7	1.3	59	1	-	-	A	D	-	D	B	-	-	-	-	A
	Fowl pox	11	91.2	8.8	40	1	-	B	A	C	B	D	C	-		-	-	A
	Hog cholera	9	98.1	1.9	24	1	-	-	A	C	D	D	B	-	-	-	A	A
	Infectious bovine rhino tracheitis (IBR)	10	100.0	0.0	89	1	-	-	-	B	B	D	E	B	-	-	-	-
E	Tick infestation	6	77.0	23.0	92	2	-	B	-	-	-	E	B	C	-	-	-	-
	Orf	3	100.0	0.0	39	1	-	-	B	D	-	E	E	-	B	-	-	-
G	*Brucella suis*	1	100.0	0.0	23	1	-	-	B	E	-	G	E	-	-	-	A	-

Table 13.2 (Contd.)

Notes:

[a]Disease impact scores (weighted by region/production system-specific relative importance of the affected species, the number of resource-poor, and the severity of poverty in the specific) were normalised to an index of 0 to 100 with 100 representing the highest impact.

[b]Portion of disease impact score (weighted aggregate across all region/production systems) contributed by each of the two main components: economic impact (Econ: incidence and herd-level productivity impact) and national livestock sector impacts (Natl: trade impacts and public expenditures).

[c]Region/production system-specific disease impact scores are adjusted to reflect the degree to which the disease affects species important to the resource-poor. Import to Poor (importance to the resource-poor) is the global weighted average importance of the species affected by the disease, where 1=all species affected by the disease are considered to be the most important to the resource-poor in the relevant region/production systems and 0=all species affected by the disease are considered unimportant to the resource-poor in those region/production systems. Species indicates the number of different species affected by the disease. The nine categories of species include: cattle, buffalo, yak, camels, horses, donkeys, pigs, shoats (combines both sheep and goats), poultry.

[d]Regions: WA=West Africa; ECSA=Eastern, Central, and Southern Africa; SEA=South-East Asia; SA=South Asia.

[e]Pastoral systems correspond to the rangeland-based systems (LGA, LGH, and LGT; which may include large-scale commercial beef production); agro-pastoral systems to the mixed crop-livestock systems (MRA, MRH, MRT, MIA, MIH, MIT), peri-urban systems to the landless farming systems (LL) (see Seré and Steinfeld, 1996).

[f]Rank groups: A=top 10 ranked diseases; B= second ten ranked (11th-20th); C= third ten ranked (21st-30th); D=fourth ten ranked (31st-40th); E= fifth ten ranked (41st-50th); F=sixth ten ranked (51st-60th); G= seventh ten ranked (61st-70th).

Source: Perry *et al.* (2002).

appropriate to the resource-poor and with clear labelling and instructions in local languages, the question of who should have access to the different classes of drugs, and the standardised registration of drugs within a region to avoid repeating the process under different sets of rules in each country. Very importantly, there is little incentive for pharmaceutical companies to develop new products to treat diseases that primarily affect livestock of the poor, and this is an issue that needs to be addressed.

Veterinary products can also be improved in terms of their economic characteristics, especially their price. Affordability can be enhanced by more efficient production and distribution through either technical or institutional improvements that lower cost. For products that provide significant benefits beyond the farm where they are used, public subsidies can directly reduce the nominal cost borne by end-users and thereby encourage uptake. Increasing the effectiveness, safety, or ease of application of veterinary products also has the effect of reducing economic risk by minimising productivity losses and wasted investment when the products do not work to satisfaction.

Photo 13.1 Intra-ocular vaccination against Newcastle disease of poultry being carried out by community vaccinators in Mozambique. (courtesy of B. Perry)

Services, institutions and policies needed for effective delivery of animal health inputs to the resource-poor

Policies and institutions modelled around animal health services provision standards in industrialised countries have rarely served the needs of resource-poor livestock-keepers to help them secure their livestock assets or enhance market access. Because of the small size and complex household goals of smallholder systems, disease control programmes designed for larger-scale livestock systems in industrialised countries are not directly transferable. Further, the lack of public services and infrastructure is an indirect constraint limiting the growth of health input markets. There is clearly a need for better economic information on the incentives for different stakeholders to participate or comply at local, national and regional levels.

The need to re-appraise modalities for the provision of livestock services has been apparent since the early 1980s when publicly provided services began to decline, and the calls for an increasing role for the private sector started in earnest. Considerable confusion resulted, due to substantial reductions in public budgets without comprehensive strategies as to how the private sector might fill the gap. The economic and institutional issues surrounding the delivery of animal health services were extensively analysed and discussed in a variety of settings, while many resource-poor livestock-keepers still went without services. The analyses showed that private inputs services have functioned only in limited regions where market-orientated producers can afford the costs of targeted quality services (e.g. the smallholder dairy producers in the highlands of East Africa).

A useful framework for classifying the full range of livestock services financing and provision options was developed in a new institutional economics setting (see for example North, 1995) and this is presented in Figure 13.4.

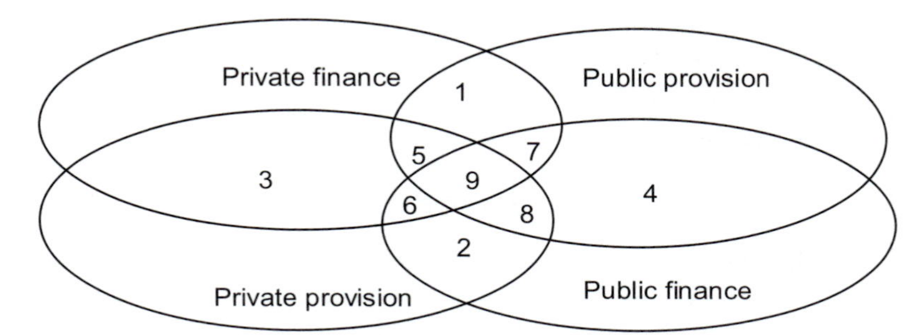

1. Private finance and public provision
2. Public finance and private provision (out-sourcing)
3. Private finance and private provision
4. Public finance and public provision
5. Private finance and a mix of private and public provision
6. A mix of both public and private finance but with private provision
7. A mix of both public and private finance but with public provision
8. Public finance and a mix of public and private provision
9. Both public and private finance and with both public and private provision

Figure 13.4 A conceptual framework for the financing and provision of livestock services

In countries where privatisation has taken place successfully, the process has broadly taken one of two forms: 'out-sourcing', i.e. contracting of private services by the public sector (2 in Figure 13.4) or privately owned services (3 in Figure 13.4). The former is used to deliver services that are either regarded as public goods (e.g. the control of epidemics and certain endemics such as vector-borne diseases), or where the farmer-perceived value of the service is below the cost, while the latter is used to deliver private goods such as clinical services. The basis for viability of fully privatised services is farmer-perception of value, which is generally driven by market-orientation of the production system. The level of farmers' market orientation underlies not only this willingness to pay for services, but also their ability to pay due to greater cash flow. This explains why such services have succeeded in peri-urban production systems but have failed in more remote and less market-orientated systems, where the costs of delivery are often exhorbitant. Though declining in its impact due to poor funding, many developing countries depend on public finance and provision of services that should ideally be privatised, such as extension (4 in Figure 13.4). Regulatory services, such as quality assurance and certification for livestock products destined for export markets, must continue to depend on public finance but could be publicly or privately provided (2 or 4 in Figure 13.4).

We briefly discuss the institutional challenges facing the delivery of animal health inputs to resource-poor livestock-keepers under two main headings: (i) balancing public and private roles; and (ii) striking a balance in regulatory standards versus benefits to poor people.

The challenge of balancing public versus private roles in inputs supply

The substantial reductions in public expenditure on livestock services in the 1980s were especially severe in Sub-Saharan Africa, where the poorest populations live. These reductions led to a drastic decline in veterinary, artificial insemination and extension advisory services. Since then, the need for institutional innovation in response to the reduced public spending has led to considerations of alternative approaches through the private sector, including strengthening of collective action (community-based organisations and co-operatives) and exploration of viability of provision of health services by semi-independent para-veterinarians and community-based animal health-care workers (CBAHWs) (cf. Chapter 5). While community action is appealing, there are often difficulties in mobilising community support, particularly when there are opportunities for individual livestock owners to benefit 'free of charge' from the disease control activities of others. Key public roles that are critical to successful privatisation include an investment in infrastructure (such as rural access roads, water supplies and electricity distribution) and the facilitation of institutional arrangements that improve the efficiency of markets, and provide incentives for market participants to invest in service delivery.

Blanket privatisation is considered to exclude vulnerable groups from accessing services, especially in areas where the evolution of the private sector to fill the gaps in service delivery is slow. Therefore, public interventions will continue to be required to fill the gaps in the short to medium term and stimulate private sector growth to take over in these areas. Unless this happens, regions experiencing economic growth will benefit from private initiatives in service delivery, while areas characterised by slow economic growth or stagnation, typically inhabited by the poor and disadvantaged and reliant on the public-sector delivery pathway, will continue to suffer from ineffective service provision. The latter scenario mainly arises from high transaction costs, which are a key barrier to effective private services and market access. The priority for the shift from public to private delivery of input services should be a managed transition, not the often-observed sudden withdrawal of operating funds to government agencies, to the detriment of the livelihoods of many poor households. Costs that could be privatised but will need to continue to be internalised for much longer are those for research, extension, information services, training and the delivery of artificial insemination services, which all have elements of public goods. The initiative in Uganda, known as National Agricultural Advisory Services (NAADS, 2000), is one potentially useful approach that includes the implicit gradual scaling-down of donor/government subsidies.

Striking a balance between regulatory standards for livestock services versus benefits for resource-poor people

Standards and quality assurance for animal health services modelled along those in industrialised countries have largely failed in many production systems in developing countries. For example, the requirement by many national authorities that only professional veterinarians can be licensed to provide animal health services has clearly rendered such services unavailable to many resource-poor and marginalised livestock-keepers, particularly pastoralists. The low off-take from such extensive systems and their limited market orientation imply that only low-cost inputs can be sustainable in the system. It is therefore clear that optimal health services should be complemented or replaced by alternative 'local equivalents' that, though provided by lesser-trained health technicians or CBAHWs, meet the need to access basic animal health-care instead of letting the livestock-keepers continue to suffer easily-preventable losses. These needs are mainly preventive medicine and simple curative measures for endemic diseases that can largely be delivered by trained para-professionals.

Muriuki *et al.* (2003), describing the Kenyan dairy sub-sector, consider that the promotion of self-regulation among service provider associations at all levels might result in better quality services than are presently available, given the failure of top-down regulation and the wide gap between policy and implementation. In countries where attempts have been made to revise laws to accommodate lesser-qualified service providers, resistance from professional veterinarians with vested interests in the *status quo* has been spontaneous. The challenge is to convince all cadres of livestock health service providers and policy makers, at both national and international levels, that trade-offs in quality of services and economic benefits are inevitable, and required. Policies that support these activities are likely to sustain competitiveness of producers and contribute effectively to improved rural livelihoods.

Further reading

Acha, P.N. and Szyfres, B. 2003. *Zoonoses and communicable diseases common to animals and man*. Three volumes. Pan American Health Organisation, Washington D.C., USA

Alders, R.G. 2001. Sustainable control of Newcastle disease in rural areas. In *SADC planning workshop on Newcastle disease control in village chickens* (ed. R.G. Alders and P.B. Spradbrow), pp. 80-90, Proceedings of an International Workshop, Maputo, Mozambique, 6-9 March 2000. ACIAR Proceedings, No. 103.

Catley, A. 1999. *Methods on the move: a review of veterinary uses of participatory approaches and methods focusing on experiences in dryland Africa*. International Institute for Environment and Development, London, UK.

Catley, A., Blakeway, S. and Leyland, T. (ed.). 2002. *Community-based animal healthcare: A practical guide to improving primary veterinary services*. ITDG Publishing, London, UK.

Coetzer, J.A.W. and Tustin, R.C. (ed.). 2004. *Infectious diseases of livestock*. Three volumes. Oxford University Press, Cape Town, South Africa.

De Haan, C. (ed.). 2004. *Veterinary institutions in the developing world: current status and future needs*. World Organisation for Animal Health Scientific and Technical Review 23: 414 pp.

Hansen, J.W. and Perry, B.D. 1994. *The epidemiology, diagnosis and control of helminth parasites of ruminants: A handbook*. International Laboratory for Research on Animal Diseases, Nairobi, Kenya.

IDL Group. 2003. *Community animal health workers: Threat or opportunity*. IDL, Crewkerne, UK.

Leonard, D.K. 1993. *Structural reform of the veterinary profession in Africa and the new institutional economics*. Development and Change 24: 227-267.

Leonard, D.K. (ed.). 1999. *Africa's changing markets for health and veterinary services: new institutional issues*. Palgrave, Basingstoke, UK.

McCorkle, C.M., Mathias, E. and Schillhorn van Veen, T.W. (ed.). 1996. *Ethnoveterinary research and development*. Intermediate Technology Publications, London, UK.

McClean, M. and Ramsey, M. 2001. *A basic course for village animal health workers in Cambodia*. International Fund for Agricultural Development (IFAD), Rome, Italy.

Murrell, K.D. 2003. *International Action Planning Workshop on Taenia solium cysticercosis/taeniosis with special focus on eastern and southern Africa*. *Acta Tropica* 87:1-189.

Norval, R.A.I., Perry, B.D. and Young, A.S. 1992. *The epidemiology of Theileriosis in Africa*. Academic Press, London, UK.

Perry, B.D. (ed). 1999. *The economics of animal disease control.* World Organisation for Animal Health Scientific and Technical Review, Special Edition 18, (2).

Putt, S.N.H., Shaw, A.P.M., Woods, A.J., Tyler, L. and James, A.D. 1987. *Veterinary epidemiology and economics in Africa,* International Livestock Centre for Africa, Addis Ababa, Ethiopia.

Singh, G. and Prabhakar, S. 2002. *Taenia solium cysticercosis: from basic to clinical science.* CABI Publishing, Wallingford, UK.

Author addresses

Brian Perry, International Livestock Research Institute (ILRI), P.O. Box 30709, Nairobi 00100, Kenya

Thomas Randolph, International Livestock Research Institute (ILRI), P.O. Box 30709, Nairobi 00100, Kenya

Amos Omore, Kenya Agricultural Research Institute (KARI), P.O. Box 57811, Nairobi 00100, Kenya

Oswin Perera, Formerly of International Atomic Energy Agency, PO Box 100, Wagramerstrasse 5, A-1400, Vienna, Austria. Current address: 40/9 Deveni Rajasinghe Mawatha, Kandy, Sri Lanka

Adriano Vatta, Onderstepoort Veterinary Institute, Private Bag X05, Onderstepoort 0110, South Africa

14

Apiculture

Nicola Bradbear[*], Uma Partap[‡], Kwame Aidoo[‡], Alice Kangave[‡]

Cameo 1 - Brazil: Honey production in Mato Grosso

Indigenous communities in Mato Grosso in the Xingu region are harvesting and processing honey that is now being sold outside the State. The communities currently market 1,500 kg of honey per month and production is increasing. In July 2003, a shipment of honey was sent to three São Paulo supermarkets. The producers are in negotiation with the Pão de Açúcar supermarket chain (with shops in twelve states of Brazil); a deal could open the door to the international market.

This honey has strong commercial appeal as it is produced by Indians and has organic certification from the Biodynamic Institute. The Certificate is awarded only to products from sustainable practices that do not harm the environment. The honey is the first indigenous product to receive a Federal Inspection Seal from the Ministry of Agriculture, which means it is produced in accordance with health and safety legislation. The Seal authorises sale of the honey in other states.

Anonymous (2003)

Cameo 2 - Nigeria: bee-keeping in Kaduna State

Zaria people, who make up today's Kaduna State, were probably the originators of bee-keeping in Nigeria. In Northern Kaduna State traditional straw hives are hung in trees, whilst in southern areas basket, clay and pot hives are used. The honey-bee *Apis mellifera adansonii* is indigenous.

---

[*] Principal author
[‡] Consulting author

Current bee-keeping training includes construction of top-bar and frame hives, baiting and capturing swarms, dividing colonies and transfer of wild colonies, the use of honey extractors and presses, and the use of protective clothing and smokers. A survey in 2000 revealed that the majority of bee-keepers in Kaduna State are men under 40 years old. Women play an active role in the processing and marketing of honey. All respondents are engaged in other farming enterprises and/or occupations in addition to bee-keeping and own an average of eight hives. Bee-keepers using top-bar or frame hives harvest 12 kg of honey per hive annually. Annual average beeswax production per hive is 1 kg. In the survey, honey sold for N 300 (US$ 3.00) per kg and beeswax for N 65 (US$ 0.65) per kg. The total annual value of bee products per hive was N 3,631 (US$ 36.31). All products are marketed locally.

Fadare (2003)

Introduction

Although many of the 30,000 known species of bees collect nectar that they convert to honey and store as a food source, it is only the large-sized permanent colonies formed by social species of bees that store appreciable quantities of honey. These bee species which collect significant volumes of honey are of the genus *Apis,* known as honey-bees. Other bees exploited by humans are the stingless bees, belonging to the genera *Trigona* and *Melipona.* Species of honey-bees and stingless bees have been exploited by man for thousands of years; until recent centuries and the advent of sugar, honey from bees was the most widely available sweetener. Other bee species are reared specifically for crop pollination; these include bumblebees (*Bombus* spp.), leafcutter (*Megachile* spp.) and other species.

The honey-bees most widely used for bee-keeping are European races of *Apis mellifera,* a species of honey-bee also indigenous to Africa and the Middle East. Honey-bees do not occur naturally in the Americas, Australia, New Zealand or the Pacific islands, and European honey-bees have been introduced to these regions during the last four centuries. Over the last 30 years, European honey-bees have been introduced to most countries of Asia. Bee-keeping technology in industrialised countries has been developed for use with European honey-bees, and most bee-keeping literature and research relate only to this bee.

Unlike the livestock species described in other chapters of this book, the bees kept by bee-keepers are 'wild' species, and in some areas (for example Europe and Africa) are indigenous species. Until recently it was true to say that any honey-bees kept inside a hive by a bee-keeper would be able to survive just as well living on their own in the wild. In recent years, humans have spread pests and predators around the world; this means that in some regions the indigenous populations of honey-bees have been killed and the only bees now surviving are those managed by bee-keepers. For example, in Europe honey-bee colonies can survive only when bee-keepers control levels of the parasitic mite *Varroa destructor*.

Honey hunting, the plundering of wild nests of honey-bees to obtain crops of honey and beeswax, is practised throughout the world wherever wild nesting honey-bee colonies are still abundant. 'Apiculture' covers this broad range of activities from the total plundering of wild bee nests for harvests of honey and beeswax, through to 'real' bee-keeping, i.e. the keeping and management of a colony of bees inside a beehive; for thousands of years it has been known that obtaining honey is made much easier and more convenient if bees are encouraged to nest inside a hive.

Tropical honey-bees

Compared with the European races of *Apis mellifera*, that have evolved in temperate climates with long, cold winters, all tropical races and species of honey-bees are more likely to abandon their nest or hive if disturbed, because they have a greater chance of survival. In some areas, tropical honey-bee colonies migrate seasonally. These are crucial factors making the management of tropical honey-bees different from the management of temperate zone honey-bees.

Africa

Apis mellifera honey-bees are indigenous to tropical Africa. Slightly smaller than the European races, and with different biology and behaviour, African honey-bees are more readily alerted to fly off the comb and to defend themselves. In most African countries, local bee-keeping methods are used, with log, bark, basket or clay hives placed in trees. The collection of honey from wild nests is carried out wherever sufficient natural resources remain.

Asia

At least eight honey-bee species, varying in biology and behaviour, occur naturally in Asia. Some of these bee species build nests consisting of single combs, in trees, bushes, or in cliffs, and a great variety of methods has been developed by human societies for their exploitation.

For example, the giant honey-bee, *Apis dorsata*, suspends its large combs (often 1 m in diameter) from tree branches and overhanging ledges on rocks and buildings. Humans obtain honey crops from this species by plundering their colonies. Throughout Asia, from Gurung tribesmen in the Himalayas, to mangrove-dwellers in the Sunderbans of Bangladesh, the rain-forest people in Malaysia, people living in the river deltas of southern Vietnam and, indeed, wherever the giant honey-bee is present, honey hunters have their own special customs for exploiting these bees.

Another species, *Apis cerana,* is known as the Asian hive bee because, like the European *Apis mellifera,* it can be kept and managed inside a hive. It is similar to *Apis mellifera* but smaller in size, both in terms of the individual bees and their colonies.

European *Apis mellifera* bees have been introduced to most of Asia, and may now be the predominant honey-bee species present in China, Japan and Thailand.

Pacific and Caribbean

Although indigenous stingless bees are present, no honey-bees occur naturally in these islands. *Apis mellifera* of European origin have been introduced to most of them and bee-keeping industries have developed using European-style bee-keeping methods. With the rapid spread of honey-bee diseases around the world, it is increasingly important that these islands endeavour to maintain stocks of disease-free bees. Caribbean bee-keepers must watch for Africanised bees that have already arrived in Trinidad.

The Americas

In the Americas, there are no indigenous honey-bees. Instead, their ecological niche was filled by many different species of stingless bees which were, and still are in some areas, exploited for their honey that is especially valued for its medicinal properties. Knowing nothing of these indigenous bees, European settlers long ago brought with them European bees, and the industry developed based on this bee. In 1956, some tropical African *Apis mellifera* queens were introduced into Brazil. These bees survived far more successfully in tropical Brazil than their European *Apis mellifera* predecessors, and quickly filled the niche. These 'Africanised' bees (dubbed 'killer bees' by the media) have spread through much of South and Central America, and are now in southern USA. In countries such as Brazil, bee-keepers have adopted new management methods and now make excellent livelihoods from these bees.

Bee-keeping

There are many ways to utilise honey-bees for their pollination services or to obtain bee products. The methods used should be determined by the types of bees available, and locally-existing skills and resources.

Bee-keeping is possible wherever there are flowering plants. Even areas that have become dry and unproductive through deforestation or inappropriate farming may support secondary vegetation that is highly valuable to bees. Common drought-tolerant plants that give good bee forage include such species as *Acacia, Azadirachta, Calliandra, Cassia, Combretum, Eucalyptus, Gleditsia, Julbernardia, Prosopis* and *Ziziphus*.

Bee-keeping does not take up valuable land; hives can be placed in trees, on wasteland, or on flat rooftops. This makes bee-keeping feasible for resource-poor smallholders and landless people. A good site should have a water source nearby, plenty of flowering plants and trees in the area, and shelter from wind and strong sunlight.

Getting started

Bees

A colony of hive-nesting bees (*Apis mellifera*, or *Apis cerana* in Asia) can be obtained by transferring a wild nesting colony (indigenous or feral) into a hive. The wild colony will already have a number of combs and these can be carefully tied to top-bars, or into frames, of a hive. Another way to get started is to set up a hive, rubbed inside with some beeswax to give it an attractive scent, and wait for it to be occupied by a passing swarm of bees. This will be successful only in areas where there are still plenty of honey-bee colonies. The best and easiest way to get started is by obtaining a colony with the assistance of local bee-keepers.

Choice of hives

A hive is any container provided by a bee-keeper inside which honey-bees can nest. Hives are of three main styles:

- local
- movable-frame
- top-bar.

Photo 14.1 A local style hive in Rwanda. The hive is a hollowed out log, protected from rain by an outer coat of banana leaves and secured with stems (courtesy of N. Bradbear)

Local style hives (fixed-comb hives)

This type of bee-keeping is practised in many countries, suitable receptacles being made from whatever materials are easily available. In forested areas, hives are made from hollowed-out logs or strips of bark; in desert areas, they are often made from date-palm trunks. Elsewhere hives are constructed variously from reeds, clay pots, baskets or odd scraps of timber, using whatever is suitable and available. Bee-keepers using local hives usually make their own, which, therefore, cost them nothing.

The purpose of the hive is to encourage bees to nest in a place accessible to the bee-keeper, and to confer 'ownership'. The bees build their nest inside the hive, just as they would build it in a naturally occurring cavity, and are more or less left alone by the bee-keeper. Bee-keepers using local, traditional methods are often highly skilled, knowing exactly where best to place the hives and when to harvest them. When nesting inside a container, bees necessarily attach the combs to the inside top of the container, hence the description as 'fixed-comb hives'. The honey crop is obtained by cutting out the combs from the bees' nest; bees may or may not be killed during this process, depending on the skill of the bee-keeper. Ideally, only honeycombs are removed, and combs with developing bees, pollen or unripe honey are left intact. Of course, this can be difficult to achieve. In obtaining honey from a local hive in this way, both the honey and the beeswax comb are necessarily removed. Local style bee-keeping methods, therefore, give a relatively high yield of beeswax. If the colony is destroyed during harvest, the hive will remain empty for a while. If there are plenty of honey-bee colonies in the area, a swarm may eventually settle in the empty hive and start building a new nest.

Local methods have evolved over a long period to suit local resources and indigenous bees. Their replacement by other methods should not be considered inevitable or necessarily desirable in every situation. For example, in the dry miombo woodlands of East Africa, local traditional bee-keeping is a major source of income, and the yields of honey and beeswax are of export quality. Bee-keepers using these local methods often benefit from assistance with access to protective clothing, smokers and good containers for the honey, training in how to harvest top quality honey and beeswax with the least harm to the colony (which is possible with this type of bee-keeping), and help in locating markets for the products. Expensive equipment is not necessary for simple bee-keeping. Good, serviceable hives can always be made from local materials, and imported goods are certainly not required.

Movable-frame hives

These hives are used in most industrialised countries and many developing countries. Bees are encouraged to build their combs within rectangular wood or plastic frames that support the combs. These frames have two major advantages:

- The ability to move combs inside frames allows the inspection and manipulation of colonies, such as moving frames of bees or stores from a strong colony to strengthen a weaker one

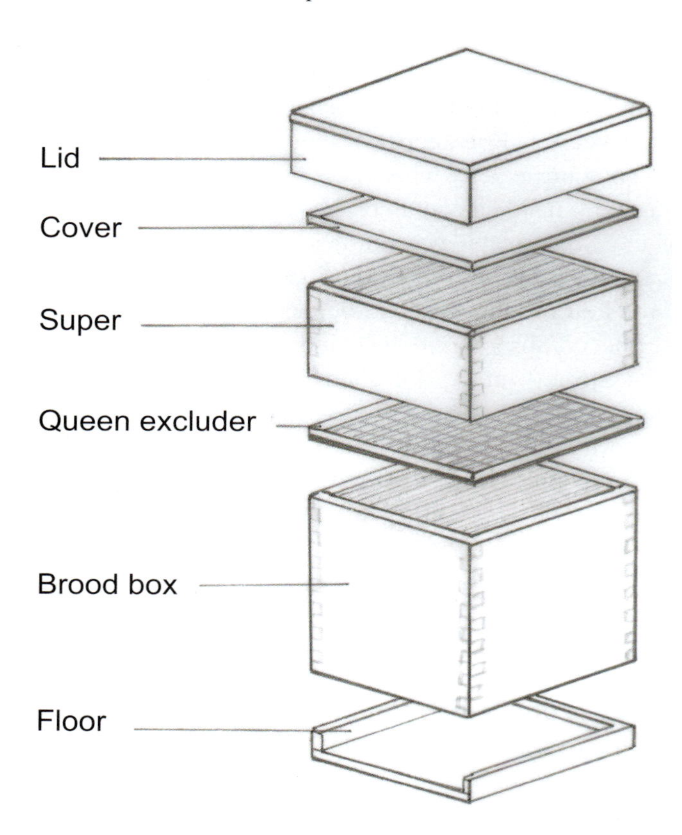

Figure 14.1 A frame hive

- The frames enable efficient harvesting of honey because the honeycombs within the frames can be emptied of honey and then returned to the hive. This allows increased honey production, as the bees do not have to use resources making wax for building fresh comb.

Frame hives consist of a series of boxes (usually wooden) stacked on top of one another. These boxes are either for the brood nest, where the queen lays eggs and young bees develop, or for honey storage. Usually the bottom box is used for the brood nest, and the queen is prevented from entering the honey storage boxes above by a *queen excluder,* a flat metal or plastic grid with parallel slots which allow the passage of worker bees, but not the larger-bodied queen or drones (male bees). The queen excluder is placed between the box with brood and the box above it. This ensures that honey alone (with no eggs or developing bees) is stored in boxes above the queen excluder. Frames are contained within all the boxes, hanging parallel to one another like files in a filing cabinet. Bees build their beeswax comb within these frames, either to contain developing brood or to store honey and pollen. The frames can be lifted easily from the box for inspection or for honey harvesting without damage to the bees. In addition to the boxes and frames, a hive stand, floor and roof are required, along with various other specialised items of equipment (see Photo 14.2).

Photo 14.2 A frame containing *Apis cerana,* the Asian hive bee (South India) (courtesy of N. Bradbear)

Frame hives must be constructed with precision. Boxes need to fit together precisely, and the spacing between frames must be the same as spacing in a natural nest. Frame hives require well-seasoned timber, planed and accurately cut, as well as other materials such as wire, nails and foundation. Foundation is a thin wax sheet, used as a base on which bees will draw out their comb. Frame hives are, therefore, relatively labour intensive to make and maintain. There must be access to replacement parts, particularly foundation and frames. The spaces between combs (bee space), nest volume and other features of standard frame hives have all been developed for use with European honey-bees (used in Europe, North and Central America, and Australasia) and are not necessarily suitable for other races and species of honey-bees. When buying equipment, it is important to have an understanding of the honey-bees to be housed and the resources available to local bee-keepers.

Honey is obtained by spinning the frames in a honey (centrifugal) extractor. The empty combs are later returned to the hive. By this 'recycling' of comb, frame hive bee-keeping contrives to make honey-bees put effort into honey production rather than comb (beeswax) production. The yield of beeswax from frame hive bee-keeping is, therefore, low compared with local or top-bar hive bee-keeping methods.

Frame hive equipment should not be used unless the infrastructure exists for manufacturing it locally. Frame hive bee-keeping is the type known to most bee-keepers in industrialised countries, and many bee-keeping projects have tried to introduce this type of bee-keeping in resource-poor countries. However, a community must have the physical, human, natural and financial assets to support this type of bee-keeping if the project is to succeed. That said, the use of movable-frame hives could have the advantage of generating 'downstream' employment for their manufacture locally.

Top-bar hives

The frame hive bee-keeping described above depends upon equipment being constructed with precision. Boxes must fit together exactly and the spacing between adjacent frames is critical in ensuring the correct bee spacing. In addition, such frame hives require carefully planed and seasoned timber, as well as nails and wire. These requirements make frame hive bee-keeping inappropriate in areas where such resources are very expensive or unavailable. To bridge the gap between the most basic receptacle-type hives and more expensive frame hives, various intermediate styles of hives have been developed. These aim to combine the low cost benefits of local hives with some of the advantages of frame hive bee-keeping. The only critical measurement in these low-technology hives is the width of the top bars, which must provide the correct space for bees to build their comb and maintain a constant temperature.

Top-bar hives are not a new invention: the Greeks were already using a basket hive with top-bars in 1682, and in North Vietnam top-bar hives were used traditionally for *Apis cerana* bee-keeping. In Africa, top-bar hives were first introduced to bee-keeping projects in Kenya in the mid-1960s. Since then projects in many countries have developed their own styles of top-bar hive, based on the resources available and the colony size of local bees.

Top-bar hives have the same advantages of manageability and efficient honey harvest as movable-frame hives, but at a lower cost, and with no recurrent costs. To allow manageability, bees are encouraged to construct their combs suspended from the undersides of a series of bars. These bars allow individual combs to be lifted from the hive by the bee-keeper. The hive body, as with local style hives, may be constructed from whatever materials are available. Many different designs have been published (for example Aidoo, 1999; Sakho, 1999; Mangum, 2001).

Many types of top-bar hive can be constructed by the bee-keeper, although village carpenters are often requested to cut the top-bars from planed timber; they need to be of a precise size because they must provide the same bee spacing for combs within the hive, as the bees would use in the natural nest. This bee spacing will depend upon the species and race of honey-bee. As a very general guide, *Apis mellifera* of European origin need top-bars 35 mm wide, African races of *Apis mellifera* need 32 mm, and *Apis cerana* need 30 mm. The best way to determine the optimum width is to measure the spacing between combs in a wild nest of the same bees.

Photo 14.3 A comb with bees from a top-bar hive (Trinidad and Tobago) (courtesy of N. Bradbear)

Another advantage of this type of equipment is that it opens up bee-keeping to new groups of people. For example, forest bee-keeping tends to be a male-only activity where bark hives are made and kept deep in forests. Groups of women bee-keepers may find it convenient to begin bee-keeping with top-bar hives that can be made and kept close to home. Top-bar hives can also be an excellent and cost-effective way of housing large numbers of colonies for pollination purposes.

Harvesting honey from top-bar hives is quite simple and can be achieved without damage to the colony. The bee-keeper selects combs containing honey but no brood: these will usually be the outermost combs. A honey extractor is not required: the honeycomb is cut from the top-bar and either cut into sections to sell as 'cut comb honey', or squeezed through a cotton bag or sieve to separate the liquid honey from the beeswax.

Protective clothing

Most bee-keepers wear clothing to protect against stings when working with bees. The most important item of protective clothing is a bee veil to cover the face: some veiling placed around a broad-brimmed hat with drawstrings to make the veil bee-proof will work well. Clothing must be loose-fitting to be cool and to keep bees away from the skin, and bees are less likely to sting light-coloured clothing. Bees can be prevented from crawling up sleeves and trouser legs by using rubber bands or string at wrists and ankles. Being adequately protected from stings gives beginner bee-keepers confidence. However, too much body armour can lead to rough handling of bees and many stings.

A bee smoker

The other important requirement is for a smoker: a device for generating cool smoke, used to subdue bees. On sensing smoke, bees rush to feed on honey and become less likely to sting. This behaviour is thought to arise because of bees' reaction to fire, when they consume honey in preparation for flight to another location. A smoker consists of a tin can with a bellows attached: useful smokers can be constructed in local workshops. Dried cow or donkey dung make ideal smoker fuels: they smoulder slowly to produce a cool, even smoke. Use of dung is nowadays considered hygienically unacceptable in some countries (e.g. Uganda), in which case dry grass is the preferred smoker fuel.

Products from bees

As described in Chapter 7, most people think of honey as the main output from bee-keeping, but this is only one of many benefits. Bees generate much more than honey; the pollination of crops and maintenance of biodiversity are the most valuable services provided by bees. Moreover, honey is just one of several different products that can be harvested.

Services rendered by bees

- The maintenance of biodiversity by the pollination of flowering plants
- The pollination of crops
- Apitherapy – medicinal use of bees' products

Products harvested from bees

- Honey
- Beeswax
- Pollen
- Propolis
- Royal jelly and venom

People in different areas of the world keep bees for different reasons. Although difficult to quantify, the financial value of the pollinatory activities of bees is much greater than the total world value of the products harvested from bees.

Honey production and value can vary greatly. Perhaps the lowest harvests are obtained from the indigenous honey-bees of Oman, *Apis florea*, and consist of very small amounts of honey, less than 1 kg, but the precious honey is then sold at up to US$ 100 per kg. The greatest honey harvests are obtained in areas where migratory bee-keeping is practised, i.e. moving colonies of bees to different crops or areas of natural vegetation as they come into flower. This is practised in some countries, for example, Argentina, China, India, Mexico and Pakistan, and bee-keepers may harvest over 100 kg of honey per colony per year. The higher honey and beeswax yields must be balanced against higher labour and transport costs.

Analysis of potentials and constraints, and suggestions for development

Diseases and parasites of honey-bees

During recent years, diseases and parasites of honey-bees have been spread around the world with serious consequences for the bee-keeping industries, and indigenous populations of bees, of many countries. This is caused by the movement of honey-bee colonies by man.

For example, the mite *Varroa destructor* is a 'natural' parasite of Asian honey-bees that can survive in the presence of the mite. However, when the mite is introduced to European *Apis mellifera* honey-bees, the whole colony will be killed unless action is taken by the bee-keeper. These mites have now been introduced to many bee-keeping countries and, for example, all populations of wild honey-bees throughout Europe have been killed during the last 20 years or so. Mites become resistant to chemicals developed for their treatment, and research is underway in many countries to find better, integrated control methods, or resistant strains of bees.

Recently another predator, the small hive beetle, *Aethina tumida,* has been spread from Africa (where it is a relatively harmless pest of bees) to honey-bee colonies in the USA, where it leads to the destruction of European honey-bee colonies.

Honey-bees and used bee-keeping equipment must never be moved from one area to another without expert consideration of the consequences. Very few regions remain without introduced honey-bee diseases, and these are mainly in developing countries. It will be highly beneficial for these countries if they can retain their stocks of disease-free honey-bees; they may in the future be able to market their disease-free stocks, and it makes opportunities for organic honey and beeswax production cheaper and easier.

Conservation of indigenous honey-bee species and races

Globalisation is taking place in bee-keeping, as in every other sector. Bee-keeping with European races of honey-bees, plus all associated technology, is being spread around the world. The consequences of competition between introduced (exotic) honey-bees and indigenous honey-bee species and races are unknown.

Loss of habitat diversity and pesticide use

This is a continuing, major problem in most regions. Loss of biodiversity ultimately leads to loss of sources of nutrition and food. Support for bee-keeping means support for the maintenance of biodiversity.

Authentication and certification of honey

Products are required increasingly to meet international standards, and to meet the demanding criteria of importing countries. The EU honey market requires imported honey to be certified free from chemical, antibiotic and other residues. These residues are most likely to be present in honey due to the use of 'medicines' (chemicals) to treat honey-bee diseases or predators, or from environmental pollution. This demand for residue-free honey opens opportunities for honey producers in the poorest countries. It is often the most resource-poor and most remote people of these countries, with few other livelihood options, who practise bee-keeping. These people can harvest honey and beeswax that are of excellent quality, and especially now, because these products are residue-free, they can achieve good prices on world markets if they are able to gain access and meet import criteria.

Technical constraints

Technical constraints include:

- lack of knowledge of appropriate methods for bee-keeping in the tropics
- lack of appropriately skilled trainers
- inappropriate technical advice and training materials
- few training possibilities
- little dissemination of new research information (especially relating to disease control).

Bee-keeping projects

Bee-keeping projects have been started in many developing countries, supported by international organisations, governments or NGOs. There are many different entry points for projects to strengthen livelihoods with bee-keeping. For example:

- including trees for bees in planting schemes to improve pollination and increase crop harvest
- assisting honey hunters

- making and marketing honey wines or beeswax cosmetics.

From a livelihoods perspective, the following are the resources necessary for bee-keeping.

Human resources: Bee-keeping skills, training and extension

Bee-keeping is a widespread activity with a wealth of existing local knowledge and skills. However addition of a little technical information can lead to harvests of honey and beeswax of greatly improved quality. There are many ways to assist honey hunters or bee-keepers as they build on their existing resources to create more income:

- by harvesting and processing honey more skilfully
- by obtaining a better price for their honey
- by saving and selling beeswax
- by making secondary products.

Bee-keepers and trainers often lack appropriate training materials (most literature discusses the keeping of European bees in temperate zone conditions). The training is often theoretical rather than practical, placing emphasis on changing the type of hive used, without providing practical guidance and follow-up. New bee-keepers need training in how to work with bees, how to maintain honey quality, how to separate honey from beeswax, how to render beeswax, how to manufacture secondary products and how to make bee-keeping clothing and equipment.

Physical resources: equipment and transport

Little or no access to transport is one reason why bee-keepers in remote areas receive the lowest prices for their products. Projects can do much to alleviate this problem. Equipment, containers and packaging can be difficult for rural people to obtain. The answer is not just to donate the items, but to give training so local people can make their own equipment, have access to good containers and packaging, and obtain the credit needed to acquire these items.

The equipment needed for bee-keeping can be very simple: the humble plastic bucket is one of the most essential items. These are very useful for bee-keepers living in remote places who need to keep their honey clean until they are able to sell it. Excellent quality honey can be harvested as long as clean buckets are available, along with cotton or baskets for sieving honey, and containers for packaging the honey.

The appropriate equipment for harvesting and processing honey and beeswax depends on the quantities to be processed and the type of product required. In some areas, bee-keeping using simple, local hives is practised on a large scale and justifies the provision of relatively sophisticated, large-scale processing equipment capable of dealing with honey in bulk.

Financial resources: credit

In resource-poor societies, lack of credit is a major constraint to all concerned with selling and buying honey. Bee-keepers with honey to sell expect to receive cash from honey collection centres or private sector traders; otherwise they prefer to sell their honey 'by the spoonful' in the market for an instant, albeit low, cash return. Those buying honey need access to credit during the honey season. The lack of credit leads to a stagnant industry, with insignificant volumes of honey available for sale, and no interest from traders.

Social resources: sector support and marketing

In resource-poor countries, there are usually government officers with responsibility for training and extension in bee-keeping. Often these people lack access to transport and other resources they need for their work. National policy is needed relating specifically to the protection of pollinators and the promotion of apiculture. A national association is desirable to represent the interest of bee-keepers, and to establish communication between producers and traders to facilitate marketing.

Photo 14.4 Honey prepared for marketing in the Gambia (courtesy of N. Bradbear)

In many developing countries, much can be done to increase retail honey sales, such as improved packaging, diversification of packaging and, especially, packaging in small volumes (see Photo 14.4). Marketing initiatives can involve media work promoting honey, interaction with consumers and traders to increase honey consumption and sales, and creation of links with packaging suppliers. Honey consumption increases according to living standards, and people are keen to buy fresh, local honey when it is well presented and they can have confidence in the product.

Project evaluation

Many bee-keeping projects have involved the distribution of hives and equipment, and provision of some technical training. Donors and local leaders might be satisfied with the outcome of such projects when shown convincing numbers of new hives installed in newly-created apiaries. However, closer examination often reveals that hives of a newly-introduced technology are not used efficiently. The true test of success in any bee-keeping development project should not be 'How many hives were distributed?' but 'Were peoples' livelihoods strengthened?'

Further reading

Bradbear, N., Fisher, E. and Jackson, H. 2002. *Strengthening livelihoods: exploring the role of beekeeping in development.* Bees for Development, Monmouth, UK.

Clauss, B. and Clauss, R. 1991. *Zambian beekeeping handbook.* Mission Press, Ndola, Zambia.

Collins, P. and Solomon, G. (ed.). 1999. *Proceedings of the First Caribbean Beekeeping Congress, Tobago 1998.* Tobago Apicultural Society, Trinidad and Tobago.

Crane, E. 1999. *The world history of beekeeping and honey hunting.* Duckworth, London, UK.

Krell, R. 1996. *Value added products from beekeeping.* Food and Agriculture Organisation of the United Nations (FAO), Rome, Italy.

Ministry of Agriculture. 1997. *Beekeeping in Botswana* (Beekeeping handbook 4th edition). Ministry of Agriculture, Gaborone, Botswana.

Morse, R.A. and Calderone, N. 2000. The value of honeybees as pollinators of US crops in 2000. *Bee Culture* 128: 3.

Ntenga, G.M. and Mugongo, B.T. 1991. *Honey hunters and beekeepers: beekeeping in Babati District, Tanzania.* Swedish University of Agricultural Science, Uppsala, Sweden.

NWRC. 1997. *Low productivity in East African beekeeping*. Njiro Wildlife Research Centre, Arusha, Tanzania.

Roubik, D. 1995. *Pollination of cultivated plants in the tropics*. Food and Agriculture Organisation of the United Nations (FAO), Rome, Italy.

Roubik, D. 2002. The value of bees to the coffee harvest. *Nature* 417: 708.

Smith, F.G. 2003. *Beekeeping in the tropics* (Reprint). Northern Bee Books, Mytholmroyd, UK.

Journal

Bees for Development Journal, published quarterly and available from Bees for Development. See www.beesfordevelopment.org for more details. If you are a bee-keeper working where payment is impossible, contact Bees for Development to find a sponsor for your subscription.

Videos

Clauss, B. 1995. *African honeybees: how to handle them in top-bar hives* (PAL/VHS) 22 min.

Wendorf, H. 1999. *Beekeeping in development* (PAL/VHS) 81 min.

Further information

Apimondia Apimondia is the World Federation of Bee-keepers' Associations. The Apimondia Congress, organised every two years, is the major international event for everyone involved with any aspect of bee-keeping. Apimondia publishes a quarterly journal, Apiacta, publishing bee research for bee-keepers, and can assist with information on many aspects of apiculture. Further information from: APIMONDIA, Corso Vittorio Emanuele II, 101, I 00186 Rome, Italy.
E-mail: APIMONDIA@MCLINK.IT
Website: www.apimondia.org

Asian Apiculture Association (AAA) AAA organises a Conference within Asia every second year (alternating with the Apimondia Congress) and operates a network between Asian bee-keepers. Further information from: Asian Apicultural Association, c/o Honey-bee Science Research Centre, Tamagawa University, Machida, Tokyo, 194-8610 Japan.
Email: hsrc@agr.tamagawa.ac.jp
Website: www.tamagawa.ac.jp/HSRC/

Bees for Development (BfD) Bees for Development offers information and advice to all concerned with apiculture as a useful part of sustainable, rural livelihoods. Bees for Development was founded in 1993 in response to the international need for a specialist organisation devoted to this sector, and functions at the heart of a network enabling coordination and exchange of information between everyone working in this field. Bees for Development manages research and development projects, organises training, provides information, and publishes the quarterly *Bees for Development Journal*. Further information from: Bees for Development, Troy, Monmouth NP25 4AB, UK.
E-mail: info@beesfordevelopment.org
Website: www.beesfordevelopment.org

Author addresses

Nicola Bradbear, Bees for Development, Troy, Monmouth NP25 4AB, UK
Uma Partap, ICIMOD, P.O. Box 3226, Kathmandu, Nepal
Kwame Aidoo, Sasakawa Centre, School of Agriculture, University of Cape Coast, Cape Coast, Ghana
Alice Kangave, Ministry of Agriculture, Animal Industry and Fisheries, Livestock Health and Entomology, P.O. Box 102, Entebbe, Uganda.

15

Giant African snails

William Oduro[*], *Tsatsu Adogla-Bessa*[‡], *Emmanuel Osafo*[‡]

Cameo - Ghana: people should be encouraged to venture into small farming

Victoria Mancell, Executive Director, Mancell Girls Vocational Institute, Kumasi, Ghana, had this to say:

"In February 2003, we requested help from the Institute of Renewable Natural Resources (IRNR), Kwame Nkrumah University of Science and Technology, to establish a snail farm at our school. After staff had received training from IRNR, we spent three million cedis (US$ 368) on materials and an initial stock of 2,000 juvenile and 288 adult snails. After six months, we realised 2,258,000 cedis (US$ 277) from the sale of 2,790 juveniles (aged six months) and 577 adult snails. We still have in stock over three thousand juvenile and adult snails. The small adult snails provide protein in the students' diet and the income generated from selling snails has been used to supplement the school's food budget. Some of our students have acquired the skills needed to rear snails and hope to start their own snail farms. Snail farming is a lucrative business and people should be encouraged to venture into it as we have done."

Introduction

The Giant African snail (*Achatina achatina*) has a geographical distribution covering several countries in West Africa, from Guinea through Ghana to Nigeria. In Ghana it is found mainly in the South and in forested areas of the Eastern,

[*] Principal author
[‡] Consulting author

Western, Central, Ashanti, Brong Ahafo and Volta Regions. The snail is normally found in the shade and undergrowth of some tree crops such as cocoa, coffee, banana, plantain and oil palm, and also under logs, stones and leaf litter.

In Ghana, Cote d'Ivoire and Nigeria snail meat is particularly popular, in the first two forming about 10 per cent of the trade in wildlife or 'bushmeat'. Most of the chop bars in Ghana sell snail meat during the peak snail season, and the prices are often higher than those charged for beef or mutton. These high prices are primarily due to customer demand and dwindling supplies, but also to the cost of transportation and processing when smoked. Among the bushmeat sold in West Africa the snail and grasscutter (Chapter 17) fetch the highest market prices per kilogram. The shells of the snails are also decorated and sold for ornamental purposes.

The snails are gathered from the forest during the wet season. In Nigeria it is estimated that the number of snails collected per farmer per month from the deciduous and rain forests is 36 and 17 respectively. Such wild snail production has declined in recent years because the wild snail populations are decreasing rapidly due to human activities. These include not only the uncontrolled exploitation of the resource but also the indiscriminate use of pesticides and bush burning during the dry season.

Commercial snail farming, such as found in Europe and other developed countries, does not exist in West Africa. However, the general scarcity of snails, particularly during the dry season, offers an opening for 'home rearing' of snails and thereby a livelihood-opportunity for resource-poor smallholders. Although the Giant African snail is slow-growing and therefore may not appear to represent a way of sustaining and improving the livelihoods of resource-poor farmers, it has been shown that with good management and careful integration into existing farming activities, snail farming can bring substantial rewards.

Snail biology

Like all snails, *A. achatina* is cold-blooded and therefore sensitive to changes in atmospheric humidity and ambient temperature (poikilothermic). Snails thrive best in areas which have moderate temperatures and high humidity. In the West African region where snails are found, temperatures do not fluctuate widely but there are significant fluctuations in humidity, which can have a profound effect on the snails. When humidity falls below 75 per cent, as is the case during the dry season (October to mid-March), *A. achatina* becomes inactive, retreats into its shell and secretes a papery covering (epiphragm) over the opening, covers itself with a white calcareous layer to prevent water loss and then aestivates. This reaction is typical of all snail species and is most common in the dry season, but snails may also aestivate if dry spells occur during the wet season. The length of aestivation depends on the conditions; in savannah regions it may be up to five months, whereas in wetter forested areas it may be only a few days or weeks. When the rains arrive, the epiphragm is broken and the snail becomes active again.

A. achatina is nocturnal, feeding at night. The Giant African snail is most active during the rainy season when conditions are well suited to its movement, breeding and growth. The soil forms a major part of the snail's habitat. The shell is made mainly of calcium, most of which, together with the snail's water requirements, is derived from the soil. The snail digs in the soil to lay eggs and to aestivate.

Snails breed mainly during the rainy season. Most snails are hermaphrodite and mate before laying, but *A. achatina* reproduces by self-fertilisation; unlike many snail species, reproduction is not preceded by coupling, although it is not unusual to find two snails in close proximity. Egg laying usually takes place in the late evening and at night, although on warm humid days eggs may also be laid during the day. The eggs are deposited in holes about 4 cm deep, depending on the softness of the soil, and covered with soil or litter. The eggs are lemon-yellow, oval in shape and slippery to touch when freshly laid. They are laid in clutches of 38-563 eggs, depending on the age of the snail. Second and even third clutches may be laid, although these later clutches are usually small. Eggs hatch after 2-3 weeks, depending on the temperature. As hatching approaches, the eggs turn white. *A. achatina* has a high hatchability and 100 per cent hatch is not uncommon.

The adult snails do not tend their young. The newly-hatched young possess a thin shell membrane, which progressively calcifies. Although this period is characterised by rapid growth, the snails are able to survive their first five to ten days without food, making their first meal of their egg shells. This is perhaps an evolutionary adaptation for an organism with poor mobility.

The juvenile phase covers the period between one to two months and sexual maturity, which is at 12-18 months (in captivity it is at 9-15 months, even with high-level management practices). At this stage the snails weigh between 100 and 450 g. Not all snails lay eggs each season, and some in captivity may not lay for three successive seasons. Average life expectancy in captivity is 5-6 years, although there are reports of snails surviving up to 10 years.

In Ghana there are several ecotypes (locally adapted populations) of *A. achatina,* which show differences in size, growth rate, aestivation pattern, colour and even flavour. Differences in size may be explained by the length of the aestivation period; a short aestivation period allowing a longer period for feeding. A study of three ecotypes known as 'Donyina', 'Apedwa' and 'Goaso' has shown significant differences between them in size and length of aestivation. The Apedwa snails have the shortest aestivation period and are largest, the Donyina the longest and are smallest (roughly half the size of the Apedwa snails). Cobbinah (1994) recommended the Apedwa ecotype for snail farming in Ghana.

According to Hodasi (1975), populations of *A achatina* have never reached epidemic proportions, as has been reported for *A. fulica* in other parts of the world. This may well be due to the high rate of human predation on *A. achatina.*

Snail husbandry

Selecting snails for home rearing/commercial farming

Before starting on 'home rearing' of snails, it is necessary to determine the social acceptance of snail meat in an area or locality. This is because the choice of meat can be coloured by culture, taboo or simply by force of habit. In the northern parts of West Africa, which are predominantly Muslim, snails do not form part of the diet and hence will be hard to sell, unless they are raised solely for export. However, in the southern forested areas snails are a delicacy and thus snail farming will be met with a resounding local response, whilst at the same time also serving as an export commodity.

Home rearing/commercial snail farming can bring higher and faster financial returns compared to traditionally-gathered stock. Until snail farms become self-sustaining, farmers will have to collect young snails from the wild or buy them cheaply in the peak season and fatten them in captivity for the off-season (dry season). Snails bought from snail-gatherers have a fairly high rate of mortality due to poor handling and adjustment to different feeds. It is best if large snails are bought rather than small ones. Large snails are likely to produce offspring with greater potential for growth, while small snails may have already attained their maximum size and therefore have no further potential for growth.

Snails with broken shells should not be selected. The broken shell makes the snail more vulnerable to attack by predators. Also snails which have any part of the shell broken, either through rough handling or by drilling a hole in the shell to allow them to be tied together for transport, require extra energy to repair the shell before breeding will commence. Snails which have been subjected to excess heat in sacks, or to starvation for long periods, take some time to recover before they can start breeding.

When selecting breeding stock from home-reared snails, it is best to do so in the wet season before aestivation. Criteria for selection include fecundity, hatchability and establishment rate, but it is important to remember that these can be affected by management as well as by genetics.

- Fecundity can be assessed on the number of eggs laid in previous years, which can be over 500 eggs per season. Fecundity differs both within and between ecotypes and, within ecotype, stress and high population density can depress fecundity.
- Hatchability is a desirable characteristic to be considered. Although 100 per cent hatchability is not uncommon in *A. achatina,* unfavourable temperatures and inbreeding can directly and indirectly affect the number of eggs likely to hatch in a particular breeding season or year.
- Establishment rate refers to the percentage of hatched snails which survive to maturity.

Housing

The Giant African snail prefers high humidity and cool areas. It is also sensitive to light, so snail houses or 'snaileries' should provide dim light conditions. The snailery should be shaded to provide the right temperature and humidity for breeding. The bottom of the snailery should be covered with moist sandy-loam soil, kept loose to allow the snails to prepare holes for egg-laying. There should be an opening at the top for feeding, and all openings should be sealed with wire netting to prevent escapes.

There are several different types of snailery that can be constructed, depending on the scale of the enterprise and the stage of development of the snails to be housed. Whatever the type, it is important that it prevents snails escaping and keeps out predators, but allows the snail-keeper access to tend the snails. Materials used can include decay- and termite-resistant timber, landcrete blocks (made with sand and mud, without using cement or fire), concrete blocks, bricks, nylon-type mosquito nets or 5 mm wire mesh.

Baskets

For small-scale fattening of snails two market-woven baskets are needed, one placed on top of the other. A basket (60 cm diameter, 45 cm depth) can contain 30-50 medium-sized snails. Plastic baskets with holes in the sides can also be used, and will last longer than woven baskets. Baskets are ideal for transporting small quantities of snails.

When using baskets, special care is needed to:

- clean the snails and the baskets frequently, under a tap or using a hose-pipe, to prevent the accumulation of faeces and rotten feed
- remove uneaten feed and replace with fresh every day
- carefully remove eggs for hatching.

Cages/hutch boxes

Cages/hutch boxes are square or rectangular boxes with lids, containing one or more chambers (Figure 15.1). The boxes are held above the ground on wooden stilts in a cool shaded area. In the centre of the lid is a mesh-covered opening. The floor of the box has several drainage holes and is covered with sieved black soil to a depth of 18-25 cm. The soil must be changed occasionally (about every three months) as an accumulation of droppings will increase the chance of disease developing. Daily cleaning is essential. A cage of the dimensions shown in Figure 15.1 can house 200 medium-sized snails.

Hutch boxes are ideal as hatchery and nursery pens as the eggs and young can be easily monitored. Mature snails should be transferred to hutch boxes to lay their eggs.

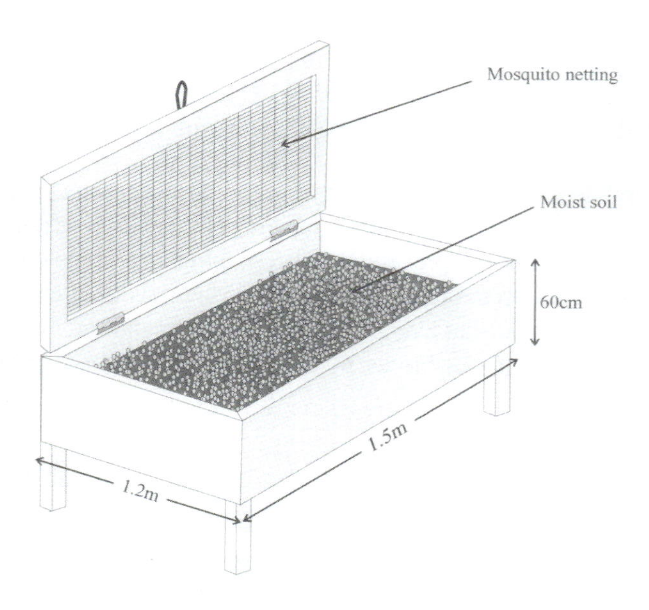

Figure 15.1 Cage/hutch box for use as a hatching or nursery pen, or for fattening snails.

Pit and trench pens

Pit and trench pens can also be used as hatchery and nursery pens but are more suitable for growing/fattening snails. However, pits and trenches are more difficult to use than hutch boxes because the snail-keeper has to stoop or bend to tend the snails. Constructing a trench involves digging a square or rectangular hole in the ground, about 50 cm deep, and then dividing it into pens. The sides of the trench pens are normally of concrete blocks and the bottoms covered with loose soil. The pens are covered with nylon mosquito mesh or with 5 mm wire mesh, nailed to wooden frames (see Figures 15.2 and 15.3). Pits and trench pens are recommended for resource-poor snail-keepers because they can be adapted from low-cost local building materials.

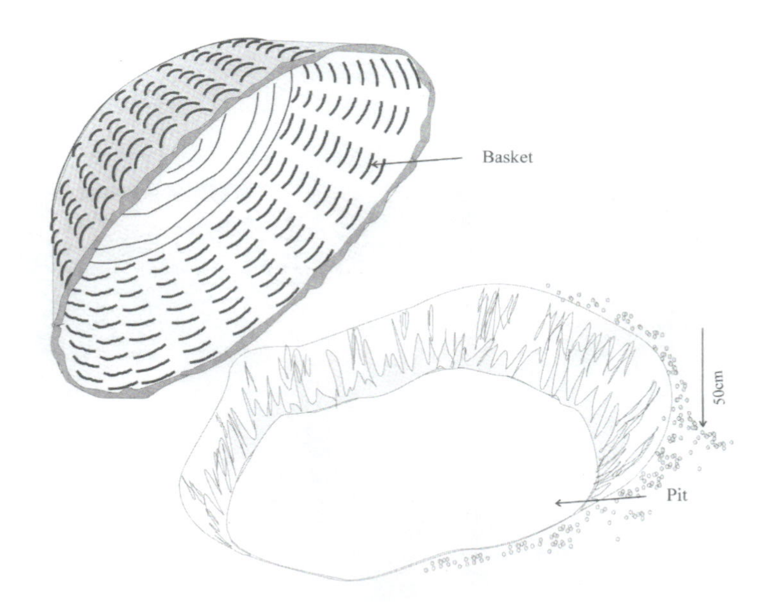

Figure 15.2 Pit pen with basket cover

Pit and trench pens have the following advantages:

- snails are able to lay their eggs directly into the soil, where they hatch easily
- snails eat the soil to obtain their mineral requirements
- snail faeces do not accumulate because earthworms and other organisms convert the waste into organic matter for inclusion in soil.

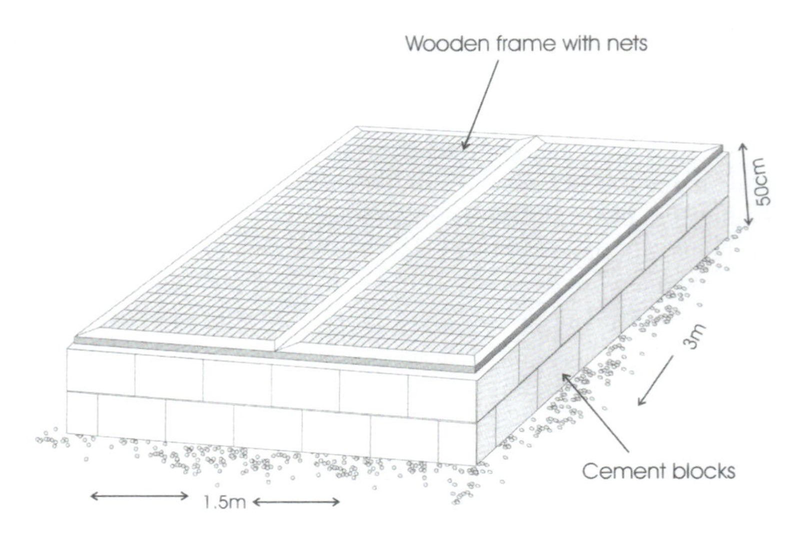

Figure 15.3 Trench pen

Photo 15.1 Cage/hutch box for use as a hatching or nursery pen, or for fattening snails (courtesy of W. Oduro)

Mini-paddock pens

The paddock is a fenced area within which the snails are confined. The paddock, of any desired size, can be constructed of simple fencing material such as 10 mm galvanised wire mesh or galvanised iron sheets, supported by wooden posts on the outside. The most common fencing materials used in Ghana are timber and bamboo, with wire or nylon mesh to prevent the snails escaping. The sides of mini-paddocks should be 50 cm high and buried 15 cm into the ground. A horizontal wooden frame covered with mesh is attached to the top of the pen fence. Mini-paddocks are ideal as fattening pens. In Ghana appropriate plants to grow in the mini-paddock to provide feed and shelter include cocoyam, banana, oil palm, cassava, sweet potato and African spinach.

The following are notable points about mini-paddocks:

- shelter and feed plants should be pruned periodically as they tend to overgrow
- paddocks should be provided with flat stones or blocks on which to step to avoid trampling on snails
- earthenware pots, pipe tubes or palm branches should be put into the paddocks to provide shelter until the feed plants are sufficiently grown
- paddocks should never be allowed to dry out.

Cobbinah (1994) states that relatively narrow rectangular pens are easier to use than square ones because they allow the snail-keepers access to the whole paddock for such tasks as distributing food, collecting uneaten food, inspecting snails and replanting shelter and feed plants, without needing to enter the paddock.

Movable pens

Movable pens are suitable as fattening pens and for housing mature snails. They can also be used as exhibition pens. They consist of wooden pens covered with wire mesh on the sides; the mesh extends over most of the top of the pen, leaving an opening for access. A fold of mesh suspended from the roof prevents the snails escaping. Movable pens should always be placed on an even surface. The main attribute of these pens is that they can be moved from one site to another as long as the framework is strong. The disadvantage is that small snails are likely to escape because the base of the pen is not buried in the soil.

Free-range pens

For a free-range snailery an area of about 18 x 12 m is planted with shelter and feed plants and enclosed by a mesh fence. The snails are allowed to move over the entire area. Although free-range pens are easier to construct than other types of snaileries, they have the following disadvantages:

- eggs and small snails are difficult to locate
- it is difficult to keep out predators.

Photo 15.2 A young snail-keeper displaying snails (*A. achatina*) collected from his free-range pens (right) behind his house (left) (courtesy of W. Oduro)

It is clearly not possible to control temperature and humidity in outdoor situations. However the magnitude of temperature and humidity fluctuations is reduced in areas of relatively undisturbed forest or fairly dense vegetation cover. Sites like those should be selected for snaileries, especially free-range pens, in preference to open grassland or farmland. During dry periods it will be necessary to water the pens regularly.

Feeds and feeding

The snail has a well-developed alimentary canal comprising a mouth, buccal cavity, pharynx, crop, stomach, intestine and anus. The mouth has a set of teeth called radulae, which help the snail to feed on vegetable matter. The snail has a well-developed sense of smell and can detect feed from a distance of about 60 cm. Snails are predominantly vegetarian and will eat the soft tender leaves and stems of most broad-leafed plants, vegetables and fruits. They do not eat grasses. Older snails show a definite preference for decaying matter, including garbage and fences. Snails also ingest soil, especially alkaline soils with a high content of calcium.

Feeding occurs mostly at night or on cloudy days when light rain is falling. Snails locate feed with their shorter pair of tentacles and feed on the green leaves of cocoyam, lettuce, cabbage, spinach, fruits and palm fruits. Gari (a fermented, partially-gelatinised granular product from cassava) with palm oil is also known to be a good fattening ration for snails. Snails may also eat root tubers and corms of cocoyam. It is advisable to offer snails a range of feeds so that they can select.

According to Cobbinah (1994), studies on *A. achatina* show that the very young snails thrive best on leafy vegetables, but for all other stages a diet made up of the following is recommended:

- Cocoyam – the leaves contribute some protein, calcium, phosphorus, thiamine (vitamin B1) and riboflavin (vitamin B2)
- Pawpaw (papaya) – the fruit provides some carbohydrate and ascorbic acid (vitamin C)
- Bokoboko (also known as 'lime flower plant') – provides some protein, phosphorus, iron, ascorbic acid and riboflavin
- Oil palm – the mesocarp (fleshy layer) provides some carbohydrate, fat and vitamin A
- Supplementary vitamins – other food plants known to contain moderate amounts of vitamins should be added, e.g. sunflower and copra cake (vitamin D), wheat germ, lettuce and other green vegetables (vitamin E), cabbage and African spinach (vitamin K)
- Supplementary minerals – placing licking stones in the pens can provide other minerals
- Supplementary calcium – if the soil is not rich in calcium a supplement is needed. Supplementary calcium can be provided by sprinkling powdered oyster or snail shells or ground limestone onto vegetables.

Captive snails should be fed twice daily – in the early morning (6-7 a.m.) and late evening (5-7 p.m.). Providing *A. achatina* with a mixture of feeds rather than only one will enhance growth. The attractiveness of feed is also important – if the feed is appetising or contains a feeding stimulant (such as pawpaw) snails will eat more and grow faster. If the feed is unattractive or lacks a stimulant, snails will eat less and grow more slowly, irrespective of how nutritious it may theoretically be. Some studies have been conducted on the growth rate of snails fed different diets, but the results are inconclusive.

Cobbinah (1994) suggests that as land pressures force people to move from extensive farming, where natural feeds are abundant, to semi-intensive farming, it may be necessary to use formulated feeds for snails. Experiments in France with *Helix* spp. showed that formulated poultry feeds reduced the period of growth of *H. aspera* from hatching to harvest by 10 months (from 27 to 17 months). Studies in Ghana in which poultry feeds were tested on snails have also shown promising results. However, for resource-poor snail-keepers formulated feeds have the disadvantage of having to be purchased. They are therefore more likely to be of value in commercial snail farming.

Good sanitation should prevail in the pens and watering, using a watering can or similar device, should be done in the morning and evening when humidity is low. Snails probably obtain sufficient water from moist soil and the feed they eat, but it is advisable to provide drinking water. For mature snails water should be in shallow containers; if the containers are too deep, the snails will not be able to crawl out and will drown. For young snails a piece of sponge or cotton wool soaked in water is sufficient.

Factors influencing reproduction

The adult snail has three phases during its active period, the timing of them dictated by the climate. In Ghana the annual cycle is as follows: the pre-spawning phase (March-April) follows aestivation and represents a fairly active period during which feed consumption is high, partly due to recovery after aestivation and partly in preparation for egg-laying; the spawning phase from April to July is characterised by egg-laying and reduced feed consumption; the post-spawning phase (July-October) is one of high feed consumption to build up body reserves for the dormant period from October to March. The aim in snail farming must be to provide near-ideal growing conditions year-round, so that the dormant period is reduced.

If faeces and left-over feed are not constantly removed it can affect snail breeding, so that a high level of sanitation is required to achieve high production. It is advisable to sterilise the soil in hutch boxes.

Pests and diseases of snails

There are many predators of snails. They include field mice, rats and shrews, frogs and toads, thrushes, crows and domesticated birds such as ducks and turkeys, lizards and snakes, drilled and curbed beetles, millipedes and centipedes. Frogs tend to take only the young snails, while reptiles eat both the eggs and the snails.

In Ghana the major parasite of the snail is the fly *Alluaudihella flavicornis*, which belongs to the same family as the housefly, which it resembles. It lays 20-40 eggs in the snail shell or on the snail. The eggs hatch in about a week and the small cream-coloured larvae start feeding on the snail's body tissue. They feed until the snail is reduced to a putrefying mass and then pupate within the shell. The adults emerge after a 10-day period. The best protection against these flies is to cover the pens with nylon mesh.

Little is known about the diseases attacking *A. achatina* in West Africa. The main disease reported is fungal disease, which is spread through physical contact by the snails licking slime from each others' bodies. However it is possible that two diseases affecting European species of snails may also affect *A. achatina* because the causal agents of these diseases occur within its natural range. These two diseases are *Pseudomonas*, which leads to intestinal infections and may spread rapidly among dense snail populations, and the fungus *Fusarium,* which parasitises the eggs of *Helix aspera.* As snail farming increases in popularity, more research on snail diseases will be needed.

Basic hygiene will prevent the spread of diseases. Pens should be cleaned frequently to remove excreta and uneaten food, as well as any other decaying/ decomposing matter that may serve as substrate for pathogenic organisms. Red ants have been observed to infest uneaten food of the snails, with dangerous consequences for the snails. The presence of the fruit fly (*Drosophila melanogaster)* is an indication of unhygienic conditions in the snailery.

Marketing of snail products

Meat and medicinal value

Snail meat is high in protein (12-16 per cent) and iron (45-50 mg/kg) but low in fat (0.05-0.08 per cent). It contains almost all the amino acids needed by humans. The meat is tasty and of good texture provided the snails are harvested at the correct age.

Snail meat is therefore highly nutritious and has been consumed by humans throughout the world, since prehistoric times. At the Imperial court in Rome, snail meat was thought to contain aphrodisiac properties and was often served to visiting dignitaries in the late evening. In the past, snail meat was recommended as a means of combating ulcers and asthma. The high iron content is considered important in the treatment of anaemia. There are also claims of its use in the treatment of hypertension and haemorrhoids. A recent study has shown that substances from edible snails cause agglutination of certain bacteria; this could be of value against a variety of ailments, including whooping cough (Cobbinah, 1994).

Edible snails have also played an important role in folk medicine. In Ghana, the bluish liquid obtained from the shell, after the meat has been removed, is believed to be good for infant development.

Processing and preserving snails

Snails reared in captivity may take 15-18 months to reach the size which meets local consumer preferences in West Africa; the size required for export is slightly smaller.

Snail meat is processed by parboiling the whole snail in its shell for about 5 minutes. When the shell cools the meat can be pulled out. The edible muscular foot is detached from the visceral contents and washed in warm water several times to remove the slime. The slime can also be removed by washing the meat in lime juice. The meat is then cut into suitable-sized chunks for further cooking.

For preservation, snails are graded according to size into juveniles, medium giants and super giants. After removing the snails from the shell, they are washed with salt, lime and vinegar and allowed to stay in the solution for about an hour to become foamy. Piped water is then used to wash the snails thoroughly. They are then put in a perforated basket to drip. In Europe the snails are passed through a nitrogen-freezing tube for 48 hours, and then a blast freezer for 15 minutes. The snails are preserved by canning, dehydrating or immersing in garlic sauce.

Snails weighing between 20 and 100 g can be canned. After the snail meat is washed with lime to remove slime, it is canned in brine or oil, and the cans then sterilised by retorting before sealing.

Snails are dehydrated by smoking at a temperature of 250-300 °C, but are not kept too long in the oven. The rest of the drying is done in the sun, usually for 2-3 days, so that the snail loses 30-90 per cent of its weight. Dried snails keep for several months.

Local markets

In the high forest zones of West Africa, particularly in Ghana, Nigeria and Cote d'Ivoire, snails form a substantial part of the meat in the diet of the local people. In Cote d'Ivoire, for example, an estimated 7.9 million kg are eaten annually. These snails are gathered in the wild, packed into bags, wooden crates or baskets and transported to selling points along main roads or in urban centres. The gatherers may sell directly to consumers, to retailers or to wholesale traders. In some cases the chain from the gatherer to the consumer may involve four or five middlemen.

The snails may be smoked and stored for sale during the off-season when prices are highest. Currently, five medium-sized snails sell for cedis 10,000-15,000 (US$ 1.2-1.8).

Snail meat is not consumed for religious reasons in the predominantly Muslim northern areas of West Africa. The climate in these areas is also unsuitable for snail-keeping.

Photo 15.3 Boys selling Giant African snails (bundled) at road-side near Kumasi, Ghana (courtesy of E. Owen)

Export markets

There is a growing international trade in snails, which are sold fresh, frozen or canned. France plays a central role, not only as a consumer but also as a processor and exporter to other European countries and to North America. The value of imports to the USA alone is about US$ 200 million per year. The annual French requirement for snails is about 15 million kg, of which over 60 per cent is imported.

The main demand in Europe is for the European snail species *Helix aspera, H. pomatia* and *H. lucorum*, but China and Thailand, as well as West Africa, supply *A. achatina*. The African species fetch only a third of the price of the European species, as the meat is considered rubbery because the snails are more mature when exported. Because of their greater size, the shell is also less suitable for presentation of the final product and European consumers generally prefer snails served in the shell.

The present product is targeted at Africans living in Europe, but recent studies in the UK have shown that juvenile *A. achatina* snails are more tender and meatier than the more-favoured European species. For West African producers, this will not only mean a greater demand for their product but also reduced costs of production because of the shorter growing period required to produce the smaller European-preferred snails.

Conclusions – snail-keeping by the resource-poor

It is unquestionable that snails can be raised in captivity in the humid tropics, such as West Africa, and the large demand for snail meat suggests that snail-keeping has a bright future. However, snail-keeping by the resource-poor in West Africa and other developing countries is still in its infancy and needs considerable research and development using modern participatory approaches (Chapter 4).

Research on the biology and production methods for *A. achatina* is needed in topics such as:

- selective breeding to improve production and increase resistance to disease
- appropriate housing
- suitable feeding regimes for the reproductive and growth stages, and also to assist disease prevention

- reduction of the aestivation period
- minimising the occurrence of *Pseudomonas* and *Fusarium* and the impact of these diseases on snail production.

There is also a need to establish communication channels so that the results of this research can be effectively disseminated to the small-scale producers.

Further reading

Baratou, J. 1998. *Raising snails for food.* (Translated from French Herb, F. 1981. *Les escargots*). Illumination Press, Calistoga, California, USA. This manual provides practical information on snail farming, particularly on indoor farming of the European species.

Cobbinah, J.R. 1994. *Snail farming in West Africa: a practical guide*. Technical Centre for Agriculture and Rural Co-operation (CTA), Wageningen, Netherlands.

FAO (Food and Agriculture Organisation of the United Nations). 1986. *Better farming series: Farming snails. Economic and Social Development Series Nos.33 and 34*. Food and Agriculture Organisation (FAO), Rome, Italy.

Sawyerr, L.C. 1995. *Short notes on practised snail farming*. CDS Press, Accra, Ghana.

Sheldon, C. 1988. *Raising snails*. Special Reference Briefs (National Agricultural Library SRB 88-04), United States Department of Agriculture (USDA), Beltsville, Maryland, USA. Report derived from review of the literature and searches of selected databases.

Journals

Snail Farming Research Journal. Biennial journal published by Scientific Committee of the Snail Farmers Association in Italy. Topics covered include taxonomy, biology, behaviour, nutrition and husbandry.

Malacological Journal and Newsletter, Online Database.

Journal of Molluscan Studies. Published three times a year. Includes articles on research on molluscs and related organisms. Oxford University Press, on behalf of the Malacological Society of London, UK, assisted by Stanford University Libraries, Wire Press, Stanford, USA.

Author addresses

William Oduro, Institute of Renewable Natural Resources, Kwame Nkrumah University of Science and Technology, Kumasi, Ghana

Tsatsu Adogla-Bessa, Agricultural Research Centre-Legon, Institute of Agricultural Research, University of Ghana, P.O. Box LG 38, Legon, Accra, Ghana

Emmanuel Osafo, Department of Animal Science, Kwame Nkrumah University of Science and Technology, Kumasi, Ghana

16

Poultry

Thomas Acamovic[*] *Arnold Sinurat*[†], *Amirthalingam Natarajan*[†], *Kaliappan Anitha*[†], *Doriasamy Chandrasekaran*[†], *Dilip Shindey*[†], *Nicholas Sparks*[‡], *Oluseyi Oduguwa*[‡], *Bartholomew Mupeta*[‡], *Aichi Kitalyi*[‡]

Cameo 1 - Tamil Nadu, India: storage of eggs for better hatchability

The desi (local) poultry-keepers in Peruganur, a village in Trichy district in Tamil Nadu, India, are progressive in their approach to poultry husbandry. Peruganur has long been noted for its contribution to a famous shandy (local market) held at the nearby bigger village, Pavithram. This shandy sells desi birds from villages from within a radius of about 20-30 km. Birds are picked up by traders and sold at premium prices in towns and cities.

Desi bird keeping is popular in Peruganur and every household is involved. Selvarani is an active desi poultry-keeper in this village; she has been involved in this for some time, keeping between 25-30 juveniles and adults, plus chicks.

Amongst the problems she encounters in keeping desi birds are: low hatchability in the eggs kept for brooding; deaths of hatchlings caused by disease and predation; and slow growth. She thought that she was unable to do much about this and accepted the situation. Sometimes, often after rain, the devastating Newcastle disease affected her birds and killed them. Many of her neighbours were also affected in the same way.

One of Selvarani's neighbours, who also rears desi birds, is Dhanabakiyam. One summer's day Selvarani told Dhanabakiyam about her problem of poor hatchability – in a recent batch of 15 eggs, none hatched.

---

[*] Principal author
[†] Co-author
[‡] Consulting author

Dhanabakiyam is a member of a self-help group (set up with funding from DFID). It was during one of their meetings that she was informed about a simple and inexpensive technique for keeping the desi eggs prior to placing them for brooding. At the meeting it was explained that if eggs were stored in a warm place the chicks might not hatch. This is why hatchability is poorer in the warm summer months. It was explained that it was important that eggs should be stored in a cool place

Dhanabakiyam explained to Selvarani that all that was required was to place some wet sand in a storage bowl, cover the wet sand with a cloth and then place the eggs on top of the cloth. The sand must be kept moist (achieved by putting the water on the sand, not the eggs), but this was all that was required and the bowl could be stored in a cool part of the house till the eggs were placed for brooding. The technique seemed to be simple, and was attractive to Selvarani. She returned to her village and started to keep the newly-laid eggs from an older hen in the way described. So that Dhanabakiyam could check that the new technique worked, another hen sat on a different batch of eggs kept in the conventional way.

After three weeks, 7 out of the 11 eggs hatched from the hen sitting on eggs kept in the old-fashioned way. However, 9 out of 10 eggs hatched from the hen sitting on the eggs stored in the new way. Selvarani could save eggs this way, and added two more birds to her flock.

Cameo 2 - Kusa Community, Kenya: small improvements in management of village chicken can transform rural poultry

Penina Gari, a local poultry farmer in Kusa Community in the Lake Victoria Basin, Kenya, has a very successful local poultry unit as a result of adopting small improvement techniques. She has a flock of over 150 birds, and in three months she can sell up to 80 birds for about KES 12,000 (US$ 158). In 2001, Penina joined a poultry common-interest group, supported by an integrated community development project funded by the Swedish International Development Cooperation Agency (Sida). As a group, they developed a Newcastle disease control programme, received training on management of poultry and discussed marketing strategy. Penina also adopted a local brooder where she raises the chicks, allowing the mother-hen to go back to laying. In that brooder she supplies the chicks with supplementary feed.

In the training, Penina was taught the importance of good feed for the birds, and how to make use of local feeds, such as fishing by-products which are readily available in the area.

Penina and the group have requested the government extension officers and other development partners to support them in this area, because they are ready to revolutionise their village chicken production system. According to the community members, this production system is now supporting many households in terms of increased food, nutrition and income security. Demand for eggs is also growing very rapidly as more households learn of the importance of a protein-rich diet, especially for those suffering with HIV/AIDS related diseases.

Photo 16.1 Penina Gari offering supplementary feed to her brooder-raised chicks (courtesy of K. Mugo)

Introduction

It has been estimated that in the developing world the ratio of poultry to human population is about 1.5:1 while in the developed world it is about 3.6:1. The rate of growth in the production of poultry is the highest when compared with ruminants and pigs (Brankaert *et al.*, 2000). Assuming that a typical average slaughter weight of a chicken for the market is about 1.2 kg (200 days old) and 2.2 kg (40 days old) in the developing and developed world respectively, then the quantity of poultry meat per person is about 2 kg and 8 kg per year. The large discrepancy between these ratios, and especially between the mass of poultry meat per head of population, demonstrates the vast gap in availability of protein from poultry for consumers in the developing and developed worlds. It also indicates that there is great scope for the increased production and consumption of poultry products in the developing world (Brankaert *et al.,* 2000). The difference between the consumption of poultry products in the developing and the developed worlds is anticipated to increase further by 2020, despite an increased per capita consumption throughout the world (Kristensen *et al.*, 1999). Furthermore, it has been estimated (Sonaiya, 2000) that up to 70 per cent of the poultry products in the developing world are produced by smallholders and in family-managed poultry systems.

Intensive farming of improved chicken breeds (broilers and layers) has spread worldwide. In the early 1990s, the population of village chickens in Indonesia was about 200 million birds and contributed about 47 per cent of the poultry meat and 18 per cent of the eggs produced in the country. In 2002, the population of village chickens was nearly 300 million birds and contributed 30 per cent of the poultry meat and 18 per cent of the eggs produced (Anonymous, 2002). The decreased proportion is, no doubt, due to an increase in the production of commercially-produced poultry meat, and is a trend seen in other parts of the world.

Nevertheless, the role of village poultry in resource-poor communities in all countries is still extremely important, especially in meeting the demand for meat and eggs for those who live in the villages (Kalita *et al.*, 2004a, 2004b; McAinsh *et al.*, 2004). Irrespective of where the resource-poor areas of the world are located and where scavenging poultry are reared, there are certain aspects that are common. The keepers of scavenging poultry usually maintain a few (2-10) birds; keepers are generally females and children and they are the poorest peoples in their communities. Smallholder production of poultry is primarily from free-ranging birds that have few or no inputs. These birds are known around the world by different names including family, scavenging, free range, desi, rural and backyard poultry. While chickens are the main species, other farmed poultry include ducks, turkeys, guinea fowl, quail and pigeons, with factors such as the environment and custom influencing the type of poultry grown. Although other poultry are important in village systems, chickens tend to predominate and thus this chapter will focus on them.

Photo 16.2 Scavenging poultry in India (courtesy of T. Acamovic)

In most family poultry systems, the native chickens have not been subjected to breeding programmes and the size, colour and productivity vary widely, with some characteristics more prized than others (Photo 16.2). Very little attention is given by farmers to managing the birds, so they tend to be kept at negligible cost. The chickens are often fed kitchen scraps and they receive no medication. As a result, the productivity of the chickens may be very low. Some studies show mortalities of chicks to be very high (50-80 per cent during the first six weeks of age) and egg production of hens to be very low (Table 16.1). With such high mortalities, villagers have little incentive to invest time or resources into managing the poultry and consequently farmers tend to obtain only a small number of eggs and therefore chicks. Some farmers provide small houses, baskets, or cages located in, or next to, their own abode to house the chickens at night. Other birds roost in trees (Photo 16.3).

Photo 16.3 Overnight accommodation for desi birds inside a farmer's house in India (courtesy of T. Acamovic)

Although scavenging poultry grow slowly and egg production is low, the price of the meat and eggs are relatively high (three to four times the price of intensively-farmed produce). This price differential reflects the preference of the indigenous people for meat and eggs that match their organoleptic (taste and texture) preferences. In Indonesia raw eggs are often used in a mixture with traditional herbal drinks because it is considered that they improve the health of the consumer.

Table 16.1 Performance characteristics of scavenging chickens

Parameters	
Body weight at day old (g)	24 – 30
Body weight at 10 weeks (g)	454 ± 220
Body weight at 20 weeks (g)	1027 ± 307
Age at point of lay (months)	6 – 7
Eggs laid (/hen/year)	30 – 80
Egg weight (g)	36 – 48
Hatchability of eggs (%)	39-86

Sources: Kingston and Creswell (1982); Gilchrist *et al.* (1994); Shindey and Pathan (2002). LPP Project R7633.

Most resource-poor families in South America, Africa, India and South-East Asia, whether they live in rural, peri-urban or urban communities, maintain a few poultry. These people can be classified according to their relative wealth and include landless people, people who have a small amount of land adjacent to their home and those that have land that is separate from their home (Conroy *et al.*, 2004a, 2004b). The latter two groups often integrate crop and animal production, both of which can benefit scavenging poultry by the provision of waste seeds for consumption and the production of a dung heap where the birds can harvest insects and worms. In households that maintain poultry with other animals, the birds often predate on insects and other parasites that affect cattle and goats, thus reducing the problems of parasites in these animals. However, the proportion of landless keepers is likely to increase, due to an increasing trend towards urbanisation in the developing world. Irrespective of the system, the poultry are frequently left to scavenge on vegetable waste, insects and whatever animal feed they can find. They are infrequently provided with supplementary feed, which amounts to a few grams of broken or waste cereal grain not used for family consumption. Where poultry-keepers also have crop land, the residual material from the crops (spilled seeds, etc.) is consumed by the poultry scavenging in the fields.

A common aspect of the sale of poultry throughout the developing world is that a high proportion is sold live. Where villagers are less organised, the birds are exchanged in a barter system (e.g. tribal areas in Rajasthan). However in some cases, e.g. in Tamil Nadu, in Southern India, there is an organised system of traders who collect poultry from villagers. The birds are then transported loose but tied to handlebars or in various containers, including baskets and crates on motorcycles, pedal cycles, buses, cars and trucks, trains and boats, to towns and sold in local markets (shandys in India). Sometimes the birds and eggs are transported further afield.

Elsewhere a more integrated village system has developed to rear and grow poultry and to produce and sell eggs. An example of such a well-developed system is one that exists in some areas of Bangladesh for the production of eggs. About 80 per cent of landed and landless households in Bangladesh maintain a small number of poultry. The rearing systems differ between the different methods of production of poultry, from scavenging, semi-scavenging, intensive smallholders to commercial production systems.

The so-called 'Bangladesh system' is integrated and highly organised, and since 1978 has been supported by the Danish (DANIDA) and the Bangladeshi Governments (Directorate of Livestock Services) and by NGOs (BRAC). The project has involved over 100,000 households in all aspects of the production of table eggs for sale and for local consumption. The system is relatively complex, consisting of trained production, marketing, supply, and service groups. The production group consists of hatchers and rearers (rearers/breeders, chick rearers and pullet rearers). The supply group consists of the parent stock farm, mixers and sellers of feeds, poultry workers (vaccinators, etc.), egg collectors and sellers. The services group includes surveyors and information gatherers, trainers, credit/saving schemes and extension provision. The system is highly dependent on micro-credit, good management practice and a well-managed system for the production of hatching eggs, chicks, layers and distribution of saleable eggs. Furthermore a supply of good quality concentrate diets is necessary for the system to function effectively. Thus, although the system is beneficial for the local population, it cannot be considered a scavenging system and requires investment, high standards of hygiene and skilled management, which is frequently unavailable in the developing world. Furthermore it is dependent upon a high population density, something which makes it difficult to transpose to some African countries, for example.

It is encouraging that various international organisations such as ACIAR, DFID, DANIDA, EU and FAO, as well as numerous NGOs in various countries (e.g. BAIF in India; BRAC in Bangladesh), have supported and have an increasing interest in the development of small-scale poultry production in parts of the developing world.

Reasons for keeping poultry

In India and other areas of the world commercial poultry production, although fairly well developed, continues to develop rapidly. Much less attention has been given to the development of the family poultry systems. In order to improve the nutritional and economic security of the people living in rural, tribal and inaccessible areas in a sustained manner, the promotion of rural poultry is appropriate and actively practised. Although, with the exception of the poorest of the poor, poultry are secondary to other agricultural activities, they have an important role in providing the local population with income and high-quality protein. The scavenging (desi) birds in villages constitute a source of ready money and they have been aptly termed 'walking money' (Mariadas, 2000). In India the women are said to be owners of the poultry and thus the role of rural women in backyard poultry production is highly important and allows them some financial freedom. In some circumstances this financial freedom can be challenged by men if poultry keeping appears to be a lucrative activity.

Table 16.2 Poultry products and services

Products	*Services*
Meat (raw, cooked, soup)	Cash income
Eggs (raw, cooked)	Gifts
Manure (crops, fish)	Loans
Nutrient supply (ruminants)	Religious rituals
	Pleasure/sport
	Medicine
	Barter material for other goods

Throughout the world, poultry provide their resource-poor owners with a range of products and services, some of which are listed in Table 16.2. The most important reasons for keeping free-ranging poultry vary between regions and countries, but the highest priorities are for meat, eggs and cash income (McAinsh *et al.*, 2004; LPP Project R7633). It is interesting that even in developed cultures, where inputs are available and affordable, there is a desire to raise 'organic' and 'free-range' poultry; the three main reasons given are identical to those of poor peoples, and also include the production of 'healthy' and 'tastier' poultry products. Poultry are inexpensive, relatively cheap to acquire and are frequently the first asset acquired through purchase or customary means by a young family, or by a resource-poor family recovering from a disaster, such as drought, flooding or war. Poultry are a valuable asset, providing security to the family as well as products such as eggs and meat.

Recent studies in villages in Tamil Nadu and in Rajasthan in India (LPP Project R7633), revealed that the priority reasons for keeping poultry differed from area to area and from village to village. There are also differences between different peoples. In Tamil Nadu, for example, the prime reason for keeping desi poultry is for home consumption or for selling in markets. In Rajasthan, especially within tribal communities, the major reason is for special occasions, including religious ceremonies, and for gifts to friends. The reason next in importance, cited by landholding farmers, is to generate income, whereas among landless people the chicken is a 'ready source of income' when they need money. Apart from that, poultry are grown for traditional rituals and for sacrificing the cocks before gods and goddesses during times of festivals and worship, gifts to relatives, entertaining guests and also it is a hobby to some of the villagers. Family flocks are also kept as a symbol of wealth, to be used in ceremonies.

Further, poultry are used in the barter-system in exchange for other livestock or items. Some of the advantages (and disadvantages) of keeping poultry are presented in Table 16.3. Obviously one of the main advantages of maintaining village poultry is to assist in reducing protein malnutrition. Indigenous birds and their eggs also fetch a price three to four times more than that of commercial broilers, because the colour and unique flavour of the meat and eggs are preferred by buyers. It has also been observed, by questioning (LPP Project R7633), that the bones (which are chewed by the consumer) of cooked desi chickens are less friable than commercial broilers and consequently are preferred.

Table 16.3 Advantages and disadvantages of poultry for resource-poor people

Advantages	*Disadvantages*
Products	*Small size*
Eggs are a compact source of valuable nutrients.	Low value often makes formal credit systems uneconomic.
Each bird is a small amount of meat (1-1.5 kg), thus reducing the problems of storage.	Susceptible to predators, trampling (chicks) and thieves.
If eggs are consumed, the source (hen) survives and continues to produce.	*Reproductive capacity*
	Reproductive capacity and viability of offspring susceptible to climate variation.
Desi poultry products command higher prices than intensively-produced meat and eggs.	*Feeding*
	Inefficient utilisation of high-fibre feeds.
Lack of religious taboos against poultry products.	More susceptible to nutrient imbalances than ruminants.
	Susceptible to toxins in feeds.
Small size	Susceptible to feed shortages if in a village system of production.
Cheap to purchase.	
Easy for women and children to handle.	
Few facilities required to keep poultry.	*Disease*
Easy to transport to market.	Susceptible to Newcastle disease, avian flu and Gumboro.
Easy to slaughter at home.	
Financial loss of a chick is not as critical as loss of larger animal.	Susceptible to internal parasites.
Low labour and input costs.	Act as reservoirs and vectors of some diseases and parasites.
Easily housed in secure premises.	
Relatively easy to get veterinary treatment (compared to large animals).	*General*
Small size enables fast movement of households in emergencies.	Larger flocks demand higher input of higher-quality feedstuffs.
	In some cases they may 'browse' young crops and reduce yields.
Reproductive capacity	Can compete with humans for food ingredients.
Rapid reproduction, thus eggs and chicks can be produced quickly.	

Table 16.3 (Contd.)

Advantages	Disadvantages
Reproductive capacity (contd.)	
Fast reproductive rate ensures early returns on investment and enables early credit repayment.	
Feeding	
Able to select and use spilt seeds at harvest.	
Able to utilise insects as a nutrient supply.	
General	
Adapted to wide range of climates.	
Suitable for landed and landless farmers.	
Manure improves soil fertility.	
During scavenging poultry cultivate soil.	
Can reduce insect damage to crops.	
Can reduce ecto-parasite burdens on other livestock, e.g. cattle.	
Provide a source of nutrients (excreta) for fish.	
Can provide a valuable supplementary income (especially for women).	
Little requirement for management or monitoring (if scavenging/free-ranging).	

In some cases it has been reported that eggs from desi poultry are nutritionally different and have higher threonine and valine contents compared to the eggs of commercial breeds (Khan, 1983). The cholesterol content in desi eggs and chicken is also significantly lower than that from commercial birds. Such differences in composition and content of some compounds are likely to be a function of the diet that the different types of birds consume, as well as the differences in breeds and size of the eggs produced. In some countries, such as the Philippines, duck eggs that have been incubated for 16-18 days, or others that have been incubated but have dead embryos, or are infertile, are cooked and eaten as a delicacy. These eggs are known as balut and penoy, and provide a variety of nutrients to the consumer.

Conventional desi poultry have other characteristics that are beneficial. The birds possess greater maternal instincts, i.e. broodiness, and can be used advantageously to incubate and rear chicks under rural conditions. Because of the coloured plumage, alertness and fighting characters, these birds are more likely to be able to protect themselves and their chicks from predators. Since the intensity of production of desi birds is low, there is no pollution problem. Indigenous birds are comparatively disease-resistant or tolerant of common infections compared to modern commercial birds, and they thrive under adverse environments such as poor housing, poor management, poor feeding (both quality and quantity), and variable temperature and humidity. The importation of higher-producing newer strains of birds has often failed for a number of reasons, including their inability to tolerate disease, heat and parasite infections. Novel or hybrid birds are also rejected because they do not have the stature, brooding characteristics and colours of indigenous poultry. As well as the characteristics listed previously, it is worth noting that in some parts of the world, e.g. Nigeria, some rare genetic conditions such as polydactyly, brachydactyly, ptiploidy and dwarfism are characteristics that enhance the value and encourage the keeping and production of village poultry. Furthermore, fancy traits, such as head spurs, crests, variation in comb colour and shape, are valuable characteristics which are sought after by villagers (Ikeobi and Oladotun, 1998; Ikeobi *et al.*, 1998; Ikeobi and Godwin, 1999; Adembambo *et al.*, 1999; Ikeobi *et al.*, 2001). In some cases, variation in feathering adds value to the birds that are kept by villagers. For example naked neck, frizzled feathers or a combination of both, are considered to be attractive characteristics (Ebozoje and Ikeobi, 1995).

Scavenging birds also help reduce the infestation of ectoparasites like ticks and lice on the cattle, sheep or goats reared by these farmers, and so reduce the incidence of diseases such as babesiosis, for which the ectoparasites act as vectors (Photo 16.4). Heavy mortality in chicks caused by disease, malnutrition and by predation are however major constraints to the development and production of scavenging birds. The feed-related problems are of greater importance for landless labourers than for landholders.

Photo 16.4 Chicken pecking ectoparasites from a calf in Zimbabwe (courtesy of B. Mupeta)

Problems faced by resource-poor people and constraints to production of poultry

The problems faced by poultry-keepers are many and reflect, probably to a greater extent, those that have been presented for other animals. Because of their low value, villagers often are ignorant of small changes that could enhance the quality, health, productivity and numbers of birds in their flocks. In the past, NGOs, donor agencies and governments have done little to educate villagers in these small changes that would benefit poultry production. Factors such as feed quality and availability are problems, although the range of feedstuffs is probably greater than for herbivores. Nutritional constraints in the use of most plant feedstuffs are probably greater with poultry than ruminants, since poultry have a very limited ability to utilise fibrous feedstuffs, and are susceptible to a greater range of phytochemicals, including anti-nutrients and toxins. Similarly, the size of seeds and their resistance to being mechanically degraded can be disadvantageous. However, poultry have the ability to be highly selective in what they consume, thereby overcoming physical deterrents such as spikes and thorns present in some potential feeds. They can then select insects and nutritious parts of plants that are inaccessible to mammals. There is a constant disease challenge to scavenging poultry because of the different ages of the birds in a flock, possible transfer from wild birds, and constant use of the land by poultry. In some instances access to markets is restricted, although there are well-developed systems for marketing poultry products in some areas of the world. Without the encouragement of outside agencies there are few inter-farmer discussions about poultry production at the village level. A major limitation to the growth of scavenging poultry is the availability of nutritious feedstuffs.

Ill-health and mortality

Since scavenging poultry are kept with minimal input by local keepers, relatively high mortality, from whatever cause, tends to be tolerated. Although this is more the case where there is very little input and less so with higher input systems, poultry-keepers aspire to have higher survivability and health in their flocks. Furthermore, commercial producers desire to ensure the good health of local scavenging birds to try and avoid the village birds acting as reservoirs for diseases such as Newcastle disease and avian influenza. Predation is a significant cause of loss in family poultry. Young chickens below eight weeks of age are extremely vulnerable to predators, which can account for more than 80 per cent of the mortality. Huque (1989) reported that 18-32 per cent of all chicks are lost due to predation during the brooding period. Recent studies in Southern India have also shown that 21 per cent of all birds died because of predation and this amounted to 73 per cent of all deaths (Natarajan *et al.,* 2004). Thus it is clear that for young birds, predation by birds of prey, crows, snakes and animals such as the mongoose and wild cat, contribute substantially to losses of scavenging poultry. Methods of protection such as the presence of bushes, areas under string, and housing hanging from trees and inside the dwellings of livestock-keepers, tend to reduce mortality from predators, and only the weaker birds are predominately predated upon. In most cases the farmers do not follow any systematic breeding programme in rural poultry, resulting in poor-quality stocks, but it is appreciated that the indigenous birds tend to have a greater resistance to various diseases.

Mortality in adult birds is mainly due to disease. In a recent survey of poultry-keepers in India (LPP Project R7633) it was discovered that the disease considered to cause most deaths was a paramyxovirus commonly known as Newcastle disease. This disease is extremely contagious and devastating to poultry flocks, causing up to 90-100 per cent mortality. When mortality is sporadic, it is unlikely that Newcastle disease (unless confirmed by diagnosis) has been the cause of death, although most scouring diseases tend to be attributed to it. Indeed unvaccinated desi birds in a village (Peruganur) in Tamil Nadu in India, had an antibody titre value below 16 in 95.6 per cent of the blood samples taken (Natarajan *et al.,* 2004), suggesting that they had not been exposed to Newcastle disease. Much excellent work has been conducted on the protection of village poultry from Newcastle disease, supported by agencies such as ACIAR and FAO/IAEA, where vaccination procedures have been developed and adopted (Spradbrow, 1999; Al-Garib, *et al*., 2003). Many other diseases of bacterial, protozoal, fungal and viral origin affect poultry. Aetiology effects, management and methods for protection and alleviation have been well documented (Calnek *et al*., 1991; Sinurat *et al.,* 1998; Spadbrow, 1999; Permin, 1997; Muralimanohar, 2004).

The most commonly observed diseases are Newcastle disease, fowl pox and fowl cholera. Since no proper vaccination is practised in most desi birds, Newcastle disease is an important limiting factor in their productivity. However, it is worth noting that considerable effort has been invested in developing vaccines and vaccination programmes for village poultry systems in Africa (Spadbrow, 1999; Alexander *et al.,* 2004). Scavenging birds also suffer from ecto- and entero-parasites. Internal parasitism (cestodes and nematodes) is very common in scavenging chickens. The presence of a few parasites does not usually cause a problem. However, large numbers can have a devastating effect on growth, egg production, and overall health. Recent work has demonstrated that more than 90 per cent of village birds examined in Rajasthan and Tamil Nadu (LPP Project R7633) were infected with intestinal parasites. In some cases the gastro-intestinal tract (GIT) was essentially blocked with parasites. Obviously the competition for nutrients, damage to the GIT and blockage of the GIT severely limit performance of the birds and increase the susceptibility to other diseases.

Some cases of ill-health also occur because of lack of feed and nutrients, as well as the consumption of feedstuffs that contain toxic or anti-nutritional compounds. Conversely, the consumption of feedstuffs with some anti-nutritional or toxic compounds may have beneficial effects because of their effects on the microflora and parasites within the GIT (LPP Project R7633; Mathias *et al.,* 1999; Acamovic *et al.,* 2004; Acamovic and Brooker, 2005). We have found that the supplementation of layers with high-tannin sorghum instead of low-tannin sorghum resulted in better egg production and viability (Acamovic *et al.,* 2004).

Feeding scavenging poultry

Water

Because of the small size of poultry, and their small numbers in the family poultry-system, there is a minimal requirement for water. Except in arid and semi-arid areas, sufficient water is usually available where humans have settled. It is important to recognise that puddles and small ponds, as well as rice paddies, are good means for the transfer of water-borne diseases and parasites. In any areas where housed birds may be trapped, they may be drowned in monsoon periods, although where there are trees, they may have the opportunity of surviving in the event of flooding.

Minerals and vitamins

In the purely-scavenging systems it is possible that mineral imbalances may occur, and also that vitamin deficiencies, especially A and E, may occur in environments where cereals are the main constituent of diets. Deficiency of vitamins is likely to cause ill-health in birds and reduced hatchability in eggs. Similarly, deficiencies of minerals, especially calcium, will cause problems with egg-shell quality and thus decrease hatchability and increase fragility of the eggs. In areas where there are high deposits of heavy metals, or where birds consume by-products from industrial sources and sewage sludge and slurry, it is possible that mineral toxicity may occur. However, during recent surveys in India, neither vitamin nor mineral deficiencies nor toxicities have been mentioned as problems.

Cereals and other crop products

In purely scavenging systems, foodstuffs are not produced for consumption by poultry. They are frequently left to scavenge seeds after harvest or from grasses and other 'weeds' that are available. Thus the feed supply is very variable and seasonal. Sometimes they are provided with supplementary waste seeds or by-products from cereal and other grains (Sonaiya, 2000, 2004). It is also difficult to convince villagers of the value of harvesting weed seeds as a supplement to their poultry's diet. This is particularly difficult if the seeds have some potential for human use, for example the use of hama grass for poultry in North-West India was not well-received by tribal peoples (LPP Project R7633).

Other feedstuffs

The use of algal material and water plants has been considered appropriate for feeding to poultry since they provide carotenoids and other fat-soluble vitamins. The carotenoids enhance the colour of the yolks in eggs and the beneficial effects of the carotenoids and vitamin E on egg quality and viability of the hatchlings has been well documented (Surai, 2002). The deeper the egg yolk colour and skin pigmentation due to carotenoids (beta carotene is a precursor of vitamin A and xanthophylls), the greater will be the supply of vitamin A to the consumer and, frequently, the greater the attraction of the meat and eggs for the consumer.

Since poultry are not vegetarian animals, they can supplement their diets by the consumption of materials such as insects, worms, larvae, pupae and any scraps of tissue from the butchering of other animals. It has been demonstrated and promulgated that termites are a good source of nutrients for poultry. However, villagers in India have indicated that the termites may damage the GIT of birds that consume them and that the termites should be killed before allowing the birds to eat them (LPP Project R7633). Thus the enthusiasm to develop systems to promote termite cultivation for poultry consumption (by scavenging) may have to be tempered with the advice to villagers to make sure that they kill the termites prior to consumption by their birds. This input of labour is a disincentive to villagers. Dung heaps from the waste of cattle, goats and sheep are also excellent sources of insect protein for scavanging birds. The birds also assist in composting by 'working' the dung heap and encouraging decomposition. However dung heaps may act as sources of parasites (e.g. coccidia) and other diseases.

Housing

During daylight hours poultry are frequently allowed to scavenge whatever food they are able to find in the local environment. In the evening the poultry are housed in a variety of environments. In some cases the birds are allowed to roost in the branches of trees or in enclosed baskets hanging from trees. However other facilities, such as baskets within the keeper's dwelling, are frequently used. These baskets can be located on the floor or in the rafter space within the dwelling (Photo 16.3). This is obviously the most secure overnight location avoiding predators and theft. Other methods involve either wooden or brick accommodation attached to the dwelling house or even stone or brick-built accommodation which is separate from the family dwelling. These houses tend to be less prevalent, primarily because of the cost of construction. Also brick-built housing tends to be difficult to clean and thus presents a potential threat due to the build-up of pathogens. In the more semi-intensive system in Bangladesh, wire netting accommodation is constructed to house a larger number of birds.

Reproduction

It is clear that egg production and hatchability are compromised in hot conditions especially when humidity is high. The lack of viability and poor quality of the eggs are due to a variety of reasons, including becoming wet and dirty, which allows the transfer of bacteria through the shell into the contents, and the high temperatures leading to premature embryo death (so-called pre-incubation).

Storage and candling of eggs

Conventionally, eggs are collected in baskets or similar containers after they are laid and the brooding hen is then allowed to incubate them. Hatchability is extremely variable, with low hatchability being accepted as the norm (Table 16.1). Storage in earthenware containers, or baskets with sand and water, to maintain eggs in a cool environment (below 27 °C) prior to incubation by the brooding hen, increases hatchability (LPP Project R7633; Sparks and Shindey, 2004). Candling of eggs is a standard method of ascertaining the quality of hatching eggs. This is achieved by shining a light (from a torch) through the egg to determine if eggs are viable, have embryos that are viable, are not fertile, have cracks, or are contaminated with bacteria. In this way eggs which are not fertile can be removed and consumed, rather than be allowed to be incubated until they should hatch, by which time they are inedible. This frequently provides families with extra protein, which would otherwise be unusable. Furthermore, eggs that are determined to be fertile can be incubated until they hatch and thus hatchability, as a proportion of eggs that are incubated, is much higher. Candling has been adopted with enthusiasm by tribal poultry-keepers in the Rajasthan area of India, where increased hatchability was obtained (LPP Project R7633). Villagers now appreciate the benefits of candling and employ a candler to test their eggs. Remuneration of the candler is often by provision of 50 per cent of the rejected eggs from the poultry-keeper.

Breeds

Typical village chickens reared in scavenging conditions have not been systematically selected for particular traits such as meat or eggs. In some areas particular characteristics (mentioned earlier) assume greater importance in the selection and keeping of poultry. In the case of the Bangladeshi system, the local Fayoumi crossed with Rhode Island Red (known as a Sonali) was selected to combine higher yields of eggs (greater than 150 p.a.) with resistance to local environmental, parasitic and disease challenges.

The appearance of village chickens varies widely, with feather colour lacking uniformity, being mixtures between red, black, grey, white and yellow. Generally the chicken has a slow growth rate and low egg productivity (Table 16.1). The hens brood after laying 10-15 eggs and nest (incubate) the eggs for 21 days. The hens normally look after the newly-hatched chicks for about 6-8 weeks before they start laying eggs again.

Improving production of poultry by resource-poor people

In the Bangladeshi system a suitable cross was selected that allowed increased production of eggs with resilience against disease. However, laboratory studies by Creswell and Gunawan (1982) showed that the productivity of village chickens could approach the performance of improved breeds when reared intensively and provided with high-quality diets. In this system, body weight of village chickens at 12 weeks of age was 575-708 g, and at 20 weeks 1203-1480 g. The age at point of lay was also reduced to 134-163 days, and the egg production increased to 119-215 /year. However, the study also showed that the overall performance of village chickens was poorer than those of improved breeds of layers. Although the performance of the chickens was improved in the research station, application of this rearing system by small-scale farmers in villages would require much development. Small-scale farmers normally have limited knowledge and skills both for rearing chickens intensively and using good management procedures. Therefore, adjustment of the methods of rearing chickens in villages is required.

A study was conducted in an Indonesian village in order to improve the performance of village chickens (Sinurat *et al.*, 1992). Comparisons were made between three management systems:

- scavenging (as normally practised by small-scale farmers)
- semi intensive (farmers were given information on how to rear chickens more effectively, extra feed was given and the chickens vaccinated against Newcastle disease)

- intensive rearing systems (farmers were trained, birds were confined, fully-fed and vaccinated against Newcastle disease).

The performance of the village chickens in the three systems is presented in Table 16.4. The performance of chickens increased as the management improved and as a result, more eggs and chickens were consumed and sold by farmers.

Table 16.4 Performance of village chickens under different management regimes by small-scale farmers in Indonesia

Parameters	*Scavenging*	*Improved (semi intensive)*	*Improved (intensive)*
Number of adult chickens owned	7.6	8.2	40.3
Mortalities to 6 weeks (%)	50.3	42.6	27.2
Body weight at 5 month of age (g)	n.a	609.0	707.0
Age at first lay (months)	n.a	8.5	7.5
Egg production (eggs/hen/year)	30.2	59.1	80.2
Number of eggs consumed (/year)	22.5	72.6	262.7
Number of egg sold (/year)	23.4	57.0	484.5
Number of chickens consumed (/year)	4.0	17.3	30.3
Number of chickens sold (/year)	2.6	16.8	57.0

n.a.: Data not available
Source: Sinurat *et al.* (1992).

The use of crop by-products as supplements to scavenging poultry is a method of improving production. In Africa, a group of ten innovative women in a dry area in North-East Zimbabwe, participated in village poultry trials (LPP Project R7524) to evaluate the performance of village chickens and commercial hybrids fed a low-cost, low-fibre sunflower-residue (SFR) poultry diet formulated with local feeds. Comparisons were made between scavenging (limited supplementation with a low-fibre SFR diet) and an intensive system (village and commercial hybrids, kept in pens and offered SFR diet ad libitum). The project was managed by farmers, after they had received training, and was supervised by research and extension staff. Each of the women reared 30 chickens (10 hybrids, 10 village chickens in pens and 10 village chickens scavenging during the day and kept in pens overnight where they received limited supplementation).

Scavenging poultry showed superior returns on money invested compared to village chickens under an intensive system (Table 16.5). Performance, measured as feed efficiency, gross margin and return per dollar, improved as the village chicken increased in age from 8 to 12 weeks. The meat from village poultry contained more protein and less fat than the hybrids at the same age, indicating a possible market for leaner meat from village chickens. However commercial hybrid birds would be sold at a younger age and lower weight than that those shown in Table 16.5, when it would be likely that the fat content would be lower. Where sunflowers are grown for oil and processed locally, low-fibre sunflower residue included in poultry diets is a viable option for increasing production from smallholder-owned poultry.

Table 16. 5 Performance of village and hybrid chickens in a farmer-managed trial in North-East Zimbabwe

Parameters	*Intensive (pens)*	*Scavenging and supplemented*	*Commercial hybrid*
Body weight at day old (g)	38 ±1.5	38 ±1.5	42 ±2.0
Body weight at 8 weeks (g)	720 ±184	700 ±215	2430 ±87
Body weight at 12 weeks (g)	1190 ±158	1050 ±260	3370 ±102
Crude protein in breast meat (g/kg DM)	743-745	739-742	689-692
Ether extract in meat at 8 weeks (g/kg DM)	338-340	334-338	491-514
Feed conversion rate, 8-12 weeks (g feed/g live-weight gain)	3.2-3.0	2.7-2.6	2.6-4.0
Return/$[1] invested 8-12 weeks	1.30-1.48	1.55-1.61	1.80-1.30

[1]Zimbabwe $

Source: Mupeta *et al.* (2003); LPP Project R7524.

Simple management, a distinct and agreeable taste of the meat and eggs, and higher prices for the products are some of the advantages in rearing village chickens. However, small numbers of chickens and poor management frequently prevent resource-poor farmers from benefiting from these advantages. Resource-poor farmers also have very limited capital, so it is inappropriate to ask them to invest in and adopt intensive rearing systems. Farmers need capital to buy feed and medication to allow the system to function effectively.

A small project was carried out to help resource-poor farmers in three neighbouring villages in Indonesia. Forty participants were selected from 75 farmers interviewed prior to the project. All participants were trained to rear chickens more effectively and were supplied with 13 pullets and two cockerels at the beginning of the project.

Information on the performance of scavenging chickens and its contribution to diet and income of farmers was collected prior to and during the 18-month project. All chickens were also vaccinated against Newcastle disease following the normal recommendations. The chickens were kept in small bamboo houses at night and allowed to scavenge during the day. Farmers were encouraged to give some extra feed such as kitchen scraps and rice bran. The results are shown in Table 16.6.

Table 16.6 Results of an 18-month project with resource-poor farmers in Indonesia to improve management of scavenging poultry

Parameters	*Before project (baseline survey)*	*During 18-month project*
Number of co-operating farmers	75	40
Number of adult chickens owned/family	4.5	12
Number of young chickens owned/family	7	41
Egg production (/hen/year)	32	53
Mortality of chicks (%)	61	37
Mortality of adult chickens (%)	40	12
Number of eggs consumed (/family/year)	22.5	128
Number of eggs sold (/family/year)	8.0	327
Number of chickens consumed/family/year	10.8	12
Number of chickens sold/family/year	12.3	36

Adapted from Gilchrist *et al.* (1994).

Improving the management improved the performance of the scavenging chickens, as shown by a reduction in mortality of the chickens, and increased productivity of the hens. As a result, more eggs and meat (chickens) became available to be consumed or sold by farmers. From the results it may be concluded that improving the rearing management of scavenging chickens could improve the welfare and health of resource-poor farmers. Similar results have been found in India, where small interventions caused disproportionately beneficial effects (LPP Project R7633).

The future

It is likely that the scavenging poultry production system will continue and possibly increase as householders see greater benefits from having extra eggs and more meat. A detraction from the continued production of backyard poultry may be the increased production of relatively inexpensive, intensively-reared birds which may reduce the market opportunities for those families that sell their produce. However there is a need to continue to encourage backyard poultry farming in rural areas to reduce hunger and protein malnutrition, especially that experienced by women and children. Poultry also provide supplemental income to the local population in developing countries. Much more effort is required to provide backyard poultry farmers with information on the benefits of maintaining poultry in good health and productive states by supplementary feeding and vaccination. The provision of information can be done through community-based, animal health-care workers (CBAHW), NGOs, and government organisations, by word of mouth and demonstrations, as well as through radio broadcasts, cameos and cartoons (see Chapters 5 and 6).

There is evidence to suggest that, in some communities, efforts currently being put into large-scale breeding programmes and similar schemes should be diverted to providing sound husbandry advice and guidance on how to manage products (table and hatching eggs, as well as the marketing of the meat). Thus activities such as organised health management programmes, with basic training covering vaccination, deworming and treatment or prevention of major diseases, should be developed further and implemented. Advice should be provided on good management procedures such as providing protective shelters/housing for the birds to reduce predator problems. Supplementary feeding should be encouraged to rectify malnutrition problems due to shortage of feeds during the period between different agricultural operations and during periods of stress, including drought.

Further reading

Alexander, D.J., Bell, J.G. and Alders, R.G. 2004. *A technology review: Newcastle disease- with special emphasis on its effects on village chickens. FAO Animal Production and Health Paper 161*, Food and Agriculture Organisation of the United Nations (FAO), Rome, Italy.

Calnek, B.W., Barnes, H.J., Beard, C.W., Reid, W.M. and Yoder Jr., H.W. 1991. *Diseases of poultry. 9 th edition*. Iowa State University Press, Iowa, USA.

Grimes, S.E. 2002. *A basic laboratory manual for the small-scale production of and testing of I-2 Newcastle disease vaccine. FAO RAP publication 2002/ 22*, Food and Agriculture Organisation of the United Nations (FAO), Rome, Italy.

Kitalyi, A.J. 1998. *Village chicken production systems in rural Africa. Household food security and gender issues. FAO Animal Production and Health Paper 142*, Food and Agriculture Organisation of the United Nations (FAO), Rome, Italy.

Kristensen, E., Larsen, C.E.S., Kyvsgaard, N., Madsen, J. and Hendriksen, J. 1999. Livestock production: the twenty first century's food revolution. In *Poultry as a tool in poverty eradication and promotion of gender equality.* Proceedings of a workshop, March 22-26, Tune Landsboskole, Denmark. At: http://www.husdyr.kvl.dk/htm/php/tune99/index2.htm (December, 2004).

Mallia, J.G. 1999. Observations on family poultry units in parts of Central America and sustainable development opportunities. *Livestock Research for Rural Development 11.* At: http://www.cipav.org.co/lrrd/

Prabakaran, R. 2003. *Good practices in planning and management of integrated commercial poultry production in South Asia. FAO Animal Production and Health Paper 159*, Food and Agriculture Organisation of the United Nations (FAO), Rome, Italy.

Rose, S.P. 1997. *Principles of poultry science.* CABI Publishing, Wallingford, UK.

Smith, A.J. 2001. *Poultry.* The Tropical Agriculturalist, Series Editors R. Coste and A.J. Smith. Macmillan Education Ltd., London, UK, in cooperation with Technical Centre for Agricultural and Rural Cooperation (CTA), Wageningen, The Netherlands.

Surai, P.F. 2002 *Natural antioxidants in avian nutrition and reproduction.* Nottingham University Press, Nottingham, UK.

Author addresses

Thomas Acamovic, Avian Sciences Research Centre, Scottish Agricultural College (SAC), West Mains Road, Edinburgh EH9 3JG, UK (Address for correspondence: Avian Science Research Centre, SAC, Ayr KA6 5HW, UK)

Arnold Sinurat, Indonesian Research Institute for Animal Production, P.O. Box 221, Bogor 16002, Indonesia

Amirthalingam Natarajan, Department of Animal Nutrition, Veterinary College and Research Institute, Tamil Nadu Veterinary and Animal Sciences University, Namakkal 637002, India

Kaliappan Anitha, Department of Animal Nutrition, Veterinary College and Research Institute, Tamil Nadu Veterinary and Animal Sciences University, Namakkal 637002, India

Doriasamy Chandrasekaran, Department of Animal Nutrition, Veterinary College and Research Institute, Tamil Nadu Veterinary and Animal Sciences University, Namakkal 637002, India

Dilip Shindey, BAIF Development Research Foundation, 969 Paneriyo ki Madvi, Near Water Tank, Udaipur 313002, Rajasthan, India

Nicholas Sparks, Avian Sciences Research Centre, Scottish Agricultural College

(SAC), West Mains Road, Edinburgh EH9 3JG, UK (Address for correspondence: Avian Science Research Centre, SAC, Ayr KA6 5HW, UK)
Oluseyi Oduguwa, Department of Animal Nutrition, University of Agriculture, Abeokuta University, Nigeria
Bartholomew Mupeta, Plan International, P. Bag 7232, Highlands, Harare. Zimbabwe
Aichi Kitalyi, Regional Land Management Unit (RELMA) at World Agroforestry Centre (ICRAF), P.O. Box 30677, Nairobi 00100, Kenya

17

Grasscutters, guinea pigs and rabbits

Emmanuel Adu[*], *Rob Paterson*[†], *Franz Rojas*[†], *Germana Laswai*[†], *Dennis Fielding*[‡], *Emmanuel Osafo*[‡]

Cameo 1 - Grasscutter farming in Ghana

Teye Ocansey, a resident of Awoshie, in Accra, Ghana, started grasscutter farming in 1994 with three wild grasscutters in his backyard. His efforts to expand the farm failed due to lack of information on proper management practices. The Animal Research Institute, Council for Scientific and Industrial Research (CSIR), Achimota, introduced him to improved management practices and he has since built his colony to over 300 breeding animals. From his sales he has built a corner shop for his wife, bought a second-hand taxi and a mobile phone! Similar results have been reported in Benin; and many NGOs and development agencies are actively promoting grasscutter farming in West Africa.

Cameo 2 - Rabbit farming in Egypt

Through raising rabbits on her husband's meagre salary Om Ahmed, a farmer in Cairo, Egypt, has been able to put her children through school. From an FAO rabbit project, she received animals from which the 30 rabbits she currently owns are descended. Om Ahmed cooks rabbit meat for her family of eight once a week. Some families on this project are now raising several hundred rabbits per year (FAO, 2003). FAO has supported and developed rabbit projects in several other countries, including Ghana, Guinea-Bissau, Equatorial Guinea, Haiti, Mexico, Rwanda, Sao Tome and Principe and the Republic of Congo (FAO, 1999).

---

[*] Principal author
[†] Co-author
[‡] Consulting author

Small mammals in livestock farming

Land is a limiting resource in many developing-country locations. Livestock-keeping is mostly on a smallholder basis, not only because of land limitation, but also due to shortage of finance for the acquisition of improved breeds, and lack of technological support for efficient livestock production. Landless people, particularly women and children, have found backyard production of small mammals an important source of part-time work (Ehui, 1999). The small size makes them easy to raise and to handle by vulnerable members of the household (women and children) in peri-urban and rural areas. The most common species kept are guinea pigs, rabbits and grasscutters.

An important feature of these species is their ability to survive on a low-quality diet due to three common factors:

- The practice of caecotrophy, i.e. the ability to recycle faecal-like pellets known as soft faeces produced in the caecum. Caecotrophy helps to recycle some of the unabsorbed nutrients as well as returning proteins and vitamin B-rich bacteria for enzyme digestion in the small intestine. About 10 per cent of the amino acid requirements of the adult rabbit can be provided through caecotrophy. Caecotrophy also increases the digestibility of tannin-containing protein sources.

- Most importantly, the rate of passage of digesta through the gastro-intestinal tract is fast, enabling the soluble components of the feed to be quickly absorbed.

- Incisor teeth for cutting, are open rooted, i.e. they continue to grow throughout life as they become worn down by the chewing.

Small mammals represent less risk and need less start-up capital for the resource-poor farmer. The cost per animal is low compared to larger species such as cattle or even goats. Shelter for these species is cheaper as they are usually kept under backyard management systems. The death of an animal is also an easier loss to manage than that of large stock. Small mammals are convenient for slaughter at home, in that the meat from an animal can be consumed at one meal without the need for storage.

Production parameters

The production parameters of the three species are summarised in Table 17.1. Weaning can be done as early as two weeks for the guinea pig, four to six weeks for the grasscutter and six weeks for the rabbit, when the young can weigh up to 150 g for the guinea pig and 900 g for the rabbit, providing nutrition is adequate.

Under traditional management systems the average period inter-partum is 190 and 211 days for the guinea pig and the grasscutter respectively. The guinea pig under such a system has 2.3 live births per litter, amounting to a productivity of 4.4 live young per breeding female per year (Paterson *et al.*, 2001), whilst the grasscutter has 3.5 live births per litter, amounting to a productivity of 6.0 live young per breeding female per year. At the other extreme is the rabbit with 5.8 live births per litter and an inter-partum period of 73 days, amounting to a productivity of 29 live young per breeding doe per year. The long inter-partum period for the grasscutter suggests that the efficiency of reproduction in the species could be improved with early weaning, but this would entail improved nutrition.

Table 17.1 Production characteristics of grasscutters, guinea pigs and rabbits

Parameter	*Grasscutter*	*Guinea pig*	*Rabbit*
Gestation length (days)	152	67	31
Litter size	4-12	3-4	4-10[1]
Weaning age (weeks)	4-6	2	6
Weaning weight (g)	285-492	150	480-900
Female age at puberty (weeks)	24-32	6-8	16-40[2]
Litters/year	1.7-2	4-5	5-7
Litters/lifetime	8-10	8	15-21
Female mature weight (kg)	2-4	0.6-2.0[3]	3-5
Male mature weight (kg)	3-6	1-2.5[3]	3-6
Productivity per breeding female/year	6.0	4.4	29.0
Slaughter age (weeks)	52	10[4]	8-16
Slaughter weight (kg)	2.3-2.5	0.8-2[3]	1.5-2
Carcass meat/breeding female/year (kg) under traditional management systems	8	2-3	24-32
Carcass meat/breeding female/year (kg) under improved management systems	15	10[4]	40

[1]Small breeds < 4, medium breeds 8-10
[2]Small breeds 16-20 weeks, medium breeds, 20-28 weeks, large breeds 32-40 weeks
[3]Common breeds are smaller than selected lines
[4]Commercial system

Growth rates in young rabbits range from 10-20 g per day in the tropics. Though this is below the average for temperate regions, it far outweighs growth rates in both the grasscutter and the guinea pig. Common guinea pigs are small (about 600 g live weight at maturity), relatively slow-growing and extremely resistant to diseases. There are, however, many lines and selections that are larger (over 800 g live weight), faster growing and more prolific, including the Tamborada lines from Peru and the Auquicuy selections from Ecuador, but these require higher standards of nutrition and hygiene to realise their potential and are more adapted to larger commercial production systems. Table 17.1 suggests that the grasscutter could play a greater role in poverty reduction, particularly in Africa where it originates, if its genotype were improved and its nutritional requirements for various physiological activities better understood.

The low production of guinea pigs, grasscutters and rabbits in the tropics is basically due to the feeding regime which, without the use of concentrates, provides excess indigestible fibre and insufficient metabolisable energy and, in some degree, to heat stress for several months of the year. Despite these difficulties, guinea pigs, grasscutters and rabbits have proven to be interesting investments for small-scale, resource-poor farmers in most tropical countries. They have the potential to play an important role in subsistence farming systems in most parts of the developing world.

Assuming low levels of losses, a group of 13 female guinea pigs and one to two breeding males will provide sufficient offspring to allow for an average consumption of one animal per week throughout the year; a female rabbit can produce at least 24 kg of meat per year, amounting to about 0.46 kg of dressed rabbit meat per week (about a third of a family's dietary need). This compares to 0.16 kg of dressed meat per week from a female grasscutter. The meat producing capacity of the rabbit is thus superior to either of the other two species. However, all three species have the potential to double their meat production per breeding female in a year under improved management systems (Table 17.1).

Grasscutters (*Thryonomys swinderianus*)

The grasscutter, also known as the greater cane rat (Photo 17.1), is one of the largest rodents of Africa, second only to the porcupine (*Hystrix cristata*). The geographical distribution is normally determined by the occurrence of preferred plant species such as *Panicum maximum* and *Pennisetum purpureum*. The animal has thus been reported in open forests and savannahs stretching from virtually all countries of West Africa to portions of East and Southern Africa, but not in desert areas.

Management systems

Grasscutter meat, a delicacy in most African countries where the animal exists, is traditionally harvested from the wild, using methods that include hunting with guns, dogs, bushfires and chemical poisons. The latter two have detrimental consequence to the environment and public health. Harvesting is usually at its peak during the dry season when agricultural activities are few. The meat commands premium prices compared to mutton, beef and poultry. For example in Côte d'Ivoire and Ghana grasscutter meat may cost three to four times that of beef on a per kilogram basis.

Photo 17.1 The grasscutter (courtesy of E. Adu)

In captivity, the animals are usually kept in strong metal/wooden and wire mesh cages or concrete/brick housing, normally constructed in three or four tiers to maximise the use of floor space (Photo 17.2). The animals are bred in colonies of four females to one male. The dimensions of a cage meant for such a family are 122 x 21 x 40-45 cm.

Most of the grasscutters being farmed now are either wild or in the transitional period of domestication and are easily stressed when disturbed. Growth rate and reproductive performance are poor (Table 17.1). However, through selective breeding, individual strains have been developed in Benin resulting in slightly improved productivity.

Male animals housed together can engage in fatal fights, particularly in the presence of females. Males are therefore housed singly or castrated on attainment of sexual maturity. Sexual maturity in the male is determined using the appearance of the ano-genital region stain. The male grasscutter is usually very aggressive towards the female during mating, particularly when the female refuses to be mounted. This may lead to casualties during mating.

Photo 17.2 A three-tier grasscutter cage (courtesy of E. Adu)

To minimise the incidence of mating-related casualties, the cages are built so that the male can be isolated from the female, but within her view, for about 14 days before they are introduced to each other. This is more important with females who might have been introduced to another male but pregnancy did not occur. The grasscutter exhibits post-partum oestrus and re-mating can theoretically be done immediately after parturition. However it is expedient to re-mate animals after weaning to allow the mother some rest before the next pregnancy.

Feeds and feeding

The grasscutter has a digestive tract analogous to that of the horse in many respects. The caecum occupies about 60 per cent of the abdominal cavity and harbours microbial organisms for efficient fermentation and utilisation of high fibre diets.

The digestibility of NDF (neutral detergent fibre) and protein in the grasscutter tend to be higher than in the porcupine, guinea pig, degu and rabbit (Van Zyl *et al.*, 1999). The grasscutter is also selective in its feeding habit, preferring the more digestible parts of graminaceous plants, making the diet consumed better than the diet on offer.

The grasscutter can be fed a combination of cut pasture, grains, tubers and kitchen waste when in captivity. Grasses such as *Panicum maximum* and *Pennisetum purpureum* are the main feeds offered by most farmers on a daily basis. Though the animals survive and reproduce on such a diet, it does not support efficient growth and reproduction. The animals are usually not provided with drinking water. Depriving the animals of drinking water leads to a higher still-birth rate (12.3 per cent vs. 1.5 per cent), a lower birth weight (98 g vs. 129 g) and a higher incidence of digestive disorders leading to enterotoxaemia, compared to the situation where drinking water is provided on a daily basis (Schrage and Yewadan, 1999; Adu, 2003).

There is a paucity of information on the nutritional requirements of grasscutters. Even diets with a crude protein of 15-18 per cent and crude fibre contents of 10-18 per cent do not appear to support efficient production of the animal. Research into the nutrition of the grasscutter in captivity is therefore needed.

Reproduction

The testes weigh about 2 g each in the adult and can first be felt when the animal is between 4 and 5 weeks of age. The penis, which is sheathed in a prepuce, is turned backwards when not erect. There is an anal gland secreting a substance which taints the ano-genital region of the male at sexual maturity.

Sex determination is by the use of the distance between the anus and the reproductive organ, i.e. penis or the clitoral sheath (referred to as the ano-genital distance). This is 2-3 times greater in the male compared to the female. A membranous substance covers the vagina, which before puberty or during sexually inactive periods presents as a slit at the base of the clitoris.

The animal is bred under polygamous pairing conditions with a mating ratio of 1: 4. The conception rate under such conditions is between 75 and 85 per cent. Successful mating in the grasscutter can be identified by the swelling of the vulva a day after mating or by mating scratches on the back of the female. Uterine bleeding 30-45 days after mating can also be used for pregnancy detection in the grasscutter. Litter size averages 4, but there is a range (Table 17.1). The young are born precocious and nibble at vegetation a few hours after birth.

Health management

The grasscutter is very hardy and requires little or no veterinary care when strict hygienic conditions and the right management practices are applied. The animal is, however, easily stressed when disturbed, especially during transportation. Mortality can be very high (ca. 60 per cent) when animals are moved during the establishment of new colonies. The causes of death in newly-established grasscutter colonies include traumatic injuries, pulmonary congestion, pneumonia and gastroenteritis, with traumatic injuries accounting for the most deaths and pneumonia and gastroenteritis the least (Adu *et al.*, 2000). During the dry season farmers, in an attempt to maximise forage harvest, uproot the whole grass stock instead of cutting the scanty foliage they find. This results in forage being contaminated with soil and can lead to worm infestation becoming a major health constraint in the grasscutter industry. Under such conditions it is recommended that forages be either wilted for 24 h or sprinkled with salt solution before being fed.

Special boxes (Photo 17.3) have been developed for transporting animals and this has led to a significant reduction in mortality due to the stress of transporting animals. These are simple boxes made of wood with perforations to ensure adequate ventilation. The box, when closed, is dark and allows the animal to settle down during transportation. It is usually filled with grass when in use.

Photo 17.3 A specialised box for transporting grasscutters (courtesy of E. Adu)

Guinea pigs *(Cavia aperea porcellus)*

Guinea pigs are native to the Andean highland areas of Latin America, where many resource-poor families keep them for meat production. Guinea pigs have also gained some significance in countries in West Africa and Asia such as Sierra Leone, Ghana, Togo, Nigeria, Cameroon, Democratic Republic of Congo and the Philippines. For example, guinea pigs are raised for food by 10 per cent of households in Southern Nigeria. They offer an alternative to chicken or small animals hunted in the forest. They are easier to prepare than poultry because they are skinned rather than plucked. They also require shorter cooking times, consuming less fuel. Information in this section has been taken mainly from Rico and Rivas (2000), with reference to South America.

Management systems

In the simplest forms of management, a small number of animals (10-30) are kept unconfined as a single group in a part of the house, usually the kitchen, or in a dedicated shed (Photo 17.4). They are fed on a mixture of household scraps such as raw vegetables and weeds, together with a small daily ration of good quality cut forage. Under these conditions, in the complete absence of vaccinations or veterinary treatment, animals of unimproved lines suffer from surprisingly few problems of health. They receive no veterinary treatment or special care, except for some protection from predators and the extremes of climate. Under such systems, they are tended by the women and children of the family and used mainly for subsistence and food security. Almost all animals are eaten on the farm. There is little trade in guinea pigs, although males are occasionally swapped between families in an attempt to avoid the problems of inbreeding. There may be occasional sales of animals in times of need, but this is not a common practice at the subsistence level, even in the poorest households.

To prevent losses due to predators, some families build simple cages for the animals, from wood and wire. These are raised on stilts, about 0.9 m above ground level, and located so that they are protected from the prevailing winds, especially in the cold season. They have slatted or mesh floors, so that the droppings can fall through onto the ground below; and roofs of palm thatch to protect the animals from rain. They are usually sub-divided, so that the adult male can be separated from the immature animals to prevent fights. Females and young animals can be kept together without signs of aggression, until the young males become sexually mature.

Photo 17.4 Guinea pigs in a typical shed in tropical lowlands, east of the Andes in Bolivia (courtesy of R. Paterson)

At the other extreme, on a commercial basis, 100-400 animals are kept, divided according to sex, age and physiological state, in small enclosures containing 8-10 animals each. These are located in a separate installation. In such systems, the guinea pigs are fed on a combination of cut pasture and purchased or home-made concentrate. Upon reaching maturity, the animals are sold live to intermediaries, who mostly supply the restaurant trade. In the highlands of Ecuador, Peru and Bolivia, the meat is generally well-liked and there is a large demand. It commands a premium price in the market, being more expensive than the normal commercial alternatives of poultry, beef and mutton.

Feeds and feeding

Guinea pigs can eat up to 30 per cent of their body weight per day in the form of green pasture or vegetable matter. Although they can live and reproduce on much poorer diets, for maximum productivity their nutrient requirements include about 18 per cent crude protein, 10 per cent crude fibre and 12.54 MJ ME/kg DM, together with a good supply of minerals and vitamins. Good quality leguminous pasture such as lucerne (alfalfa, *Medicago sativa*), tropical kudzu (*Pueraria phaseoloides*) or perennial peanut (*Arachis pintoi*) will meet the need for protein, but will contain too much fibre and too little energy to give optimum results. To obtain the highest yields of meat, up to 40 per cent of the diet can be provided as concentrate feed, including cracked grain. The change from one major feedstuff to another should be done gradually, to avoid diarrhoea. The daily requirements for water are about 30, 80 and 100 ml/day for sucklers, growing animals and adults respectively. As the proportion of dry feed in the diet increases, so does the need for a constant supply of clean drinking water. Mortality rate is significantly increased when the animals are deprived of water, with pregnant and lactating females being the most affected, followed by young, rapidly-growing animals.

Reproduction

Litters can contain up to seven young, although three or four are commonly born at a time. They are born fully-formed and will suckle immediately. They are capable of eating cut pasture within a few hours of birth, so clean, high-quality forage should be provided at all times. The mothers produce milk for a period of about 14 days, after which the young should be fully weaned. When well-fed, females can reach puberty at 6-8 weeks, although the males take about two weeks longer.

In commercial systems with improved lines, slaughter weight is reached in about 10 weeks, while those chosen for breeding should be first mated at about 12 weeks, when they will have achieved about 60 per cent of their maximum adult weight. Females show heat at intervals of about 16 days, with the first occurring immediately after parturition. The gestation period is about 67 days, so with continuous mating, 4-5 litters can be produced in a year. Under this system, the females are culled after producing 5-6 litters. If the males are separated from the females for a period of three weeks starting about one week before birth, four litters per year can be produced, and the females can give up to eight litters before they need to be culled. One male can service 6-8 females. Under optimum conditions of feeding and health, one breeding female of an improved line can produce about 12 live offspring per year.

Health management

The optimum temperature for young animals is over 12 °C. Adults can generally tolerate cooler temperatures better than extremely high ones, the optimum for breeding animals being 18 °C. Temperatures outside the range from 3-34 °C can adversely affect breeding or lactating females. Although guinea pigs are susceptible to draughts, good ventilation and natural lighting are important to maintain health.

Diseases are not usually an issue with the rustic common types but, as with other species of livestock, the more productive lines and selections show less resistance to health problems. Common infectious diseases include salmonella and pneumonia, while fungal attacks can cause skin problems. Guinea pigs can be infested with ectoparasites such as fleas, mites and ticks, together with endoparasites, including coccidia, nematodes and liver flukes, so they should not be allowed close contact with other domestic livestock, or family pets. To avoid passing on endoparasites, other livestock should not previously graze the pasture areas cut for guinea pigs. Bacterial conjunctivitis can occur, particularly under dusty conditions, while fresh pasture herbage can produce bloat. Most of these problems can be treated with the veterinary products or home remedies commonly used for other animal species. Bloat can be avoided by allowing freshly-cut pasture-herbage to wilt in the shade for at least an hour before offering it to the animals. Guinea pigs show marked resistance to endemic diseases and are not susceptible to myxomatosis that prevents the raising of rabbits in certain areas.

Rabbits *(Oryctolagus cuniculus)*

Small-scale rabbit production has been advocated as a viable income-generating activity, which could emerge as a low-cost answer to food insecurity in rural communities. In countries such as Ghana and Mexico, raising rabbits has been accepted socially on the combined basis of the low space requirement, high reproductive rate, low competition with humans for the same foods, minimal zoonotic health hazard and minor capital investment. However, there are some social taboos that may affect the consumption of rabbit meat.

Rabbit meat is highly nutritious, easily digested, rich in protein (14-25 per cent), low in fat (3-6 per cent), low in cholesterol (135 mg/100 g) and a good source of B-group vitamins, with only a small amount of uric acid formed during metabolism. In addition, rabbit meat is low in sodium, which is advantageous to people with heart problems.

Management systems

In developing countries a semi-intensive production system is commonly practised, whereby small rabbit units are constructed around the backyard of residential houses, using locally available resources. The animals are kept in hutches normally built with bamboo, raffia palm, bush sticks, woven wood, paddled soil and sometimes soil or cement bricks. They are placed one metre above the ground with a height of 60 cm at the front, 50 cm at the back, width of 50-60 cm and length of 90-120 cm (Photo 17.5). The hutches are equipped with water and feed troughs, forage rack and kindling box made of various local materials. In other units, rabbits are mixed together with rural chickens and both are used to supplement the family diet and income. In contrast to backyard systems, top management rabbit units, e.g. in institutions, involve keeping 20-50 rabbits in well-constructed buildings equipped with wire mesh cages, feed hoppers, bottle drinkers and wooden or plastic kindling boxes.

Photo 17.5 Backyard rabbit keeping in Tanzania. Rabbit-keeper moving a pregnant doe from the common hutch to a kidding place (courtesy of G. Laswai)

Feeds and feeding

Rabbits in developing countries are mainly raised on forages, supplemented with agro-industrial by-products. The protein and energy requirements for animals in various physiological states are presented in Table 17.2.

The palatability of the forages fed is important, particularly in situations where the forages provide a major part of the daily nutrient intake. With the exception of gliricidia (*Gliricidia sepium*) tropical legumes are preferred over grasses and agro-industrial by-products. For efficient production, about 50 g of concentrate per animal should be offered per day. This is to meet the mineral and vitamin requirements of the animals. The concentrate can either be purchased or home-made, though commercial rabbit concentrates are rarely available in most developing countries.

Table 17.2 Protein and energy requirements of rabbits in different physiological states

Physiological state	*Nutritional requirement*	
	Crude protein (%)	*Energy (MJ ME/kg DM)*
Pre-weaning	15	10.03
Young animals (4-12 weeks)	16	10.45
Mixed category	17	10.66
Lactating does	18	11.08

Adapted from Lebas *et al.* (1997).

Reproduction

The age at sexual maturity, litter size and other reproductive parameters are all influenced by the breed (Table 17.1). Oestrus is induced, enabling the doe to conceive even on the day it kindles. The gestation length is 31 days. The young are born naked, very fragile and with their eyes closed. Therefore kindling boxes are provided from about 10 days before parturition, to keep the newly-born rabbits in the nest and together during the first critical days of life.

Exposure to light for 8 hours out of 24 hours favours spermatogenesis and sexual activity in bucks, with 14 to 16 hours of light favouring female sexual activity and fertilisation. During mating it is advisable that does are taken to the buck's cage in order to avoid fighting and delayed mating. When a buck is taken to the doe, mating becomes difficult as does tend to express possessive behaviour regarding their cage. Successful mating is indicated by the buck 'falling off' the doe on completing copulation. Pregnancy can be diagnosed by palpating the abdomen with the thumb and index finger to feel the developing foetuses in the uterus. This is done when the doe is relaxed, so that the abdominal muscles are not tense.

Health management

One distinct attribute of rabbit farming is the relatively low incidence of epidemic diseases when a high standard of hygiene and careful management is practised. They are also resistant to low temperatures, and can therefore be raised under a range of different climatic conditions. The rabbit's normal body temperature ranges between 37 and 39.5 °C. Heat stress can occur at body temperatures above 40 °C. High temperatures in the tropics result in lower feed intake, reduced growth rate, reduced reproductive rate and increased water intake. It is well established that high ambient temperatures can cause infertility in breeding rabbits, bucks being more sensitive than does. Prolonged exposure to a critical temperature in excess of 30 °C is considered the threshold point at which infertility may result. Rabbits are more sensitive to low humidity (below 55 per cent) than high humidity. Abrupt changes in humidity, from low to high and vice versa, affect rabbits more than the level of humidity per se. Ventilation must be adequate to remove excess heat, ammonia and moisture, which may predispose the rabbits to diseases.

Diseases such as myxomatosis, pasteurellosis, coccidiosis, worms and pneumonia are important health constraints to rabbit production. Various sulphur-based drugs have shown good results in controlling rabbit coccidiosis. Sanitation is a critical determinant in the control of episodic incidences and morbidity levels due to coccidiosis outbreaks. Therefore it is imperative, where large rabbit operations exist, to maintain stringent levels of hygiene and culling of diseased animals, as well as implementing proper quarantine measures.

Conclusion

Though commercialisation of smallholder agriculture, which in most developing countries involves backyard production of small mammals, is one of the strategies for addressing rural poverty, the potential of guinea pigs, grasscutters and rabbits for alleviating poverty under backyard systems cannot be generalised. The dietary importance, for humans, of most of these species is localised. Whilst guinea pig meat is widely accepted among highland people in Latin American countries, the grasscutter is mostly known and accepted in West Africa. Large-scale commercial production of these species is also limited due to a combination of factors, viz: low genetic potential, localised consumer demand, insufficient promotion, variable product supply, unreasonable prices, competition from other meats, lack of product diversification and poorly-developed marketing channels (Lukefahr and Goldman, 1985).

Rabbit production suffers from inadequate marketing opportunities in most developing countries. Where there are favourable market prices for rabbit meat, such as in Kenya, Trinidad and Cameroon, rabbit production could improve the livelihood of many resource-poor farmers (Wanjaiya and Pope, 1985; Lukefahr and Goldman, 1985). However, in projects aimed at greater commercialisation of small mammal production, the first step is to investigate the people's needs and resources, using participatory methods (cf. Chapter 4). Furthermore, projects should be introduced to a particular area only when viable and well-established markets exists, e.g. such as for the grasscutter in West Africa.

Further reading

Adu, E.K., Alhassan, W.S., and Nelson, F.S. 1999. Smallholder farming of the greater cane rat, *Thryonomys swinderianus* Temminck, in southern Ghana: A baseline survey of management practices. *Tropical Animal Health and Production* 31: 223-232.

Board on Science and Technology for International Development, National Research Council. 1991. *Microlivestock. Little-known small animals with a promising economic future*. National Academy Press, Washington, D.C., USA.

Fielding, D. 1991. *Rabbits.* The Tropical Agriculturalist, Series Editors, R. Coste and A.J. Smith. Macmillan Education Ltd., London, UK, in cooperation with Technical Centre for Agricultural and Rural Cooperation (CTA), Wageningen, The Netherlands.

McNitt, J.I., Patton, N.M., Lukefahr, S.D. and Cheeke, P.R. 2000. *Rabbit production. 8th edition.* The Interstate Printers and Publishers Inc., Danville, IL, USA.

Schrage, R. and Yewadan, L.T. 1999. *Raising grasscutters.* Deutsche Gesellschaft für Technische Zusammenarbeit (GTZ), GmbH, Eschborn, Germany.

Wagner, J.E. and Manning, P.J. (ed.). 1976. *The biology of the guinea pig,* Academic Press, New York, USA.

Author addresses

Emmanuel Adu, Animal Research Institute, Council for Scientific and Industrial Research (CSIR), Achimota, P.O. Box AH 20, Achimota, Ghana

Rob Paterson, Formerly of Natural Resources Institute (NRI), University of Greenwich at Medway, Chatham Maritime, Kent ME4 4TB, UK

Franz Rojas, Formerly of Centro Internacional de Agricultura Tropical (CIAT), Santa Cruz, Bolivia

Germana Laswai, Department of Animal Science and Production, Sokoine University of Agriculture, P.O. Box 3004, Morogoro, Tanzania

Dennis Fielding, Formerly of Centre for Tropical Veterinary Medicine, Division of Animal Health and Welfare, University of Edinburgh, Easter Bush Veterinary Centre, Roslin, Midlothian EH25 9RG, UK
Emmanuel Osafo, Department of Animal Science, Kwame Nkrumah University of Science and Technology, Kumasi, Ghana

18

Pigs

David Holness[*], *Rob Paterson*[†], *Brian Ogle*[‡]

Cameo - Bolivia: simple improvements to pig production

Don Fecundo Araúz lives with his family in the community of Potrerito in the tropical lowlands of Bolivia. The family grows a range of crops, including maize, cassava, sugar cane and bananas, together with many types of vegetables, while native fruits are collected from the forest. There is seldom much of a surplus for sale, because the operation is aimed more at subsistence than at commercial production. They keep several types of poultry and also a couple of free-ranging sows, but in 1999 they were ready to sell off the pigs, because of the problems that they caused with the neighbours when they strayed off the farm during the cropping season.

CIAT, the nearby international agricultural research and development organisation, started a programme of work with small animal species at that time. The team encouraged Don Fecundo to plant some extra cassava for the pigs and then helped him to design and build a simple pen of poles and palm thatch cut from the farm, to confine the animals during the critical periods of the year. The animals were treated against internal parasites and a boar was borrowed from a neighbouring community. When the sows farrowed and the piglets were two months old, all of the animals were again treated for worms. In the past, it was usual for at least half of the piglets from each litter to die from parasites and accidents. With these simple improvements, only three of the 17 piglets from the two litters were lost before weaning. In the next 7-8 months, the family consumed three animals and reserved the best two females for breeding. The sale of the remaining

[*] Principal author
[†] Co-author
[‡] Consulting author

12 animals, concentrated in the periods of Christmas and carnival, brought in about half as much money as Don Fecundo could expect to earn in a full year by doing the casual labouring work that was available in the district. Although they still have no boar of their own, the Araúz family now has three breeding sows on their farm. They see them as important in sustaining their livelihood and have no intention of getting rid of them.

Introduction

On a world-wide basis, more meat is produced from pigs than from any other domestic species. It is estimated that over 65 per cent of the world's pig population is in the poorer developing regions, representing over 500 million pigs. Within the developing regions the vast majority of the pigs are situated in Asia, especially China, while the Americas and Africa account for less than 15 per cent (FAO, 2002). This difference is reflected in consumption patterns, where in Africa per capita annual consumption of pig meat is around 2 kg. Part of the explanation for this uneven distribution is due to religion, as believers in Islam and the Jewish faith are forbidden to eat pork. Also, in contrast to parts of Asia and China where pig meat predominates in the national diet, African and South American people have traditionally shown a preference for meat from cattle or from small ruminants (sheep and goats). Nevertheless, pig meat consumption in the developing world has shown a steady increase at the expense of beef during the past decade, in part due to the lower fat content and healthier image of pig meat.

Pigs are monogastric omnivores and will eat a wide range of feedstuffs. However, although they possess a caecum, which is part of the large intestine and houses a population of microbes which breaks down fibre, their ability to digest high-fibre diets is limited. This contrasts with the situation in the ruminant, which derives the majority of its nutrients from microbial breakdown in the rumen. Nevertheless, pigs fed high-fibre diets have been shown to develop an enlarged caecum. At one time it was thought that indigenous-type pigs may have evolved a bigger large intestine and an enhanced ability to digest fibre, but this was found not to be the case because, when fed high-quality concentrate diets, the caecum and colon were in the same relative proportion to the intestinal tract as that in modern exotic pig breeds (Chigaru *et al.*, 1981).

Because pigs are a litter-bearing species with a relatively short gestation period, their potential for meat production far exceeds that of the other main domestic species, and yield of meat per unit live weight of breeding female can be up to six times that of cattle.

The domestic pig is believed to be derived from the European wild boar (*Sus scrofa*), which is adapted to live in a cool forest environment. Hence pigs possess a sub-cutaneous layer of fat to help insulate against heat loss, and possess sweat glands only on the snout. As a consequence, they are not very efficient at dissipating body heat, and when the ambient temperature approaches body temperature, they will soon suffer hyperthermia and die if they do not have the means to cool themselves, such as wallowing in water or mud.

Advantages of pig production

For the resource-poor farmer in the developing world, the advantages of pig production compared to other forms of livestock production are:

- Pigs can be confined and reared in relatively small areas. As a consequence they are not subject to the same problems which confront cattle, sheep and goat producers in the many regions where communal land tenure is common.
- For the same reason that they only require a small area, pig production is particularly appropriate in densely-populated areas. Accordingly, as red meat becomes more expensive, pig production is growing in these areas.
- If pigs are kept in pens, they do not contribute to erosion and land degradation, a trend which continues to expand in the developing world.
- Pigs will convert a variety of crop waste, kitchen waste and livestock offal into high quality meat.
- Pigs are very efficient converters of concentrate feeds to meat when compared to ruminants. However, on low-quality high-fibre diets they are less efficient than ruminants.
- Pigs give a relatively rapid return on investment as, even on low planes of nutrition, a piglet is ready for slaughter at twelve months of age.
- Pigs are often considered in small-scale farming communities as 'living banks' which can be slaughtered at times of particular financial need, e.g. for medical or school fees.
- Pigs have a higher dressing percentage than any other livestock species, i.e. the carcass forms a higher proportion of the body at slaughter.

- The size of pigs compared to cattle makes slaughter and marketing a more flexible and easier process. Young pigs are often in big demand at holiday periods and at times of family celebrations.

- Pigs produce relatively rich manure, and a sow can produce up to 750 kg annually. This becomes a very important resource to the farmer when the price of inorganic fertilisers is prohibitive.

Pig breeds

In a large part of the developing world, involving Central and South America and Africa, there is a clear dichotomy between modern domestic breeds (*Sus domesticus*) and so-called indigenous breeds. In Sub-Saharan Africa, for instance, most commercial pig production units are based on introduced modern breeds, such as the Large White and Landrace, and latterly the Duroc and Hampshire. The modern breeds have been subjected to relatively intense selection for growth rate, feed conversion efficiency and carcass conformation; the indigenous breeds have selectively adapted in order to survive in their local environment. Consequently, the two types are very different in size, shape and productivity. In China, where pigs have been domesticated for centuries, the situation differs. There are over 100 different breeds, mainly derived from the *Sus vittatus* type, with short, broad, fat bodies, often sway backs, short legs and a dished head. This type of pig is now widely distributed over the rest of Asia and surrounding territories.

In Sub-Saharan Africa it appears that the bushpig (*Potamochoerus porcus*) and warthog (*Phacochoerus aethiopicus*) are the only truly indigenous members of the *Suidae*. European settlers probably introduced the so-called indigenous pigs in the nineteenth century. Livingstone and Livingstone (1865), for instance, reported finding Portuguese settlers at Senna on the Zambezi river in possession of "horrid long-snouted greyhound-shaped pigs wallowing in foetid mud pools". There is now considerable diversity between the breeds or types of pig in Africa, ranging from the relatively small and slight build of the Ashanti Dwarf in Ghana, and the Bakosi in Cameroon, to the larger, Windsnyer-type in Zimbabwe, which is longer in the leg and razor-backed, and the Kolbroek-type in South Africa, which is short and fat, with a dished face (Mason and Maule, 1960).

While it is possible that some of the indigenous pigs in the Americas originated from Chinese pigs introduced in the sixteenth century, there was large-scale introduction of domestic livestock into South America from Spain and Portugal in the seventeenth and eighteenth centuries. The unimproved descendants from these imports are known as 'creoles' and they are common in all parts of Latin America (Wilkins and Martinez, 1983).

Photo 18.1 Indigenous Mukota sow in Zimbabwe (courtesy of A. Allen)

Indigenous or creole pigs form the basis of the pig populations in the resource-poor regions of the world. The extent to which they have been cross-bred with modern improved breeds varies from region to region. Productivity is generally low, but this is strongly affected by the environment, especially intermittent feed supplies. Average performance figures derived from data from four countries in Africa, where pigs were well-managed, and an example of performance figures for Asian (Vietnam) pigs, are shown in Table 18.1. Similar data are available from South America (Paterson *et al.*, 2001), although without very good levels of management it is hard to keep mortality levels below 10-15 per cent. The pre-weaning mortality (<5 per cent) for indigenous pigs in Africa appears low. However, trials in Zimbabwe with indigenous sows run extensively and farrowed in arks, with no farrowing aids, confirmed these values. Table 18.1 also shows some comparative figures for exotic pigs raised in commercial units in the tropics.

Table 18.1 Average performance figures for well managed indigenous pigs in Africa and Asia, and comparative figures for exotic pigs raised in commercial units in the tropics

	Africa indigenous	*Asia indigenous*	*Commercial exotics*
Number of piglets born per litter	7.5	8.3	10.5
Pre-weaning mortality (%)	<5	20.5	11.0
Weaning weight (8 weeks) (kg)	7.9	8.0	18.0
Live-weight gain from weaning to slaughter (kg/day)	0.42	0.50	0.65
Food conversion rate (kg feed/kg live-weight gain)	4.10	4.50	3.20

Compared to exotic breeds, indigenous or creole pigs are better-adapted to 'harsher' environments and poor management systems. They are more mobile and better equipped to scavenge and root. They are considerably less susceptible to heat stress and more resistant to most local diseases and parasites. These characteristics contribute to hardiness and survivability when crossed with an exotic breed.

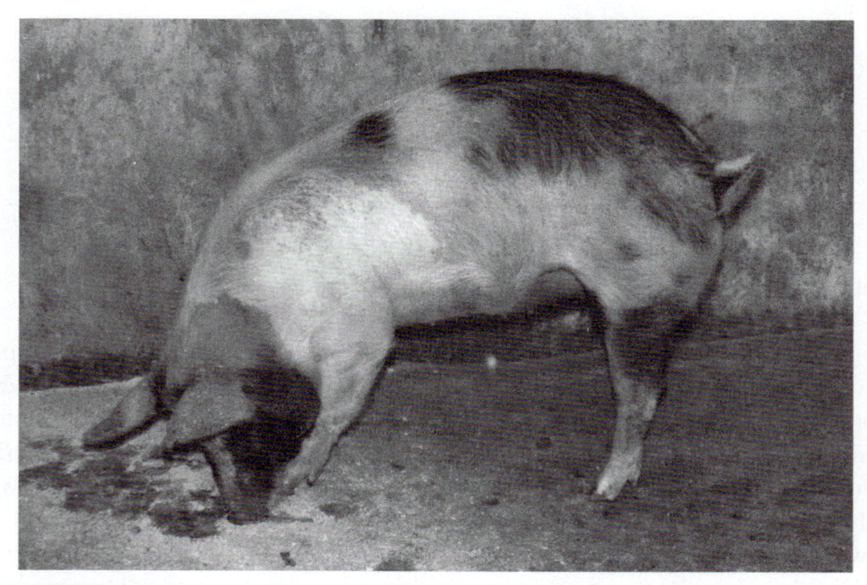

Photo 18.2 Indigenous cross Duroc sow which performs well in the tropics (courtesy of A. Allen)

Indigenous pigs are earlier-maturing than exotic breeds, and hence will start depositing fat in the carcass at an earlier age than their exotic counterparts. As a consequence, indigenous pigs reared throughout on good diets to slaughter at conventional ages for exotic pigs (e.g. 5 to 6 months), are invariably down-graded as over-fat. In comparative trials in Zimbabwe (Chigaru *et al.,* 1981), over a range of ages at slaughter (4, 6 and 8 months), dressing percentage was consistently 2-3 units higher and the percentage of fat in the carcass 4 units lower in the Large White compared to the indigenous pigs.

Feeds

The pig requires access to feedstuffs which will provide sufficient of the essential nutrients, broadly protein, energy, minerals and vitamins, to promote adequate growth and productivity, and to generate an end-product which can be sold. In addition, other factors of importance are the quality of the protein, i.e. the make-up of amino acids in the protein, which determines the pig's ability to convert the protein into meat, and the digestibility and quantity of the fibre in the diet, which will affect intake and utilisation of the total ration.

Feeds that are commonly used and their suitability for feeding pigs are noted below.

Protein feeds

Oilseed meals are the most widely used sources of protein. These include soyabean meal, sesame meal, cottonseed meal, groundnut meal, sunflower meal, coconut oil meal, safflower meal and palm kernel meal. Only soyabean meal has a high protein quality, but the others tend to be deficient in different amino acids and can therefore be blended to improve overall quality. Some may contain toxins, e.g. groundnut meal is susceptible to mould infection and aflatoxin, which will depress pig growth, and cottonseed meal can contain a toxin called gossypol, which may be fatal. As a consequence, cottonseed is not generally included above 10 per cent of the ration. High-fibre contents restrict the use of some of the meals, e.g. sunflower meal, coconut oil meal and palm kernel meal. The fibre in safflower meal is also highly indigestible, which severely limits its use.

Various peas and beans, e.g. cowpeas, chickpeas, pigeon peas, kidney beans, and Jack beans can be grown and fed. They are lower in protein than the oilseed meals and tend to contain anti-nutritional factors which need to be destroyed by cooking before feeding to pigs. Also the foliage of crops such as cassava and sweet potato can be useful sources of protein, although some varieties should first be sun-dried or ensiled to reduce levels of anti-nutritional factors. Although generally high in fibre content, the leaves of some forage trees and shrubs can be fed. However, care must be taken due to the presence of anti-nutritional factors. The commonly grown *Leucaena leucocephala,* for example, contains mimosine, a toxic amino acid which necessitates the leaves being soaked and dried before feeding.

Animal by-products, such as meat and bone meal, blood meal and offal, are excellent sources of high-quality protein and can be included at up to 10 per cent of the diet. If not properly processed, there is a risk of disease (e.g. foot and mouth disease and anthrax) transmission from offal.

Fish-processing wastes, sometimes available to pig producers adjacent to inland fisheries, are a source of high-quality protein and are also rich in calcium and phosphorus. They can be very effective in the diet if included at a low level. In island and coastal territories, marine scrap-fish or fish offal are preserved as silage and used as pig feeds.

In Bolivia, earthworms and small animals, usually found along the banks of streams, form a major source of protein for scavenging pigs.

Energy feeds

The cereal grains and their by-products are the most-widely used and are all good sources of energy. These include maize, sorghum, millets and, to a lesser extent, wheat. Some varieties of sorghum contain tannins which reduce the protein digestibility in the diet. Bulrush millet is susceptible to ergot fungus which will cause agalactia (no milk) in sows.

Rice, and particularly rice bran and 'polishings', which are the outer layers of the grain, is widely used in pig diets in large parts of Asia. Energy content is lower than in the cereals due to the higher fibre content. Rice husks contain a high level of silica and are, therefore, of limited value.

Cassava (manioc or tapioca) roots, when dried and ground, provide a high-energy feed for pigs. It is widely grown in Central and South America, parts of Africa and South-East Asia. The roots can also be fed fresh or chopped and sun-dried, especially to sows, although some bitter-tasting varieties can cause prussic acid poisoning and must be avoided.

Potatoes are a rich source of starch energy, but contain a toxin, solanin, and a trypsin inhibitor, and therefore should be cooked before feeding. Similarly yams, which are widespread in West Africa, are rich in starch but require to be cooked to destroy toxins and tannins.

Sweet potatoes, common in small-scale farming systems, can be fed fresh, or dried and ground, and have an energy content equivalent to maize.

Cane sugar by-products, especially molasses, are also very palatable and will enhance intakes of the total diet. Molasses is a good source of sugars, minerals and the B vitamins, although high levels of 'C'-molasses ('final' or 'blackstrap' molasses) can result in diarrhoea. The maximum recommended level in growing pig diets is 30 per cent (Perez, 1997).

Oils are a rich source of energy for pigs, and in smallholder systems can be provided as whole oilseeds. Whole soyabeans, for example, if boiled to destroy the anti-nutritional factors, are an excellent source of both energy and protein. Various residues containing palm oil have also been used in pig diets in South-East Asia, the west coast of Africa and South America. For free-ranging pigs in tropical Bolivia, the fallen fruits of a range of native trees, including the motacú and cupesí palms, are important sources of both energy and protein.

Bulkier feeds

Brewers' grains are generally available and widely-used as pig feeds in small-scale areas and villages. They are very palatable and, although relatively high in fibre, do provide some protein.

Kitchen waste or swill, although very variable, can be nutritious. Because of the danger of infectious diseases, it should be boiled before feeding to pigs. The whey, left over from the production of cheese, is a low-density source of nutrients.

Pumpkins and melons, often grown in the same lands as arable crops in small-scale systems, are commonly fed to pigs and provide some energy, succulence and gut-fill.

Bananas and plantains, when ripe, are palatable and useful sources of energy.

Water hyacinth, very common in waterways in tropical areas, is fed to pigs in South-East Asia. The plants are chopped, boiled slowly for a few hours until they turn into a paste, and then fed. Other water plants that are commonly fed to pigs in the humid tropics are Azolla and Duckweed (*Lemna* spp.), both of which are good sources of protein, vitamins and minerals.

Formulation of diets

One of the major drawbacks to productivity in the small-scale situation is the feeding of unbalanced and poor-quality rations. Once the producer has assessed the amount of feed which is likely to be available for his pigs over an annual period, an attempt can then be made to optimise the quality of the ration to be provided.

The main considerations are the amount and quality of the protein, and the protein: energy ratio, as the efficient utilisation of protein is dependent on the amount of available energy. The fibre content of the diet is important in relation to digestibility and the release of other nutrients.

Minerals and vitamins are also essential components of the diet of the pig. Of the minerals, calcium, phosphorus, magnesium, sodium, chlorine, copper, iron, zinc, manganese, iodine and selenium are considered essential. These elements may already be present in the feed ingredients, or may need to be supplemented. A particular problem is iron deficiency in piglets without access to soil, which will quickly lead to anaemia. Similarly, fourteen vitamins are normally considered to be required by the pig. Including some green feed in the diet can often help to supplement the vitamin supply.

Systems of production

Free-range or scavenging systems

These traditional systems predominate in large areas of the developing world and are based on the rural village. Indigenous or local pigs will forage for the bulk of their food around homesteads, kraals and adjacent areas, and receive some form of supplementary feed later in the day, often in the form of cassava, cracked cereal grain or household scraps. Productivity of these village pigs is generally low, with litter sizes of three to five piglets and low growth rates (less than 120 g per day). Past efforts to increase productivity by the introduction of improved pig breeds into the traditional village system have not been successful (e.g. Walters, 1981), largely due to poor management, parasites and diseases. The potential of these basic production systems for wealth creation is limited, but they can make a significant contribution to the livelihoods of resource-poor people. The important role that the pig plays in the social life of the village, in traditional and family rituals, and the exchange of meat, as well as in the provision of high-quality protein in the diet, should not be underestimated.

Interventions to improve productivity by small-scale producers

The first of these is the enclosure of pigs in pens or yards. Pens can be relatively simple and cheap to construct, particularly if poles and palm thatch are available for cutting on the farm. In the simplest case, the pens can be used to confine normally free-ranging animals only when they might damage crops in the field. Full confinement is often preferable, but this demands that the owner provides all of the feed that the animals require on a year-round basis. Benefits that result are as follows:

- Animals conserve energy by not having to forage for generally low-quality feed sources.
- Both the quality and quantity of feed provided can be controlled.
- Clean drinking water can be made available.
- It is easier to ensure a routine programme of vaccinations and parasite control.
- Pigs can be protected from the climate, particularly the sun in tropical and sub-tropical climates, by the provision of shade. Cooling facilities, e.g. wallows, can be provided.
- Access to human faeces is prevented, avoiding infection with human tapeworms.
- If solid floors are regularly cleaned, or a deep litter system is practised, a build-up of roundworm infestation can be avoided.
- Animals are protected from direct contact with sources of infectious diseases, e.g. foot and mouth disease, anthrax, African swine fever, classical swine fever, swine dysentery, hog cholera and tick-borne diseases.
- A controlled breeding programme can be implemented. Sow oestrous periods can be monitored and boars can be introduced to sows at set times in order to optimise the number of litters per sow per year.

Secondly, during periods of confinement, the producer can attempt to feed a balanced ration. Feed is invariably the first limiting factor and the highest cost item for the small-scale producer. The biggest cause of waste tends to be the feeding of rations that are unbalanced for the purpose of achieving effective growth and reproduction. A feeding programme which has been successful in parts of Africa involves the producer purchasing small quantities of a concentrated protein source (which often also contains some essential minerals) from a commercial supplier. An amount calculated to provide the majority of the protein requirements is fed daily, and pigs then have ad libitum access to lower-density energy feeds produced on-farm.

Thirdly, the improved environment and feeding allows the producer to introduce improved boars in order to generate cross-bred progeny for fattening and future female breeding stock. Cross-breds will grow faster, convert concentrate feed more effectively, and produce a carcass which is more acceptable in the market place. At the same time, some of the adaptability and hardiness characteristics of the indigenous breeds are retained.

Peri-urban production

Because of their small requirements for space, pigs are suitable for keeping in a suburban environment, in pens in gardens and backyards. In parts of Asia, pens are often made of bamboo and elevated, or alternatively pigs can be restrained by a chest-tether and rotated around larger yards.

In the developed world, peri-urban pig production is largely prohibited due to problems with odour, noise, flies and manure disposal. However, no such regulations exist in large parts of the developing world, and the problems of manure disposal are often overcome by the use of digesters to produce biogas for fuel, a practice common in the Philippines and other parts of Asia. Major advantages which accrue to the urban producer are the ready availability of kitchen wastes as a feed source, and the proximity of relatively-affluent markets for the end product.

Centralised systems of production

As a means of stimulating pig production in small-scale communities, some centralised systems have been established. These consist either of a centralised pig fattening unit, which contracts to buy weaners from out-growers with small breeding herds, or a centralised breeding unit, which sells weaners to out-growers for fattening. Benefits to the individual producers are that the centre can serve as a source of extension information, help procure and sell feedstuffs and veterinary inputs, in certain circumstances provide breeding stock and semen, and assist with transport. These systems are not common in tropical Latin America, and in Africa they have proved very vulnerable to variations in feed supplies, especially in times of drought.

Intensive units

In a number of instances, small-scale, specialist producers progress to more intensive systems of production. These involve larger herds (20-30 sows), higher-quality buildings with concrete floors providing greater protection and hygiene, and feed grown or purchased specifically for the piggery. Modern breeds or cross-breds are used and the progeny grown to produce a specific product for the commercial market.

Integrated production systems

Tropical regions in Asia have been at the forefront in developing systems which link pig production with other productive enterprises, thereby enhancing the output for the whole operation.

The use of pig effluent to fertilise fish ponds is probably the most common system, especially in tropical and sub-tropical areas where temperatures allow for a long growing season for the fish. Effluent is channelled from the pig sties into suitable ponds, or in some cases pig sties are constructed over ponds so that the faeces and urine drop straight through into the water. The enriched water serves to generate algae, which can be converted by fish into edible protein. In Zimbabwe, trials have shown that effluent from 60 pigs is sufficient for one hectare of pond per year and, if stocked with 50,000 *Tilapia* fingerlings per hectare, edible fish yields of 4 t/ha can be realised (Holness, 1985). Fish production is increased if the pond water is aerated. As they fill up with sediment, ponds can be dried-out in rotation and the sediment forms rich manure which can be used on gardens or arable lands, or alternatively vegetable gardens can be established in the pond beds. Other systems establish water hyacinth on the ponds to utilise the nutrients, and this can then be fed back to the pigs. Ducks can also be introduced to contribute to fertilising the ponds and provide an alternative meat source.

Furthermore, manure can be fermented anaerobically to produce biogas, which can be used for domestic cooking and lighting. Requirements are a tank (concrete or steel) for digestion, a condensation/safety trap, a reservoir to provide a constant pressure and a piping system to transport the gas to the user. In Vietnam, a cheaper, polyethylene-tube digester system has been developed, and more than 20,000 have been installed over the last decade (Ogle and Preston, 2004). The daily manure/ effluent from 12-15 growing pigs (each producing 1.5 kg fresh weight) produces enough gas (about one cubic metre) for 4-5 hours cooking per day.

Management

As outlined earlier, the traditional, unimproved village system of pig keeping will not lead to significant wealth creation, because it is uncontrolled and inherently inefficient. When pigs move from scavenging systems of production into some form of simple pens or housing, the producer has the means to begin managing his animals and small changes or improvements in the system can lead to significant increases in productivity.

The basic aim of good management is to provide conditions and impose procedures which optimise the efficiency with which feed is converted to meat by the pig. Some of the fundamental aspects were considered earlier, but other points include management of pig comfort, reproduction, growth and disease.

Pig comfort

Particularly in hot climates it is essential to provide sufficient space for the animals. Recommendations (per animal) are one square metre for growing pigs and three to four for sows in groups. This minimises problems of social interactions and fighting. It also allows room for positional and postural changes to alleviate problems of high temperatures. Overcrowding is a common cause of depressed performance in pigs.

Mixing of batches of pigs should be avoided if possible, as this will invariably cause fighting and damage. If mixing is unavoidable, pigs to be introduced into an existing group should be smeared with a strong smelling substance (e.g. old engine oil) so that they have no recognisably-different smells.

Provision of bedding, that is some form of roughage, grass or wood shavings, is very beneficial for pigs and will help prevent fighting and tail biting. It also serves to insulate the pig against the cold in winter.

Good ventilation in pens and buildings is also very important in allowing pigs to lose heat in hot weather. In addition, pens and buildings should be built to face north-south to minimise direct exposure of the pigs to the sun.

Reproductive management

One of the critical factors influencing the profitability of a pig unit of any size is the number of pigs reared and sold per sow per year. To achieve optimum results, a sow should produce two good size litters (according to breed) per annum, which involves weaning the litter on or before eight weeks of age, whereupon she can be re-served as soon as she comes on heat (normally four to seven days after weaning).

Boars should be kept in good body condition, exercised regularly, and not overworked (more than seven services per week) or underworked (more than 14 days interval between services). They should be housed within sight and smell of the sows. This will ensure good semen quality, maintainance of libido and help optimise fertility levels in the herd. Sows should be served in the cool of the day to avoid any heat stress. Before farrowing, sows must be housed in a separate pen and provided with some bedding for nest building. Care of the sow and her litter at and around farrowing is very important to ensure a good milk flow and minimise piglet mortality. Sows should be well-fed during lactation to promote milk production and prevent major loss of weight before weaning, but requirements for nutrients in gestation are lower, and low-energy fibrous feeds can be given. Piglets should be fed some solid feed before weaning to assist in the weaning transition. After weaning and service, sows can be transferred, in groups, to a dry pen until the next farrowing.

Growing pigs

Once weaned, pigs can be group-fed in pens. Care should be taken to ensure that weaners receive the highest-quality feed that is available, so that growth is not restricted at this stage. Growth should be linear through to slaughter weight, without periods of stasis. Indigenous pigs have a limited potential to deposit lean meat and are also early maturing. Thus they will tend to start depositing fat at a relatively early age and small size. The use of feed to produce live weight then becomes very inefficient. Slaughter weight and age should, therefore, be tailored to slaughter animals when the lean: fat ratio is optimum. Waste of feed is common in small-scale pig pens and troughs or drums should be provided to minimise spillage and spoiling of feed.

Disease

In common with all domesticated animals, pigs are subject to a variety of infectious diseases, of which the most important are African swine fever (when the most common cause of infection is contact with bushpigs and warthog), foot and mouth disease, anthrax, rabies, brucellosis, erysipelas, enteric colibacillosis, hog cholera and swine dysentery. Preventing contact with vectors and good hygiene are two very important preventative measures.

Planning

Lack of an adequate supply of feed is a critical factor in rural pig production, and supplies are often both erratic and unreliable. Particularly is this the case in regions where there is a long dry-season, which often results in animals going through a period of sub-maintenance feeding. This has a negative impact on both reproductive and growth performance, causing marked, and sometimes prolonged, reductions in animal growth and annual output.

To combat this situation on an individual unit, the farmer must plan ahead. This involves a simple recording system so that the necessary information is available. Calculations on an annual basis to determine what the total feed requirements are for the unit, what feeds can be produced or are available on the farm, and the amounts to be purchased or acquired, are necessary. In addition, it is necessary to assess storage capacity so that some feed can be stored in case of unpredictable events, e.g. drought or outbreaks of crop disease. If the calculations show that it is not possible to provide adequate amounts of feed, then the size of the pig herd may have to be reduced to bring it into balance with the available feed resources.

The way forward

For the resource-poor, small-scale producer, the way forward must be to up-grade the operation and profitability through step-by-step improvements in the management of the pigs, particularly with regard to shelter, nutrition and health care. A portion of the extra money generated by these improvements can then be re-invested in further up-grading, until a plateau is reached in terms of performance and productivity. When this occurs, genetic improvement can be introduced further to improve pig production and carcass quality. Alternative approaches to improve the wealth-generating ability of the small-scale producer, such as introducing modern technology from the developed world, are usually unaffordable unless aid-funded and, even then, are generally unsustainable.

A pre-requisite for the development of small-scale pig production is access to information by the producer. All the necessary information is available in academic and commercial circles, but the problem relates to the dissemination of this information to the resource-poor farmer. Taking Africa as an example, the main vehicles for disseminating agricultural information and advice are the government extension services, which are generally under-funded and without sufficient specialist staff. Some regions are overcoming the problem by involving the private sector in rural development. In Zimbabwe, small-scale dairy producers have overcome the problem, to a certain extent, by forming producer associations for an area. Similar pig-producer associations could provide focus for the transfer of information and could invite specialists to give training courses in efficient pig production.

Further reading

Devendra, C. and Fuller, M.F. 1979. *Pig production in the tropics*. Oxford University Press, Oxford, UK.

English, P.R., Burgess, G., Cockran, R.S. and Dunn, J. 1992. *Stockmanship, improving the care of the pig and other livestock*. Farming Press, Ipswich, UK.

Eusebio, J.A. 1980. *Pig production in the tropics*. Longman, Harlow, UK.

Holness, D.H. 1991. *Pigs*. The Tropical Agriculturalist, Series Editors R. Coste and A.J. Smith. Macmillan Education Ltd., London, UK, in cooperation with Technical Centre for Agricultural and Rural Cooperation (CTA), Wageningen, The Netherlands.

Muirhead, M.R. and Alexander, T.J.L. 1999. *Managing pig health and the treatment of disease*. Nottingham University Press, Nottingham, UK.

Perez, R. 1997. *Feeding pigs in the tropics. FAO Animal Production and Health Paper 132*, Food and Agriculture Organisation of the United Nations (FAO), Rome, Italy.

Author addresses

David Holness, Dr David Holness (Pvt) Ltd., P.O. Box BW 943, Borrowdale, Harare, Zimbabwe

Rob Paterson, Formerly of Natural Resources Institute (NRI), University of Greenwich at Medway, Chatham Maritime, Kent ME4 4TB, UK

Brian Ogle, Department of Animal Nutrition and Management, Swedish University of Agricultural Sciences, Box 7024, 750 07 Uppsala, Sweden

19

Goats

Christie Peacock[*], *Canagasaby Devendra*[†], *Camillus Ahuya*[†], *Merida Roets*[†], *Mujaffar Hossain*[‡], *Emmanuel Osafo*[‡]

Cameo - Bangladesh: goats, a pathway out of poverty

Sufia Khatun of Charbogra village of Mymensingh district, Bangladesh, had this to say:

"I am one of the focus group members taking part in 'Action Research' for a UK DFID project (LPP Project R8109) on livestock to alleviate poverty. In January 2004 I received a goat doe of 20 months of age with two male kids aged two months (cost Taka 1800 [US$ 31.1]). I fed the three goats by grazing and tethering. Kitchen waste like vegetable peel and left-over rice was also fed to the goats. My husband is a daily labourer. At night I keep the goats in the dwelling house. On 17th April 2004, the doe gave birth to two new kids. I am collecting the droppings of the goats in a small pit near my house and use them for my vegetable garden. Now I am happy and busy keeping five goats.

Photo 19.1 Sufia Khatun with her five goats (courtesy of M. Hossain)

[*] Principal author
[†] Co-author
[‡] Consulting author

I am very poor, but I can now dream of selling two male kids of the first generation during the Ramadan Eid Festival in November 2004 for Tk 1500 (US$ 25.9) for each goat. I can imagine goat rearing will improve my daily food consumption and also generate a continuous source of income."

Role of goats in the livelihoods of resource-poor people in developing countries

Roughly 95 per cent of the 765 million goats in the world are found in the developing countries of Asia, Africa and South America (Table 19.1). For too long the role of goats in the economies of developing countries has been grossly underestimated, and their contribution to the livelihoods of people, particularly resource-poor people, has been ignored. There are many reasons for this. Goats are difficult to count and so their numbers are nearly always underestimated. Most goats and goat products seldom enter a formal marketing system, and they are not taxed, so their contribution to the national economy is not recognised. Furthermore, goat-keepers are often resource-poor people who are economically and politically marginalised. As a result, politicians, development planners and academics give goats a low status.

Table 19.1. Goat population and geographical distribution

Region	*Population (million)*	*Percentage of total population*
Asia	477	62.3
Africa	220	28.7
Latin America and Caribbean	27	3.5
Europe	18	2.3
North & Central America	14	1.8
Former Soviet Union	8	1.0
Oceania	1	0.4
Total	765	100.0

Source: FAO (2003).

However this past neglect is starting to change. The goat's role in the support of people who are poor is starting to be appreciated and their potential to play a role in enabling resource-poor people to find a pathway out of poverty is, at last, starting to be exploited.

Goats provide their owners with a vast range of products and services, some of which are listed in Table 19.2.

Table 19.2 Goat products and services

Products	*Services*
Meat (raw, cooked, blood, soup)	Cash income
Milk (fresh, sour, yoghurt, butter, cheese)	Security
Skins (clothes, water/grain containers, tents, thongs, etc.)	Gifts
Hair (cashmere, mohair, coarse hair tents, wigs, fish lures)	Gifts
	Loans
Horns	Religious rituals
Bones	Judicial role
Manure (crops, fish)	Pleasure
	Pack transport
	Draught power
	Medicine
	Control of bush encroachment
	Guiding sheep

Goats are relatively cheap and are often the first asset acquired, through purchase or customary means, by a young family or by a poor family recovering from a disaster, such as drought or war. Goats, once acquired, become a valuable asset providing security to the family as well as products such as milk and dairy products. The advantages and disadvantages of goats relative to resource-poor people are shown in Table 19.3.

Table 19.3 Advantages and disadvantages of goats of relevance to resource-poor people

Advantages	*Disadvantages*
Able to use fibrous feeds, especially browse, efficiently enabling use of marginal land.	Susceptible to predators and thieves.
Efficient use of water.	Small value often makes formal credit systems uneconomic.
Wide climatic adaptation.	Small value makes formal insurance systems difficult to administer.
Relatively cheap to purchase.	Susceptible to broncho-pneumonia.
Security from several low value goats being less risky than one high value cow.	Susceptible to internal parasites.
Suitability to small farms.	Less easy to control than other species.
Relatively drought tolerant.	
Fast reproductive rate quickly builds up herd.	
Fast reproductive rate ensures early returns on investment and enables early credit repayment.	
Small size enables easy and quick movement of households in emergencies.	

Table 19.3 (Contd.)

Advantages	*Disadvantages*
Easy for women and children to handle.	
Few facilities required.	
Lack of religious taboos against goat meat which often commands higher price than other meat.	
Small size allows easy home slaughter.	
Potential for integration into perennial tree crop systems.	
Relative trypanotolerance.	
Manure improves soil fertility.	

Classification of major production systems in developing countries

Goat production systems fall into four categories:

- Extensive grazing or browsing systems
- Systems using feed from:
  - by-products of arable cropping
  - roadside verges, communal grazing/forest areas through tethering or cut-and-carry feeding
- Systems integrated with tree cropping
- Urban goat keeping.

Goat populations tend to be higher in arid and semi-arid areas; for example, in Sub-Saharan Africa 65 per cent of goats are found in these areas. However, goats are highly adaptable animals, and may be found at the highest altitudes of the Himalayas and in the humid coastal areas of Africa and Asia. They are an integral component of agro-pastoralism and range-based livestock systems in Sub-Saharan Africa, West Asia and North Africa, parts of Central Asia, Northern Brazil, Peru, Mexico and the Indian sub-continent (see Chapter 3). Goats may be kept by themselves, or in mixed-grazing systems with sheep or other livestock. In more humid and sub-humid regions of Asia, Latin America and the Caribbean, resource-poor people integrate goats with arable cropping systems. The integration of goats with perennial tree crops, e.g. coconuts, oil palm and rubber, has largely been under-exploited. Attention is also drawn in Chapter 3 to the relevance of goats in the higher-order livestock systems and the opportunities for increasing their contribution in the developing countries.

Problems faced by resource-poor farmers and constraints to production

Resource-poor goat farmers face many problems which are common to keepers of other species. As indicated in other chapters, these problems include:

Physical

- Low quality and shortage of available feed
- Limited and declining access to communal grazing
- Limited access to water
- High disease challenge
- Poor infrastructure, e.g. markets, slaughter houses, watering facilities.

Economic

- Very limited cash for investment
- Limited access to land
- Limited access to relevant support services, e.g. veterinary, breeding, marketing etc.
- Limited access to new technology, e.g. vaccinations, and information, e.g. market information
- Limited access to credit.

Policy

- Weak political representation at local and national levels
- Few farmer organisations meeting needs of goat farmers
- Negligible allocation of national resources to developing goat production.

In the past, there have been very few goat development programmes established by national governments, or by international or non-government organisations (NGOs). As a result, goat farmers have received little attention and support. However, as indicated earlier, the potential of goats to improve the lives of the resource-poor of the world is beginning to be acknowledged.

Strategies of resource-poor livestock-keepers and rationale for current husbandry practices

Resource-poor goat-keepers in societies where there is no formal welfare system, and in harsh tropical environments, are driven by the short-term need for survival and to ensure the basic necessities of life. Goats often form a vital part of a family's strategy to reduce risk and diversify its sources of income. As such, goat-keepers are unsurprisingly averse to risk-taking, and struggle to plan very far into the future. For example, the marketing of goats is usually driven by an immediate need for cash, rather than to sell a male goat at its optimum weight. Being cash-poor usually results in goat-keepers struggling to pay for expensive veterinary or breeding services. Furthermore, goats are valued because of the multiple benefits they offer. Goat owners may not necessarily be motivated to develop specialised systems of production, e.g. dairying, but will tend to be interested in improving several aspects of production simultaneously, providing it does not increase risk.

Goat farmers adopt a variety of strategies to manage goats, and especially to cope with the harsh environment of the arid and semi-arid tropics, mainly very high temperatures (35-45 °C), low annual rainfall (250-600 mm), lack of feed and water. Husbandry practices are therefore aimed at mitigating these elements; the rationale is the avoidance of risk (Devendra, 1998). Strategies may include diversification (crop cultivation practices, other income sources, other livestock species), reallocation of resources (e.g. increasing cropping area, mixed cropping), adjustment in dates of sowing, transplanting, etc.), increased investment in goats ('banks on hooves'), household survival mechanisms (asset disposal, land mortgage, credit) and off-farm activities and employment.

These factors should be kept in mind when assessing levels of production from goats managed by resource-poor livestock keepers.

Improving goat production as a pathway out of poverty

Understanding production problems of goats

Before goat production systems can be improved, they must be accurately described and their problems properly analysed and understood. It is most important that the motivation and aspirations of the goat owners are known so that appropriate improvements can be developed *with* them. Chapter 4 and Peacock (1996) describe several tools that may be used to assess goat production problems in the field.

Most production problems encountered are not simple, but are caused by a complex of factors. For example, it is often the case that there is relatively high pre-weaning mortality among goats, but this is seldom caused by any one factor. Poor nutrition of the dam, combined with her genetic potential, may cause her to produce little milk, which will undernourish the kid, making it susceptible to disease and harsh environments.

Peacock (1996) describes most of the major goat production systems in the tropics and analyses their main production problems, and factors causing or contributing to the problems.

The improvement of the production and marketing of goats kept by resource-poor livestock-keepers can be the first step out of poverty. The basis of this improvement has to be a thorough understanding of the existing system and a genuine engagement of the goat-keepers themselves. Without their involvement and motivation, outside interventions are doomed to fail.

Different strategies that can be adopted

There are many different strategies that can be adopted to improve goat production:

- *Small/big improvements to existing system,* e.g. reducing kid mortality through better supervision of suckling, supplementary feeding of young males for fattening, growing of forage trees to provide dry-season feed, etc.

- *Fundamental change of system*, e.g. moving to housing goats for intensive production (dairying and/or fattening), perhaps also including cross-breeding and market orientation.

- *Stocking/re-stocking people with goats*, e.g. goat credit to stock people with goats or re-stock families that have lost their stock through some sort of disaster. In addition to supplying credit or grants for goat purchase, husbandry skills may have to be taught to new goat-keepers.

Specific problems of goats and opportunities to improve production

There will always be opportunities to improve the existing system. Many of these improvements will have a cost associated with them but many will not. A series of small adjustments to husbandry practices can have as great an impact on production as one expensive improvement.

Feeding

Feeding goats well is the major factor underpinning all goat enterprises. Good nutrition is a prerequisite for good health, good reproduction, high milk yields and high growth rates, all necessary for a successful goat production system. Most goat farmers make use of natural vegetation or crop by-products and try to feed their goats as best they can with what is available. This might be termed 'supply-driven' feeding as goats are fed according to the supply of feeds available, which may vary seasonally and over which the goat-keeper may have little, if any, control. Goat-keepers can reduce the seasonality of feed supply by growing out-of-season forage crops or by conserving forage and other locally occurring resources, for example protein-rich tree fruits.

There are many problems commonly encountered in feeding goats in the tropics. These include:

- Fibrous feeds causing low intake rates and production
- Seasonal fluctuations in quantity, digestibility, protein and water availability
- Low levels of protein for growth and milk production
- Specific mineral deficiencies
- Poor presentation of feed to confined goats
- Poor access to water
- Poor nutrition of lactating dams, leading to low milk yields and low rates of growth and survival among kids
- Low quality of feeds for kids at weaning, causing a sharp drop in weight and possible death.

Key improvements

The key principles that should be considered when improving the feeding of goats are:

- Types of feeds available and seasonality of supply
- Improve the balance of nutrients reaching the rumen and small intestine
- Increase the amount of feed eaten
- Target goats with special needs, e.g. pregnant or lactating females
- Reduce the seasonal fluctuation in feed supply
- Reduce the burden on women and children.

In order to consider ways of improving the feeding of goats it is vital to know:

- Seasonality of feed supply
- Seasonal methods of feeding
- Seasonal labour supply
- Seasonal patterns of production
- Likely periods of feed deficit
- Availability of supplements, minerals and water.

Options to improve feed supply and diets are shown in Table 19.4. Factors affecting intake are shown in Table 19.5.

Stall-feeding goats

As the areas cultivated increase and encroach on grazing lands throughout the developing world, there is often pressure to confine livestock and stop them from grazing freely. In densely-populated areas of the highlands of Africa and most of South-East Asia there is an increasing trend towards confinement of goats through tethering or housing and stall-feeding in some way. Unless farmers are careful in their feeding, confinement can seriously reduce the quality of the diet consumed

Table 19.4 Options to improve feed supply and diet

	Free grazing	*Tethered*	*Stall-fed*
Feed supply	Select grazing area. Develop forage crops. Supplement diet with energy, protein, minerals.	Select best site. Develop forage crops. Supplement diet with energy, protein, minerals.	Select quality feeds. Mix feeds. Develop forage crops. Supplement diet with energy, protein, minerals.
Treatment of feed	Conserve feeds.	Chop unpalatable feeds. Wilt wet feeds. Mix feeds. Treat with urea. Conserve feeds.	Chop unpalatable feeds. Wilt wet feeds. Mix feeds. Treat with urea. Conserve feeds.
Presentation	Increase total grazing time. Allow time for ruminating. Ensure presence of shade. Select best time to graze.	Ensure comfort and safety. Move frequently. Ensure presence of shade. Allow sufficient time to graze.	Feed at correct height. Present feeds in an accessible manner. Ensure adequate space for all goats. Feed little and often. Clean up waste feed.
Water	Allow frequent access.	Allow frequent access.	Allow continuous access.

Source: Peacock (1996).

by goats. For this reason it is important to understand how goats naturally like to eat so that, when they are confined in some way, their natural habits can be copied as closely as possible so they are comfortable, consume their feed readily and are therefore productive. The goat is a natural browser, feeding by preference on tree leaves, flowers and seed pods (tree-fruits) 20-120 cm above the ground. They very often find it difficult to eat directly off the ground. Goats tend to move quickly over the available vegetation, selecting the best available plants and parts of plants thus ensuring a diet of reasonable quality.

It is sometimes wrongly thought that because goats eat a wide range of plants they will eat anything. Goats are quite fastidious in their tastes, preferring a wide range of vegetation, quickly becoming bored with the same feed unless it is one they particularly relish.

If goats are stall-fed, feed should be presented in an accessible manner. Ideally it should be raised off the ground and presented to the goats within their preferred browsing height. A high feed rack, trough or even hanging bunches of forage will ensure good access and that the feed is kept clean. This will reduce wastage. If there are many goats feeding from the same rack there should be enough space for them all to have easy access to the feed. If there are small, weak goats they should be fed separately.

Table 19.5 Factors affecting intake by goats

Feed factors	*Presentation factors*	*Goat factors*
Taste	Feeding time	Appetite
Smell	Frequency of fresh feed	Preference
Variety	Quantity offered	Size
Moisture content	Competition from other goats	Pregnancy
Digestibility	Temperature	Growing
Size and form of feed	Humidity	Lactating
	Method of feed presentation	

Water

Goats are second only to the camel in the efficiency with which they use water. However they do need regular access to water and this is particularly important for lactating females. They need an allowance of 1.3 litres water per litre of milk produced. The lack of water is often a major factor reducing milk production. Generally the lower the quality of feed the more water goats need. Free-grazing goats should ideally be offered water every day and stall-fed goats should have access to water at all times.

Minerals

Mineral deficiencies can be difficult to detect, but can lower production considerably. Some mineral deficiencies will have specific symptoms, and in extreme case may cause kids to be born deformed. Each area will have its own deficiencies and it is important to know what these are. Goats may often be observed eating soil as they crave a deficient mineral. Ideally, goats would have mineral supplements but this may be too costly. Locally-sourced mineral supplements are often traded in markets and simple cooking salt can be useful in very hot temperatures.

Forage production

The development of forage crops, including multi-purpose trees (MPTs), is one option for making affordable improvements to the diet of goats. Forage crops can benefit goats, as well as the rest of the farm and the wider environment, and can:

- Improve the total supply of feed available, e.g. growing elephant grass
- Improve the quality of feed, e.g. undersowing crops with legumes as a protein supplement
- Compensate for seasonal fluctuation in feed supply, e.g. growing tree legumes for dry-season feed
- Reduce soil erosion
- Provide a source of green manure for food crops
- Provide firewood and building materials
- Reduce labour required to feed goats
- Provide shade.

Key improvements

Experience of introducing forage-growing to farmers has shown that it is most successful in systems of production that offer a good monetary return, such as dairying or fattening. Successful forage development requires there to be:

- Well-motivated goat-keepers
- Good initial extension support
- A good forage strategy that fits well into the existing system
- Stock control
- Availability of planting material.

To develop an effective forage strategy it is must be clear where and when forage can be grown. The areas that might be considered for growing appropriate forage for goats are:

- The area immediately around the house, e.g. tree legumes cut as a hedge
- Along the edges of fields, including bunds in rice fields, e.g. herbaceous legumes or grass strips
- A strip of land in the field

- Underneath an established crop, e.g. undersowing with a herbaceous legume
- Underneath a perennial crop, e.g. undersowing with a grass/herbaceous legume mix
- Communal grazing areas, e.g. over sowing with a hardy legume.

Permanent pastures

While the nutritive value and cultivation methods of individual forage species may be known, very little research has been carried out on designing appropriate permanent pastures for goats that gives sufficient consideration to their browsing habit.

Housing and restraint

The reduction in grazing land available for goats to graze freely means that they are increasingly likely to be confined, housed or restrained in some way, for all or part of the year. Goats are very difficult to restrain into a particular plot and seem to be able to escape through most fencing. This means that they will normally be housed or tethered in some way. Housing can be expensive for resource-poor goat-keepers to construct and is only justified in systems of production offering high returns, e.g. dairying.

Goats are usually tethered with a rope tied to their leg or around their neck. Both sites can have their problems. Ropes tied around the neck will usually rub the skin there, which can pre-dispose the goat to sarcoptic mange. This may also occur if the rope is tied tightly around the leg, where the blood supply can be constricted. Ideally goats would be tethered by attaching the rope to a leather collar and to a pin placed in the ground. It is also important that goats tethered to graze, are allowed enough time to maximise their intake. A recent UK DFID project in Tanzania showed this to be 8 hours per day (LPP Project, R5194).

Reproduction

It is important that goat-keepers are able to manage effectively the reproduction of their goats according to their own objectives, the availability of feed and the requirements of the market. Reproduction determines the rate of expansion of the flock, the number of goats that can be sold and the availability of milk.

Puberty is reached at about seven months, but it can be at as young as three months or as old as 12 months. Successful matings can occur as young as three months, so control needs to be exerted early. The age at which females should first mate will vary according to breed and stage of maturity. A female should never be mated

Photo 19.2 Goats tethered on edge of fallow maize field awaiting feed in Kitui, Kenya (courtesy of FARM-Africa)

before one year. Ideally, she would have one pair of permanent incisors, i.e. be 14-17 months. Ideally, bucks should not mate until they are one year old.

Oestrus occurs in healthy goats every 19 days (range 17-21 days). The main signs are bleating, tail wagging, seeking out a male, standing to be mated, swelling of the vagina and mucus discharge. In the tropics and sub-tropics oestrus occurs all the year round. However, in areas with a pronounced and lengthy dry-season, during which females lose weight and suffer nutritional stress, they may not show oestrus; this is known as nutritional anoestrus.

Mating management

If fertile males are kept with fertile females the males will easily detect oestrus and mate.

Poorer goat-keepers owning small numbers of goats may not necessarily have a breeding male in their flock and may have to seek the use of a neighbour's. It is important that oestrus is detected in good time and, with small numbers of goats kept in close proximity to the owner's family, the change in behaviour will be relatively easy to spot. Once oestrus is identified the female should be taken to the male or vice versa, and mating should take place as soon as possible to ensure the greatest chance of conception.

If goat-keepers want to control the time of breeding, for example so that kids are born at a favourable time of year, there are various options they can use to achieve this:

- Separation of males from females (suitable in housed goats)
- Buck apron (suitable in large flocks)
- Buck penis string (suitable in large flocks)
- Castration of unwanted males.

Artificial insemination

Artificial insemination (AI) is one method of introducing new and better genes into a flock. However, in the developing world there have been very few successful AI programmes except on commercial farms or large breeding stations. While the AI technique in goats is simpler than in cattle, in practice in the field, ensuring ideal conditions for conception is very difficult, resulting in low conception rates (cf. Chapter 12).

Breeding

Over 350 breeds of goats are recognised, all of which originated from the original goat stocks domesticated in the Middle East and Central Asia over 10,000 years ago. Tropical goat breeds are the result of hundreds of years of pressure by the tropical environment through natural selection, combined with some selective breeding by their owners. As a result, goat breeds are well adapted to surviving in tropical environments with high temperatures, low-quality feeds, limited water and a high disease challenge. The environment does not allow these goats to perform at a high level of production and they rarely have the potential to do so. The multi-purpose nature of the role of goats in the tropics also means that they have a broad range of useful traits, but with few of them having been developed for specialised economic purposes. Many goats are referred to as 'nondescript' or 'local' breeds simply because they have never been properly described and named. In general, little effort has been put into the systematic description, classification and evaluation of goats, so the potential of most tropical breeds is poorly understood and exploited (FARM-Africa, 1996).

In some societies it is clear that pressure exerted by owners can be both positive and negative. For example, the needs of the owners mean that fast-growing males are often sold ahead of their slower-growing cohorts. In harsh environments goat-keepers may select against twinning, reasoning that a single kid has a much higher chance of survival than twins.

Characteristics of tropical goat breeds

- Size: tropical breeds are usually smaller in size than dairy breeds bred in temperate conditions. Some of the desert types, such as the Jamnapari or Sudan desert type, may be tall and leggy, while breeds in the humid tropics tend to be short and small.

- Colour: varies widely, from pure white or black, and from solid colours to patches and speckles.

- Coat type: most breeds in the tropics have a short coat to reflect light and keep the goat cool. In colder environments breeds may have a longer coat, such as the Kashmiri (Pashmina) goat of the Himalayas.

- Growth: generally slow, partly as a result of poor nutrition, as well as genotype. Nutritional studies show a generally poor response to improved nutrition by tropical breeds.

- Milk production: yields are generally low (200-1,000 ml per day) and lactation length is short at 3-5 months.

- Prolificacy: most breeds regularly give birth to twins, and triplets are not uncommon.

- Disease resistance: most breeds build up their own resistance to diseases of the area in which they are kept. There is great variability, both within and between breeds, in resistance to disease. This is a trait that could be exploited through breeding programmes.

- Water use: tropical breeds use water efficiently and some breeds can last for several days without water.

- Survival: normally good, although pre-weaning mortality can be a problem in some systems.

Key improvements

Breed improvement should only be considered if the standard of management can be improved sufficiently to take advantage of the greater genetic potential. It is pointless to waste resources improving the breed potential of goats unless the 'improved' goat is fed well and kept healthy. However, in practice it is often found that owning an improved goat will stimulate owners to improve their feeding and management. Owners will quickly learn that the improved genetic potential, expressed as milk yields or growth rates, brings greater rewards to better management (see later Section 'Goats as a pathway out of poverty – examples').

As indicated earlier, there have been very few successful goat-breeding programmes in the developing world. This does not reflect any lack of potential, but rather the history of neglect by government agriculture departments and researchers. Often when governments have promoted new breeds, they have not put sufficient effort into management improvements, leading people to perceive the new breed to have failed. All breeding improvement requires a long-term commitment, over 10-20 years, and this is often hard to achieve in developing countries. New practice concerning the provision of agricultural services presents opportunities for 'community-based' or 'farmer-managed' breed improvement (cf. FARM-Africa cross-breeding programme in Kenya discussed in Chapter 12)

Very little selection *within* tropical goat breeds has ever been undertaken. There is clearly potential to improve traits of economic value with high heritability through selection, but this would require the long-term commitment of resources in a systematic manner (cf. Chapter 12).

Health

Keeping goats healthy is obviously important for all livestock keepers and particularly for poor ones. Goats may make up a large part of the family's total assets and so steps need to be taken to ensure they do not die or become sick needlessly. Keeping goats healthy does not involve the use of expensive drugs and highly-trained veterinary staff. In most situations, the majority of the important diseases can be controlled through simple preventative measures such as good feeding, clean water, clean housing, vaccination, drenching, spraying/dipping and foot trimming. If these basic measures are done when appropriate, 80-90 per cent of the important diseases affecting goats can be controlled. Efforts should be directed at establishing what the common and important diseases are in any area, and efforts should be focused on controlling them.

Smith and Sherman (1994) remains the best text on goat medicine and Table 19.6 (from Peacock, 1996) provides a useful introduction to the causes of the common disease syndromes of goats.

Table 19.6 Common disease syndromes of goats

Disease syndrome	*Likely causes*
Kid death	Coccidiosis; colibacillosis; colostrum deprivation; enterotoxaemia; internal parasites; suffocation; malnutrition.
Diarrhoea and loss of condition	Acidosis; bloat; coccidiosis; enterotoxaemia; internal parasites; Nairobi sheep disease (NSD); peste des petits ruminants (PPR); Rift Valley fever.
Respiratory problems and fever	Anthrax; contagious caprine pleuro-pneumonia (CCPP); lungworm; melioidosis; Nairobi sheep disease; PPR; pneumonia; goat pox.
Skin diseases and swellings	Caseous lymphadinitis; streptothricosis; goat pox; mange; orf; ringworm; warts.
Poor condition/anaemia	Anaplasmosis; babesiosis; coccidiosis; internal parasites; teeth problems; trypanosomiasis.
Lameness	Akabane disease; caprine arthritis encephalitis (CAE); contagious agalactia; foot and mouth disease (FMD); foot rot; mastitis; melioidiosis; mineral deficiencies; navel ill; ticks; physical injury; vitamin deficiencies.
Diseases affecting the nervous system	CAE; copper deficiency; enterotoxaemia; heartwater; listeriosis; melioidosis; navel ill; pregnancy toxaemia; rabies; scrapie; tetanus.
Male and female infertility	Brucellosis; intersex; metritis; physical damage; sperm granulomas; trypanosomiasis
Abortion	Brucellosis; chlamydial abortion; FMD; listeriosis; malnutrition; NSD; poisoning; Rift Valley fever; salmonellosis; shock and stress; toxoplasmosis; trypanosomiasis.
Udder problems	Mastitis; orf; physical damage; warts.

Source: Peacock (1996).

Goats are particularly susceptible to the following diseases.

Internal parasites

Goats are more susceptible to internal parasites than sheep or cattle, perhaps because they are browsers, normally consuming vegetation above the height at which infective larvae are found. Goat farmers all over the world have found that controlling internal parasites is a key determinant of successful goat production. There are four groups of gastro-intestinal parasites that affect goats:

- Nematodes: *Haemonchus contortus, Ostertagia* spp., *Trichostrongylus* spp., *Nematodirus* spp., *Strongyloides* spp., *Oesophagostomum* spp., *Trichuris* spp.

- Cestodes: *Moniezia* spp.

- Trematodes: *Paramphistomum* spp.

- Protozoa: Coccidia, including *Eimeria*

Of these, *H. contortus* is by far the most important species and should nearly always be suspected if there is a problem. They attach themselves to the wall of the abomasum and can consume a great deal of blood, making the host goat anaemic and unthrifty. Parasites such as *H. contortus* can be controlled through changes to management and/or use of appropriate anthelmintics. The feeding of forage with high-tannin content has also been found to help control internal parasites (LPP Project, R7424, cf. Chapter 10). Management controls include:

- Grazing kids separately from adults on 'clean' pasture or ahead of the adults

- Avoiding wet/swampy areas

- Selecting bushy areas

- Consider cut-and-carry feeding and wilting wet feed.

Mange

Mange is an underrated disease of goats. If left unchecked it can kill goats and in certain circumstances can sweep through a flock causing high mortality rates. The most important mange in goats is sarcoptic mange caused by *Sarcoptes scabei*. It is vital to treat the disease early with an effective acaricide vigorously scrubbed into the affected sites. An infusion of the leaves of the castor bean plant (*Ricinis communis*) has been found to be effective against mange (Peacock, 1996).

Tick-borne diseases and tick control

There are four main diseases of goats transmitted by ticks: heartwater, anaplasmosis, babesiosis and Nairobi sheep disease. Of these heartwater, caused by *Cowdria ruminantium* transmitted by the *Amblyomma* spp. of tick, is probably the most important and is a common disease in Africa and the Caribbean. As immunity can be developed to many tick-borne diseases, expensive tick control using acaricides should be practised only if there is a clear economic benefit.

Contagious caprine pleuro-pneumonia (CCPP)

Contagious caprine pleuro-pneumonia is an acute pneumonia of goats which causes high mortality rates. The disease is widely distributed in North and East Africa, the Middle East, Eastern Europe and some parts of Asia. It is a highly contagious disease that can kill whole flocks of goats, devastating their owners' livelihoods. Effective vaccines are available and should be given to susceptible flocks wherever possible. Treatment using long-acting tetracyclines can be effective if started early.

Peste des petits ruminants (PPR)

Peste des petits ruminants (PPR) is a highly contagious viral disease similar to rinderpest in cattle. It is widespread in West and Central Africa and has recently been identified in East Africa and the Middle East. Devastating outbreaks can occur, causing mortality rates of 70-90 per cent and it is a major factor inhibiting goat development in West Africa. There is no treatment for PPR. Control is by movement restriction and vaccination. The cattle rinderpest vaccine has been found to be effective.

Pneumonia

Respiratory problems are relatively common in goats, particularly housed goats. It is often difficult to identify the cause of the problem. Pneumonia is mostly caused by poor ventilation and may be triggered by stress. It is most important that if goats are housed they have plenty of fresh air. In highland areas where the temperature may drop considerably at night, there is sometimes the belief that goats should be kept warm by shutting them into a small space with poor ventilation. It should always be remembered that the rumen produces a lot of heat, which is quite sufficient to keep goats warm at night. In temperate locations goats are kept in the snow without any ill effects!

Community-based goat health systems

The small size and relatively low value of goats means that they are unlikely to attract the attentions of qualified veterinarians. Resource-poor livestock-keepers are unlikely to be able to afford to treat individual sick goats. This means that preventative care by owners is vital and they should be trained to be able to manage simple diseases as they arise. Community-based animal health-care systems have been found to offer a cost-effective approach to providing basic health care to livestock kept by resource-poor people (see Chapters 5 and 13).

Marketing and processing goats and goat products

The incomes of resource-poor goat-keepers can be boosted significantly through the better marketing of goats, as well as through adding value to goat products, e.g. processing milk into cheese or yoghurt. However, resource-poor farmers are often in a weak bargaining position in relation to traders and may need to group together to strengthen their position to ensure that the producer retains the bulk of the 'value-added'. There are many marketing opportunities, including international markets, which could be exploited by goat farmers, e.g. processing of hides into quality leather and leather products, sale and processing of cashmere, export of carcasses into Middle Eastern markets, etc. However the collection, handling, processing and marketing of goats and their products is disorganised in many developing countries, severely constraining the more 'market-orientated' goat farmer.

Goats as a pathway out of poverty - examples

The recent increase in knowledge and understanding of goats and the people who keep them has led to better-planned and -implemented goat development programmes in many parts of the developing world. There is a growing awareness of the life-changing role goats can play in assisting some of the poorest people out of poverty and onto a pathway to prosperity.

Example from Meru, Kenya, of poverty-focused dairy goat development

Smallholder farmers in Meru district, Kenya, have few options to increase their incomes and improve their lives. Landholdings are small, 0.5-2 ha, on which maize, beans and coffee are grown. Livestock are an important asset and source of income, with wealthier families owning one to three cattle, while poorer farmers, if they have ruminant livestock, may own a few goats. Goats are traditionally milked by people in Meru and improving their production offered one option to improve the lives of people in the district. The British NGO FARM-Africa, worked in partnership with the Kenya Ministry of Agriculture to enable farmers to up-grade local goats by cross-breeding. Farmers had access to the British Toggenburg breed through community-managed buck stations, supplied with breeding stock from small private breeders who were part of farmer groups. The poorest families, without any livestock, were supplied with local goats on credit. One such family was the Kinoti family, who did casual labouring and owned a small plot of land. Mr Angelo Kinoti received two local goats on credit and became a buck keeper. He managed the buck so well that it became champion at the Nairobi International Show. The cross-bred goats he bred were in demand and commanded a high price.

Through good management and hard work Mr Kinoti was able to buy oxen for ploughing, which he rented out for contract ploughing. He acquired land and is now able to build a stone house and pay the fees to send his eldest daughter to secondary school.

Photo 19.3 Collecting manure from housed dairy goats in Meru, Kenya (courtesy of FARM-Africa)

Example from Ethiopia of role of goats in food security in very food-insecure areas

After years of war and regular droughts, and resulting famine, there were many impoverished, women-headed households on the edge of survival. Typically these households had a small plot of land but no assets, not even a chicken, nor realistic way of improving their lives. The British NGO FARM-Africa started a Dairy Goat Development Project to assist women and their families in this difficult situation. Two local female goats were provided on credit to each selected woman. Women were organised into small groups of 25-30 who were responsible for managing the repayment of the credit, which was repaid in kind by returning a kid to the group, who would then select another woman to benefit. Women were trained in forage production, improved management and basic health-care and were given the opportunity to up-grade their goats through cross-breeding with Anglo-Nubian bucks. Two women per group were trained as community-based animal health-care workers (CBAHW) providing basic veterinary care for a small fee to members of the group and their neighbours.

Women quickly built up their flocks of goats and in some cases were able to sell goats to buy oxen for ploughing or a milking cow. During the severe drought of 1999-2000 households with livestock to sell were able to buy grain and survive the drought without resorting to food aid handouts. Selling one goat could buy enough grain to feed a family of 5 for 2-3 months.

A market-oriented example of goat commercialisation in South Africa

National and international trade opportunities and globalisation have created a niche for the commercialisation of indigenous South African goats. Product innovation, the opening of global markets, the shift in South African population demographics, and the increase in accessibility of information to non-commercial farmers, have created opportunities for this under-utilised resource. Certain institutional arrangements, comfortable and culturally-acceptable to non-commercial farmers, had to be created, while at the same time addressing the global challenges of quality, consistency and high standards. This has required attention to formal (contracts, organisations, markets) and informal (traditions, customs) institutions, both at macro (legal) and micro (organisational form) level. The businesses (some in the process of development) link farmers with processors and markets, through vertical integration, and ultimately to export markets with their high demands for safety, quality and consistency.

The Kalahari Kid Corporation (KKC) was registered in 2002. This company has as its core functions the branding, brand management, quality control and marketing of goats and goat products. It is a joint initiative between private-sector commercial partners, several government stakeholders, and emerging and commercial farmers in several provinces of South Africa. Kalahari Kid's initial emphasis is on the development of (especially) the non-commercial goat sector to supply animals of the correct quality. This entails the organisation of small-scale farmers into 'Goat Interest Groups', undertaking a contractual relationship with KKC to deliver a chosen number of goats of a pre-determined quality per year, the transfer of knowledge and skills regarding the animals and their management, and, finally, delivery. The process of contracting, and the limiting of the purchase of stock only to goats grown by 'contract growers', serves to establish the traceability system that is required by international standards.

Having secured its supply base, the KKC has placed further emphasis on the design of market-oriented products. This has been achieved through product development and consumer testing. Marketing has emphasised goat meat as a healthy, interesting, and 'naturally reared' meat alternative. Entry into both the local retail market and foreign (especially Middle-Eastern) markets has been achieved.

The development of these institutions has created the possibility of entry of small-scale goat farmers into the formal national and international markets.

The future

At last the goat is coming into its own as a species of livestock worthy of serious academic and development interest. The new poverty-focused development agenda of national governments, supported by multi/bilateral development institutions, demands that the goat be placed at the centre of any serious development effort that addresses the needs of the resource-poor. The goat has been a servant of mankind for generations and today it offers a huge potential for transforming the lives of some of the poorest people in the world.

Further reading

Bath, G. and de Wet, I. 2000. *Sheep and goat diseases*. Tafelberg Press, Cape Town, South Africa

Devendra, C. and Burns, M. 1983. *Goat production in the tropics. Second edition*. CABI Publishing, Wallingford, UK.

Devendra, C. and Chantalakhana, C. 2002. Animals, poor people and food insecurity - opportunities for improved livelihoods through natural resource management. *Outlook on Agriculture* 31:161-175.

Devendra, C. and Haenlein, G.F. 2002. Goat breeds. In *Encyclopaedia of dairy sciences* (ed. R. Roginski, J.W. Fuquay and P.F. Fox), pp. 498-495. Academic Press, Oxford, UK.

FARM-Africa. 1996. *Goat types of Ethiopia and Eritrea. Physical description and management systems*. Joint publication FARM-Africa, London, UK, and International Livestock Research Institute (ILRI), Nairobi, Kenya.

Peacock, C.P. 1996. *Improving goat production in the tropics. A manual for development workers*. Oxfam/FARM-Africa, Oxford, UK.

Porter, V. 1996. *Goats of the World.* Farming Press, Ipswich, UK.

Smith, M.C. and Sherman, D.M. 1994. *Goat medicine*. Lea and Febiger, Philadelphia, USA.

Steele, M. 1996. *Goats*. The Tropical Agriculturalist, Series Editors R. Coste and A.J. Smith. Macmillan Education Ltd., London, UK, in cooperation with Technical Centre for Agricultural and Rural Cooperation (CTA), Wageningen, The Netherlands.

Author addresses

Christie Peacock, FARM-Africa, 9-10 Southampton Place, London WC1A 2EA, UK

Canagasaby Devendra, Formerly of the International Livestock Research Institute (ILRI), P.O. Box 30709, Nairobi, Kenya. Current address: 130A Jalan Awan Jawa, 58200 Kuala Lumpur, Malaysia

Camillus Ahuya, FARM-Africa, Kenya Country Office, P.O. Box 49502, Nairobi 00100, Kenya

Merida Roets, c/o Kalahari Kid Company, 4th floor, 356 Rivonia Boulevard, Rivonia, Sandton, South Africa

Mujaffar Hossain, Department of Animal Science, Bangladesh Agricultural University, Mymensingh 2202, Bangladesh

Emmanuel Osafo, Department of Animal Science, Kwame Nkrumah University of Science and Technology, Kumasi, Ghana

20

Sheep

Tsatsu Adogla-Bessa[*], *Alan Carles*[†], *Ruth Gatenby*[‡], *Tim Treacher*[‡]

Cameo 1 - Peri-urban sheep production in Ghana

Mr Tetteh is a low-paid junior employee at the University of Ghana and lives in South Legon, a suburb of Accra; he keeps livestock to generate additional income in order to survive in this expensive suburbia. He has about 20 sheep, which he houses at night and for part of the day. He releases them occasionally to scavenge on the almost non-existent pasture and anything else edible that they can find. The basal diet offered during confinement is mainly cassava and plantain peels, which he obtains from a nearby restaurant, and sometimes grass cut from neighbouring undeveloped plots of land. Browse is scarce due to land clearing for residential and related developments.

At the onset of the rains, he has to confine his sheep to prevent damage to crops being grown in the backyards of homes around him, otherwise the complaints from neighbours are unbearable. Preventive health-care against internal and external parasites is non-existent but when an animal shows clearly visible signs of illness, he brings in a nearby veterinary assistant to help. Annual lamb mortality is as high as 40 per cent, and of those surviving, up to a third are lost due to theft and motor accidents.

During the Muslim Eid-ul-Fitr festival of sacrifice, he gets very high prices from selling surplus rams. He also sells to a nearby restaurant when in need of cash. Being a Christian, he always slaughters an animal for family consumption at Christmas and another at Easter. Mr Tetteh is seen as

---

[*] Principal author
[†] Co-author
[‡] Consulting author

moderately well-off because of the living bank of 20 or so sheep that are noticed by everyone in the community.

Improving the husbandry of the flock would go a long way to improving the profitability of the enterprise and increase his income.

Cameo 2 - Subsistence pastoralism for a Maasai widow and her family in Kenya

Nashiluni Nkurruna is a Maasai who was widowed some 20 years ago; she has six children and the family live in southern Kenya. The whole family is solely dependent on the livestock, which consists of a flock of up to 90 Red Maasai sheep, up to 30 Small East-African zebu cattle and some 10 goats.

Nashiluni's cattle provide the family with milk, but for the rest of their needs they depend on the flock of sheep. The annual off-take from the flock is about 20 per cent; three quarters of this is sold for cash for the purchase of cereal foods, medicines, household materials and school fees. Just a few sheep are slaughtered for home consumption.

Mortality rates for her sheep are about 20 per cent per annum, over half of which occur between weaning and one year of age. Only about 40 per cent of the breeding ewes lamb each year. Natural pasture is the only source of feed for the livestock, but this is rapidly diminishing due to the expanding human population. Her finances do not permit any additional expenditure, for instance on feed supplements that would increase flock productivity.

In addition to the normal homestead chores, the livestock need to be tended. Four of Nashiluni's children are still of school age, and she has to choose between their going to school or helping to tend the livestock. For richer owners the hiring of a herdsman is an option. The inflation in the cost of the goods she has to buy has greatly exceeded that of the prices she has been able to obtain for her livestock, so her level of poverty has worsened over time.

Despite her many hardships Nashiluni is well aware of the potential of her sheep. The easing of only one of the constraints to the flock's productivity would produce a major improvement in the family's livelihood.

The value of sheep in the livelihoods of resource-poor farmers

The diversity of sheep

Sheep are important domestic animals in tropical livestock production systems. There are breeds of sheep to suit different environments from desert to humid rainforest regions, from snow-covered hills to lowland swamps (Devendra and McLeroy, 1982). No other domestic animal species is adapted to as wide a range of environments. In all of the environments of the world inhabited by humans, sheep are present and they supply food, fuel, power, clothing and socio-economic benefits to humans. The range of products and services to humans provided by sheep is similar to that of goats (cf. Chapter 19, Table 19.2).

Sheep vary in size, conformation, coat cover, productive ability and adaptability to different conditions. Numerous methods have been used to classify sheep, such as tail type, coat cover, function, and body size, but they have all defied complete categorisation (Devendra and McLeroy, 1982; Gatenby, 2002). The application of analytical techniques used for goats in Ethiopia and Eritrea (FARM-Africa, 1996) would greatly assist in identifying significantly different types of sheep with respect to performance in different regions of the world, as indeed would the use of DNA analysis.

Why sheep are important in production systems of resource-poor farmers

Over half of the world's sheep are in developing countries. Sheep are involved in the simplest, as well as the more sophisticated crop-livestock agricultural systems (cf. Chapter 3). Sheep have unique advantages to the resource-poor farmer:

- Feeding behaviour. Sheep are highly selective, with greater preference for grass than browse, and are amenable to herding. In contrast, goats are also very selective, but prefer browse to grass. Sheep are thus more efficient at extracting all the available nutrients from a forage base in times of scarcity. There is therefore advantage in keeping both sheep and goats in terms of maximising the utilisation of available grazing and browse.

- Sheep lay down fat depots more readily from surplus energy than goats, and so are better able to utilise surplus feed in times of plenty (Photo 20.1).

- In cooler climates, where wool is the desired product, sheep are more appropriate than goats.

- There are certain social and religious functions where sheep are specifically required. In this regard (e.g. as sacrificial animals) the live animal has a value exceeding the meat or cash value.

The above are also advantages of sheep over goats. In addition, sheep (like goats) are suited to resource-poor systems because of the following:

- Small size allows easy handling and management by family labour, especially women and children.

- Smaller land holdings are required compared to cattle.

- Low capital investment per animal; very few facilities are required to rear sheep.

- Sheep are easier to confine than goats.

- Manure, presented as pellets, is easy to handle and valuable for crop production.

- Higher survival rates than cattle under drought conditions and, because of their higher reproductive rates, numbers can be restored more rapidly after disasters such as drought.

- Their short reproductive cycle and high incidence of multiple births in many breeds give higher annual off-take rates than cattle.

- Their small size and early maturity make them especially suitable for meeting subsistence needs for meat and milk.

- They use fibrous feeds and water efficiently.

- There are no major religious taboos against consumption of sheep meat.

The role of sheep in wealth creation for resource-poor farmers

Sheep supply food and at the same time generate income for the direct benefit of resource-poor people. In the subsistence sector, farmers depend on them for much of their livelihood, often to a greater extent than on cattle because sheep are generally owned by the poorer sectors of the community. Therefore, any intervention that improves the productivity of sheep is an important route to creating wealth and improving the standard of living of the resource-poor farmer.

Photo 20.1 Somali sheep grazing an arid-zone pasture just after the rains. Sheep are well adapted to maximising nutrient intake and storage at times of abundance, and drawing upon their energy reserves in their fat deposits at times of scarcity (courtesy of A. Carles)

Sheep are relatively cheap to buy. Due to their rapid reproductive turnover (early puberty, short gestation) and, in some breeds, high incidence of multiple births, a farmer can quickly build up a sheep flock as a major capital asset. With increased production, the farmer has a surplus for sale, and with the cash income can buy inputs to increase his crop production. Like goats, sheep can be sold for household needs, school fees, medical bills and other emergencies; indeed, they serve as a living bank.

In many developing countries, ownership of sheep also confers social status. This may be as important as food and cash, but is difficult to quantify. This is particularly important in Sub-Saharan Africa where livestock hold an important place in many cultures. For example, in the Maasai community sheep have a special function in the first months after the birth of a child and when someone is sick. After milk and blood, fat and deep-fried mutton are the foods of choice for lactating mothers, patients and special guests. Thus even a small livestock holding can boost the morale of the poor, with a multitude of benefits. However, it may be difficult to place monetary value on the personal satisfaction and social prestige that come with owning livestock; economists should seek to quantify such values.

Unfortunately some local sheep are endangered. For example populations of the Bonga breed of Ethiopia, known for their large sizes, and the Nungua Blackheads of Ghana, known for their prolificacy, are rapidly declining. This may ultimately threaten the livelihood of resource-poor farmers in those locations.

Major sheep production systems in developing countries

Sheep production varies from extensive to urban/peri-urban systems. In extensive systems animals obtain most of their feed requirements from grazing. A large amount of land is needed but inputs per animal of labour, housing and equipment are low. Prevalence of infectious disease is generally also low. In urban and peri-urban systems sheep are housed and are largely stall-fed, but scavenging around dwellings is allowed; losses from thefts and accidents are high.

Typical features of production systems for resource-poor farmers are:

- Small numbers of animals.
- Grouping of animals from several owners for grazing.
- Owners have little technical expertise.

Livestock systems in developing countries are discussed in Chapter 3 (see also Wilson, 1995). Sheep production systems can be briefly classified as follows:

- Traditional pastoral systems, which are principally range-based. These systems are typified by the absence of crops.
- Agro-pastoral systems, which are extensive grazing systems with a low level of crop-livestock integration.
- Crop-based livestock systems that involve combining livestock with annual or perennial crops.
- Landless systems that include urban and peri-urban, and rural landless production systems. These involve cut-and-carry feeding and grazing of roadside verges.

Sheep production problems and how to alleviate them

There are two common features of livestock production by resource-poor farmers:

- Levels of inputs are low, resulting in low production.

- Livestock yield below their genetic potential (in consequence of low inputs).

Devendra and McLeroy (1982) wrote "Subsistence farmers pay little attention to the feeding, management and health of their sheep even though relatively low inputs, of both feed and veterinary requirements, could lead to relatively high gains in productivity". For example, when West-African Dwarf Sheep were put on an improved feeding regime, lamb mortality was reduced dramatically (Table 20.1). Such simple interventions would presumably bring about significant improvements in farmers' income. However information, such as that in Table 20.1, should be accompanied by an assessment of the economic benefit, having also considered the practical feasibility of adopting the intervention by resource-poor farmers.

Table 20.1 Productivity of West-African Dwarf sheep given supplementary feeding when grazing natural savanna rangeland in the Ivory Coast

Supplement (g/day)	*Pre-weaning lamb mortality (%)*	*Number of weaned lambs, per ewe, per year*	*Total weight of weaned lamb, per ewe, per year (kg)*
0	32.4	0.98	9.6
350[1]	16.8	1.37	13.4
650[2]	2.5	1.51	24.0

[1]Supplement composition was 72 % rice flour, 14 % molasses and 14 % cottonseed meal.
[2]Supplement composition was 47 % wheat bran, 37 % molasses and 16 % cottonseed meal.
Source: Charray *et al.* (1992).

Feeds and feeding

Feed supply is the most pervasive constraint, and has both quantitative and qualitative dimensions. As sheep are selective feeders, quantity is often the greater constraint. Options available to the resource-poor farmer are to plant forage crops (often difficult due to limited land), make greater use of crop residues (timely harvesting and storage [LPP Project R6993], allow selective feeding [LPP Projects R6619, R7955]), and improve basic soil and water conservation practices.

Feeds commonly available to sheep are:

- Natural grasses/legumes.

- Shrubs and trees.

- Cultivated legumes.

- Cultivated grasses.
- Crop/industrial by-products (molasses, pulp etc.).
- Crop residues (straws, stovers etc.).
- Domestic kitchen waste.

Water intake

The availability of water can be a primary constraint, although sheep are much less demanding of water than cattle, especially if the forage does not have a very high dry matter content. Different types of sheep have different water requirements and differ in tolerance of both water-stress and water-quality. For example, Desert Sudanese sheep are normally watered at three-day intervals, and the Indian Marwari and Chokla breeds are known for their ability to consume saline water. Immature and lactating sheep require most water. The recommended water intake is 2 kg/kg feed DM consumed. This means sheep on a good diet will require a total (in feed and drinking water) of 2-3 litres of water daily, or 4-5 litres of water every 2 days (Baudelaire, 1972) (Chapter 11).

Common feeding problems

Seasonal fluctuations in feed and water supplies are problematic because, although they are abundant in the wet season, inability to conserve them leads to shortages in the dry season. The impact of these shortages in lowering reproductive rate is significant. Other common feeding problems are:

- High-fibre feeds causing low intakes, resulting in low levels of overall production.
- Seasonal fluctuations in quality of available forage, particularly protein content and digestibility.
- Low protein intake restricting growth and milk production, leading to low lamb growth and survival.
- Specific mineral deficiencies.
- Poor access to water.

Key feeding improvements

The specific solution to the problem of feed supplies will depend on the production system and ecological zone. Attending to these constraints generally results in large increases in production. Increases in live weight of up to 50 per cent, at comparable ages, have been reported (Charray *et al.*, 1992). Overcoming the constraints will require two basic approaches.

Firstly, farmers could make the following changes to increase feed production:

- Production of forages and use of by-products to bridge the feed gap.
- Improvement of rangeland pastures.
- Cropping systems that include forage legumes (intercropping, relay cropping, rotations), or use of multi-purpose crops for feed and food.

Secondly, farmers could make changes to flock management to make better use of available feeds, for instance:

- Flock segregation to feed females and young animals separately.
- Earlier off-take of growing stock.

The low-quality forages (e.g. crop residues) available to feed to sheep and other ruminants on smallholder farms are bulky, high in fibre, degraded slowly in the rumen (<3 per cent/hour), and low in nitrogen (<70 g/kg crude protein) and minerals such as sulphur (Osuji, 1994). These characteristics result in low forage-intake. As resource-poor farmers are unlikely to be able to afford chemical and physical treatments to improve such forages, feeding strategies involving supplementation will be more appropriate, and will often result in significant improvements in forage intake and utilisation. Supplementation is often simpler to apply at farm level, as it requires less labour than treatment of crop residues with urea (Preston and Leng, 1984).

The use of multi-nutritional feed blocks (usually with a molasses base) can be a cost-effective way of supplementing energy, protein and minerals (see Chapter 11). They have proved to be effective and acceptable to resource-poor farmers in India and other parts of the world (Sansoucy *et al.*, 1988).

The importance of leguminous and other forage trees as supplements to low-quality diets should not be underestimated. Most smallholder farmers can grow trees/shrubs to feed as supplements. In Nigeria supplementation of a grass- and cassava peel-based diet with a mixture of Leucaena and Gliricidia significantly increased the growth rate of lambs, from 25 g/day to 50 g/day, up to 24 weeks of age, and survival increased from 50 to 100 per cent over the same period (ILCA, 1988). In Ethiopia, Horro sheep supplemented with Leucaena attained better growth rates than when supplemented with other herbaceous legume hays and seed cakes (Lemma Gizachew, 1993). Growing forage trees offers an opportunity to enhance nutrition at least cost to the resource-poor farmer.

Housing

The type of housing needed depends on the system of production. Sheep prefer environments that are neither too hot nor too cold. In extensive and agro-pastoral systems where animals are not confined during the day, only a simple open or unroofed fold may be required at night. In intensive systems, the design of the house is more important because animals may be confined for most of the time. For all housing, however, it is important to keep the floor dry to prevent a build-up of disease and parasites.

Low-cost, locally-available construction materials, such as poles, bamboo, thatch and mud are not very durable. Concrete, timber and metal roofing sheets are more expensive, but last longer.

An exceptionally good traditional design has been reported in Java, where the climate is hot and humid (Gatenby, 2002). The roof is made of thatch, is steeply sloping and has large overhangs so that even in heavy rain the sheep remain dry. The walls are made of well-spaced pieces of wood or split bamboo, and animals stretch through the walls and eat from troughs attached to the outside of the walls. The floor is raised 60 cm above ground and is made of slats of split bamboo. Typical floor space is 4 x 1.5 m, to house eight sheep.

Key housing improvements/features

Important features that are necessary include:

- Slatted floors that stay dry and clean.
- High roofs to encourage air movement in the house.
- Long overhangs to protect against sun and rain.
- Partially open walls to enhance ventilation.

Housing is one of the more costly interventions and therefore particularly difficult for the resource-poor. Construction costs can be decreased by taking advantage of some parts of the domestic dwelling (a leeward wall with extended overhang) or shade trees (in drier climates).

Reproduction

An understanding of the reproductive parameters is crucial in reproduction management. The most important parameter is the annual lambing percentage, defined as the number of lambs per flock raised to marketable age (3-5 months). This corresponds roughly to the percentage weaned per flock per year. Some ewes may lamb more than once a year, some will have multiple births, some ewe lambs conceive and lamb before maturity; all these factors can lead to an increase in breeding efficiency. A higher annual lambing rate results in a higher overall flock production and off-take, and, therefore, more income for the farmer. Reproductive performance varies, particularly with environment (e.g. breeding is seasonal in temperate zones and year-round in the tropics), but some improvement is possible by adopting better management practices.

The onset of oestrus varies from 5-12 months of age. Breed, season of birth, level of nutrition, disease incidence and general physical development all influence age at onset of oestrus. Length of the oestrous cycle is 15-19 days. If no conception occurs the ewe will return to heat. Gestation is about five months (150 ± 5 days). Ewes are not receptive to the ram whilst pregnant, but in the tropics return to heat 2-3 weeks after lambing. The reproductive performance of the ewe reaches a peak at 5-6 years of age and declines thereafter.

Farmers should allow first mating when females are sufficiently developed to withstand the stress of pregnancy and lactation, usually at 8-12 months of age, but mating should not be delayed beyond 18 months. The younger the female is at first pregnancy, the better must be the feeding and overall management, otherwise the lamb may be born weak with reduced chances of survival, and the growth of the mother may be stunted.

Ram lambs reach sexual maturity at 5-10 months of age and can be used for limited service at 12-18 months, but after two years they can undertake full service. Mating is often uncontrolled and the ram is available year-round; while it is common in these circumstances for the ram to run with 10-25 ewes, it could be four to five times as many. Alternatively, mating can be controlled by running the ram with ewes continuously for 50-60 days. Uncontrolled mating will usually give a higher conception rate than restricted mating, but also higher lamb mortality due to some lambs dying because of being born when feed supplies are low. The larger the sheep enterprise, the greater is the advantage of controlled mating.

Key management improvements

Management practices that improve the annual lambing percentage are:

- Disease control.
- Good nutrition.
- Mating at a suitable age and frequency.
- Paying particular attention to health of ewe (udder, teeth, feet).
- Ensuring good health of rams used for mating.
- Culling ewes more intensively after 5-6 years of breeding.
- Maintaining the correct ratio of fertile rams and ewes (one ram to 20-25 ewes).

Control of mating (particularly post-partum and for the maiden ewe) in order to avoid lambing in adverse climatic or nutritional conditions is a particularly powerful management tool. Low-cost options include:

- Housing ewes and rams separately to prevent mating.
- Use of a leather apron to give a physical barrier to penetration, as used in Maasai flocks in East Africa.
- Use of the Kunan, which is common in Islamic regions (Gatenby, 1986). This is a cord tied round the neck of the scrotum and looped over the prepuce to prevent extrusion of the penis.

Breed development/selection

The breed types available in Africa are not limiting factors except where nutritional and disease constraints have been overcome. Improved breeds that have been developed include the Israeli improved Awassi, Nungua Blackhead, Dorper and Desert Sudanese (Table 20.2). These breeds have improved productivity indices compared to the original indigenous breeds.

Table 20.2 Some developed sheep breeds

Type/Breed	*Country*
Meat breeds	
Blackhead Persian	South Africa
Dorper	South Africa
Dorsimal	Malaysia
Nungua Blackhead	Ghana
Santa Ines	Brazil
Vogan	Togo
Wiltiper	Zimbabwe
Meat and wool breeds	
Avikalin	India
Dormer	South Africa
Rabo Largo	Brazil
Meat, milk and wool breeds	
Assaf	Israel
Israeli Improved Awassi	Israel

Source: Gatenby (2002).

Key breeding improvements

As discussed in Chapter 12, strategies to improve breeding efficiency and productivity should be developed to suit the particular production environment. Tools available include the well-established traditional methods of selection and planned mating, and the new technologies, including artificial insemination.

In applying improvement tools, the technical and economic feasibility of changing the genotype, as opposed to changing the environment, needs to be considered. For example, if disease is an overriding constraint, which will be best: developing preventative measures or treatment, or genetic improvement for resistance/ tolerance?

Three relatively-inexpensive approaches can be adopted to develop and improve genotypes for resource-poor farmers:

- *Development of new breeds* Two or more breeds are used to synthesise a new breed, combining all or most of the favourable characteristics of the original breeds. Examples are the cross of the Blackhead Persian and the West-African Forest Type to give the Nungua Blackhead sheep in Ghana, or the Blackhead Persian and Dorset to give the Dorper in South Africa. Such new breeds have been developed for meat, milk, wool or multi-purposes (Table 20.2). Productivity is also usually better than the original indigenous breed. The Nungua Blackhead showed a 72 per cent gain in birth weight, and a 51 per cent gain in ewe mature body weight, over the original West-African Forest Type parent (Table 20.3).

- *Introduction of improved breeds* Breeds noted for high levels of productivity have often been introduced into exotic areas. Instead of transferring such breeds (often from temperate regions), emphasis should be on transfer of superior genotypes that have evolved or been selected under similar environmental conditions. Principal obstacles to such introductions include lack of well-characterised stocks from which to select animals to be transferred, and the potential for spreading disease. The establishment of sheep breeding and disease evaluation centres in developing countries can help overcome this problem.

- *Selection within indigenous breeds* Rams and ewes with outstanding traits such as growth rate, lambing interval or twinning, are chosen and used as parents for the next generation. This selection is continued systematically from generation to generation, but care must be taken to avoid inbreeding. Selection should be done in an environment similar to the conditions in which the offspring will ultimately live, otherwise the offspring will not be adapted and their performance depressed.

Table 20.3 Improved growth performance of the Nungua Blackhead sheep compared with the West-African Forest Type from which it derived

Trait	*West-African Forest Type*	*Nungua Blackhead*	*Per cent gain*
Birth weight (kg)	1.3 ±0.23	2.3 ±0.45	72
Average daily gain to weaning (g)	50 ±9	90 ±20	67
Weaning weight (12 weeks) (kg)	5.8 ±0.77	9.8 ±1.9	69
Ewe mature body weight (over 4 years) (kg)	21	32	51

Source: Ngere (1973).

Some indigenous breeds have become well-recognised for particularly valuable productive traits, e.g. resistance to haemonchosis (Red Maasai), resistance to trypanosomiasis (West-African Dwarf), large litter size (West-African Dwarf, Priangan).

Because one ram generally mates with several ewes, selection of rams is the most efficient way of increasing productivity. For example, a ram lamb may be selected because its mother always produced twins. Again, sheep breeding and evaluation centres can be used for this process. For the resource-poor, access to a ram is probably more of a problem than choice of breed.

Health

As discussed in Chapter 13, animal health problems are complex and closely interrelated with other biological and socio-economic constraints. Constraints imposed on sheep (and goats) by diseases and parasites are substantial. The technology for prevention and treatment of many animal health problems exists. However, the means to deliver such technology is generally lacking due to shortage of veterinary personnel, poor roads and the cost of drugs. There is an unfortunate perception with respect to veterinary assistance; when the assistance is confined to the treatment of sick sheep it is seen as usually not cost effective, due to their small size and output, and so veterinary help is avoided. As a result, prophylaxis is even more important. The saying 'prevention is better than cure' is very applicable in sheep (and goat) production.

Health problems of resource-poor farmers' sheep usually fall into two broad categories:

- Lowered resistance caused by poor nutrition leading to death from disease and parasites.
- Control measures for transmissible diseases normally controlled by direct intervention (e.g. vaccination, vector control, or other prophylactic measures) being too expensive for resource-poor farmers.

Important diseases and parasites are outlined in Table 20.4. Overcoming major constraints require the following:

- Providing adequate nutrition, which decreases susceptibility to disease and parasites.
- Use of disease resistant animals.
- Improved parasite control.
- Improved control of endemic disease.

Table 20.4 indicates where vaccination, probably the most cost effective method of control of infectious disease when it is available, may be an option. Because the resource-poor are so much less able to control circumstances, thus greatly facilitating the spread of infectious disease, potential zoonoses are much more serious; for this reason potential zoonoses are also indicated in Table 20.4.

Table 20.4 Principal diseases of sheep in developing countries

Disease type [1]	*Disease*	*Vaccine* [2]	*Zoonosis threat* [3]
Alimentary diseases	Bloat		
	Coccidiosis		
	Johne's disease		
	Liver flukes		Yes
	Parasitic gastro-enteritis		
	Taeniasis		
Blood diseases	Anaplasmosis	Yes	
	Babesiosis	Yes	
	Haemonchosis		
Nervous diseases	Coenurosis		
	Listeriosis		Yes
	Rabies	Yes	Yes
	Scrapie		
	Tetanus	Yes	Yes
Reproductive diseases	Akabane		
	Brucellosis	Yes	Yes
	Chlamydial abortion	Yes	Yes
	Mastitis		
	Toxoplasmosis		Yes
Respiratory diseases	Enzootic pneumonia	Yes	
	Lungworms		
	Maedi-visna		
	Pulmonary adenomatosis		
Skin and feet diseases	Contagious pustular dermatitis	Yes	Yes
	Dermatophilosis		
	External parasites: mange, fleas, lice, ticks		
	Foot rot		
	Infectious keratoconjuctivitis		

Table 20.4 (Contd.)

Disease type [1]	*Disease*	*Vaccine* [2]	*Zoonosis threat* [3]
Systemic diseases	Anthrax	Yes	Yes
	Bluetongue	Yes	
	Caseous lymphadenitis		Yes
	Clostridial diseases – enterotoxaemias, blackleg, black disease, braxy, malignant oedema	Yes	
	Foot and mouth disease (FMD)	Yes	
	Heartwater (cowdriosis)	Yes	
	Hydatidosis		
	Melioidosis		Yes
	Nairobi sheep disease	Yes	
	Pasteurellosis	Yes	
	Peste des petits ruminants (PPR)	Yes	
	Rift Valley fever	Yes	Yes
	Sheep pox	Yes	
	Trypanosomiasis		Yes
	Wesselsbron	Yes	
Nutritional diseases	Deficiencies/imbalances of energy, protein, fibre, minerals, vitamins		

[1]Principal body system affected; if more than one then disease is designated Systemic.
[2]Yes = Vaccine is available and is usually a principal means of control, but there are exceptions e.g. FMD, enzootic pneumonia.
[3]Yes = Significant zoonosis threat if disease present.

The introduction of the 'community-based animal health-care workers (CBAHW)' concept (Chapters 5 and 13), where a local resident is trained in basic animal husbandry and health practices can help alleviate the problem of delivering technology, but it needs investment and the will of governments to train and pay personnel.

The cost of treating diseases is usually beyond the resource-poor farmer and emphasis should be placed on disease prevention; cost-effective methods of disease prevention and control available to the resource-poor are listed in Table 20.5.

Table 20.5 Cheap forms of disease control for resource-poor sheep-keepers

		Problem	*Action*
Genotype	Adaptation/resistance		Select animals from healthy flocks within the same environment
Vaccination	Specific disease		Veterinary advice on disease incidence
Hygiene	Grass/soil	Excess excreta (urine and faeces)	Reduce infection level and exposure of sheep
	Enclosures and wetlands	Excess moisture, vector/parasite habitats	Effective drainage
	Water source	Excreta contamination	Proper disposal
	Atmosphere	Inadequate ventilation	Proper housing
	Structures/equipment	Excreta, discharges	Clean, disinfect
	Carcass material	Infected material, e.g. Canid tapeworms	Proper disposal
	Abortions	Foetuses, placentas	Proper disposal
General care	Feet	Overgrowth, infection	Trim, disinfect
	Mouth	Teeth missing	Deselect, cull
		Jaw deformation	Deselect, cull
	External genitalia (ram)	Abnormalities	Cull
	Udder (ewe)	Abnormalities	Cull
	Wounds		Cleanse and disinfect
	Ill health	Early detection	Isolation
	Fleece	Contamination (urine and faeces)	Shear and cleanse
Stress evasion	Physical	Trekking	Minimise stress
	Temperature		Shade and ventilation
	Nutritional	Deficiencies, imbalances	Dietary management
	Exposure to foreign microbes	Movement to new environments	Maximise other forms of protection

Table 20.5 (Contd.)

		Problem	*Action*
Positive health	Nutrition	Quality and quantity	Attempt to balance diet
Vector control	Ticks		Spot-on applications or hand removal

Notes:

1. The groups have been put in order of usually increasing cost.
2. Adapted genotypes are probably the most cost effective long term way of disease control; recent findings of tolerant genotypes for haemonchosis, trypanosomiasis, heat tolerance, and water deprivation are examples.
3. Vaccines are extremely cost effective.
4. Improved hygiene often costs little.
5. Positive health can play a major role but may be rather expensive for the resource-poor.

Source: After Carles (1992).

Measuring productivity

The level at which the resource-poor operate their livestock, and the constraints they face, would suggest to many that any measurement of productivity is unwarranted and impractical. In the early stages of interventions, when production improvements are likely to be large, qualitative evaluation may well be adequate. Later, however, adequate evaluation of changes in productivity can be made only if some quantitative measure is used.

Productivity is defined as quantity of output(s) per unit of (one or more) input(s), standardised for a unit of time. While there are many outputs (potentially), the main ones for most resource-poor sheep farmers are meat and livestock (for breeding or sale). The position is more complex with inputs. The breeding ewe has often been used as an appropriate unit of input. Thus a simple index, incorporating the main outputs, a key input and the levels of fertility and mortality, is suggested as*: weight of (carcass) off-take (live or dead) per ewe mated per year.*

Unfortunately, the flock dynamics of most resource-poor sheep-keepers complicate the estimates of these parameters, because females are usually entering and leaving the breeding-ewe component throughout the year, and animals are usually sold at widely differing ages (over and under a year of age) and weights. Because of these complications, the estimates of these parameters for the standard unit of one year have to be adjusted accordingly. Also, mating is usually uncontrolled, thus breeding occurs throughout the year, and a significant level of discipline is required to make a few simple records (unless the flock size is very small). These difficulties may be insurmountable, but if they can be overcome the records are extremely valuable.

More detailed discussions of the estimation of sheep productivity are given by Carles (1983) and Gatenby (1986); the latter also reviews levels of production (both biological and economic) of a number of production systems from several developing countries. James and Carles (1996) describe a method for the estimation of energy efficiency of total flock productivity, incorporating all components of the flock and the effects of flock dynamics.

Future perspectives

Three critical areas need to be improved to enhance productivity and to increase income of the resource-poor sheep farmer:

- *Supply of adequate feed throughout the year.* Feed availability can be enhanced primarily through sustainable forage development, especially forage trees. The feed deficit could also be met through the development of cropping systems which meet animal feed requirements without reducing food or cash-crop yield; this includes harvest and feed preservation strategies to maximise nutritive value of crop residues. Efforts must be made to identify crops which, when intercropped or grown in rotations, increase the yield of human food as well as provide feed for animals; such dual-purpose crops are more readily adopted by resource-poor farmers.

- *Improved health.* Lamb mortality can be reduced through better health-care such as improved mothering, endoparasite control, ectoparasite control and vaccinations against epidemic diseases such as peste des petits ruminants (PPR). The development of low-cost flock health programmes (low cost, low labour) which are acceptable to farmers can enhance lamb survival and increase productivity.

- *Production of improved sheep genotypes.* A great diversity of sheep exists. Strategies have to be developed for combining the superior traits of different breeds, with particular attention to breeds that have evolved in the tropics. In order to meet demand for superior, adapted genotypes, centres to produce performance-tested, disease-free stocks need to be established, preferably by governments. Government stations should be enabled to supply superior rams to villages or producer groups on a sale or loan scheme.

Concluding remarks

It appears that in most pastoral areas sheep (and goats) are becoming more widely distributed, and are representing an increasing proportion of the grazing livestock biomass, as documented for eastern Maasailand by de Leeuw *et al.* (1991) and Grandin (1991). Rightly, many welcome this as a broadening of the environmental niches being utilised to support the increasing human population. However, frequently the main reason for this extension is the degradation of the forage resource from overgrazing by the less selective cattle population. An additional reason in most pastoral areas could be increasing poverty, with its associated trend toward small stock (Thornton *et al.*, 2002).

This increase of the small ruminant component provides an important, but temporary, boost to the animal products required by the human population, at least in pastoral areas. But if nothing is done to redress the underlying excessive grazing pressure, then the degradation of this resource will continue and next time there will be no alternative ruminant livestock species to provide further relief. It is of great importance that this short time of reprieve is not squandered, but that the opportunity is taken to develop more sustainable grazing systems. This will include a better balance, than in the past, of the grazing livestock species, but must also include some relief of the pressure on environmental resources. This process will be assisted as poverty is alleviated and the resource-poor have a wider range of options available to them in the management of their livestock.

Further reading

Brocklesby, D.W. and Sewell, M.M.H. 1990. *Handbook of animal diseases in the tropics. Fourth edition*. Bailliere Tindall, London, UK.

Carles, A. B. 1983. *Sheep production in the tropics*. Oxford University Press, Oxford, UK.

Charray, J., Humbert, J.M. and Levif, J. (Translated by Alan Leeson) 1992. *Manual of sheep production in the humid tropics of Africa*. CABI Publishing, Wallingford, UK in collaboration with the Technical Centre for Agricultural and Rural Co-operation (CTA), Wageningen, The Netherlands.

Devendra, C. and McLeroy, G.B. 1982. *Goat and sheep production in the tropics*. Intermediate Tropical Agriculture Series, Longman Scientific and Technical, Harlow, UK.

Gatenby, R.M. 1986. *Sheep production in the tropics and sub-tropics*, Longman, Scientific and Technical, Harlow, UK.

Gatenby, R.M. 2002. *Sheep*. The Tropical Agriculturalist, Series Editors R. Coste and A.J. Smith. Macmillan Education Ltd., London, UK, in cooperation with Technical Centre for Agricultural and Rural Cooperation (CTA), Wageningen, The Netherlands.

Hunter, A. 1994. *Animal health. Volume 2 Specific diseases*. The Tropical Agriculturalist, Series Editors R. Coste and A.J. Smith. Macmillan Education Ltd., London, UK, in cooperation with Technical Centre for Agricultural and Rural Cooperation (CTA), Wageningen, The Netherlands.

Muchaal, P. K. 2002. *Urban agriculture and zoonoses in West Africa: an assessment of the potential impact on public health. Report 35*. Cities Feeding People Series, International Development Research Centre (IDRC), Ottawa, Canada.

Schiere, H and van der Hock, R. 2001. *Livestock keeping in urban areas – a review of traditional technologies based on literature and field experiences. FAO Animal Production and Health Paper 151*. Food and Agriculture Organisation of the United Nations (FAO), Rome, Italy.

Author addresses

Tsatsu Adogla-Bessa, Agricultural Research Centre – Legon, Institute of Agricultural Research, University of Ghana, P.O. Box LG 38, Legon, Accra, Ghana

Alan Carles, Formerly of Department of Animal Production, The University of Nairobi, Faculty of Veterinary Medicine, P.O. Box 29053, Kabete, Nairobi, Kenya

Ruth Gatenby, Formerly of Small Ruminant Collaborative Research Support Programme, University of California, Davis, USA, and Centre for Tropical Veterinary Medicine, Division of Animal Health and Welfare, University of Edinburgh, Easter Bush Veterinary Centre, Roslin, Midlothian EH25 9RG, UK

Tim Treacher, Formerly of International Centre for Agriculture Research in the Dry Areas (ICARDA), P.O. Box 5466, Aleppo, Syria

21

African camels and South American camelids

Chris Field[*], *Jonathan Rushton*[†], *Rommy Viscarra*[†], *Bessie Urquieta*[‡], *Hichem Ben Salem*[‡]

Cameo 1 - In praise of camels - comments from Tunisia

We learn many things from our parents, grandparents and from old people in general. The experience they acquired is often not documented. Some time ago I came across three aged farmers in an arid zone of Tunisia, escorting their camels. I simply asked them whether the camel played an important role in farmer livelihoods.

Mr. Hédi Jaouadi said, "The camel is a miracle animal created by God to help people living under severely-dry conditions. It was even mentioned in the Koran. Until the 1970s, the camel was everything for Tunisian people, as it was used for transportation and numerous agricultural activities. All products and by-products of the camel are valuable, especially meat and milk. Traditionally all farmer-families had at least one camel which was also a status indicator. The car and tractor have now substituted as status indicators. But the car cannot replace the multipurpose function of the camel".

Mr. Belgacem Hanzouli added, "The camel, in contrast to sheep, goats and cattle, can go without food and water for several days. It can eat almost anything that grows in the desert, including salty plants and convert them to milk and meat for humans. A young male camel is a valuable source of income, as it is now sold for 700 Tunisian dinars (about US$ 600). The meat is appreciated by many people and has almost the same value as beef. Although motorisation has caused a significant decrease of camel herd-sizes in many parts of Tunisia, the camel is still very important for the

---

[*] Principal author
[†] Co-author
[‡] Consulting author

livelihood of people in the desert. It is an important beast of burden, but also valued for its meat, milk, hair and skin. It is also sacrificed in local ceremonies".

Mr. Mohamed Lamine Hanzouli says, "Camel meat and milk are considered by rural people as 'medicaments', therefore as healthy products for humans. The camel is now involved in tourism activities, and so farmers are selling this animal at high prices, sometimes reaching US$ 2,500 per head."

The three farmers agreed that the camel will continue to represent the symbol of human survival in the desert. However, efforts should be made to rehabilitate camel husbandry in the other zones.

Cameo 2 - A new opportunity for resource-poor llama keepers in Bolivia

A regional association of camelid producers (ARCCA) was established in 1997, in the Department of Potosi, Bolivia. ARCCA has 1,500 members, who own and manage nearly 180,000 llamas in one of the harshest environments in the world. This region has low rainfall, a high percentage of days when the temperature falls below freezing and much of the land is 4,000 m above sea level. The only livestock opportunity for the people in this region is the llama, which has traditionally been important for the production of pack-animals for transporting goods and for producing meat for home consumption. The region is also recognised as being one of the poorest in Bolivia.

ARCCA began with a vision of providing a new economic opportunity for its members, which involved shearing their llamas and selling the fibre. The association provided advice, at producer-level, on animal health, production and methods of shearing. The association also searched for a spinning factory that was willing to process llama fibre, a product that has, until recently, been ignored by the textile industry. A Peruvian factory agreed to process the llama fibre. Over a three-year period the production and collection of llama fibre from the members tripled (see table below).

Year	Greasy fibre (kg)	Value of the production Boliviano ($b)	US$	Value per member (US$)
1997	4,723	31,170*	5,667	3.78
2000	14,848	97,995**	15,076	10.05

*1997 US$=$b 5.5; **2000 US$=$b 6.5

Whilst the income per member may appear relatively small, it has to be placed in the context of an average income per capita in the region of between US$ 150 and 200 per year. Therefore this intervention signifies an increase of between 5 and 7 per cent in income levels and an important economic opportunity in the livelihoods of some of the poorest people in South America.

Source: UNEPCA (1999, 2002).

Introduction and origins

The Horn of Africa is arguably the most vulnerable region in the world to drought, overuse and poverty. It is far from government resources and development assistance.

The key to survival in this area is the camel. By providing food security during drought, the camel enables humans to live and prosper under exceptionally difficult conditions in rangelands accessible only to desert wildlife.

In South America four species of related, hoofed mammals have survived and are uniquely adapted to rangelands associated with the high altitudes of the Andes. These are collectively known as New World or hump-less camelids.

Camelids are mammals of the order Artiodactyla (even-toed ungulates) and sub-order Tylopoda (pad footed). The hooves protect the leading edge of the thick fatty pads on the soles of the feet, while the latter spread their weight in soft soil and sand to prevent sinking.

Fossil records suggest that camelids evolved in North America 50-60 million years ago, but were no larger than hares. Some 45 million years later they had become much larger, some with humps and others without.

The latter migrated to South America where they are found today as llamas, alpacas, guanacos and vicuñas. The arrival of the first ancestral forms of camelids in South America occurred during the Pleistocene period, around two to three million years ago (Hoffmann *et al.*, 1983). Four genera evolved during this period, two of which still exist today, the Lama and the Vicugna. The genus Lama has three species: *Lama pacos* (alpaca), *L. glama* (llama) and *L. guanicoe* (guanaco). The genus Vicugna has only one species: *Vicugna vicugna* (vicuña) (Bustinza, 1984).

Another branch of camel stock crossed from North America into Eastern Asia by a land bridge. From there they spread westwards into the cold deserts of China, where the Bactrian or two-humped camel evolved, and then to the hot deserts of the Middle East and Africa, where the one-humped camel or camel lives today.

African camels (one-humped/dromedaries)

Numbers and distribution of dromedaries

According to FAO (2001) there were 11.7 million dromedaries in Africa in 2001, amounting to 78 per cent of the world's population. The largest population, amounting to 6.2 million camels, occurs in Somalia, with 3.2 million in Sudan, and about a million each in Ethiopia and Kenya.

During the past two decades camels have expanded their range southwards into Maasailand, both in Kenya and Northern Tanzania in response to the failure of cattle to provide household needs during drought. Cattle cultures, such as the Boran of Ethiopia and the Samburu of Kenya, have adopted camels for similar reasons.

Biological and physiological characteristics of dromedaries

The unique physiology and biology of camels enables them to prosper under conditions unfavourable for other livestock.

Direct radiation

Camels have long, narrow bodies, which can be orientated towards, or away from, the sun to minimise the surface-area exposed to radiation.

Photo 21.1 One-humped camel and cactus in Tunisia (courtesy of H. Ben Salem)

Reflected heat

Their long legs lift their bodies well above the hot, reflecting surfaces of the ground. Cooler air can pass underneath because, even when camels are lying, the sternal pad lifts much of the abdomen above the ground so that air can flow beneath.

Loss of metabolic heat

Deep body heat is lost through convection by the blood circulating through a network of capillaries under the skin. Localisation of the fat in the hump minimises its effect as a heat insulator.

Role of the skin and hair

Camels' hair and skin act as a good insulator against incoming radiation; surface temperatures may reach 70 °C while skin temperatures are 30 °C lower.

Respiration

Respiration rates of camels remain low even at high temperatures. They lose less water through respiration than a cow, which may breathe twice as fast.

Water loss and circulation

The blood plasma of humans and camels contains 16 per cent of total body water and must remain fluid to conduct heat from the body. If 25 per cent of human weight is lost in water, the blood becomes viscous and cannot conduct heat, which then increases, causing heat stroke. By contrast if a camel loses 25 per cent of its weight in water, its blood volume decreases by only 10 per cent and remains fluid.

Drinking and thermolability

Weight-for-weight camels drink less water than other livestock. When a camel drinks 135 litres at one watering, this amounts to only 25 ml per kg per day, or one third of that consumed by cattle, sheep and goats under the same dry conditions. After rain, camels obtain all their moisture from their forage and do not need to drink.

Camels are thermolabile, allowing their body temperatures to fluctuate with the ambient temperature of their surroundings. Low night temperatures enable camels to start the day quite cool but they heat up by 6 °C or more, before cooling mechanisms are triggered. For a camel weighing 500 kg, this saves 5 litres of water. They store heat during the day and dissipate it during the night when less energy is required.

The red blood cells of a camel can absorb large quantities of water without rupture and may swell to several times their normal size. Camel-owning nomads recognise this and, after initial watering, return the camels to top up at the well several times over six hours, before departing for up to another two weeks without drinking.

Water losses in urine and faeces

The kidneys of a camel are very efficient at absorbing water and excreting concentrated urine. Urea is recycled as a source of nitrogen.

In dry seasons, water losses in urine of camels are only one tenth that lost by sheep, when compared per unit body-weight. During the rains they may produce 4-6 litres of urine daily, but during dehydration this is reduced to 0.75 litre per day. In hot conditions camels urinate down their hind legs, to extract heat from the skin when the urine evaporates.

Normally a camel produces hard, dry faeces containing 1-2 litres of water daily, while a cow may lose 20-40 litres of water a day in its faeces.

Anatomical adaptations

Camels tolerate dust by having long eyelashes to protect the orbit and nostrils which can be closed. Their large pad-like feet do not break up soil like the hooves of cattle, sheep and goats, and cause less erosion, whereas in dry seasons herds of cattle can easily be seen from afar by the clouds of dust following them.

Some of these attributes occur in other livestock but none, save the camel, combine them all. While other livestock in dry seasons are pinned to the vicinity of permanent water, and overgraze their pasture, the camel has almost unlimited access to rangelands beyond. Camels travel for seven days, or 40 km, from water, which gives them access to three times the area available to other livestock. This explains the ability of camels to continue to produce milk during droughts, when all other livestock have dried-off, and may even be dying.

Reproduction and breeding

Camels mature slowly, having their first calf at four to five years. They are induced ovulators, being mated several times before conception. Gestation period is one year, and the calving interval about two years.

There are many breeds of camels, which have developed according to the needs of the people. Usually smaller camels are found in the extremely arid areas, where feed may be limiting. Larger, higher-yielding animals are found where there is better forage or crop residues. When introduced to more arid areas these animals perform less well and may not survive droughts. However, their F1 hybrids with local breeds yield more, as well as retaining some of the hardiness.

Feeding habits

The average composition of the diet of camels in northern Kenya has been found to be dwarf shrubs 50, trees 25, herbs 14 and grasses 11 per cent respectively. Camels feed over a considerable height-range, from ground level to 3.5 m, which gives them an advantage over other livestock. Perennial woody-plants comprise three quarters of the diet, the remainder being herbs and annual grasses, which are eaten in dry seasons as standing hay. Perennial grasses, eaten by cattle, have a high fibre content and low digestibility and nutritive value, and are avoided by camels.

Camels need minerals for their physiology, so prefer halophytic (salt loving) shrubs such as *Sueda monoica* and *Salvadora persica*.

In semi-arid areas where cattle can survive, pastoralists such as the Samburu and Rendille keep a combination of both cattle and camels to utilise the available spectrum of vegetation more fully, and so produce more food.

During the rains and shortly afterwards, when nutrition is good, a camel may have surplus metabolisable energy in its diet which is converted to fat in the hump. Young males are sometimes castrated so that they will accumulate more fat. When slaughtered, such a camel may provide a household with enough fat for its needs for a year (i.e. 60-80 litres).

Production and marketing

Camel production concerns mainly its use as a food resource, but also for draught, and as a source of clothing and shelter. In Rajasthan, India, camels are used mostly for pulling carts and very little for food, as the majority of people are vegetarian (Hindu).

Milk production

The most important camel product is milk, which is often the staple diet for nomads. In Kenya camels provide 60-70 per cent of the nomadic Gabra and Rendille people's diet.

Camel milk-yields may be six times those of local cattle under the same conditions, but vary, due to many factors which affect production.

Wet-season daily yields are often double those of the dry-season. Climate affects forage condition and water availability. Camels watered more frequently produce more milk. Somali camels produce more milk than Turkana camels, under identical conditions. Increasing milking frequency may increase milk output by 30 per cent. In Kenya, pure-breed camels from Pakistan had higher milk yields than indigenous breeds, provided there was adequate forage. Milking regimes vary from twice daily in many parts of Kenya, to six times a day among the Afars of Ethiopia.

Several diseases may affect milk yield, the most important being trypanosomiasis. A herd receiving veterinary treatment yielded 67 per cent more milk than a herd under traditional management, due to the combined effect of improved health of the dam and increased survival of calves. Camels with calves surviving to weaning produced 2.9 times more milk than those whose calves died. Camels with a reduced calving interval produced 20 per cent less milk per lactation. Age and parity also affect yield.

Daily milk yields in Kenya vary from 2.4 litres under traditional management, to 4 litres under improved ranch-management.

Comprehensive veterinary inputs improve milk yield by just under a half a litre per camel per day. The treatment of helminthiasis, alone, can increase milk yields by more than 11 litres spread over five weeks, which is cost effective, because the increased milk is more valuable than the cost of the drug.

Peaks in lactation occur 6-10 weeks after parturition, when some camels can yield more than 12 litres per day.

The duration of lactation in a camel varies considerably. While 12 months is average, it may be prolonged to 30 months during drought. Once conception occurs, a camel normally dries-off within the following three months. Prolonged lactation is considered advantageous to subsistence-orientated camel owners. However, if one wishes to improve one's herd, and alleviate poverty, one should breed from the good milkers. Possibly, the tradition of prolonging lactation prevents the best genes from increasing, which may be a constraint to improving livelihoods.

The quality of camel milk is similar to cow milk, but the butterfat does not separate easily. However, Vitamin C in camel milk may be as much as six times higher than cow milk. This is important in arid and semi-arid land (ASAL) regions as fruit and vegetables are not available as a source of Vitamin C. Camels' milk is invariably found to be better than any relief food.

Meat production

Camel meat is eaten infrequently by nomads, because it is a large animal which yields more meat than can be handled comfortably by a family. Camels are reserved for ceremonial occasions, such as weddings and funerals, or during drought when other animals are emaciated and unfit for slaughter.

Increasingly, however, camel meat is popular in butcheries, because it is cheaper than beef and therefore more affordable. It is estimated that 7.5 per cent of a camel population could be slaughtered, or exported, annually without causing a decline in the population, but the actual off-take is lower. In Kenya, it was estimated that an annual off-take of about 6,000 camels would be possible. With an average live weight of 500 kg, and a 55 per cent carcass, they would yield 1.65 million kg of carcass valued at US$ 1.76 million, thereby helping to alleviate poverty.

Young camels grow slowly as they may be affected by climate, forage, disease, breed, sex, whether castrated or not, and the maternal milk supply. Birth weights vary from 25 to 50 kg (Pakistan x Somali hybrid calves are heaviest). Male calves are normally heavier than females. Weaning occurs naturally at 8-15 months, soon after the mother's next conception, which may cause a growth check. In countries where camels are produced primarily for meat, calves are weaned at a later age.

Mature body weights are reached at 6-7 years and are affected most by the breed. Disease and management may delay a camel reaching mature body weight. In Kenya the Somali breed is almost twice the weight of other breeds, with a maximum of 970 kg.

Because of their importance in milk production, females are rarely slaughtered. Surplus males are slaughtered at 8-10 years, although a more optimum age in terms of tenderness of meat is 4-5 years old.

The carcass averages about 55 per cent of the live weight, depending on the sex and the nutritional status of the camel. Despite their long legs, camels have a higher dressing percentage and meat: bone ratio than range-cattle, possibly because the stomach of a camel is relatively small as it is not a roughage feeder.

The meat of a camel up to 5 years of age is similar to beef, after which it may become tough. It has 22 per cent protein and only 1 per cent fat, as this is stored in the hump, and is low in cholesterol. Camel fat has a high melting point, possibly because of the high ambient temperatures reached and the use of fat for insulation. It can be stored for at least a year without becoming rancid, which is convenient for nomads.

Blood

Blood is withdrawn from camels and consumed by non-Muslims, for whom it forms an important part of their dry-season diet. Up to 5 litres of blood may be taken at a time and as much as 35 litres a year from a single camel. Lactating females and breeding bulls are not bled as it can weaken them.

Hides

Camel hides are strong and durable and have many traditional uses: rope, tying building poles, water and milk containers, sandals and floor or bed mats. During prolonged famine camel hides are roasted, boiled and eaten.

Although shoe manufacturers claim that camel hides produce very good leather for boots, they still have a poor market. The main reason for this is that the hides are often damaged through fighting among males, skin infections, mange, putrefaction where fat remains on the skin, and cuts from flaying. Nevertheless in North Africa, the Middle East and the Indian Sub-Continent camel leather is used for shoes and boots, luggage and a variety of ornaments.

Bones

Camel bones have been used for carving ornaments, being not unlike ivory in appearance. They may also be burnt, ground in a mill and mixed with other minerals as a supplement for livestock.

Urine and dung

Camel urine is used traditionally as an ethno-medicine. It is highly concentrated and induces diarrhoea and vomiting.

In some countries, especially in desert areas where there is little vegetation, camel dung is used for fuel. Nearer to cultivation, the dung may be harvested from the night enclosure and sold for fertiliser.

The camel as a draught animal

Camels are large, strong animals, which move slowly and deliberately, making them suitable for traction. With the advent of the motor vehicle, camels have lost much of their importance as draught animals since vehicles are faster and can carry bigger loads. However, in Rajasthan camels regularly pull carts in the face of competition from trucks. Camel owners cite the economic advantage of the camel, which does not require expensive fossil fuel but can be powered by cheap range-forage and grain, which it transports along with the merchandise. It thus helps resource-poor people, with little capital, to set up in business.

In many areas where there is difficult terrain, camels are used as baggage animals.

Camels have a useful working life of about fifteen years, undergoing their first training as three year olds. The traditional saddle is a complex system of ropes and poles, which hold protective layers of sacks and hides firmly to the back of the camel and create a frame for the attachment of loads. They are most frequently used for carrying water containers and portable houses. The Gabra and Rendille, in Kenya, use two camels to carry the components of one traditional house and its contents. They have also been used to transport merchandise, the oriental 'Silk Route' being the best-known example. Salt is carried from the Sahara southwards in the Sudan and from the Danakil depression to the highlands of Ethiopia, while frankincense, myrrh and gum Arabic are all transported widely in Ethiopia and Somalia.

Agricultural produce is carried from the market to the homestead, usually in the form of sacks of grain such as maize meal, or sugar, while milk is back-loaded from the camp to market.

Average loads are 100 kg, which may be carried over distances of 25 km at speeds of 4 kph. The heavier loads are carried by larger Somali camels.

Camel saddles may be used as the frame for the transport of the sick, elderly or young children and to carry books for mobile libraries. Some Eritreans believe they helped win the war of secession from Ethiopia since, when all other supply lines were cut, camels were still able to get through with food and weapons.

In regions surrounding the Sahara, camels take the place of motorcycles, being more efficient in negotiating soft sand. They have also been used for military and security purposes, being cost-effective when compared with modern alternatives, and are still being used in some areas of the Indian Sub-Continent.

Recently camels have been trained for recreational purposes, principally eco-tourism and racing. The former may involve safaris of several weeks, where riding camels are combined with baggage camels and the tourists alternately ride and walk. Apart from the Sudan where camel riding is traditional, racing has not taken-off in the rest of the Horn of Africa, as it has in the Arabian peninsula.

Camels have been used for ploughing, milling, pulling carts and, in parts of Kenya, Ethiopia and Somalia, for de-silting dams by pulling a ripper followed by a scoop. However, none of these activities has taken-off on a large scale, although they have been promoted extensively by NGOs.

Camel health, sickness and productivity

A camel-owner knows when his camels are healthy and when they are sick. A great deal can be determined from the appearance and behaviour of the animal or the herd.

A sick camel may be thin, with a dull, erect coat. Ears are laid back and the eyes dull or opaque. Faeces may be loose and urination may be difficult. It may not feed or drink well, but stands around listlessly, or even lie down for extended periods. It may seek shade and, if in pain, grind its teeth. Blood samples may reveal parasites and have a packed cell volume of less than 20 per cent. Faecal samples may have parasites.

The productivity of a camel may decline when it falls sick. A young camel may lose weight and cease to grow. A pregnant female may abort and lactating females may show an abrupt decline in milk production. Male camels may cease to rut and become emaciated.

A recent study lists 95 diseases, which may affect livestock of the resource-poor, but only one third of these infect camels. There are seven key diseases which affect camels.

Trypanosomiasis

Trypanosomiasis is caused by infection with a blood protozoan and is the most important health hazard for camels over most of their habitat, leading to debility, sometimes death, and considerable economic loss. Camels lose condition, appetite and become emaciated. Pregnant camels abort, lactating camels show a dramatic loss in production and bulls stop rutting. Diagnosis is firstly through the detection of anaemia and then identification of the parasite in the blood. Prophylaxis and treatment is by injection of trypanocidal drugs.

Mange

Mange is caused by a minute tick-like parasite called a mite, which burrows into the skin and causes intense irritation. The camel first shows signs of mange by rubbing the affected parts against hard surfaces. There is loss of hair and, in the chronic stage, emaciation and thickening of the skin. Camels with mange cannot feed and rest properly, while productivity declines in milking females. Treatment is through the external application of acaracides (such as Amitraz) and a subcutaneous injection of Ivermectin. Old engine oil is often applied externally, as it is cheap and can be effective in the early stages. Salt is important in the diet. Mange is economically important as many calves die, and their mothers then invariably cease lactation. The estimated combined loss is US$ 467 per death.

Calf diarrhoea

Characteristic symptoms are loose stools, emaciation, reluctance to suckle and sunken eyes due to dehydration. It affects mostly calves in their first month.

It is possible that diarrhoea may have more than one causative agent, including salmonella, which responds to certain antibiotics. Since death is due to dehydration and the loss of electrolytes, it is important to ensure that re-hydration therapy is given. This involves giving several litres of water daily, in small doses, with a little sugar and a pinch of salt dissolved in it.

Helminthiasis

Helminthiasis (or worms) commonly occurs in camels and other livestock, and can be tolerated at low levels. When levels increase, camels of any age, but especially weaners, develop loose stools and oedema (swelling) under the skin.

Although worms do not normally cause death, they appear to predispose camels to Clostridium infections, which produce toxins and may kill. Treatment is with any anthelmintic containing albendazole, administered orally.

Camel pox and orf

These two diseases, which are caused by viruses, have similar appearances and usually affect young animals. Once recovered, the camel acquires prolonged immunity. Symptoms of pox are the development of lesions around the mouth. Usually there is spontaneous recovery, but if the lesions develop internally a weak calf may die.

Orf involves swelling of the head due to impeded drainage of the lymphatic system. The camel may become temporarily blind. Severe cases should be treated with injectable antibiotics to prevent secondary infections. However there is now the possibility of vaccination against camel pox. Two injections, separated by one month at the age of 6-8 months, provide lifetime immunity.

Mastitis

Infections of the udder of lactating camels are widespread. They cause considerable economic loss in terms of milk production and also, in some cases, milk spoilage. Chronic cases usually lead to the loss of the infected quarter. Because of the narrow double milk ducts, normal intramammary applicators for antibiotics do not function and an applicator with a finer nozzle is needed.

Many of the diseases affecting other livestock, such as foot and mouth and rinderpest, do not infect camels.

Research and development for poverty alleviation

The following cameo illustrates the benefits of farmer-training.

A farmer in Samburu District, Kenya, had several problems:

- during dry-seasons his animals travelled far to graze, forcing him to buy milk
- he had insufficient milk for home needs
- there was little income from the sale of sheep and goats
- up to 50 sheep and goats died annually from disease.

After training and joining a local camel-improvement group, he now uses his camels in several ways:

- in dry-seasons they transport camel milk over 30 km from his herds to his home
- contract ploughing and supplying cereals and vegetables to Maralal town
- contracting camels to supply sand and ballast to building sites
- hiring his camels out to tourists
- using his training in animal health-care, he treats camels and small stock in the community, for which he is paid. He has also opened a shop.

Research and development in the following areas would help alleviate poverty:

- community-based milk and meat production, processing and marketing
- adding value to hides by improved processing
- export marketing of live animals
- the economic cost of camel diseases, in particular mastitis, calf diarrhoea, helminthiasis and trypanosomiasis
- research and extension on camel harness for various types of traction.

Conclusions

The camel, by virtue of its many unique attributes, commands a very special position in the life of pastoral nomadic societies, which becomes greater as the habitat becomes drier. Camels alone continue to lactate during severe drought and their milk is highly nutritious and sustaining, which is why they are called desert dairies.

There is a potential for adding value for the nomad to both milk and meat products and also hides. Although considerable advances have been made in our knowledge of camel diseases, its practical application through community-based animal health-care workers (CBAHW) (cf. Chapters 5 and 13) is still inadequate.

The use of camels for transport where there are no roads, and harnessing their great strength for traction, deserves more attention.

South American camelids

General characteristics

The camelidae are unique among mammals in that their red blood cells are oval shaped. They have an ability to survive and produce at high altitudes and cope with the extreme variations in temperature that are common in the high Andes. All species have the same number of chromosomes (74). It is possible to cross-breed all types of South American camelids, with all resulting offspring being fertile (Bustinza, 1984).

The alpaca and llama are domesticated and are socio-economically the most important species. Guanaco and vicuña are found only in the wild. The key characteristics of each species are shown in Table 21.1.

Distribution of the South American camelids

Table 21.2 shows the estimated population of South American camelids in Ecuador, Peru, Bolivia, Argentina and Chile. The data for the vicuña and guanaco are based on estimates, and are probably not very accurate. The populations of llamas and alpacas are based on census data for Ecuador, Peru and Bolivia for the years 2002, 1994 and 1997 respectively.

The main concentration of alpacas can be found in the Departments of Puno and Huanacalica, in Peru, and the main concentrations of llamas are also in these Peruvian Departments, and the Department of Oruro, in Bolivia. The main population of vicuña is found in specific regions of Peru, Bolivia and Chile. Guanaco are concentrated in the southern tip of the continent and are the only species found naturally from 0 to 4,000 m above sea level.

The distribution of the alpacas and llamas in the main camelid-producing countries, Bolivia and Peru, is related to the ability of these species to compete with alternative domestic animals. In harsh conditions with low feed-quality alpacas and llamas have a better conversion index than sheep or cattle (note that goats are not normally found in the high Andes), but this advantage disappears when feed-quality improves.

Table 21.1 Key characteristics of the four South American camelid species

Species	*Mature weight (kg)*		*Shoulder height (m)*	*General characteristics*
	Male	*Female*		
Llama[1]	116 (66-151)	102 (70 – 150)	1.1 – 1.5	There are two llama breeds: • T'hampulli which is a wool breed • Q'ara which is a meat and pack breed. Llamas are mainly kept for meat production and as pack animals
Alpaca[2]	64	62	1.0	Alpacas have ears that are short and a face covered with wool. They have a rounded rump and a straight tail. The two most important breeds are Sury and Huacaya, the latter producing the finest wool.
Vicuña[3]	37		0.90	Vicuña are generally golden brown in colour, with wool that is the finest of the four species. Vicuña are very excitable and capable of running at 60 km/hour for short distances.
Guanaco[4]	80 – 120		1.1 – 1.5	Guanaco have a light brown upper body and whitish underbelly. They have an unpredictable temperament.

[1]Sumar (1977); [2]Condorena (1980); [3]Jessup and Lance (1982); [4]Bustinza (1984)

Table 21.2 Population of South American camelids, by country

Country	*Alpaca*		*Llama*		*Guanaco*		*Vicuña*	
	Population	*%*	*Population*	*%*	*Population*	*%*	*Population*	*%*
Peru	2,454,718	84.1	999,440	27.5	3,810	0.7	118,600	56.7
Bolivia	416,952	14.3	2,398,572	66.0		0.0	34,500	16.5
Argentina	400	0.0	135,000	3.7	550,000	95.9	36,000	17.2
Chile	45,282	1.6	79,363	2.2	20,000	3.5	18,000	8.6
Ecuador	1,561	0.1	20,873	0.6		0.0	2,000	1.0
Total	2,918,913	100.0	3,633,248	100.0	573,810	100.0	209,100	100.0

Sources: Various Government (Bolivia, Peru and Ecuador) and CITES censuses between 1994 and 2002

Social and economic importance

Rearing of alpacas and llamas is an important activity for the indigenous families living in the High Andes (Sumar and Garcia, 1986). Due to the high altitude and harsh climate in this region, other domesticated animals do not thrive, and there are limited opportunities for crop activities. Therefore, South American camelids are a key component in the livelihood strategies of some of the poorest people in the continent.

Alpacas are kept mainly for fine fibre production, although meat is also an important product. The annual fibre production from an adult animal is estimated to be around 2 kg. The dressing percentage is high, at 55 to 59 per cent, and does not vary with age or sex. The carcass weight is about 20 kg at two years old, and alpaca meat has a high protein (20.3 per cent), but a low fat content (1.3 per cent) (Bustinza, 1984). The llama is an important pack animal, which has remarkable endurance, being able to carry loads of 40-50 kg over rough terrain for long distances. However, transporting goods is becoming less important with the introduction of motorised transport, and meat from this species is rapidly becoming the most important output. In addition, llamas also produce fibre, but this requires more careful processing than alpaca fibre, due to a higher proportion of coarse fibres.

In Peru it is estimated that 2.9 million people are involved in camelid production. The Peruvian camelid population is estimated to produce 4,000 tonnes of fibre and 7,000 tonnes of meat. In Bolivia, 53,000 families have camelids, which produce 300 tonnes of fibre and 12,800 tonnes of meat.

Of the wild South American camelids, the vicuña gives a biennial wool yield of 250 g and an adult may be sheared four times during its lifetime (Torres, 1984). The hunting of guanaco calves, which have very soft, valuable pelts, is permitted in limited numbers in the southern areas of Argentina. In Chile, guanacos have been protected since 1963 and the hunting of all animals, regardless of age, has been prohibited.

The sustainable use of wild vicuña and guanaco is ruled by the Convention for International Trade of Endangered Species of Flora and Fauna (CITES) in areas where their population size allows exploitation within conservation.

The countries where the vicuña is endemic signed the 'Convention for Management and Conservation of the Vicuña' during the seventies. Protection efforts, through national and international inputs, increased the numbers of vicuña and guanaco, and generated information about their management, with a result that fibre utilisation from live animals is now permitted. Systems of wild management, and management in confinement in natural environments, have been established, and Argentina, Chile and Peru are now disseminating these techniques from government agencies to the indigenous people. People have been trained in capture manoeuvres, shearing techniques, animal and fibre management, and given guidance in business administration for export. British and Italian textile industries are processing vicuña fibre into luxury products. The quality (13 μm diameter) and price (US$ 300-500/kg) of vicuña fibre compensates for its small quantity (12 tonnes/year). The revenue from fibre export is returned to indigenous communities, thus increasing livelihoods (Galaz and González, 2001). These policies have the added benefit of maintaining indigenous identity, and reducing migration to urban areas with the subsequent loss of culture.

Guanaco governmental research centres and private breeding farms are developing scientific research and fibre production in Argentina and Chile (González *et al.*, 2000). Local participation is focused on people with ovine wool-handling skills, living in depressed areas. The guanaco coat has a greater proportion of coarse fibre (20 per cent) than vicuña coat (10 per cent) which demands a de-hairing process, but the fine fibres (16-23 μm) are included among the special animal fibres. Brut fibre is commercialised in the countries of origin (US$ 100-150/kg) and its processing by the European textile industry is increasing. Controlled harvests of guanaco from wild populations for meat production has been authorised recently in Southern Patagonia; the appearance of guanaco has provided competition for ovine producers. Guanaco meat has had a good reception in internal, exclusive markets.

Production systems

Llamas and alpacas are found in extensive production systems, with very few inputs, and family management of grazing and reproduction. As mentioned above, the main focus of alpaca production is fibre, whereas llamas are principally raised for meat. The majority of camelid-keepers does not maintain production records and generally focuses on herd, rather than individual, animal management.

Reproductive aspects

Seasonality

In their natural habitat South American camelids normally breed during the wet, warmer months (late November to early May) when forage is most abundant.

Photo 21.2 The winners of the T'hampulli breed of llamas at the 5th South American Camelid Show in Turco, Oruro, Bolivia, June 2003 (courtesy of J. Rushton)

Puberty

The onset of puberty in South American camelids reportedly occurs around one year of age in the females, and two years in males.

Mating behaviour

Alpacas, llamas and vicuñas show a similar pattern of mating behaviour. It can be divided into:

- courting phase, which lasts only a few minutes, the female is receptive, and the male chases and attempts to mount her
- copulatory phase, which is longer but of variable duration.

A mean copulatory time of 18 minutes has been reported (Vivanco *et al*., 1985). In the llama the duration of coitus ranges from 3 to 65 minutes (England *et al*., 1971). In vicuña, it has been estimated to be around 30 minutes (Hoffmann *et al*., 1983) and 15 to 20 minutes in guanaco (FAO, 1985).

Ovulation

South American camelids are induced ovulators. Ovulation has been described as an event normally triggered by the act of copulation.

Gestation

The gestation period is about 11 months for all species. The presence of a corpus luteum appears necessary for the maintenance of pregnancy and it appears to be the main source of progesterone throughout gestation (Sumar and Garcia, 1986). The camelids generally give birth in the standing position and dystocia and placenta retention are uncommon.

Post-partum

Uterine involution takes about 15 days in alpaca and 20 days in llama. This means that for optimum breeding efficiency females should not be bred until at least 15 days post-partum. The young suckle for a period of 8 months and the milk is not removed for human consumption.

Main health problems

Internal and external parasites are the main health problems in South American camelids. The important external parasites are mange and lice. Of the internal parasites, roundworms are important and other helminths, such as hydatid disease (*Echinococcus granulosus*), fasciolosis (*Fasciola hepatica*), tapeworms (*Moniezia benedeni* and *M. expansa*) and lungworms (*Dictyocaulus filaria*) have been reported. Within the protozoa coccidiosis and toxoplasmosis (*Toxoplasma gondii*) have been reported, but sarcosporydiosis (*Sarcocystis aucheniae*) is of the greatest economic importance, especially in systems that focus on meat production.

The most important infectious bacterial diseases are enterotoxaemia (*Clostridium perfringens* type A and C), diarrhoea (*Escherichia coli*) and 'Fiebre de Alpaca' (*Streptococcus* spp.). The latter disease causes high levels of losses in young animals and is of particular importance in alpaca systems in Peru.

Key interventions

An important intervention to improve the livelihoods from domesticated South American camelids has been the control of external parasites and alpaca fever in the alpaca wool-producing systems. This intervention has improved fibre quality and yields, and helped to reduce mortality in young animals.

Further reading

Bravo, P.W. 2002. *The reproductive process of South American camelids.* Seagull Printing, Salt Lake City, UT, USA.

Evans, J.O., Simpkin, S.P. and Atkins, D. 1995. *Camel keeping in Kenya. Range management handbook of Kenya* Volume III, 8, Republic of Kenya, Ministry of Agriculture, Livestock Development and Marketing, Nairobi, Kenya.

Fernandez-Baca, S. (ed.). 1991. *Avances y perspectivas del conocimiento de los camélidos sudamericanos* (Advances and perspectives of the knowledge on South American camelids). Food and Agriculture Organisation of the United Nations (FAO), Santiago, Chile.

Field, C.R. 2005. *Where there is no development agency. A manual for pastoralists and their promoters. With special reference to the arid zone of the Greater Horn of Africa*. Natural Resources International, Aylesford, Kent, UK.

Fowler, M.E. 1993. *Medicine and surgery of South American camelids. Third Edition*. Iowa State University Press, Ames, Iowa, USA.

McCorkle, C.M. (ed.). 1990. *Improving Andean sheep and alpaca production. Recommendations from a decade of research in Peru*. Small Ruminant Collaborative Research Support Project, University of California, USA. Published by University of Missouri-Columbia, Columbia, USA.

Schwartz, H.J. and Dioli, M. 1992. *The one-humped camel* (Camelus dromedarius) *in Eastern Africa: A pictorial guide to diseases, health care and management*. Verlag Josef Margraf, Weikersheim, Germany.

Author addresses

Chris Field, Kenya Camel Association, P.O. Box 485, Nanyuki, Kenya

Jonathan Rushton, CEVEP, Casilla 10474, La Paz, Bolivia

Rommy Viscarra, CEVEP, Casilla 10474, La Paz, Bolivia

Bessie Urquieta, Universidad de Chile, Fac Cs. Vet. y Pec., Casilla 2, Correo 15, Santiago, Chile

Hichem Ben Salem, INRAT, Laboratoire des Productions Animales et Fourragerès, Rui Hédi Karray, 2049 Ariana, Tunisia

22

Cattle

David Mwangi[*], *Joseph Methu*[†], *Mujaffar Hossain*[‡], *Samad Khan*[‡], *Siboniso Moyo*[‡], *Anne Pearson*[‡], *Dannie Romney*[‡]

Cameo 1 - Smallholder dairy farmers in Kiambu, Kenya

"On my farm I grow maize, beans, potatoes and vegetables, but also keep dairy cows. The most important enterprise on my farm is the dairy cattle. This is because from the milk I sell, I am able to educate my children and buy food. I have four children who depend on me, and the dairy animals are my main source of income" said Margaret Nongari, who is a widow with a one hectare farm in Kiambu district, Kenya. This is the story of approximately 800,000 smallholder dairy farmers in Kenya who depend on dairy cattle for their livelihoods. The dairy enterprise generates double the income of other rural income opportunities.

Source: Unheard Voices Video (Smallholder Dairy Project and Action Aid Kenya)

Cameo 2 - Smallholder dairying in India has transformed the lives of millions of people

"It was the success of the Kaira District Co-operative Union, popularly known as Amul, which inspired the co-operative dairy programme in India over the last five decades. It has also led to the creation of organisations like the Gujarat Co-operative Milk Marketing Federation Ltd., responsible for marketing the products of all the co-operative's dairies in this particular state. The Tribhuvandas Foundation, which is Asia's largest NGO, works in over 600 villages in the State, in the field of maternal and infant care.

---

[*] Principal author
[†] Co-author
[‡] Consulting author

What is unique about the programme of the Foundation is that it rides on the back of milk. It is the village milk co-operative that appoints a village health worker and pays an honorarium to undertake the work. So it is milk paying for health. We hope you will share our belief that smallholder dairying can transform the lives of millions of people. Most of the co-operatives, health care and education in this area of Gujarat result from smallholder dairy farmers joining together to create resources, not only benefiting themselves, but also creating a better community for all."

Source: Speech by Amrita Patel, Chairman National Dairy Development Board (NDDB) of India while opening the South-South workshop in Anand, March 2001 (Rangnekar and Thorpe, 2002).

Introduction

Cattle and other ruminants convert forage, including low-quality crop residues, into valuable products. The relative importance of these products will vary from one livestock system to another. In large ranching companies, in Africa and Latin America, meat is the main product. Among resource-poor smallholders and most pastoralists, manure, milk and draught power are the important products from cattle. Among the Maasai of East Africa, where it has always been assumed that the sale of animals is the driving factor in cattle rearing, milk is ranked as the number one commodity. In areas where a market exists, these products are sold to provide income. Cattle are a very important source of high-quality proteins, energy and micro-nutrients essential for growth and good health, especially among young children. Research in Kenya indicated a lower level of malnutrition in young children among households with dairy cattle than those without (Nicholson *et al.,* 2003).

In mixed farming systems, cattle play a major role in household food security. In some of these systems manure is more important than milk or meat, as it helps to improve or maintain soil fertility and hence food production. As resource-poor households in most of the developing world cannot afford chemical fertilisers, manure is generally the only source of nutrients applied to the crop fields. As draught animals, cattle also allow households to plough and cultivate more land and remove labour bottlenecks during weeding. Cattle have been used traditionally as draught animals in many parts of the developing world, especially in Ethiopia, parts of East and Southern Africa, and Asia.

Photo 22.1 A happy boy drinking milk in front of the family's zero-grazed, Ayrshire cross cow in Kiambu district, Central Kenya (courtesy of Smallholder Dairy Project, 1999)

Cattle, and indeed other livestock, help resource-poor livestock-keepers accumulate wealth without their having to invest a large amount of money. In the intensive dairy system practised in central Kenya the male calf is considered of low value, and is therefore sold for as little as KES 3,000 (US$ 39.47) at weaning (8-12 weeks of age). Resource-poor families will purchase such calves, and feed them for 2-3 years and then sell them for KES 25,000-30,000 (US$ 329-395), which is enough to purchase a dairy heifer. In an area where average household monthly incomes are estimated to be KES 5,831 (US$ 76.72) (Staal *et al.*, 2001), the purchase of a dairy heifer to start a dairy unit is beyond the reach of most households. However, by starting with small stock (e.g. goats) or male calves, it is possible to accumulate enough funds to purchase a dairy heifer. Cattle are also an inflation-free form of banking for poor people and can be sold to meet family financial needs, for example school fees or medical bills.

Cattle also provide fuel. In Ethiopia and parts of Asia dried faeces is (regrettably) used as fuel. In parts of East Africa and also Asia biogas produced from manure of housed cattle is beginning to contribute to the energy needs of the family whilst at the same time generating a by-product fertiliser. Cattle are also important in social functions and as insurance against natural calamities like drought (cf. Chapter 2).

Photo 22.2 Cattle manure being applied prior to planting rice in Mymensingh, Bangladesh (courtesy of E. Owen)

Cattle production systems in Sub-Saharan Africa (SSA)

In Africa most of the cattle are in or near the Sahel, the higher-potential areas of East Africa, (including the Ethiopian Highlands), Zimbabwe and South Africa (Thornton *et al.,* 2002). The arid and semi-arid zones, which comprise 55 per cent of the area of SSA, are characterised by extensive pastoral livestock systems, with varying degrees of mobility (nomadic or transhumant practice). The Borana pastoral system of Southern Ethiopia and Northern Kenya, and the Maasai pastoral system of East Africa are some of the traditional systems in this category. These systems were based almost exclusively on livestock production, with little or no integration with crops. This is now changing rapidly due to increasing human population and growing pressure on land, with a resulting decrease in number of cattle per person. Thus crop production now features in many households among the Borana, the Maasai and other pastoral communities.

The pastoral system is based on indigenous zebu cattle grazed extensively in dry lowland zones. The main product is cull animals sold for beef. Milk is used in the household and is rarely sold. Milk off-take from the cattle is therefore low. With an improved marketing infrastructure, including milk collection, off-take from this system can be improved, as has happened in Tanzania. De Haan *et al.* (1997) reported a 95 per cent increase in the meat produced per hectare and 47 per cent in the meat produced per head in five Sahelian countries, marking the importance of this cattle production system. Botswana derives a large share of its GDP from cattle production from semi-arid lands.

The agro-pastoral systems found in semi-arid and sub-humid zones are a result of an increasingly sedentary lifestyle among the pastoralists. Varying levels of intensification are found in this system, ranging from the more extensive grazing systems, practised by the Fulani community in West Africa, to the more-or-less controlled grazing systems in the sorghum-millet belt of Eastern and Southern Africa. In the agro-pastoral system, herd sizes are much smaller than in the pastoral system. The animals are mainly grazed and supplemented with crop residues or forage crops. In this system cattle are used for draught, and the production of manure and milk. There is a general increase in productivity per animal as you move from the agro-pastoral systems to the crop-livestock systems (Table 22.1). As the milk off-take is based on the demand for milk by the household, production per km^2 and per animal are much lower where milk is not sold, compared to a situation where a market exists. In the smallholder farms in Kenya, where the access to market is much better compared to the situation in the communal lands of Zimbabwe or in Guinea Bissau, the off-take per animal and per km^2, and cattle density, are much higher. However, intensification in this system depends on improving the feeding, either through purchasing feed off-farm or through improved forage production and conservation (cf. Chapter 11), as well as improving the animals, through cross-breeding with exotic breeds and selection within the indigenous breeds (cf. Chapter 12).

Table 22.1 Selected examples of major land-based systems producing milk in Sub-Saharan Africa (1993)

	Agro-pastoral			*Crop-livestock*		
	Kenya	*Gambia*	*Nigeria*	*Zimbabwe*	*Guinea Bissau*	*Kenya*
Cattle density (cattle/ km^2)	35	30	32	15	10	78
Milk (kg/animal/year)	44	85	44	25	8	820
Milk (tonnes/km^2)	1.5	2.6	1.4	0.4	0.1	63.7
Human density (humans/km^2)	5	80	70	29	21	322
Milk availability (kg/ human/year)	308	32	20	13	4	197

Source: Falvey and Chantalakhana (1999).

In Zimbabwe, ownership of cattle is associated with enhanced food security, and improved livelihoods. Other reasons for keeping cattle have been discussed at the beginning of this chapter. Smallholder-owned cattle depend on natural grazing and crop residues for the bulk of their feed. Almost 90 per cent of Zimbabwe's indigenous cattle are in the smallholder sector. In several comparisons with exotic beef breeds and their crosses, indigenous cattle have proved superior across a range of production indices (Table 22.2).The results in Table 22.2 indicate that indigenous cow breeds (Mashona, Tuli and Nkone) were the most productive per cow unit weight, despite the low weaning and low 18-month weights of their calves. This was due mainly to the high reproductive rates and the small mature weight of these indigenous cows and thus low maintenance needs. These attributes have resulted from long evolution within these harsh environments. When choosing a breed to suit an environment, adaptation characteristics play a key role. Similar results from a beef cattle cross-breeding study in Tanzania were reported in Chapter 12.

Smallholder cattle are more likely to be regarded as multi-purpose than single-purpose, with cows providing draught animal power (DAP) as well as producing the next generation. Gemeda *et al.* (1995) reported delayed conception in non-supplemented working cows, compared to those not working and supplemented. Pearson and Dijkman (1994) suggested that work required 1.8 times maintenance energy, usually at a time of year when the animals are in poor condition at the end of the dry season, and feed resources are at their lowest. The smallholder dairy industry in Zimbabwe has grown in recent years and, because of the decline in the number of registered herds in the commercial sector (from 383 herds in 1994 to 198 in 2004), will need to expand still further if market requirements are to be met.

The tropical highlands, which are mostly found in Eastern Africa, carry 20 per cent of the cattle in SSA. Cattle production in this zone is managed under mixed farming, with intensive crop production, using draught oxen, and a range of crop residues and crop by-products for livestock feeding. Ethiopia has the largest share of the African highlands, and the largest cattle herd. This zone is best for dairying because the high rainfall and cool climate support a high production of forage and a low incidence of animal diseases. However, the presence of dairying also depends on there being a well-developed market and whether cultural consumption patterns include milk or dairy products.

The intensive, market-oriented smallholder production system in Kenya will be used as the main example of dairy development in SSA.

Table 22.2 Productivity of beef breeds evaluated in Zimbabwe (1978-89)

Cow genotype	*Number of cows*	*Cow weight[1] (kg)*	*Annual calving rate (%)*	*Weaning weight (kg)*	*18-month weight (kg)*	*Productivity indices*			
						Per cow mated[2]		*Per cow unit weight[3]*	
						Weaning	*18-month*	*Weaning*	*18-month*
Mashona	253	351	77	176	261	153	235	206	346
Nkone	147	388	71	188	279	160	246	203	335
Tuli	349	400	72	187	275	160	240	193	318
Afrikaner	447	398	59	189	273	138	208	169	281
Brahman	106	440	71	207	299	166	248	197	311
Sussex	134	418	59	180	275	139	222	162	281
Charolais	116	480	67	188	276	160	241	172	284

[1]Weight at peak time (May) three months before weaning calves; [2]Weight of calf per cow exposed to the bull; [3]Weight of calf per 100 kg live weight$^{0.73}$ of cow exposed to the bull.
Source: Moyo (1996).

Intensive smallholder dairying in Kenya

Breed of dairy cattle kept

The dairy herd consists of Friesian/Holstein, Ayrshire, Guernsey and Jersey cattle, and their crosses with indigenous breeds. The crosses make up 50 per cent of the herd, with the Friesians and Ayrshire dominating the exotic dairy breeds. The dairy cattle population has increased from approximately 0.8 million in 1960 to 3.2 million currently. This major increase can be attributed to the breeding policies adopted by the Kenyan government since independence. In 1966 a breeding system was established to provide AI services to smallholder farmers in an effort to improve the milk production potential of the indigenous zebu. Cross-breeding was preferred so as to combine the milk potential of the exotic breeds and the hardiness of the indigenous animal. As a result, Kenya has the largest population of improved dairy animals in East and Southern Africa (Figure 22.1).

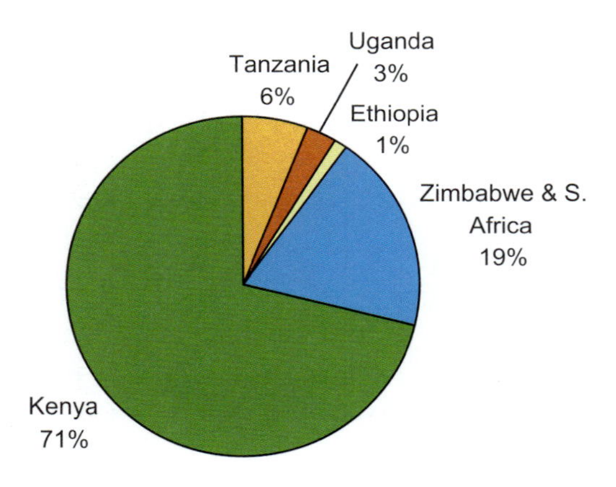

Figure 22.1 Distribution of improved dairy animals in East and Southern Africa (source: Thorpe *et al.*, 2000)

Feed source

During the last decade or so the smallholder dairy production system has changed tremendously. In the initial years the animals were mainly grazed in paddocks, a system that still exists in parts of the country but which is relatively rare in central Kenya. Here, due to the cultural practice of land division and therefore diminishing land holdings, the 'zero-grazing' system has developed. The animals are confined in stalls and fed a diet largely consisting of (planted) Napier grass (*Pennisetum purpureum*) and crop residues, which are supplemented with grass from communal land or roadsides. Napier grass is the major cultivated forage crop on these farms and supplies about 40 per cent of the feed, while the maize crop supplies a further 24 per cent (Figure 22.2), with considerable variation between farms.

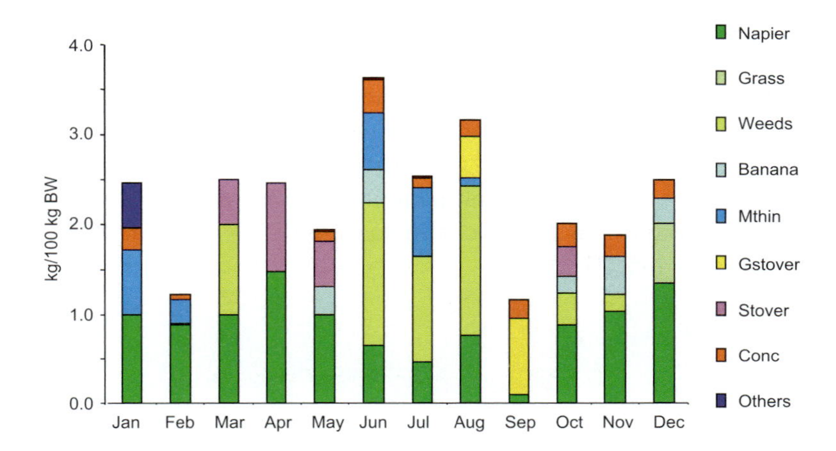

Figure 22.2 A typical seasonal feeding pattern on smallholder dairy farms in Kenya. BW = body weight; Mthin = maize thinnings during the growing period; Conc = concentrate feeds (dairy meal); Gstover = green forage from maize harvested at roasting stage (source: Smallholder Dairy Project, 1999)

Ruminant production in the tropics is constrained by the low crude protein (CP) content of dry-season forage (cf. Chapter 11). In central Kenya the CP content of Napier grass is between 64 and 118 g per kg dry matter (DM), depending on age at cutting, while that of maize stover is between 56 and 61 g per kg DM (Abate and Abate, 1991; Methu *et al.,* 1996). Maize stover and other crop residues are the main source of dry-season feeds and contain less than the critical level of 60-70 g CP per kg DM needed to maintain efficient rumen function (cf. Chapter 10). Therefore milk yields decline during the dry season and animals suffer large weight losses. In general, the CP content of the main feeds on smallholder dairy farms is lower than the 120 g per kg DM required for moderate production by dairy cattle (AFRC, 1993). To bridge this protein gap many attempts have been made to introduce forage legumes (both herbaceous and shrubby). However, the cases of successful adoption on smallholder farms have been few. The low adoption can be attributed to many factors, including lack of planting material and a dearth of clearly demonstrated benefits.

As smallholder dairy farms intensify, the proportion of feed purchased from outside the farm tends to increase. A study in central Kenya (Romney *et al*., 2004), where zero-grazing is the main production system, showed that more than 25 per cent of the feed was purchased. In contrast, less than 15 per cent of feed was purchased where extensive paddock grazing was used. Of the feeds purchased, concentrate supplements could represent 85 to 95 per cent of the cost. The main concentrate supplements were dairy meal and other cereal milling by-products (maize germ, pollard/wheatfeed and wheat bran, and oil seed cakes). The supplements were fed at a flat rate regardless of the quality of the basal diet or the stage of lactation, resulting in a collapsing curve of milk production ('pre-trial' in Figure 22.3). On-farm research (Romney *et al.*, 2000) has shown that encouraging farmers to lead or challenge-feed with dairy meal in the early part of the lactation (first 4 months), rather than feeding at a flat rate throughout lactation, and assisting farmers to access concentrate through credit, can increase milk yield by 20 to 40 per cent ('during trial' in Figure 22.3).

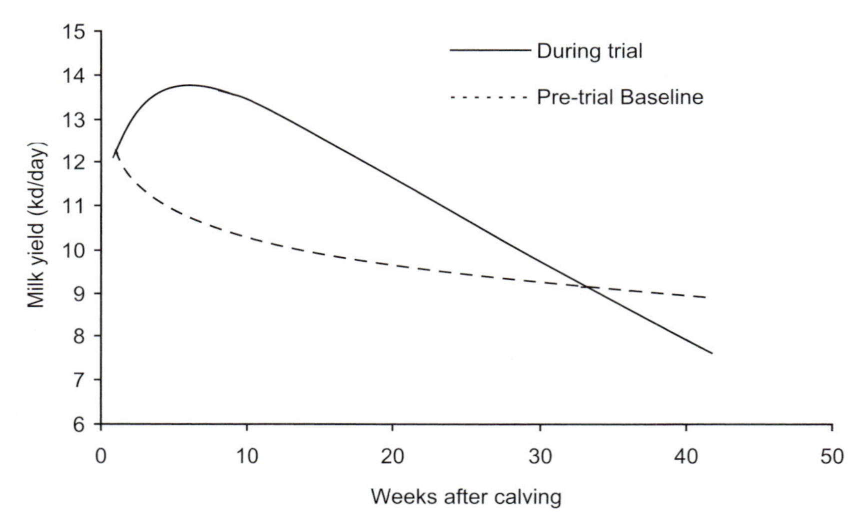

Figure 22.3 Lactation curves during a study where farmers were offered additional credit through the cooperative to allow feeding higher levels of concentrate during the first 4 months of lactation (source: Romney *et al.*, 2000)

Production parameters in smallholder farms in Kenya

A general feature on all smallholder dairy farms is the low growth rate of replacement heifers. On most farms weight gains are below 200 g per day between birth and weaning. This results in heifers calving at over 3 years of age. Calf mortality is high, at over 20 per cent, on smallholder farms. The high mortality and low weight gains are attributed to both malnutrition and undernourishment. The age at weaning decreases as farms intensify (Table 22.3).

Table 22.3 Production parameters across selected cattle production systems in Kenya

Production parameter	*Production system*			
	Large-scale commercial (exotic/crosses)	*Small-scale semi-intensive*	*Small-scale semi-intensive (exotic/crosses)*	*Small-scale intensive (exotic/ crosses)*
Grazing management	Free	Free (agro-pastoral)	Semi-zero	Zero
Age at first calving (years)	3.5	3.5	3.0	3.0
Pre-weaning calf mortality (%/year)	8	20	15	20
Age at weaning (days)	>120	>200	90	90
Lactation length (days)	>300	>200	450	450
Milk off-take (l/cow/year)	3,700	250	1555	2000
Mature weight (kg)	450	300	350	350

Source: Peeler and Omore (1997).

Calving intervals are long, e.g. 471-598 days in central Kenya and up to 1290 days in western Kenya (Peeler and Omore, 1997; Staal *et al.,* 2001). This is attributed mainly to undernutrition, which leads to long anoestrous periods. It is not uncommon to find animals that have not cycled up to 200 days after calving.

Although most smallholder farmers in Kenya keep improved dairy breeds capable of producing more than 15 l per cow daily, the average production is 4-5 l. Well-managed smallholder farms can produce up to 20 l per cow per day, which is similar to the production on well-managed commercial farms. Average lactation yield on many smallholder dairy farms is about 1,500 l per lactation, compared to over 3,000 l on commercial farms and well-managed smallholder farms. Lack of persistency, due to undernutrition, results in rapidly-declining milk yields and short lactations in many smallholder-owned herds (Figure 22.3; Romney *et al.,* 2000).

The low milk yields and long calving intervals so characteristic of exotic cross-breds, such as used in Kenya dairying, have led to an increasingly held view that national strategies should be based on improving local breeds rather than importing exotics. There is increasing evidence (e.g. King *et al.,* 2005) that exotic dairy cattle, such as Holsteins, are less productive than local animals, especially when compared on a lifetime basis and where ambient temperature and humidity are high.

Cattle production in Asia

Three major livestock production systems exist in Asia; pasture-based; rain-fed crop-livestock; and irrigated crop-livestock. The extensive pasture-based production covers about 70 per cent of the land. Milk production is low as the animals are kept mainly for beef production. About 15 per cent of the beef consumed in Asia comes from this system. The rain-fed crop-livestock system supports higher milk production per animal unit (AU) of cattle or buffalo in tropical Asia, followed by the irrigated crop-livestock system. Milk availability per person follows the same trend (Table 22.4)

Table 22.4 Livestock and milk production in tropical Asia

	Pasture-based	*Rain-fed crop-livestock*	*Irrigated crop-livestock*
Cattle (AU/human)	0.14	0.17	0.11
Buffalo (AU/human)	-	0.07	0.05
Grazing (ha/AU)	16.7	0.8	0.8
Crop land (ha/AU)	0.7	1.6	1.3
Milk			
kg/AU/yr	20	509	324
kg/ha/yr	1	211	155
kg/person/yr	2	56	24

Source: Seré *et al.* (1996)
AU = Animal unit of 400 kg

Landless, marginal and smallholder producers occupy over 84 per cent of the holdings but they own or control less than 35 per cent of the total farm land (Birthal and Parthasarathy Rao, 2002). Livestock ownership is more evenly distributed with landless, small and marginal farmers owning approximately 64 per cent of the dairy animals and producing 65 per cent of the milk (Table 22.5).

Table 22.5 Distribution of dairy animals and milk production among different classes of farmers in India

Type of farmer	*Farmers (%)*	*Dairy animals (%)*	*Milk production (%)*
Landless (<0.002 ha)	26	22	23
Small and marginal (0.002-2 ha)	49	42	42
Medium and large (>2 ha)	25	36	35

Source: de Jong (1996).

Over 70 per cent of the cattle kept in Asia are indigenous and are low producers. The number of cross-bred cattle is increasing. For example, in India the number of cross-bred cattle doubled between 1982 and 1992. These cross-bred cattle are higher yielding compared to the indigenous animals (Table 22.6).

Table 22.6 Population of indigenous and cross-bred cattle and their productivity in India

Animal category	*Cattle population (x 10^6)*		*Lactation yield by region 1992 (kg/cow)*			*Daily yield 1992 (kg/cow)*		
	1982	*1992*	*East*	*North*	*West*	*East*	*North*	*West*
Indigenous	180.3	189.3	452	658	638	3.01	3.29	3.19
Crossbred	8.8	15.2	1746	2121	2340	5.82	7.07	7.80

Source: Shukla and Brahmankar (1999).

As elsewhere in the developing world, crop residues and pastures are the major sources of feed in the smallholder system in Asia. Crop residues supply approximately 70 per cent of feed on smallholder farms with 15-30 per cent coming from grasses and grazing (Birthal and Parthasarathy Rao, 2002). Cultivated forages and concentrate feeds contribute only a small proportion of the total feed supply.

Smallholder dairying is a major source of household income, especially among the landless farmers. The impact study on Operation Flood in India showed smallholder dairying accounting for more that 50 per cent of the household income of landless farmers and that the importance of dairying as a source of household income decreased as the size of farm increased (Table 22.7). As land is not equitably distributed, this indicates that development of smallholder dairying would be a path out of poverty for the resource-poor in the society.

Table 22.7 Sources of household incomes in rural areas of India

Household	*Dairying*	*Crop husbandry*	*Others*	*Total*
Landless (<0.002 ha)	53.1	0.0	46.9	100
Marginal (<1.00 ha)	30.1	46.6	23.3	100
Small (1.00-1.99 ha)	29.7	53.8	16.5	100
Semi-medium (2.00-3.99 ha)	26.3	59.0	14.8	100
Medium (4.00-9.99 ha)	25.3	62.8	11.9	100
Large (>10.00 ha)	19.0	71.5	9.5	100

Source: Shukla and Brahmankar (1999).

However, in Asia, as elsewhere, dairying is not the only reason for keeping cattle.

In many areas off-take for meat production is a major source of income, as typified by experience of small-scale fattening schemes in Bangladesh where indigenous cattle are mainly reared by rural farmers for draught power, meat, milk and calves. Cattle fattening before Eid-ul-Azha (Muslim festival) for beef production has become an important business for small farmers. Of the calves born, very few are reared as bulls for natural breeding. Most of the male calves are therefore available for rearing as beef animals. If the following steps are undertaken, small-scale cattle fattening in rural areas in developing countries like Bangladesh will be facilitated:

- provide access to credit from banks, NGOs etc.
- develop specialised breeding programmes for beef from local breeds
- provide subsidy (from government) for the purchase of animal feed
- provide support for vaccines and other veterinary treatments for cattle
- develop marketing systems for buying and selling bull calves.

Many researchers have investigated the factors associated with cattle fattening by rural farmers in Bangladesh, and a summary of their findings is presented in Table 22.8.

Table 22.8 Farmer survey of small-scale cattle fattening in Bangladesh

Question	*Answer*
Education level of farmers	More than 50% received primary education
Sex of calves	More than 70% uncastrated males
Age of bull calves at start of fattening	2-3 years
Number of cattle fattened by each farmer	More than 70% fattened 2 animals
Duration of cattle fattening	6-7 months
Source of finance for cattle fattening	More than 80% farmers invested own resources
Training on cattle fattening	90% of farmers received no training
Housing facilities	More than 80% of farmers provided separate houses for fattening cattle
Deworming of cattle	About 60% of farmers dewormed cattle at start of fattening period
Health treatment of cattle	More than 76% reported using unskilled village doctors
Feed sources used	Household by-products and locally-available feed resources
Feed technologies used	Urea molasses blocks, urea and molasses rice straw
Average net return per animal, within 6 months	Taka 2000 (US$ 34.5)
Expected outputs of rural farmers through small-scale cattle fattening schemes	• Cash income generated • Beef production increased • Local feed resources used • Employment opportunities created • Availability of organic manure increased • Foreign currency earnings generated through exporting hides

Cattle for draught power

The most common system for keeping cattle (and buffalo) in Asia and much of Africa too, on which most resource-poor people depend, is the mixed crop-livestock farming system. Cattle and buffalo play an important role, along with other livestock, as a source of family savings and economic security in these systems. Small-scale farmers rarely have surplus income to bank, but by keeping a range of livestock, from poultry through pigs or small ruminants to the larger cattle and buffalo, they have from short term (poultry) to long term (cattle and buffalo) savings accounts. Despite the increase in availability of small and two-wheeled tractors, throughout Asia in particular, the need for animal power continues to be an important reason for maintaining cattle and buffalo on many small farms, particularly where the terrain is unsuitable for tractors. Additional benefits to farmers in these systems include manure, calves and, where markets are opening up, milk, and finally meat when the animal is no longer useful for work.

Cattle and buffalo in these farming systems are now rarely kept for manure and work alone. The draught animal is truly multi-purpose. In the highlands of Vietnam, for example, there is still almost 100 per cent ownership of 1-2 cattle or buffalo on small farms, although animal power has been almost entirely replaced by tractor power in the rice growing areas of the Mekong delta, and on some of the flatter lands in the midlands and on the Red River delta. Most of the working animals are female and produce a calf at least every two years. Similar situations exist in many highland areas throughout the tropical and sub-tropical areas, where it is difficult to see the draught animal being replaced on mixed farms.

However, changes are occurring within these mixed farming systems, in some areas. The price of cattle and buffalo meat is increasing in South-East Asia and some small-scale farmers are being tempted to move into more intensive systems of meat production, with a move away from rain-fed rice production to more specialised forage production alongside their livestock. In Northern Thailand, for example, many small-scale farmers in 2004 are getting better returns from beef production from planted forage than they are from rain-fed rice production, although this means that they move away from being self-sufficient in rice production for the household. The cattle and buffalo remain the focal point of the farming system and key to the economic security of the family. In Latin America the draught cattle on small mixed farms are tending to be replaced by horses and mules, which people find more versatile than cattle to use for pack, riding and draught, as well as being faster.

In most mixed farming systems where cattle are used for draught power, changing labour availability is now a key factor in influencing activities. With increasing education opportunities, many young people are leaving the poorer rural communities and the average age of the smallholder farming community is increasing. In the Altiplano of Mexico, for example, the average age of farmers is over 50 years, whereas in South Africa many farmers in Eastern Cape Province are over 65 years old. Mechanisation can help reduce labour demands in these situations, but in reality, in resource-poor communities where mechanisation is not an option, it is the women who pick up the burden of livestock management. In parts of Asia and Africa it is now the women who are increasingly taking responsibility for working the draught cattle or buffalo, as well as the daily management of the livestock themselves.

Photo 22.3 Cattle cultivating field for planting rice in Mymensingh, Bangladesh (courtesy of E. Owen)

Conclusion

Cattle on smallholder farms produce milk, manure and meat that are used to improve the food security of the farm household. Cattle also provide draught power, allowing the smallholder farmer not only to expand the amount of land cropped but also to prepare it on time, which is of essence, especially in rain-fed production systems. As indicated above, most resource-poor farm households keep livestock. Therefore, based on the two cameos at the beginning of the chapter and the broader discussion that followed, it can be concluded that cattle keeping can be used as a path out of poverty. The main constraints to this are the availability of feeds, and animals with appropriate genetic production potential to match available resources and the environment. As demonstrated by the Kenyan case, the improvement of the genetic potential of the animals through cross-breeding can be feasible, but must be made in conjunction with improvements in the availability and quality of feeds. However, as noted earlier (and in Chapter 12), cross-breeding with exotic cattle may not be the most appropriate strategy for sustainable improvement in milk or beef production

Non-monetary outputs, like organic manure for use in crop production, dung for fuel and insurance against contingences, are important in smallholder livestock production. These outputs are often ignored when calculating profitability, yet recent work has shown that they can contribute as much as 20 per cent of the annual profitability of these systems.

Further reading

Hill, D.H. 1988. *Cattle and buffalo meat production in the tropics*. Intermediate Tropical Agriculture Series, General Editor, W.J.A. Payne. Longman Scientific and Technical, Harlow, Essex, UK.

Kitalyi, A., Miano, D., Mwebaza, S. and Wambugu, C. 2005. *More forage, more milk. Forage production for small-scale zero grazing systems. Technical handbook No. 33*. Regional Land Management Unit (RELMA in ICRAF)/ World Agroforestry Centre, Nairobi, Kenya.

McIntire, J., Bourzat, D. and Pingali, P. 1992. *Crop-livestock interaction in Sub-Saharan Africa*. The World Bank. Washington, D.C., USA.

Mathewman, R.W. in collaboration with Chabeuf, N. 1993. *Dairying*. The Tropical Agriculturalist, Series Editors R. Coste and A.J. Smith. Macmillan Education Ltd., London, UK, in cooperation with Technical Centre for Agricultural and Rural Cooperation (CTA), Wageningen, The Netherlands.

Pearson, R.A. and Dijkman, J.T. 1994. Nutritional implications of work in draught animals. *Proceedings of the Nutrition Society* 53:169-179.

Peeler, E.J. and Omore, A.O. 1997. *Manual of livestock production systems in Kenya*. Kenya Agricultural Research Institute (KARI)/UK Department for International Development (DFID), National Agricultural Research Project, Kikuyu, Kenya.

Phillips, C.J.C. (ed.). 1989. *New techniques in cattle production*. Butterworths, London, UK. Now Nottingham University Press, Nottingham, UK.

Preston, T.R. and Leng, R.A. 1987. *Matching ruminant production systems with available resources in the tropics and sub-tropics*. Penambul Books, Armidale, Australia.

Smith, A.J. (ed.). 1976. *Beef production in developing countries*. Centre for Tropical Veterinary Medicine, University of Edinburgh, Edinburgh, UK.

Stobbs, T.H. and Thompson, P.A.C. 1975. Milk production from tropical pastures. *World Animal Review* 13: 27-31.

Author addresses

David Mwangi, Kenya Agricultural Research Institute, P.O. Box 57811-00200 City Square, Nairobi, Kenya

Joseph Methu, Land O'Lakes, International Development, P.O. Box 45006, Nairobi, Kenya

Mujaffar Hossain, Department of Animal Science, Bangladesh Agricultural University, Mymensingh 2202, Bangladesh

Samad Khan, Department of Dairy Science, Bangladesh Agricultural University, Mymensingh 2202, Bangladesh

Siboniso Moyo, Department of Livestock Production and Development, P.O. Box CY 2505, Causeway, Harare, Zimbabwe

Anne Pearson, Centre for Tropical Veterinary Medicine, Division of Animal Health and Welfare, University of Edinburgh, Easter Bush Veterinary Centre, Roslin, Midlothian EH25 9RG, UK

Dannie Romney, International Livestock Research Institute (ILRI), P.O. Box 30709, Nairobi 00100, Kenya

23

Buffalo

Oswin Perera[*], *Harischandra Abeygunawardena*[†], *William Vale*[†], *Charan Chantalakhana*[‡]

Cameo - Projects to help resource-poor buffalo keepers in Sri Lanka and Brazil

A resource-poor farmer in the semi-arid rice growing region of Sri Lanka was looking for ways to overcome the problem of keeping a herd of indigenous buffalo for use in ploughing his rice fields. New irrigated settlement schemes in the area had made traditional extensive grazing systems untenable due to communal lands being no longer available. "I was fortunate to join a group of farmers who were participating in a project to develop an innovative buffalo production system called 'Smallholder Intensively Managed Buffalo Units' (SIMBU). We received assistance to construct a shed and purchase four dairy buffalo. We were trained to test various improved management, feeding, breeding and disease control practices and to adopt those that were most applicable. We managed our SIMBU with household labour and used low-cost feed resources that were available on or around the farm. We also learnt how to convert milk to curd, thus earning more per litre of milk. In addition to meeting our daily cash needs, the SIMBU provides us with draught power and manure for crops. The size of my herd has doubled within four years, enabling me to repay the assistance received".

Resource-poor smallholder farmers in the flood plains of the Amazon river basin in Brazil are faced with similar problems of managing large herds under extensive systems. The State Government is therefore providing special credit to them for purchase of a herd of 10 improved females and a male. The repayment is done in the form of animals rather than cash, commencing three years after receiving the animals. This consists of four animals in the fourth and fifth year

---

[*] Principal author
[†] Co-author
[‡] Consulting author

and three in the sixth and seventh year. These farmers are making use of the assistance to increase their incomes through multi-purpose use of buffalo.

Why keep buffalo?

Since its domestication some 5,000 years ago in the Indus Valley, the domestic water buffalo (*Bubalus bubalis*) has become an integral part of rural agriculture in most countries in South and South-East Asia. Together with cattle (see Chapter 22), it provides milk, meat, draught power, fertiliser, fuel and hides. In many rural farming systems it is also a source of economic security and social status. Buffalo have spread into many countries outside Asia (Table 23.1). However, it is mainly in Asia and Latin America that the buffalo is important in the livelihood of resource-poor rural smallholder farmers. This chapter will therefore briefly review the role of the buffalo in rural farming systems in these two regions and examine its future potential as a means for food security and poverty alleviation.

Table 23.1 Population of water buffalo in different geographical regions and countries

Region/Country	*Population (million)*	*Region/country*	*Population (million)*
South Asia	123	*Central and West Asia,*	4.0
Bangladesh	0.83	*North Africa and Europe*	
Bhutan	0.01	Azerbaijan	0.30
India	95.10	Bulgaria	0.02
Nepal	3.70	Egypt	3.55
Pakistan	24.00	Iran	0.52
Sri Lanka	0.72	Iraq	0.09
		Italy	0.17
South-East Asia	38	Kazakhstan	0.10
Cambodia	0.62	Romania	0.20
China	22.24	Turkey	0.14
Indonesia	2.30		
Laos	1.06	*Latin America and*	4.0
Malaysia	0.15	*the Caribbean*	
Myanmar	2.55	Argentina	0.12
Philippines	3.12	Brazil	3.50
Thailand	2.10	Colombia	0.08
Vietnam	2.80	Cuba	0.02
		Peru	0.03
		Trinidad-Tobago	0.01
		Venezuela	0.20

Total world population ~ 170 million
(Adapted from: FAOSTAT (2003) and other sources.

Comprehensive coverage on all aspects of buffalo husbandry, physiology and productivity are found in the books cited under Further Reading. Reviews on current global statistics and trends on buffalo populations and production are available in Singh and Dhanda (2003). A website containing information on water buffalo can be accessed at: http://ww2.netnitco.net/users/djligda/waterbuf.htm.

The domestic buffalo is believed to be derived from the Asian wild buffalo, *Bubalus arnee*. It is referred to as 'water buffalo' due to its characteristic of wallowing in water, which is a mechanism for dissipating body heat in hot humid environments. The wild buffalo of Africa (*Syncerus caffer*) is called the Cape buffalo and belongs to a separate genus. Although the American Bison (*Bison bison*) is also called 'buffalo', it is more closely related to cattle.

In South Asia agriculture is the livelihood of a high proportion of rural people and in the majority of cases this comprises mixed farming, with crops and livestock performing complementary roles (cf. Chapter 3). In this region milk and its derived products (ghee, yoghurt, sour milk and cheese) are important in the diet of children as well as adults, and the buffalo is the major producer of milk. In South-East Asia, too, agriculture is still the dominant occupation and although milk is less important, mixed crop-livestock farming forms the backbone of rural agriculture. In both regions buffalo provide draught power for the cultivation of rice and other annual crops, manure for use as fertiliser or fuel and, usually towards the latter part of its productive life, meat. They also serve as capital and savings for the poor, confer social status and have cultural or religious roles in rural communities.

In Latin America and the Caribbean (LAC), where buffalo were first introduced through Brazil in the late 19th and early 20th centuries, subsequent importations of live animals as well as semen for use in artificial insemination (AI) have resulted in their spread to most countries in the region. They are raised under a variety of farming conditions, ranging from smallholder to large-scale ranching systems. Buffalo milk has a high demand for making mozzarella cheese, which has a premium local and export market. Buffalo meat is also finding a niche as a more lean and healthy alternative to beef. Buffalo farming is perceived as a viable option for providing principally milk and meat, but also draught power, in situations where climate, feed availability or diseases limit cattle production.

The buffalo is therefore favoured by resource-poor farmers in Asia and Latin America due to its ability to utilise low-quality forage, survive under harsh environmental conditions and provide economically important products with the minimum of external inputs. Indeed, the buffalo is considered a part of the family in some farming communities and is essential for their survival under difficult conditions. Thus the question 'Why keep buffalo?' has many answers, depending on the region and farming system. As one poor farmer in Asia put it, "If I die my family will be sad, but if my buffalo dies they will starve".

Types of buffalo kept

Water buffalo were classified into three main 'types' by McGregor in 1939 (Cockrill, 1974), based on their habitat, external (phenotypic) characters and behaviour. The buffalo of the Indo-Pak subcontinent were called 'river type' due to their origin around river valleys and preference for wallowing in clear running water. The working buffalo of South-East Asia were called 'swamp type' due to their habitat in swampy areas and preference for wallowing in stagnant pools or mud holes. The Mediterranean buffalo, which are now considered a distinct type, are thought to have originated from the river type.

Photo 23.1 River type buffalo bull in the Punjab, Pakistan (courtesy of O. Perera)

Some characteristics of river and swamp buffalo are given in Table 23.2. The river types comprise a number of distinct dairy breeds, such as Murrah, Surti and Jafarabadi in India and Nili-Ravi and Kundi in Pakistan. There are no specific breeds among swamp buffalo, but differences in body size are evident between some countries. Swamp buffalo in Indonesia, Malaysia and the Philippines are called 'carabao'. White (albino) as well as black and white (piebald) animals occur in some countries. The indigenous buffalo of Sri Lanka were originally classified as swamp type based on phenotype and behaviour, but they have 50 chromosomes and are therefore genetically river type.

Table 23.2 Characteristics of river and swamp buffalo

Parameter	*River buffalo*	*Swamp buffalo*
Percentage of world population	70	30
Wallowing habits	Running water (rivers)	Stagnant water (swamps)
Number of chromosomes	50	48
Main uses	Milk and meat	Draught and meat
Body colour	Darker	Lighter
Body hair	More	Less
Chevrons on neck	Absent	Present
Horns	Curled, swept back from forehead	Sickle shaped, in line with forehead
'Bos' on forehead	Present	Absent
Growth rate (kg/day)	0.8-1.0	0.5-0.7
Weight at puberty (kg)	250-350	200-300
Adult weight (kg):		
Male	450-750	400-600
Female	400-650	350-450
Milk production:		
Litres/day	5 to 15	1-3
Litres/lactation	1,500-4,500	200-750
Lactation length (months)	9-10	5-8
Fat content of milk (%)	6-10	6-10

In South Asia the river type predominates, although nondescript indigenous animals (called 'deshi'), which resemble the swamp type, are also found in India and Bangladesh. The majority of farmers in South-East Asia keep the swamp type, but improved breeds of the river type have also been imported to many countries for use in cross-breeding. The main objective of such breeding programmes is to upgrade the milk potential of the swamp type, while retaining its draught ability. Genetic improvement of the swamp buffalo through performance testing and selective breeding is also being done in some countries (Chantalakhana, 1992).

The LAC region has all three types of buffalo. The original stock of swamp buffalo which landed on Marajó Island in the delta of the Amazon River in 1895 became feral and multiplied rapidly. Subsequent imports included Mediterranean buffalo

Photo 23.2 Swamp type buffalo ploughing a rice field in Sri Lanka (courtesy of H. Abeygunawardena)

to Brazil and river types from India to Trinidad-Tobago. More recent importations of live animals, as well as semen for use in AI, have been done from India, Pakistan, Egypt, Italy and Bulgaria to Argentina, Brazil and Venezuela (Vale, 1996). The most common breeds of river buffalo in this region are the Murrah, Jafarabadi and Nili-Ravi, while crosses between them and the Mediterranean and swamp types also exist. A cross-bred called the 'Buffalypso' was developed in Trinidad and is used in the USA and Caribbean countries for beef production.

Biology and physiology

Growth rate and adult weight

Growth rate and adult weight (see Table 23.2) are both influenced by interactions between the genotype of the animal and the environment in which it is raised, which includes climate, management practices, feeding and the incidence of parasites and other diseases. Buffalo usually reach their mature weight at about 4-5 years of age.

Nutritional characteristics

The nutritional requirements of dairy buffalo and guidelines for feeding calves, heifers, pregnant and lactating cows have been published (Ranjhan, 1998).

Common feeding practices for ruminants and methods of supplementing low quality feeds, including the use of urea molasses multi-nutrient feed blocks (UMMB) have been described in Chapters 10 and 11.

A unique advantage of the buffalo is its ability to utilise low-quality forages more efficiently than cattle (Sebastian *et al.*, 1970; Abdullah *et al.*, 1990). This is due to a combination of factors, including higher intake of dry matter and longer retention time of feed in the digestive tract than cattle. The ecology of the rumen, which consists of bacteria, protozoa and fungi, is more favourable than in cattle to the digestion of coarse fibrous feeds. Coupled with these factors is the buffalo's ability to utilise a wide range of plant types and stressful environments. In northern Brazil, for example, during the wet season the heat and humidity cause a reduction in feed intake of cattle, but not of buffalo. They feed on a wide variety of vegetation, including thorny bushes, sedges, reeds, floating grass and aquatic weeds, and are even known to dive underwater to graze on submerged vegetation.

Resource-poor farmers can benefit from these characteristics of the buffalo by combining different low-cost feed resources available on or around their farms to produce economically important products. An example of such a system is described later (see Comparative advantages and potential for improvement - South and South-East Asia).

Reproduction and fertility

The buffalo has been traditionally regarded as a poor breeder. This arises from the observation that in the majority of conditions under which buffalo are raised (i.e. smallholder farming systems in harsh environments with low-quality feed and minimal managerial inputs), they have low fertility. The latter is manifested mainly as late maturity, long post-partum anoestrus, poor expression of heat (oestrus) signs, low conception rates (CR) and long calving intervals (CI). However, a review of more recent studies shows that buffalo can have excellent fertility provided that genotypes are matched to the environment and they are managed and fed properly. Comprehensive reviews on reproductive functions and fertility of buffalo, including physiology and endocrinology, have been published (Perera, 1999a; Jainudeen and Hafez, 2000). A brief overview is given below.

The female buffalo

Buffalo are capable of breeding throughout the year and are termed polyoestrous. However, in many countries a seasonal pattern of calving has been observed. This is due to the influence of climate and feed supply, which result in seasonal fluctuations in ovarian activity and conception, and therefore in calving. In some locations, where annual changes in rainfall determine the availability and quality of feed, ovarian activity commences some two to three months after the onset of rains, followed by conceptions that result in a peak calving season 10 months later.

Buffalo heifers usually attain puberty when they reach 55-60 per cent of adult body weight (Table 23.2), but the age at which this occurs can be highly variable. This is influenced by genotype, nutrition, management, social environment, climate, year or season of birth and diseases. Under optimum conditions buffalo heifers can conceive as early as 16 months of age, but more commonly this occurs around 26 to 30 months. Although buffalo attain puberty later than cattle, they have a longer reproductive life, which compensates for the early disadvantage.

Reproductive characteristics of buffalo are summarised in Table 23.3. The external signs of oestrus are less obvious in buffalo than in cattle. A major difference is that homosexual behaviour between females, which is common in cattle, is rarely seen in buffalo. The main behavioural signs are restlessness, bellowing and frequent voiding of small quantities of urine. Externally detectable physical changes include swelling of the vulva and reddening of the vestibular mucosa. Mucus secreted from the cervix during oestrus is less conspicuous than in cattle. In hot climates the duration of heat may be shorter and the signs may be exhibited only during the night or early morning. The most appropriate time for breeding, whether by natural service or AI, is during the latter part of the oestrus period. Thus the AM/PM rule which is applied in cattle, where females detected in oestrus in the morning are bred in the afternoon of the same day and those detected in the afternoon or evening are bred the next morning, also applies in buffalo.

Table 23.3 Reproductive characteristics of buffalo

Parameter	*Mean*	*Range*
Age at puberty (months)	30	18-46
Length of oestrus cycle (days)	21	17-26
Length of oestrus (h)	10	5 to 27
Time of ovulation:		
After onset of oestrus (h)	34	24-48
After end of oestrus (h)	14	6-21
Length of gestation (days):		
River type	310	300-320
Swamp type	330	320-340
Birth weight of calves (kg)	26	22-36
Involution of uterus (days)	30	25-35

The period of anoestrus or acyclicity after calving is longer in buffalo than in cattle under comparative management conditions, and under good conditions ranges from 30 to 90 days. However, factors such as nutrition, suckling management and climate (which also influences nutrition through feed quality and availability) influence the duration of post-partum anoestrus. For example, indigenous buffalo in Sri Lanka, raised under harsh conditions with free suckling by the calves, commenced ovarian activity between 150-200 days after calving. Conversely, when raised under free-grazing conditions with abundant natural feed and suckling restricted to once per day, they resumed cyclicity by 30-60 days (Perera *et al.*, 1987).

The male buffalo

Under good feeding and management, young buffalo bulls can be used for breeding from about 26-30 months of age, but sexual maturity or full potential is usually reached about one year later. Although they can reproduce throughout the year, seasonal changes are usually observed in libido and fertility and are related to factors such as rainfall, which influences nutrition, and to temperature and humidity, which influence testicular functions.

Artificial insemination (AI) and cross-breeding

The procedure for collection, evaluation and preservation of buffalo semen for use in AI is similar to that in cattle, with the major difference being in the semen diluents, where Tris buffers, skim milk and coconut water are used more commonly as ingredients (Vale, 1994a). The two main methods of preserving semen are the chilled form (+ 4 °C) and the deep frozen form (– 196 °C). The former can be used for up to three days after processing, while the latter can be used for over 10 years.

AI using semen from improved breeds can be an effective method for rapid gains in milk production, particularly for resource-poor farmers who own draught buffalo and wish to turn to dairying to increase their income. However, the major limitations are lack of awareness among farmers on the advantages of AI and inadequacies in infrastructure for efficient delivery of the service. Addressing these deficiencies will permit the benefits to percolate down to the poorer farmers.

A factor to be considered in cross-breeding is the difference in number of diploid chromosomes in swamp and river types (Table 23.2). The F1 generation will have a proportion of animals with an unbalanced karyotype of 49 chromosomes. Although this does not affect fertility in females, a proportion of males can have lower fertility and such programmes should include measures to detect and cull them.

Reproductive efficiency

Under optimum conditions buffalo can achieve a first service CR of over 65 per cent with natural service and 40-50 per cent with AI, chilled semen usually giving higher CR than frozen semen. In well managed herds, overall annual pregnancy rates of over 80 per cent can be achieved, with CI of 14-15 months. However, in many smallholder systems, particularly where females are also used for draught, the annual pregnancy rates are around 50-60 per cent, with CI of 18-24 months.

In modern dairy cattle production systems the optimum CI is considered to be one year and, since the average gestation length in cattle is 280 days, cows must become pregnant by 85 days after calving. With the longer gestation length of 300-340 days in buffalo, a one year CI is often impossible to achieve. Furthermore, although a short CI is desirable in purely dairy systems, it may not be so where draught power and other outputs are also important. Indeed, in conditions with limited and highly seasonal feed resources, buffalo may calve only every other year. Attempts to reduce this interval could result in losses due to deaths of calves born during unfavourable seasons or to low survivability of the dam herself. Thus a more realistic and achievable set of criteria needs to be developed for such production systems, based on the optimum economic benefits that can be derived by resource-poor farmers using their available resources in a sustainable manner.

Improving reproductive efficiency

Two major factors that need to be addressed to improve buffalo fertility and thereby the economic benefits to smallholder farmers are oestrus detection and post-partum anoestrus (Perera, 1999b). Improving oestrus detection needs education and motivation of farmers, proper identification and records for individual animals and regular close observation for the occurrence of heat signs. Oestrus synchronisation is an alternative that could be applied in some farming systems. However, its use requires a full evaluation of the fertility status in order to select the most suitable hormonal protocol, and careful planning to ensure success.

Prevention of prolonged post-partum anoestrus is best tackled by providing farmers with information on how to manage, feed and breed their buffalo, in order to avoid deleterious effects of harsh climate, seasonal fluctuations in feed availability, parasites and diseases. Under smallholder systems this demands an holistic approach from researchers and livestock support services, working closely with farmers to develop and transfer a package of appropriate technologies that are acceptable to them. The use of hormones and other therapeutic interventions to overcome anoestrus are unlikely to be successful unless these basic constraints are resolved.

Milk production

The milk production capability of river and swamp type buffalo is given in Table 23.2. Cross-bred animals have intermediate yields of 4-8 litres per day. Buffalo cows can continue to produce calves and milk up to around 20 years of age and therefore have a much longer economically-useful life than cattle. In most smallholder systems buffalo are milked by hand, often with the presence of the calf to stimulate milk let-down. In such situations three of the four quarters of the udder are milked and the fourth is left for the calf to suckle. Specialised dairy breeds of buffalo can be trained to let down milk without the presence of the calf and in large commercial operations weaning is done at birth and milking is by machine.

The fat content of buffalo milk (Table 23.2) is higher than in cow milk (see Chapter 7). The concentrations of lactose, protein, tocopherol, calcium and phosphorus are slightly higher, while that of cholesterol, sodium and potassium are lower. Buffalo milk fat is white in colour due to the absence of ß–carotene, which is converted to vitamin A and passed into the milk. These characteristics of buffalo milk make it ideal for processing to products such as clarified butter (ghee), yoghurt and cheese. Ghee is an important component of the human diet in South Asia, while yoghurts are popular in South-East Asia. In Sri Lanka a high proportion of buffalo milk is converted by rural small-scale farmers and cottage processors to a type of natural yoghurt called 'curd', which has a high demand in urban markets, adding value to the milk and providing employment to resource-poor families.

In Latin America, buffalo milk is popular for making cheese and butter. One kg of cheese requires only 5 kg of buffalo milk, compared to 8 kg of cow milk. In this region buffalo milk is used mainly for producing mozzarella cheese, which has a high demand. It offers particular benefits to resource-poor farmers in areas where refrigeration is not widely available, transportation of liquid milk is difficult, and there are seasonal fluctuations in milk supplies.

Meat production

Both river and swamp buffalo are used in many countries as a source of meat, but they are often slaughtered at an advanced age after being used for milk or draught. Hence the quality of meat is invariably poor. However, when raised with adequate feeding and slaughtered at around 2-3 years of age, the tenderness and eating quality of meat is comparable to that of similarly raised cattle, and has the added advantages of containing less fat, cholesterol and saturated fatty acids.

Farmers in Brazil and Venezuela use swamp type bulls to breed dairy buffalo cows and rear the male offspring for meat, attaining a weight of 400 kg at 18-22 months age. In Asia, male calves of dairy buffalo are often neglected and die early in life or are sold for low prices. Resource-poor farmers can obtain added income by rearing male calves under low cost feeding systems, utilising crop residues, tree leaves and farm by-products, and selling them for meat when they reach mature weight.

Work capacity

The swamp type is the major contributor to draught power, but river types and crosses between the two are also used in some locations. The predominant usage is for work on rice fields, where they have an advantage over cattle due to larger hooves and greater flexibility of the lower joints (fetlock and pastern) in the legs. They are used for ploughing, harrowing and levelling of fields before planting rice, for threshing the harvested rice and for hauling carts to transport agricultural produce. Other uses include extraction of oil from copra and oil seeds, drawing water from wells and mixing clay for making bricks or pottery.

In South and South-East Asia the working buffalo is most important in farming systems involving lowland rain-fed rice cultivation. Farmers prefer males to females due to their larger size and higher work capacity, but some also use females, including pregnant animals, when there is a scarcity of draught animals. In South Asia ploughing is done by a pair of animals per plough, whereas in South-East Asia a single animal is used. Farmers who own draught buffalo have an advantage over their neighbours, being able to cultivate fields in a timely manner to benefit from seasonal rainfall patterns. They also earn additional income by hiring out buffalo, with payment being made in the form of cash or produce, usually at the time of harvest.

In Latin America buffalo are seldom used for work, but there is a growing interest in such applications. Studies in Brazil have established that an adult male buffalo can draw a cartload of 500 kg a distance of 25 km in a day and can also pull boats in shallow water. Resource-poor farmers in flooded, muddy and forest areas, and those in the pottery and brick industry in the Amazon valley, are being encouraged to use buffalo for such purposes.

Buffalo are capable of generating a draught force of around 0.5 kilowatt, corresponding to about 0.75 horsepower. A mature buffalo can develop a tractive power equivalent to 10-15 per cent of its body weight. Studies in Thailand show that buffalo ploughing different soil types generate draught forces of 19-32 kg, travel at speeds of 2.7-3.5 km/h and plough an area of 390-800 m^2/h (Skunmun, 2000). In general, two working buffalo can supply the draught needs for cultivating 2 ha of rice per year.

In most situations the methods of harnessing used for buffalo and the work implements are of traditional design based on historical preferences. Some of these harnesses are uncomfortable for the animal and can even cause injury. Studies have shown that improvements in the design of harnesses and implements can increase the speed and output of work, as well as the safety of the animal and the operator (cf. Chapter 25). Such improvements should be tested under different local conditions and proven methods disseminated to resource-poor farmers to facilitate more efficient utilisation of buffalo draught power.

Stress tolerance and adaptation

The density of sweat glands in the skin of the buffalo is only one tenth of that in cattle. Thus the buffalo has a poor capacity for dissipating body heat through sweating. Its darker skin also absorbs more solar heat. The buffalo has therefore developed the strategy of wallowing in water in order to dissipate body heat and maintain homeostasis when environmental temperature is high. It has a relatively labile body temperature, whereby absorbed heat is tolerated until body temperature reaches a critical level, and then the heat is dissipated quickly by wallowing.

Buffalo have a higher rate of daily water turnover (i.e. intake and loss from the body) than either *Bos taurus* or *B. indicus* cattle. This is due to greater loss of moisture from the skin by cutaneous evaporation rather than true sweating, coupled with loss from the respiratory tract. This enables the buffalo to handle larger quantities of roughages of high water content, which are common under humid tropical conditions.

When buffalo are raised under confined housing systems, access to wallowing is often unavailable. In such situations farmers can provide relief from heat stress by sprinkling or spraying water on the body of buffalo. The optimum regime is to sprinkle for periods of 5-10 minutes every hour, on 3-4 occasions during the hottest period of day.

Disease tolerance and susceptibility

Although buffalo are generally regarded as being more resistant to diseases than cattle, there are some conditions to which they are more susceptible. Infestation by the parasite *Toxocara vitulorum*, if not controlled, causes high mortality rates in buffalo calves. Treatment with a simple anthelmintic drug such as Pyrantel, which is also used in children and hence available in most village pharmacies, controls the parasite. Treatment must be done at the critical stage of 10-14 days after birth. Growing and adult buffalo are particularly susceptible to haemorrhagic septicaemia, which is an acute disease caused by the bacterium *Pasteurella multocida*. It causes respiratory symptoms and high mortality, particularly in animals that are stressed or debilitated. Foot and mouth disease affects buffalo in much the same way it does cattle, causing mostly distress, loss of condition and inability to work, but not high mortality. Both diseases can be prevented by vaccination, which has to be done at least annually. Other diseases that are common in buffalo include brucellosis, tuberculosis, trypanosomiasis, filariasis and mange. Buffalo are less susceptible than cattle to mastitis, foot rot and metritis.

Major production systems in different regions, their problems and limitations

Buffalo are reared in different regions and countries under a wide variety of environments and management systems, either alone or in combination with other on-farm and off-farm activities (Cockrill, 1974, Mahadevan, 1992). The farming system is determined by a matrix of several interacting factors. These include climate (tropical or temperate, humid or arid), location (rural, peri-urban or urban), integration with cropping systems (rain-fed or irrigated, annual or perennial crops), type of operation (small or large farm, subsistence or commercial), and the primary purpose (milk, meat, draught, capital or mixed). These farming systems have been classified in various ways (FAO, 2001) but, for the purpose of this chapter, the four basic systems described in Chapter 3 will be used: (a) landless (rural, urban and peri-urban); (b) crop-based mixed (annual crops and perennial crops); (c) agro-pastoralist; and (d) range-based.

South Asia

This region has 72 per cent of the world buffalo population (Table 23.1) and is the home range of the best known buffalo breeds. The majority of animals are of the river type, with some swamp types and crosses. The landless and crop-based mixed farming systems are the most common, while range-based systems exist in the more arid areas. The majority of farms are small (1-5 animals) and at subsistence level, but medium sized (5-20 animals) semi-commercial farms and some large specialised dairy operations are also present.

In India and Pakistan the buffalo is the major producer of milk. In the irrigated and rain-fed areas with intensive crop production, resource-poor farmers keep a few improved dairy buffalo under confinement, with stall-feeding on cut grass, crop residues and concentrates. The latter are based on oilseed cakes, coconut cake (poonac) and bran from rice, maize and wheat. In areas with less intensive cropping, farmers keep larger herds under extensive management, with grazing on communal lands supplemented by crop residues. Milk production from buffalo is also important in Nepal and the southern part of Bangladesh. In most other regions of Bangladesh and in Sri Lanka the majority of buffalo are nondescript indigenous animals, raised in mixed farming systems. Resource-poor farmers use them mainly for draught power associated with the cultivation of rice and other annual crops. Management is usually semi-extensive with tethered grazing or extensive with grazing on communal lands. A few systems integrating buffalo with perennial crops such as coconut also exist.

In mixed farming systems the major problems faced by resource-poor farmers are shortage of land and feed resources. Pressures from growing human populations and crop production have resulted in decreased herd sizes, forcing farmers to adopt more intensive systems of production. Intensification requires more inputs, particularly in terms of labour, and will be successful only if economic returns are worthwhile. In the case of specialised dairy buffalo, animals with high genetic merit have been bred within institutional farms and disseminated through AI to farmers. However, this genetic potential is not fully exploited in many situations, due to inadequate nutrition and management. Farmers in these situations require knowledge and skills on proper utilisation of crop residues and other locally-available feed resources in order to be economically viable.

Areas around some major cities in South Asia have large herds of buffalo raised purely for milk production under a unique intensive landless system, where all feed is bought and animals are in total confinement. High producing cows are purchased from rural areas and brought in to such systems, kept for the duration of their lactation and then sold. These are usually controlled by businessmen and do not benefit resource-poor farmers.

South-East Asia

This region has 22 per cent of the world buffalo population. The swamp type predominates and is reared mainly by resource-poor farmers engaged in mixed crop-livestock husbandry. Both irrigated and rain-fed mixed systems, as well as range-based systems are common, with tethered or free grazing on communal lands and confinement near the homestead at night. Little or no supplementary feed is provided, except rice straw and leguminous tree leaves during periods of drought. The buffalo has long been part of the rural socio-economic structure of this region and many mixed smallholder systems exist, involving the integration of crops with one or more livestock species (cattle, buffalo, goats, sheep, pigs, chickens or ducks) and fish farming (see Chapter 3). On larger holdings perennial crops, such as coconut and oil palm, are integrated with cattle or buffalo.

The problems faced by resource-poor farmers in this region include low genetic potential of indigenous buffalo for growth and milk production, seasonal shortages in feed and poor marketing channels for livestock products. Most farmers therefore keep buffalo under a minimum-input system with low use intensity. They need advice and services for genetic improvement and skills to feed and manage improved buffalo, in order to derive cash income through milk production and raising surplus animals for meat. In peri-urban and urban areas, where a demand for milk exists, breeds of the river type have been established and cross-breeding the swamp type is being done to obtain dual- or even triple-purpose (dairy/draught/meat) animals. Good examples of resource-poor farmers moving from subsistence to sustainable production are seen in such areas, where cross-breeding has been followed-up with improved feeding and development of marketing channels, leading to thriving dairy industries. Small-scale 'backyard' fattening units for male buffalo also exist, where resource-poor farmers use cut grass, crop residues and by-products such as straw, banana stems and soyabean waste for fattening small batches of buffalo calves each year. These systems have the potential for replication by other resource-poor farmers, through provision of information, support services and improved marketing.

Latin America and the Caribbean

Buffalo production has expanded dramatically over the past decade in this region, particularly in areas with poor soil or periodic flooding, where they are able to thrive on conditions which even zebu cattle cannot cope with (Vale, 1994b). Production systems vary depending on the purpose, but herd sizes are generally large compared to those in Asia. Beef production is usually under extensive ranch conditions, while milk production is under confinement or semi-extensive systems. Resource-poor smallholder buffalo farmers exist in many countries of this region and their main problems are limitations in land, forage and pasture. An example of how they can be assisted to intensify buffalo production and earn higher income is given in the next section (see Latin America).

Photo 23.3 Buffalo cows capable of producing up to 4,500 kg milk per lactation in Brazil (courtesy of W. Vale)

Comparative advantages and potential for improvement (moving from subsistence to sustainable production)

South and South-East Asia

Mixed crop-livestock farming is the predominant means of livelihood for the rural poor in this region. Whereas income from crops is seasonal and often associated with risks of failure due to uncertain weather conditions, livestock provide a regular source of income, minimise risks and help overcome the rural agricultural debt cycle. Furthermore, due to the ability to survive under harsh environmental conditions with minimal inputs and provide a range of economically-important outputs, buffalo offer opportunities for harnessing family labour and biomass that would otherwise be unutilised. In order to move from subsistence to economically viable enterprises, resource-poor farmers need the knowledge and skills required to utilise these resources optimally and sustainably. This involves the development and testing of appropriate improved methods for managing, feeding, breeding and disease control, together with procedures for value addition and marketing of the products.

A successful example is available from the predominantly mixed crop-livestock smallholder farming system in the rain-fed and irrigated rice growing regions of Sri Lanka. Although rice is the major source of income for resource-poor farmers, buffalo play a vital complementary role through provision of draught power. In the past they were managed extensively with minimum inputs and fed on communal scrub lands. The sustainability of this system has been threatened due to intensification of land use for rice and other crops. Intensified dairy production under these conditions, on the other hand, is uneconomical due to the unfavourable ratio between farm-gate price of milk and cost of compounded feeds. A project was therefore conducted, initially using Participatory Rural Assessment (PRA), to identify constraints and opportunities, followed by a series of steps undertaken with selected groups of farmers using the Farmer Participatory Approach (FPA), to develop and test potential improvements, their costs and benefits (Abeygunawardena *et al.*, 1996). This resulted in the development of a model for 'Smallholder Intensively Managed Buffalo Units' (SIMBU), incorporating improved technologies and greater integration with crops. The model was then expanded to other farmer groups and refined based on location-specific needs. The characteristics of SIMBU included the following:

- Each unit comprised of 3-4 adult buffalo and their calves, either pure-bred (Murrah, Surti, Nili-Ravi) or cross-bred, with milk yields of 4-6 litres per day
- Low-cost housing with hygienic floors, feed and water troughs, and storage space in the loft for crop residues
- Feeding based on grass and leguminous tree forage, the latter grown as alleys or as perimeter fence
- Utilisation of rice straw supplemented with urea molasses multi-nutrient feed blocks (UMMB)
- Cooling by sprinkling of water as an alternative to wallowing
- Improved oestrus detection, timing of AI service and timely pregnancy diagnosis
- Suckling of calves restricted to twice a day, at milking time only
- Improved health care of calves, including strategic deworming and vaccination
- Body condition scoring using a 1-5 scale, as an aid to appropriate feeding decisions
- Hygienic milk production and value addition by conversion to yoghurt (curd) and other products

- Recycling of animal and crop wastes to provide mutual benefits
- Maintenance of simple farm records and their use for decision making.

The total cost of establishing a SIMBU, including construction of housing and purchase of buffalo, was US$ 1,000. The average daily milk yield per household was 4.3 litres, which if sold in liquid form would yield US$ 1.00. However, when converted to curd the income increased to US$ 2.00 per day. Since the cost of family labour and energy for heating the milk using firewood was zero, this supplementary income was sufficient for meeting the daily cash needs of the family. The average herd size doubled within four years, resulting in a gain in capital assets which offset the initial investment for the SIMBU. Other benefits included use of non-lactating buffalo for draught in rice fields and use of dung as manure in home vegetable plots.

Latin America and the Caribbean

Although the major expansion of buffalo production in this region has been in the larger more commercially-oriented farming systems, the situation in some areas, like the Amazon valley, Chaco region of Bolivia and Paraguay through the Paraná river and Llanos of Orinoco Basin, as well as in other flooded or savannah areas, is quite different. New strategies for sustainable development using buffalo as a suitable animal in these areas must be focused on smallholder farmers. In some areas of Brazil, State Governments have provided special credit to resource-poor farmers for the purchase of a herd of 10 females and a male. The repayment is done in the form of animals rather than cash, commencing three years after receiving the animals. This consists of four animals in the fourth and fifth year and three in the sixth and seventh year. The project aims to support the development of the water buffalo industry, particularly through a breeding programme to enhance the potential for meat, draught power and milk production for smallholders. Such programmes are suitable for extension to other countries in the region, using national and international involvement.

Suggestions for development and contribution to poverty alleviation

The majority of traditional farming systems which have evolved over the past millennium in Asia have used indigenous genotypes which are well adapted to the environment and can reproduce, as well as provide commodities, at a level which is in harmony with the available resources. However, higher demand for livestock products coupled with decreasing farm land and communal grazing areas have necessitated profound changes to traditional farming systems in many regions (Seré and Steinfeld, 1994). Livestock farmers have usually responded to these pressures, particularly in urban and peri-urban areas, by intensifying their production systems (Ghirotti, 1999), often without adequate assessment of the costs to natural resources and long-term sustainability. These emerging farming systems pose a challenge as well as an opportunity to research workers, to develop appropriate methods to assist resource-poor farmers to move from subsistence to semi-commercial and sustainable enterprises. The challenge is to determine how the limited resources available to poor farmers, which include indigenous and improved genotypes, underutilised family labour, locally-available feed resources and access to support services and markets, can be optimally utilised to obtain higher outputs of livestock products (Perera, 1999a). A further aspect that will inevitably become increasingly important during the next decade is the quality and safety of livestock products, which will be dictated by the needs of more discerning consumers and the growing international trade in these commodities (cf. Chapters 7 and 8).

Meeting these challenges will require interdisciplinary adaptive research, undertaken with full participation of all stakeholders including the farmers and their service providers. The essential steps involved in such initiatives are now well documented (Abeygunawardena *et al.*, 1998) and include:

- Identification of strengths and weaknesses of the specific production system
- Formulation and on-farm testing of appropriate improvements and interventions
- Extension and dissemination of proven technologies in the form of an integrated package.

Research and development strategies based on this approach have been promoted by international and bilateral donor agencies during the past decade and have resulted in the availability of valuable baseline information on livestock farming systems in Asia, Africa and Latin America. The need now is to use this information, together with recent knowledge gained from technological developments in advanced countries, to the benefit of a wider range of resource-poor rural farming communities. This will require working with the farmers, understanding their needs and limitations, improving their knowledge and skills, and assisting them to develop and apply their own solutions.

Further reading

Cockrill, W.R. (ed.). 1974. *The husbandry and health of the domestic buffalo*. Food and Agriculture Organisation of the United Nations (FAO), Rome, Italy.

FAO. 2001. *Mixed crop-livestock farming: A review of traditional technologies based on literature and field experience. FAO Animal Production and Health Paper 152*. Food and Agriculture Organisation of the United Nations (FAO), Rome, Italy. (Full document available at: http://www.fao.org/DOCREP/004/Y0501E/Y0501E00.HTM)

Ranjhan, S.K. 1998. *Text book on buffalo production. 4th edition*. Viskas Publishing House, New Delhi, India.

Tulloh, N.M. and Holmes, J.H.G. (ed.). 1992. *World animal science, Volume. C6: Buffalo production*. Elsevier, Amsterdam, the Netherlands. (Contents page available at: http://www.elsevier.com/wps/find/bookdescription.cws_home/522234/description#description).

Website containing information on water buffalo available at: http://ww2.netnitco.net/users/djligda/waterbuf.htm

Author addresses

Oswin Perera, Formerly of Animal Production and Health Section, Joint FAO/IAEA Division of Nuclear Techiques in Food and Agriculture, International Atomic Energy Agency, Wagramer Strasse 5, P.O. Box 5, A-1400 Vienna, Austria. Current address: 40/9 Deveni Rajasinghe Mawatha, Kandy, Sri Lanka

Harischandra Abeyguawardena, Faculty of Veterinary Medicine and Animal Science, University of Peradeniya, Peradeniya, Sri Lanka

William Vale, Institute of Animal Husbandry and Health, Amazonian Federal Rural University-UFRA Belém 66.077-530, Pará, Brazil

Charan Chantalakhana, Suwanvajokkasikit Animal Research and Development Institute, University of Kasetsart, Bangkok, Thailand

24

Yak

Gerald Wiener[*], *Han Jianlin*[‡]

Cameo - Qinghai-Tibetan Plateau, China: a livelihood from yak on 'the roof of the world'

I am Nyima Tsering. My wife, Gesang Zhuoma, and our family herd about 70 yak, 25 of them cows, and a small number of sheep, in Zenqin township of Chengduo county of Qinghai, China, at an altitude of 3,780 m.

Life was harder than usual last year. Because of a dry summer there was not enough grazing for the animals to get into good condition before the six-month-long winter. And then for the first time in four years we had a lot of snow for several weeks and we lost 22 yak, including 10 of the cows. We never have much hay for the winter, as it is difficult to make. We keep hearing about feed blocks, but we could not afford to buy them. With the cows in such poor condition we had even less of such little milk as they give in winter - milk that we keep mostly for the children. But we did not go hungry because there was still enough stored meat.

This year, green-up began late, at the end of April, and the herd ate the ground bare before we could move them. Zhuoma relies on the milk to make butter and cheese. We use milk in milk tea and usually sell a little butter locally - but with fewer cows this year there may be almost none to sell. I wonder if the people from the College think I might obtain, or breed, a couple of hybrid yak from those black and white cattle - it would get us some more butter to sell - but I bet they won't do well up here.

Milking needs an early start when the cows are still grazing near our campsite. We keep the calves enclosed overnight for safety and warmth but

[*] Principal author
[‡] Consulting author

also to prevent them taking the milk from the cows. At the start of milking we have to let them out because the cows won't let down their milk without their calf starting to suckle. And then it's a day of driving the herd up to the tops of our range to graze and later down, keeping our eyes open for wolves, before the next milking at night. We might finish by 9 o'clock.

We do have a bit of fun each year at the county festival, especially the horse racing. But Zhuoma will be disappointed this year if we can't afford some new jewellery for her to show off to her friends. My two sons want more than just to continue with such a harsh life, but that is all they have ever known and they have not been given the chance to learn any other skills. We are not alone in this; we'll have to decide about our future.

Background

The vast rangelands of the Qinghai-Tibetan Plateau, the 'roof of the world', and adjacent regions, as well as the interspersed alpine areas of high mountains and deep valleys, support mostly livestock production. Livestock is the mainstay of life for the people, and the yak, a species of bovid, is the single most important animal. The yak provides for most of the essential needs including milk, meat, clothing, tents, transport and, of vital importance, dung for use as fuel. Sheep and goats have a supporting role in the economy, to a greater or lesser extent depending on the climate and topography of the area. Yak keeping, however, is also a way of life for the herders and bound up with their traditions and culture. Much yak keeping is at a subsistence level. Some governmental policies and interventions aim to improve living conditions through settlement within a largely nomadic society and to increase economic returns from the livestock. It remains to be seen how successful some of these policies are in securing a future for the people and a sustainable exploitation of the natural resources of the rangelands.

The Qinghai-Tibetan Plateau of western China extends over 250 million ha. Of the 14 million or so yak in the world, at least 13 million are on the plateau and in the mountainous areas of western China, about half a million in Mongolia and most of the rest in countries bordering on the Himalayas and to the north into Russia. Small pockets of yak are found in other countries of Western High Asia. Yak keeping is therefore associated, for the most part, with harsh climatic conditions and short growing seasons at high elevations. At somewhat lower elevations and in better conditions, hybrids of yak and cattle are also kept.

In the context of this book it needs to be asked firstly, and most importantly, what might be done to raise the productivity of the animals and the living standards of the people. A second intriguing question is whether the special qualities of the yak, which, among large ruminants, is almost uniquely adapted to these harsh conditions, may offer something useful for other regions of the world that have similar environmental conditions but where yak do not at present occur. Small populations of yak in other parts of the world, notably North America, provide a useful insight into the adaptability of yak to different conditions and to their potential performance. Thirdly, as will appear later, there are attributes of the yak which, if genetic, might in the future become transmissible to other bovine species.

An overview of yak production and of the characteristics of the animals is a necessary preliminary to discussing the prospects for improvements and is outlined in what follows. However, much of the detail, with the supporting evidence, and a large bibliography of published literature relating to yak and their management and environment, can be found in a recent book (Wiener *et al.*, 2003).

Environment

Typically, yak are kept at elevations of between 1,000 and 4,500 m, with the lower elevations at the more northerly latitudes (Russia). On the Qinghai-Tibetan Plateau elevations are generally above 3,000 m and in Mongolia between 1,500 and 4,000 m. The climate is cold and harsh, with short growing seasons of around 100 days followed by some 3 to 4 months when wilted herbage provides a near-adequate but decreasing supply of feed for maintenance. Then, from late winter into early spring, shortage of feed and variable snow cover lead to frequently severe malnutrition and weight loss of the animals. There is said to be no totally frost-free day at any time of the year.

Climate and topography have played major roles in determining the soil types and vegetation of the rangelands. At one extreme is desert steppe (rainfall around 50 mm per year) with *Ceratoides* spp. predominant. At intermediate rainfall (precipitation around 200 mm) are areas of alpine steppe with a preponderance of *Stipa* spp. among the grasses. In areas of highest precipitation (up to 600 mm per year), alpine meadows support sedges and low shrubs along with other vegetation. These factors influence stocking densities, management systems and the relative importance of yak to other species, such as sheep and goats. There is a lack of legumes in the alpine vegetation. A detailed account of the rangeland ecosystem and its management is given by Ruijun (2003a).

Breeds and hybrids

In 1766 Linnaeus classified the yak as *Bos grunniens* (because of the characteristic grunting noises made by the animal). Later, several investigators listed the yak as a separate sub-genus, *Poephagus grunniens,* on account of its morphological distinctions from other cattle and from bison, and this classification is supported by recent molecular evidence (Jianlin, 2003). Both classifications, however, remain in use.

There are two distinct populations of the yak species - the wild yak and the domestic yak. Only 15,000 wild yak were thought (in the 1990s) to remain in remote mountain regions, from among the millions of former times, because of hunting and the encroachment on its territories by humans and their beasts (Schaller, 1998). The wild yak are larger and more ferocious than their domestic counterparts, but it is not clear to what extent this difference is genetic - though there is some evidence to that effect from the allegedly better growth and performance of crosses between wild and domestic yak.

There are 12 recognised breeds of domestic yak distinguished on the basis of colour, conformation, history and distribution, among other factors. The best known and numerically most important of these are the Jiulong, Maiwa, Tianzhu White, Plateau and Huanhu. In other countries yak are generally referred to by the name of the area where they are found or have originated. Most of the named breeds live in different parts of the vast territories and rarely mix or interbreed. So, apart from obvious characteristics of colour or strongly inherited morphological traits, there is no clear evidence of how much these breeds differ in fitness or performance traits. This consideration has important implications for potential genetic improvement strategies.

For centuries hybrids of yak and other cattle have been produced at altitudes intermediate between the highest occupied by pure yak and the lower-lying territories where other cattle predominate. These hybrids were, and still are, for the most part created between the local cattle and the local yak and can be produced reciprocally - although the use of cattle males on yak females is more common. The hybrid males are sterile and cannot be used for further breeding, but the females are valued for what is generally found to be better performance (size, milk production, etc.) than either the yak or the local cattle. From a strictly genetic point of view, doubt remains about how much of this superiority is due to actual heterosis (hybrid vigour) (chapter 7 in Wiener *et al.,* 2003). This matter is of critical importance in any consideration of breeding plans involving these animals and has particular relevance when assessing the greatly improved performance from hybrids of yak and 'improved' exotic breeds of cattle (for example the Holstein or the Simmental). Such 'improved' hybrids have been promoted since the late 1930s. These hybrids are larger and the females produce much more milk than the parental yak. Two factors, however, prevent them from playing a role in the economy of all herders. Firstly, the 'improved' hybrids need better treatment and a better environment than the typical yak. Secondly, they have to be produced by artificial insemination of yak females with semen from the chosen cattle breeds. This makes its own technological demands and requires rapid access to yak cows in heat, which often is difficult in remote areas.

The reciprocal local hybrids are generally given different names (varying with country) and there is an absolute plethora of names for the various types of back-cross hybrids (of local cattle and yak) that can be produced, but rarely are. Performance of back-crosses declines relative to the first generation hybrids and for that reason they are not much favoured in practice (chapter 7 in Wiener *et al.*, 2003).

Characteristics of yak and performance

Typically, adult females (six years old and above) weigh around 200 to 300 kg, depending on breed, at the end of the summer grazing season and perhaps 25 per cent less at the end of winter. Bulls are substantially heavier. Calves at birth, after a gestation period of only around 258 days, weigh about 5-7 per cent of the maternal weight.

In their native environment most yak females do not show oestrus before two years old and then generally only once a year. A majority of the cows calve first at four years old. Calving once in two years is the norm but two calves in three years is not uncommon. Four to five calves are produced in the average lifetime. Twins are very rare.

Castrated males are not generally slaughtered before the age of four years and then at the end of the summer grazing period when they are in their best condition. By that age the animals have passed through several cycles of seasonal weight gain and loss. This should be a consideration in the overall efficiency of the management systems and in assessing opportunities for change. In areas where meat from yak is the main, or only, product, slaughter at earlier ages is preferred and a different rearing system is needed. There is some documented and some circumstantial evidence that the cost of producing meat from yak is less than that from other cattle or from sheep. Smirnov *et al.* (1990), reporting on yak in the Caucasus, quote a tenfold lower cost of a unit of live-weight gain for yak, on account of the yak's almost exclusive use of mountain pastures relative to more costly feeds needed by cattle. Katzina (2003) notes that the cost of producing meat from yak in Buryatia and Yakutia is about a third of that from other cattle, whilst several ranchers keeping both yak and cattle in North America claim (personal communications reported by Wiener, 2003) that yak have a better efficiency of feed conversion. It is not known whether there have been further comparative studies to define more precisely the reasons for the reported species differences in the economics of meat production.

Lactation is seasonal and averages from 150 to 500 kg (varying with breed and location) with a fat content between 5.4 and 7.5 per cent. Cows rarely dry off completely during the winter and lactation can recover in a second year without calving again. Such a second-year lactation produces between one half and two thirds of the quantity of milk of the first year (and is popularly termed the 'half-lactation') but with a higher fat percentage.

Yak have an outer coat of long hair and in winter (and from birth in calves) a fine undercoat of down hair that has a quality similar to that of cashmere. Total fibre yields vary with breed and location from 0.5 kg to 2.9 kg. The down represents from 60 to 70 per cent of the total fleece in calves, but can decline to as little as 20 per cent in mature animals.

Yak are susceptible to most of the diseases reported for cattle and none that is unique (Dorji *et al.,* 2003). Incidence of some diseases tends to be high because of lack of preventive measures due to cost. In addition, mortality from starvation and near-starvation over winter, and particularly in periodic 'snow disasters', is a hazard of the region. Wild predators, especially wolves, also cause losses, but herdsmen try to prevent these through their herding methods.

The yak is adapted physiologically and in terms of physical characteristics and temperament to high altitude (low oxygen), cold, shortage of feed over winter, precipitous terrain and danger from predators (chapter 4 in Wiener *et al.,* 2003). However, the presence of yak herds in some other countries, even at quite low elevations and in temperate climates, suggests that yak can also be more versatile in their environmental tolerance than is traditionally assumed.

Products from the yak

The relative importance of milk and meat production from the yak varies between countries and districts and with market demands, though in many areas milk products provide the largest economic return. The fleece has, in general, less economic importance, but exploitation of the down fibre for textiles is receiving increasing attention and will be referred to again later. As is well-known in yak circles, nothing from the yak is wasted and by-products, including the hides, the viscera and the blood, find many uses (chapter 10 in Wiener *et al.,* 2003). Traditionally, the yak has had an important role as a pack and draught animal and for riding (though this role is in decline). Special mention must be made of the fact that in the economy of yak-keeping, most especially on the Qinghai-Tibetan Plateau at these tree-less elevations, the dung is the main, and often the only, source of fuel. This, perhaps as least as much as the other attributes of the yak, allowed the development of human life and of civilization in this relatively remote region of the world.

Photo 24.1 Pack yak, near Xia La Xiu township, China (courtesy of P. Horber)

Milk as a raw product is used primarily by the herdsmen and their families in 'milk tea'. The main product from milk is butter, used both as food and for other purposes, including its use for oil lamps both in the family and in temples, and for temple sculptures. Different types of soft cheeses and yoghurt are made mostly for local consumption. In Nepal a Swiss-type hard cheese has been produced for some years in small factories located in the yak territories (Joshi *et al.*, 1999). This cheese finds a market among tourists in particular.

Meat from animals slaughtered mostly before the onset of winter is preserved for consumption dried, frozen (in nature's 'deep freeze') or salted. It is also made, mostly locally by the herdsmen, into different types of sausages and other products (chapter 10 in Wiener *et al.*, 2003). Most slaughtering is done locally and the meat and products consumed by the family or locally. Little finds a ready market in towns and cities. By contrast, in Mongolia, where there is a large demand for meat in urban areas, emphasis is placed on slaughter of yak in approved slaughter houses. This trade and that in dairy products from yak are constrained, however, by the fact that markets are often far distant from the areas of production and herdsmen may need to trek as much as 1000 km to reach the markets and slaughter facilities (Magash, 2003).

Photo 24.2 Milking a yak in Tibet (courtesy of O. Hanotte and ILRI)

Management

Traditional management follows a transhumance system ruled by season, climate, topography and socio-cultural factors. In summer and early autumn herds are kept on pastures at the higher elevations and moved as necessary, depending on the availability of grazing. In winter and early spring yak are kept nearer to the permanent homes of the herders at lower elevations. Various types of shelter are provided, especially for calves (Chapter 8 in Wiener *et al.,* 2003).

In several provinces of China, things have changed and are continuing to do so through a policy of 'Household Responsibility', whereby herders own their animals and have rights (though not ownership) to parcels of rangeland. Some of this rangeland is now fenced through government intervention and support. Accordingly, animals of several families are less often herded together than formerly and movement of the animals across summer and winter pastures is more constrained. Some of the incentive for these changes and the benefits from them are social, but it has yet to be shown whether the new systems are equally effective in utilising the natural resources of the rangeland to best advantage.

For most yak the naturally existing vegetation is the only, or almost the only, feed available. In summer, following green-up (typically around April but possibly later) animals gain weight rapidly. In late winter and early spring animals can be close to starvation and deaths are common, especially in years of heavy snow. Supplementary feeds, such as hay or crop by-products, are not widely or sufficiently available but efforts to improve that situation are widespread and the use of urea-molasses and multi-nutrient blocks is finding some acceptance (Ruijun, 2003b). But yak are not readily compliant. Yak may reject supplementary feeding (even hay which is the most favoured) if adequate amounts of wilted herbage remain accessible on the ground in winter, or during green-up in early spring. This may occur even though intake falls short of requirements. However, this is not the experience in non-traditional areas of the yak.

Mating normally occurs in the autumn. In traditional practice, the bulls run in groups and fight for possession of cows. The strongest, generally older, bulls get most mates. Because the successful bulls are likely to be followed in due course by their sons in the same herds, there is an inbuilt potential for inbreeding, unless counter-measures are taken. Inbreeding reduces nearly all aspects of performance. Controlled breeding practices, including the use of artificial insemination, have been introduced in some cases and, of necessity, for producing hybrids with exotic cattle breeds and for crossing with the wild yak. In the very different environments of North America and Europe, year-round breeding of yak has been recorded.

Calves are born in the spring and early summer. Their mothers are generally in poor or very poor condition at that time after their often serious weight loss over winter. Calves thus have a difficult start to their life. Those born late in the year or those that have not put on sufficient weight by autumn may themselves not survive the following winter. April and May are generally the peak months for calving, but this will vary with location.

In the areas where milk is the most important product for the herders, the calves are allowed to suckle all the milk of their mothers for only the first two to four weeks after calving. Thereafter calves are given only limited access to their dam's milk by keeping them apart from the cows for long periods daily. Cows are tethered during milking and the calf is allowed to suckle briefly to stimulate milk let-down. After the hand milking of the cow, the calf is allowed to strip out the remainder. This practice restricts calf growth. Calves grow and survive best when allowed constant access to their dams. Meat and milk production, therefore, compete. This consideration is especially important in areas where meat production is, or could become, a major aim.

The grazing habits of yak are unusual in that they can graze long grass like other cattle by wrapping their tongue around it and can graze short grass, roots and creeping vegetation in the way of sheep using their incisors and lips. Nonetheless mixed grazing by yak and sheep appears to utilise the grazing lands to better advantage than with single species stocking. Sheep are the most common companion grazing animal alongside yak but the proportions of yak to sheep vary with climate and topography. Yak represent the lowest proportion of the total livestock population in the driest, coldest areas. Goats also are companion grazers, but to a lesser extent than sheep. Wherever there are herders there are horses, for riding and for enjoyment, and these, alongside a variety of wild species of grazing animals, add to the pressures on the rangelands from the domestic livestock. Increasing demands on the rangelands and some inappropriate grazing practices, superimposed on the disequilibrium between the supply of feed and the requirements of the stock, have led to significant degradation of some of the rangeland (Ruijun, 2003a).

Social and cultural considerations

Yak keeping in the traditional areas is intertwined with the social, cultural and religious life of the people. As described by Wu Ning (2003), complex forms of social organisation have developed in yak-keeping societies which impinge on the utilisation of the rangelands and influence herding practices. Not least among the social factors is that yak represent a source of wealth (social capital) and insurance for the owner, and because of the use of yak as a dowry when a herder's daughter marries. There is thus an incentive to keep larger numbers of yak than might be appropriate by production criteria alone.

Constraints and opportunities for change

Constraints

Constraints on any improvement of animal performance or of income, in cash or kind, from the land include the harshness of the environment, the seasonal nature of herbage availability with a prolonged period of feed shortage each year, and the relatively low productivity and reproductive rate of the animals themselves. This is compounded by the effects of large seasonal weight gains and losses and traditionally sub-optimal herd structures (a less than optimum ratio of the milking cows to other yak stock, delaying the slaughter of steers to relatively mature ages, etc.). Any change of system or its components needs also to be considered within the traditional context of yak keeping. In one sense, this, in itself, represents a constraint. Moreover, it has to be established which opportunities for change provide a worthwhile return for any extra investment.

But perhaps the single most important constraint is that of the sustainable carrying capacity of the rangelands. In that context there is an argument, though it may be a counsel of despair, that any increase in livestock production is undesirable as it would lead to further degradation of an already fragile ecosystem. Large increases in stocking rates, following the private ownership of the stock (in China), coincided initially with the retention of communal grazing, which created a further potential for serious overstocking of rangeland - although traditional property arrangements among herders provided some control (Wu Ning, 2003). The more recent allocation of land to individual herders may have brought these problems into even clearer focus for the herders themselves. Recognition of the overriding problem of rangeland management by herders and scientists alike, the development of appropriate stocking and grazing strategies and the parallel development of cropping and supplementary feeding, may yet turn the tide toward greater economic returns for the herders from their livestock, whilst restoring the integrity of the rangelands. Increased output without increased stocking should be the aim wherever productivity of the rangeland is at its limit.

Intervention

What are some of the opportunities for intervention? The environment cannot be altered, but its worst effects might be. Additional shelters might be provided in winter but this is limited by the construction materials available, not only by cost. Providing more winter feed appears as an obvious candidate. However, making hay from natural pastures, and during a moist summer, is difficult, as is the setting aside of grazing land for that purpose. The introduction of artificial pastures for winter feed provides some hope. The use of forages is feasible close to agricultural areas but growing of, for example, oats or other cereals for forage, may become more widespread as cold resistant varieties, adapted to short growing seasons, are developed (cf. Ruijun, 2003b). Feed blocks as a supplementary source of winter feed are becoming available, but the costs and returns will be uppermost in the thoughts of herders. As discussed earlier, however, account must be taken of some apparent reluctance by yak to accept supplements.

Tradition and change

Many traditional practices may already represent 'best practice' for the area, while others act as constraints on change. The traditions are a signal that the pace of change needs to be measured. Some of the arguments for or against change are circular. For example, if the number of animals owned is regarded as conferring status and wealth, it increases pressure on the grazing lands with the risk of serious degradation already referred to. Greater stocking density also reduces individual animal productivity and, eventually, herd output. Culling the less productive animals and slaughtering steers at younger ages may appear obvious solutions but involve marketing less weight of steer and hence less immediate return in cash or kind. The expected extra productivity and more calves from cows that have to compete less for feed, because of lower stocking, is less immediate but more long-term in its effects. The herder may also feel that extra animals provide some insurance against heavy losses in, or after, particularly harsh winters, quite apart from the status argument. On the other hand, overgrazing leads in turn to animals in poor condition that are more prone to die. 'Problems' and 'solutions' are caught in a vicious circle.

Grazing patterns to minimise overgrazing require rotational access to the grazing lands and the transhumance system allows for this. It remains to be seen whether the allocation of 'private' parcels of formerly communal grazing lands, in some parts of the region, and the fencing of land for private use, may not run counter to the optimum use of the rangeland. The arguments were well rehearsed by Ruijun (2003a). Even well-intentioned proposals for intensification of milk production for use in processing plants or in urban areas, carry a risk that this may place undue strain on the winter pastures as these are, in general, the closest to the roads for the collection of milk. A scheme involving such measures is being introduced by a large commercial concern in China to at least one area where a quite large yak population is relatively easily accessible by road. Evidence of sustainability and of economic viability of such 'industrialised' yak production should therefore become apparent in future years.

Low productivity

The relatively low productivity of yak is without doubt a function of the environment, particularly in relation to growth and reproductive rate. The much greater output in some non-traditional environments and management systems provides some of the evidence for this assertion. There are no equivalent clues in respect of milk yield because in more intensive rearing areas, where meat is the principal output, the milk is used only for calf growth and not directly measured. But as a consequence of the greater calf growth, first calving occurs earlier with more regular calving thereafter and lower mortality, as well as an earlier age at maturity.

Genetic change

The extent to which genetic selection strategies, well established in specialised cattle breeds, could be applied to yak is largely a matter of organisation and infrastructure, but the problems in remote areas would be substantial. Cross-breeding with wild yak is claimed to provide a benefit but certainty that hybrid vigour is involved remains to be proven. The counteracting of potential inbreeding in yak herds is a likely benefit, but this alone could also be achieved by more controlled breeding practices and by keeping mating records.

The production of hybrids with cattle, improved or otherwise, is constrained by the environment in which the hybrids are expected to live, but in the right conditions they provide more milk and meat than pure yak, or the local cattle in these areas. Hybrids also make good draught animals. But a system of hybridisation can become complex as the male hybrids are sterile and back-crosses are less productive. There is also a limit on the extent to which the hybrids can be produced, which is set by the reproductive rate of the yak females. In theory, only females surplus to the need for the pure yak population to reproduce itself are available for hybrid production.

Infrastructure

An obvious but little exploited way of improving living standards of relatively poor yak herders is to improve the infrastructure of the region and the access to markets, so as to allow value-added products from the yak to be produced, provided these are not allowed to increase grazing pressures. The establishment of cheese factories in Nepal is an example of such a development. A prospect might be to improve the harvesting and processing of yak down for the production of valuable cashmere-type garments which fetch high prices in western markets but bring, at present, relatively little reward to the local community since the processing and manufacture is far removed from the villages. Major improvements in hide quality (preventing parasite and other damage) and processing could improve income. Improved slaughter facilities and access to them, to supply a niche market in yak meat for urban areas, is another possible way of bringing cash flow to yak herders, as well as the commercialised marketing of milk referred to earlier.

Research and development

Research and development work with yak proceeds at various centres in China and most other countries with significant numbers of yak. But relatively little of it is co-ordinated across centres and a large proportion of the work is small-scale or concerned with micro matters. Production-oriented investigations on a statistically adequate scale are inevitably more costly and need more facilities, for example the evaluation of breed differences, the effects of cross-breeding among yak breeds, the value of the effects of alternate grazing systems or of the enclosure of rangeland. An insight into the range of research undertaken can be seen from papers published in the Proceedings of successive international yak congresses (Zhang *et al.,* 1994; Yang *et al.,* 1997; Jianlin *et al.*, 2002).

A place for yak elsewhere?

Outside the Asian Highlands there are also regions with harsh, cold climates and seasonally variable availability of natural forage. Parts of the Andes in South America, northern parts of Canada, and possibly parts of Greenland are among potential regions where yak or its hybrids might be considered. Such introductions need not involve pure yak populations. Thus, J.J. Rutledge proposed a scheme (personal communication, 2002) whereby surplus Holstein eggs from slaughter cows in the USA could be fertilised *in vitro* with yak semen, the resulting embryos sexed and the hybrid female embryos transplanted to recipient cows in South America for ultimate use of the hybrids in the Andes. They would be expected to provide the local population with much more milk than is presently obtained from local llamas. The main deterrent to investigating such wider potential uses of the yak stems from the current over-supply of livestock products from rich parts of the world and from the reluctance of administrators, and not from biological considerations. Also, in more distantly futuristic terms, it may be envisioned that the yak's characteristics of survival in hostile climates, or its allegedly superior efficiency in converting roughage to meat, will have their genetic origin identified and transferred to other bovine species.

For the present, these ideas among others serve to underline the need to conserve the yak as a valuable genetic resource, pending changes in the needs and aspirations of a growing world population.

Conclusion

The indigenous knowledge of yak keeping and the culture, social context and traditional practices of yak-keeping society are the framework on which changes should be built if living standards in these relatively impoverished regions are to be improved. Already in peripheral areas of the yak's distribution, such as North-East India, Nepal and even Mongolia, yak populations are in decline. A younger generation of herders is less content than their parents with harsh living for inadequate reward. A decline in yak numbers is not yet obvious in the main areas of the Qinghai-Tibetan Plateau, even though the aspirations of the young people may be no different. It may just be a matter of time. The yak is irreplaceable as the principal source of livelihood in most of these areas - with other species, tourism or alternative employment making, at best, smaller contributions. Thus, in order to preserve viable societies in these regions and, most importantly, to use the natural resources to good effect, a progressive scheme of improvement in productivity from yak and the rangelands is desirable. The means to that end need to come to a large extent from the will of the policy makers and through the provision of funding to improve infrastructure, marketing, extension and veterinary services, research and development projects and education.

An overriding requirement is, however, that degradation of the rangelands due to overstocking is halted and reversed and that economic improvements are achieved within a sustainable ecosystem. Any increased output should be achieved, in general, without increasing the stocking rates.

Further reading

Wiener, G., Han Jianlin and Long Ruijun (ed.). 2003. *The yak. Second edition.* Food and Agriculture Organisation of the United Nations (FAO), Bangkok, Thailand.

Author addresses

Gerald Wiener, Roslin Institute (Edinburgh), Roslin, Midlothian EH25 9PS, UK (Formerly Animal Breeding Research Organisation)

Han Jianlin, International Livestock Research Institute (ILRI), P.O. Box 30709, Nairobi 00100, Kenya

25

Equines

Anne Pearson, Brian Sims‡, Ali Aboud‡*

Cameo 1 - Sub-Saharan Africa

Lack of animal power is a problem for many resource-poor smallholder farmers in the drier areas of Sub-Saharan Africa, who cannot afford tractors and have lost cattle due to drought or disease. Tandiwe Dube in Zimbabwe already uses her donkey to fetch water and fuel but, along with other farmers, is wondering whether she can use her donkey to cultivate the fields. Is her donkey strong enough for ploughing or would she need to borrow others; would she need to feed it more if it was ploughing; what harness would she need; would the ox plough she still has be too heavy for donkeys; could she still produce enough cereals to feed her family using donkey power? Researchers working closely with Tandiwe and other farmers like her have been able to provide the information needed. Using her donkey with others, Tandiwe is now planting a larger area of land than she had been able to do since she lost her draught oxen. She is not the only farmer in the area to have benefited from the research, and others are now successfully using donkeys to assist them in crop production.

Cameo 2 - Latin America

Many smallholder farmers on the steep hillsides in Latin America rely on animal power on their farms, as tractors are too expensive and impractical on the slopes and terraces. Olegario González in La Era and Antonio Herrera in Altzayanca in Mexico keep mules and horses respectively, which they use in crop production and transport. In Mexico the main problems these hillside farmers have is soil erosion and providing enough feed for their animals in the dry season. Similar problems are found in the mid-Andean

---

* Principal author
‡ Consulting author

hills of Bolivia but, in addition, the implements used are often too heavy for the small horses found at these altitudes. Researchers have been working closely with these farmers and others to help them. Cultivation techniques such as contour barriers, tine cultivation, and cover crops and sown pastures which can be conserved as feed, have been tested. Spring and winter vetch (*Vicia sativa* and *V. villosa*) are now sown with forage oats or with the maize crop in Mexico. In Bolivia, in the mid-Andean hills, a high-hitch harness has reduced the draught force needed to pull implements, resulting in equines being used for a wider range of tasks than before.

Cameo 3 - Asia

In many towns and cities in Asia equines carry loads or pull carts to transport people and goods, sometimes coming in from the nearby rural areas. Many of the owners have no land on which to grow crops and live in very poor homes. In Gadaipuri, close to the centre of Delhi, Nizamuddin relies entirely on his six donkeys to earn a living for himself, his wife and six children. He uses the donkeys to transport materials to and from construction sites inside Delhi each day. In Afghanistan, Gulam Sadiq has five horses. The income these animals generate supports his entire extended family of 11 adults and 45 children. In Gujranwala, Pakistan, Sadiq Masih's family eat well if he has had a good day with his Tonga horse ferrying people around the town. If passengers are in short supply then the family go hungry. These people in Asia are not alone in relying solely on their working animals for a living. There are many people elsewhere who live from day to day on the income that their horse or donkey earns. Veterinarians and husbandry specialists, funded largely by equine charities, have been working with horse and donkey owners in the poorer areas to help them improve their husbandry and keep their animals fit and healthy.

Introduction

Horses (*Equus caballus*) and donkeys (*E. asinus*) and their crosses, the mule and hinny, belong to the Genus Equus and the family Equidae. They were domesticated about 5,500-6,000 years ago, mainly for draught and transport, rather than food. Most of the domesticated donkeys and mules (over 95 per cent), and horses (75 per cent) in the world are kept in developing countries (Table 25.1). In the tropics horses are usually more common in cooler, less-arid areas and at the higher altitudes and donkeys in the drier semi-arid areas. Equines make a significant contribution to farm power on small farms and to local transport, as a low-cost option to motorised power. Transport is largely over short distances in both rural and urban areas, moving commodities and people. Virtually all of the equines kept in developing countries are used for work at some point in their lives. They are rarely kept just for breeding and sale for meat is generally only when the animals have come to the end of their useful working life. Except where horse and donkey meat are traditionally eaten, such as in the Northern parts of Ghana and small areas in Eastern Africa, consumption of horse and donkey meat is often a last resort when other protein sources are not available, and is not often acknowledged. Mare's milk is consumed by the pastoralists in Mongolia and donkey's milk is drunk for medicinal purposes in some African societies. See Table 25.1 for the relative advantages and disadvantages of each equine species for resource-poor livestock-keepers.

Table 25.1 Estimated number of horses, donkeys and mules (millions) in developing countries compared to numbers elsewhere (FAO statistics 2002), and relative pros and cons of keeping these animals

Number of equids	*Horses*	*Donkeys*	*Mules and hinnys (horse x donkey)*
Africa	3.4	13.7	1.3
Asia	15.9	18.2	5.2
Latin America and Caribbean	24.1	7.8	6.5
Developing countries	41.6	39.1	13.1
Developed countries	14.8	1.4	0.3
World total	56.4	40.5	13.4
Relative advantages	Can be heavy Powerful Medium/high work capacity Fastest speed of travel	Cheap to buy Low maintenance Easy to train	Powerful for size High work capacity Require less maintenance than a horse Can have a long working life if well managed
Relative disadvantages	Expensive to buy Need supplementary feed if working Most susceptible to disease and poor management	Small low power unit Slow to variable speed	Expensive to buy Work best for one owner Can be difficult to train

In some places the numbers of equines are increasing. In agriculture equines are tending to replace draught oxen on many smallholder farms where they can survive in the environment and tractors are not a viable option. An example is the Gambia where there were few horses kept before the 1960s and much of the cultivation was done using hand labour and the small trypanotolerant N'dama cattle. The latest livestock statistics for the Gambia, for 2002, show that there are now 30,000 horses working on the land. In semi-arid areas of Africa, where successive droughts and diseases have reduced the cattle population, many resource-poor farmers have had to turn to donkeys to provide power in crop production. Numbers are impossible to determine accurately as many people do not regard the donkeys as worth mentioning in surveys or on census forms, thinking of them as a temporary measure until they can use oxen again. However, in many semi-arid areas of Sub-Saharan Africa more donkeys are now seen at work than in the past. In places where HIV/AIDS has reduced the available labour force, then the smaller, low-maintenance donkeys are seen by some as a more realistic provider of farm power than draught oxen.

Table 25.2 Introduction of donkeys in Mgeta Division, Tanzania, to solve local transport problems for cash crops

Location	Tchemzema and the surrounding villages in the Uluguru mountains
Climate	Mild temperatures, good rains
Terrain	Mountainous, few roads
Households	6,000 households of Luguru people
Agriculture	Intensive production of temperate fruit and vegetables
Start of project	1994
Transport for produce before 1994	Head loads to nearest road, then trucks
Number of donkeys at start of project	2
Number of donkeys ten years later	168
Outcomes	People now using donkeys instead of head loading to transport produce to roadsides Association of Donkey Users has 250 members, more than half are ladies Group bank account for buying donkeys and drugs Value of donkeys has increased

In some areas equines have been introduced to help solve transport problems, largely due to the activities of research and development projects (e.g. Table 25.2). In most other parts of the developing world, equine numbers are relatively static, but can be high. For example, The Brooke Hospital for Animals (BHA), which works with equine owners in urban areas in Egypt, India, Pakistan and Jordan, estimates that there are 100,000 working equines in Cairo, at least 30,000 across Delhi, Jaipur and Hyderabad and 60,000 in and around the five major cities of Lahore, Multan, Gujranwala, Mardan and Peshawar (BHA, 2003). These numbers are just in the places in which BHA operates. China has over eight million donkeys and a similar number of horses working in rural areas and in towns.

Characteristics of equines

Horses are often not as heavy as oxen, and therefore not able to generate as much draught force during work; their main advantage is their speed of working. This is an advantage in the tasks requiring a relatively low draught force, such as seeding and weeding (often no more than 270-350 N [newtons] on sandy soil) and transport; whereas size is more important in the primary cultivation tasks (ploughing or ripping) which require a high draught force (sometimes more than 1 kN). Horses, donkeys and mules are often regarded as being more versatile than oxen, being better for riding and transport. As a result, they are often used for more tasks than a pair of draught oxen during the year. This is a consideration when selecting a work animal, since the working animal requires feeding every day, unlike the tractor. Where large draught horses of 500-600 kg are found, it is generally recognised that one large horse can do the work of two smaller 400-450 kg oxen (Photo 25.1). Horses come in many different sizes and the best one for each situation depends mainly on the feed resources available to maintain it, as well as on the power needed to do the work. Animals can be teamed up, shared or hired to make up deficiencies in power, but again this adds to the feed 'bill'. Efforts to reduce the force required to undertake the tasks can also be important, enabling smaller or fewer animals to be kept. Feed deficiencies are difficult to make up on smallholder farms, so choosing the right animal can be important. In North Vietnam, for example, where feed resources for horses are limited, small 150-200 kg ponies provide sufficient power for most local transport needs for people who cannot afford to use motorised power, and they complement the larger swamp buffalo which do the field work (Photo 25.2).

The attributes of the donkey are that it is cheap to buy and to maintain. It is one of the few domesticated animals that appear to do rather well with minimal management. In many parts of the world donkeys seem able to obtain sufficient nutrients from grazing in all but the most extreme of dry seasons. With the exception of the camel, they can go longer without water than any of the other large domestic animals used for work. The Zimbabwean donkey, for example, is able to maintain relatively high levels of voluntary feed intake of conserved forage despite being deprived of access to water for periods up to 72 hours in moderate ambient temperature conditions (maximum about 26 °C) (Nengomasha *et al.*,1999a, LPP Project R5926). Donkeys also seem to be less affected by some of the common diseases that debilitate or kill other working animals, notably trypanosomiasis and African horse sickness.

Photo 25.1 Horse and oxen working near Chillán, Chile (courtesy of A. Pearson)

Photo 25.2 Contract transporter with his pony and cart in Song Cong, Thai Nguyen Province, North Vietnam (courtesy of A. Pearson)

Equines generally do not do well in very humid areas where they tend to succumb to insect-borne diseases, respiratory problems and foot problems. The working equine rarely has a resale value as meat at the end of its working life. This can result in some people giving it a lower priority, compared with other livestock, when resources are short. In Ethiopia, for example, when asked to rank their animals in order of economic importance, many livestock owners regarded their donkeys as economically the most important or second only to their work oxen, more important than their cows and small ruminants. However, when asked to rank their animals in the order in which they would spend money on improving health and management, then the donkey was at the bottom of the list (Pearson *et al.*, 2002).

Performance of working equines

The productivity of livestock kept for beef or dairy production is relatively easy to measure; that of a working animal is more difficult, since there is no simple objective criterion that can be used. When an animal uses energy to pull an implement through the soil or to carry a load over a distance, work is produced. The amount of work an animal can do depends on the speed at which it works, and the draught force generated. For a particular draught force, it is the speed which will determine the power output of the animal, i.e. the rate at which the animal does the work. A large animal can sustain a higher draught force than a small one; but for any animal, as draught force increases so speed decreases. For instance, an animal ploughing a heavy clay soil (high draught force) will walk more slowly than when it is pulling a well-balanced cart over a level sealed road (low draught force).

Performance tests and field trials (e.g. Table 25.3) provide guidelines to help match animals to implements and tasks on a farm, and help to predict outputs (e.g. area cultivated, distance travelled) in a working day. However, they are often carried out with well-trained animals in good condition. Obviously animals that are inexperienced, underfed or sick will be incapable of producing as good a performance as animals which are well-fed and healthy. Adverse working conditions, such as heat stress, badly fitting harness and difficult ground conditions, will all reduce the rate of work and therefore, the amount done in a day. Unfortunately these are some of the problems faced by many resource-poor farmers and equine owners when they lack the knowledge or resources to manage their equines well.

Table 25.3 Average draught forces (expressed as kg draught force/100 kg live weight) which can be sustained over a 4+ hour working day

Species	*Average individual live weight (kg)*	*Hitching*	*Average draught force (kgdf/100kg live weight*[1]*)*	*Country*	*Reference*
Donkey	140	Single	12.9	Niger	Betker and Kutzbach (1991)
Ox	392	Single	12.0		
Cow	330	Pair	12.0		
Ox	372	Pair	12.0		
Donkey	100-150	Single	10-16	Cameroon	Vall (1996)
Horse	200-300	Single	10-15		
Ox, cow	300-450	Single	10-15		
Donkey	170	Team of four	13-16	Zimbabwe	Nengomasha *et al.* (1999b); LPP Project R5926
Ox	330	Pair	13		
Horse	500-600	Single	11-12	Chile	Pérez *et al.* (1996)

[1](1.0 kg draught force = 9.81 N [newtons])

Attitudes and approach to equines

An important issue to consider in improving husbandry of equines is that the equine means different things to different people. Some people hold the horse, mule or donkey in high regard, whilst to others (more common when considering the donkey) it is an expendable item. Some people using equines are experienced animal keepers, well aware of the problems in feeding and managing livestock, others are relatively new owners (or users) and may have had no experience of animal management before buying (or borrowing) an equine.

The diversity of situations and attitudes among equine users offers a considerable challenge to those aiming to improve management. It is not enough to develop one solution or approach for all situations. A more readily acceptable (and likely to be more successful) technique is the 'participatory' approach, which is now being widely encouraged and used in other disciplines in development. This involves close exchange and collaboration between users and advisers, and between new and experienced users. Consultation takes place with the users to identify their specific problems and discuss the options that might be appropriate, drawing on the experiences of good equine managers within the community, if they are available. The choice of input or modification to try is ultimately the decision of the users, who can decide on a course of action, for whatever reason, to suit their own situation. Most scientists and development experts and NGOs working with the resource-poor who keep equines, now adopt this approach.

Owners are encouraged to help themselves and prevent injuries and ill-health in their equines through improved husbandry, rather than, as Pablo Torrealba, a hill farmer in Mexico said, "relying on the veterinarian to fix the animal when it cannot work". Training and education now play an important part in most operations to improve equine husbandry and welfare.

What is often forgotten is that improvements in management do not always have to involve expensive or time-consuming activities. Some of the possibilities available are discussed in the following sections. The emphasis is on possibilities at the practical level, rather than at the policy level of decision making, although it is recognised that changes of attitudes at the policy level can have an important impact on attitudes elsewhere.

Nutrition and feeding

Comparative studies of the intake and digestion of forage feeds by donkeys and horses have shown that, while differences between the two species are small on good quality, highly-digestible forages (such as lucerne), on low-quality forages (such as barley and oat straws) donkeys are better able than horses to digest the high-fibre, low-protein material (e.g. Pearson, *et al.*, 2001, LPP Project R5198). This may account for their ability to maintain weight in the dry season compared to the horse.

The problems that owners have in feeding their equines vary from area to area (Table 25.4). Good feeding does not just benefit the working capacity of equines. An adult horse or donkey in good condition is better able to resist disease challenge, lives longer and has a higher rate of reproduction to provide replacement animals. There is a range of options available to improve the nutrition of the equine, some more expensive than others (Table 25.5).

In summary, for some equine users concentrate feeding is an acceptable way of improving the nutrition of their animals, while for others low-cost more time consuming, labour-intensive options, will be more acceptable. The key is to identify which of these options are likely to be the most acceptable to any target group.

Table 25.4 Potential differences in nutrition and feeding practices for equines kept in rural and urban areas

Rural areas	*Urban areas*
Good grazing	Poor or no grazing
Seasonally variable feed supply	More regular feed supply
Less buying power for supplementary feeds	More buying power for supplementary feeds
Better knowledge of livestock feeding	Less knowledgeable of livestock feeding
Potential to underfeed when seasonal shortages occur	Potential to feed unbalanced ration through incorrect purchase

Table 25.5 Methods to improve the nutrition and feeding of equines

Need money	*Need time*
Purchase concentrate (e.g. maize, beans, barley mixtures)	Offer more forage to improve selection
Purchase forage (e.g. alfalfa, cowpea hay, groundnut hay)	Collect more (e.g. weeds, pods, roadside grasses)
	Grow better quality forages (oats/vetch mixture)
	Prolong feeding time
	Feed individually
Purchase water	Offer water more frequently

Reproduction and breeding

The reproductive characteristics of the donkey and horse are given in Table 25.6. The mule is usually infertile.

Many people in urban areas prefer to use geldings or stallions rather than mares to avoid having to reduce or stop work in late pregnancy and early lactation. Mares are more common in rural areas, where a foal may be produced every 1-3 years. Donkey breeding is notoriously unplanned in most developing countries. In Sub-Saharan Africa there are localised instances of good male donkeys being kept and sought-after for breeding. However, very often the larger more dominant animals are castrated to prevent fighting and so mating of females is with smaller less aggressive males. Breeding is often better organised in Latin America where cross-breeding with the horse to produce mules is more widespread, and some donkey stallions are kept specifically for this purpose. Owners of horses will more often make some effort to select the stallion to use for breeding, but this is not always the case.

Table 25.6 Reproductive characteristics of the horse and donkey

	Horse	*Donkey*	*Remarks*
Males sexually mature	In year 2		Can vary considerably
Females sexually mature	1-3 years		Depends on nutrition
Breeding	Seasonally polyoestrous in temperate areas in late spring and summer; in tropics breeding all year round		
Oestrous cycle length	21-25 days	20-26 days	
Oestrus	2-7 days	2-10 days	
First heat (foal heat) days after parturition	7-8 days	17-18 days	Can vary in donkey from 6-69 days after parturition
Gestation length	336-340 days	360-375 days	
Length of reproductive life	18-20 years	20-25 years	

There are few instances of planned breeding programmes for working equines in resource-poor areas. Work is of primary importance and the ability of the female to produce a foal is regarded by some as a bonus and by others, usually where resources are inadequate, as a drain on the mare, reducing its condition and work output and increasing feed requirements. In some areas of Southern Africa, where animals graze communally for much of the year and breeding is unchecked, numbers of donkeys can reach epidemic proportions, leading to increased grazing pressure and more animals grazing at roadsides, thus increasing the chances of road accidents. In these places donkey owners have been asking for assistance in castration programmes to control numbers.

For donkeys in many parts of Africa, inhumane castration methods are used to crush the testes and spermatic cord, which involve considerable pain to the animals for some time, and may not result in successful castration. Veterinarians recommend the surgical method, cutting open the scrotum and removing the testes and tying-off the spermatic cord under local anaesthetic. The use of the burdizzo is believed to be unsuitable for castration of donkeys as it produces a significant local reaction and swelling. However the burdizzo method of castration is being practised on young donkeys in several parts of Sub-Saharan Africa, with some apparent success. Because there are no wounds, the risk of infection, a problem in surgical castration, is minimised. Until veterinary methods are accessible and affordable to everyone, then the burdizzo may offer an alternative to the inhumane local castration methods practised on donkeys. For this reason a research project is currently underway by veterinarians to determine the feasibility of using the burdizzo as a method of castration in young donkeys.

Disease and injury prevention

A major problem for resource-poor people owning equines is access to veterinary services. Many working animals are kept in remote areas where it is not easy to find veterinary assistance. A second problem, even when these services are available, is the cost of treatment. Owners are often more willing to spend money or time on treating the 'visible' problems (harness sores, lameness, ticks and skin diseases) rather than the 'less visible' ones (endoparasites and infections, the largest group, including worms, viruses, bacteria and protozoa). Because of the financial cost there is a tendency to treat the symptom, often too late, rather than prevent the disease in the first place. People are more willing to spend money on mules and horses than on the lower value donkey.

Some effort to prevent at least the acute diseases would seem to be justified, so that replacement animals do not have to be found at short notice, especially in the middle of the cultivation season. Farmers can often be persuaded to vaccinate their animals against the acute diseases, when the vaccinations are available, but control of the chronic or less visible diseases can be more difficult to justify. Control of intestinal worms provides a good example. In many places the 'high cost' prevention of worm infestation with drugs, although convenient, is not acceptable economically. This problem has been recognised and alternative 'low cost' measures of worm control have been tested in developing countries. These can involve a management approach or use of local, indigenous plants. *Artemisia maritima, Caesalpinia crista, Melia azedarach, Mallotus philippensis, Chrysanthemum* spp, and *Matteuccia orientalis* have all been said to have anthehelmintic properties in livestock and warrant further investigation (Hammond *et al.,* 1997). Use of local indigenous remedies can also provide an alternative low-cost option to treat other health problems. While some are effective (e.g. aloe and honey in wound treatment), others are definitely harmful (e.g. battery acid in wound treatment).

It is common to see equines in developing countries with saddle and harness sores, or scars where sores have healed. For example, in Ethiopia donkey owners using their animals for transport reported that 40-50 per cent of their donkeys had skin sores 'all the time', and most animals had a sore some time in the year (LPP Project R7350). The fact that these sores are preventable, provided harnesses are made out of suitable material and are correctly fitted, makes the situation more frustrating to those aiming to improve the management of the equine (Table 25.7). Again people will spend more time and money on a harness for a mule or a horse than for a donkey. But in general few equine users adopt preventive measures against harness sores, dealing with the problem only when it becomes apparent.

Success in reducing harness-related injuries lies in education and encouragement of users to invest in harnessing, as they would invest in an implement or cart, and to encourage the adoption of preventative measures as part of good management, rather than waiting for a problem to develop. Equine charities and other NGOs are investing in training both of owners and of artisans who can produce and repair harnesses that fit the animals and will not rub when the animals are working hard. For example, the National Society for the Prevention of Cruelty to Animals (NSPCA) in South Africa, has been running a training programme funded by the BHA in which, as well as training in husbandry, donkey owners can participate in workshops where they make harnesses for their own donkeys, under supervision, using materials supplied by the project. The result is a better harness for the donkeys and one which the owners value more because they have invested time and effort in making it (Photo 25.3). The Kenyan Network for Draught Animal Technology (KENDAT) has been running similar self-help workshops outside Nairobi with encouraging results (Kaumbutho, 2003).

Photo 25.3 A group of donkey owners making their harnesses at a NSPCA workshop at Mayaeyane, NW Province, South Africa (courtesy of A. Pearson)

Table 25.7 Harness and equipment – causes of injury and preventable measures

Location	*Cause of injury*	*How to prevent*
Harness	Incorrect size and not properly fitted to animal	Make back straps/saddle straps, traces and breeching straps adjustable
	Too narrow or thin	Use wide bands or straps, not ropes or narrow bands
	Sharp edges	Sew with strong thread, rather than bolts or wire
	Stitched joints/bolts	Use natural materials, leather where harness in contact with the animal, or webbing
	Unsuitable material	
Pack saddle	Poorly designed and fitted	Ensure weight rests on ribs, not backbone
		Measure on animal when making
	Made of unsuitable materials	Use light materials
		Use good padding underneath
Halters	Attached incorrectly	Avoid using bits and blinkers where possible
Bridles and bits	Incorrect size	Use wide soft straps, not narrow ropes or wire, no sharp edges
	Unsuitable materials	Make sure the halter or bridle is not tight around nose or throat
		Make adjustable
Hobbles	Unsuitable material	Use on front legs only
	Not properly fitted	Use wide straps, no sharp edges
		Should not constrict blood flow to the feet
		Attachments should have easy release
		Should be easy to adjust

Table 25.7 (Contd.)

Location	*Cause of injury*	*How to prevent*
Neck ropes or collars	Unsuitable material Not properly fitted	Use wide straps, no sharp edges Should be loose around neck and not be able to slide to tighten Attachments should have easy release
Tillage implements	Too heavy for the animal Incorrectly set for depth or width of operation	Must be suitable for job and soils Add more animals in pairs if necessary Try lighter implement and/or high hitch harness
Carts	Too heavy for the animal No brakes (for carts) Poorly designed and hitched No breeching straps on harness Shafts too short	Loads should be well balanced Wheels should be the same size Wheel bearings in good condition/wheels turn easily Breeching strap to prevent cart hitting animal Saddle to enable animal to take weight of shafts on the back, not the neck Hitching points in right position Allow space for swingletrees and evener
All injuries on body	Poor body condition Beating	Give water before feeding Feed well Avoid heat stress Do not overload Train animal well Do not beat

Source: Pearson *et al.* (2003b)

New technical solutions have been developed to assist people who have found it more expedient to use their smaller donkeys or light-weight horses for field work than to rely on the more expensive draught oxen. The smaller animals have lower feed requirements, but cannot produce as much power as the larger animals. Scientists have, therefore, studied ways of reducing the power needed in tillage. Research by Inns (2003) has shown that the draught of an animal-powered tillage implement can be reduced by using a lightweight implement and/or by pulling it at a steeper angle of up to 40° (a function of harness design and adjustment). A high-lift harness combined with a lightweight implement allows the lighter equines to prepare the soils for planting crops. The system has been introduced into Tanzania, Kenya and Bolivia (Inns, 2003). In Bolivia a range of implements is now available, manufactured, sold and extensively used on a commercial unsubsidised basis (Sims and O'Neill, 2003, LPP Project R6970). A lightweight plough has also been developed in Zimbabwe, and is now sold in a number of neighbouring countries by a commercial manufacturer to reduce the draught required to plough the soils there. This is advertised as a 'donkey' plough, and enables donkeys to be more easily used for ploughing (where they have replaced oxen in field operations) than if they use the traditional Zimbabwean ox-plough (LPP Project R5926).

Housing and foot care

While horses and mules are generally housed at night, and fed supplements, in many parts of Africa donkeys have traditionally been allowed to roam free to graze. This practice is becoming less common as people worry about theft and crop damage. This can have an effect on the nutritional status of the donkey when it is grazing. In trials in Ethiopia and Zimbabwe, cattle were able to compensate for restricted feeding time by spending more time grazing per hour, and increasing bite rate. Donkeys, however, ate less if grazing time was restricted and also selected a poorer diet. When given the opportunity, donkeys used the hours of darkness to feed, but cattle did not graze at night (Smith, 1999; LPP Project R6166). This emphasises the need to supplement donkeys when they are kraaled or stabled at night and have been working during the day.

Water requirement increases during work as heat load increases. In urban areas and at markets, some success has been achieved by providing shaded areas, and water troughs, where animals can rest and be watered. At markets in countries as far apart as Mexico, Pakistan and Southern Africa, permanent water troughs have been provided by local organisations, NGOs or authorities. These have an immediate impact on improving the welfare of the working animals (Photo 25.4) and, because they are seen, can increase awareness of the importance of shade and water in those equine users who may not normally give it a thought. These shade and water points are often backed up by educational campaigns by the equine charities operating in the areas to increase awareness.

Equines working mainly on fields and dirt roads in rural areas rarely need to be shod, provided their feet are cleaned regularly and checked for stones. Trimming is sometimes necessary and should not be left until the hooves are too long and distorted. Equines working daily in urban areas on tarred roads, need shoeing regularly to prevent excessive wear of the hooves. In some places farriers are available, and in others the owners do their own trimming and shoeing. The quality of the shoeing varies considerably, as does the quality of the shoes.

Photo 25.4 Watering a horse on University Road, Lahore, at BHA wayside station in Pakistan (courtesy of A. Pearson)

Conclusions

The main difficulty to be overcome in applying options to improve management of equines is one of information dissemination. Farmers, urban users of equines and extension officers are very often not aware that options exist, and are ignorant of the beneficial effects that improvements in management can have on performance, longevity and reproduction. In the case of donkeys, users themselves are in many cases the initiators of improvements in feeding and management practices, as the value of their donkeys to them has increased. There is a need in the future to improve the dissemination of information on the management options that are available to those involved in development and extension in areas where equines contribute to the economy and quality of life. The equine charities and other NGOs do a good job of supporting the equine populations in the areas in which they operate, with an increasing emphasis on education and training. However they cannot be everywhere. There is a need for governments to recognise the value of the equine to the poorer members of society, and to include them along with the 'food animals' in their livestock activities, rather than ignoring them. Benefits from training, education and improved support are not necessarily immediate or dramatic in nature, but they can result in an improvement in health and welfare of the equine, which in turn results in improved productivity in work and reproduction, and an increased life-span.

Further reading

Animal Traction Network for Eastern and Southern Africa (ATNESA) website: www.atnesa.org

Draught Animal News a biannual publication, website: www.vet.ed.ac.uk/ctvm

Hadrill, D. 2002. *Horse healthcare – A manual for animal health workers and owners*. ITDG Publishing, London, UK.

Payne, W.J.A. and Wilson, R.T. 1999. *An Introduction to animal husbandry in the Tropics. Fifth Edition*. Blackwell Science, Ltd., London, UK.

Pearson, R.A., Lhoste, P., Saastamoinen, M. and Martin-Rosset, W. (ed.). 2003a. *Working animals in agriculture and transport. A collection of some current research and development observations*. EAAP Technical Series No 6, European Association of Animal Production, Wageningen Press, The Netherlands.

Pearson, R.A., Simalenga, T.E. and Krecek, R.C. 2003b. *Harnessing and hitching donkeys, horses and mules for work*. Centre for Tropical Veterinary Medicine, Division of Animal Health and Welfare, University of Edinburgh, Easter Bush Veterinary Centre, Roslin, Midlothian EH25 9RG, UK.

Author addresses

Anne Pearson, Centre for Tropical Veterinary Medicine, Division of Animal Health and Welfare, University of Edinburgh, Easter Bush Veterinary Centre, Roslin, Midlothian EH25 9RG, UK

Brian Sims, Engineering for Development, Bedford MK1 7EG, UK
Ali Aboud, Department of Animal Science and Production, Sokoine University of Agriculture, P.O. Box 3004, Morogoro, Tanzania

26

Wildlife

Ken Campbell[*], *Martin Loibooki*[‡]

Cameo 1 - Saiga antelope in Central Asia

Covering roughly 260 million hectares, the vast rangelands of Central Asia form the world's largest contiguous area of grazing and are home to the nomadic Saiga antelope (*Saiga tatarica*). *S. t. tatarica* occurs in Kazakhstan and in Kalmykia, Russian Federation, and *S. t. mongolica* in Mongolia. Horns are borne by males only and are highly valued in traditional Chinese medicine. Following heavy hunting pressure during the 19th century, a complete and well-regulated ban on hunting allowed populations to recover. Controlled hunting for meat commenced during the 1950s and continued for the remainder of the Soviet period. Following the collapse of the Soviet Union, state and collective livestock herds were divided among small landowners. Privatisation was accompanied by an end to input subsidies, disintegration of marketing networks, lack of winter forage, and poor maintenance of infrastructure, all factors leading to a decline in livestock numbers. Rural poverty, unemployment, loss of livestock, and a lack of funding for Saiga management have resulted in uncontrolled and large-scale poaching for Saiga meat, which formed an important contribution to many people's livelihoods. By 1998 Saiga population numbers had collapsed to about 5 per cent of numbers ten years previously. Selective hunting of adult males for their horns has resulted in over 100 females per adult male. It is likely that a lack of males has caused a serious reduction in conception rates which, in addition to the high mortality from unregulated hunting, has resulted in a population collapse. Saiga antelope are now listed as Critically Endangered on the 2002 IUCN Red List and are also listed on Appendix II

---

[*] Principal author
[‡] Consulting author

of the Convention on Migratory Species (Milner-Gulland *et al.*, 2001; 2003). With the collapse of the Saiga populations, dependence on Saiga is clearly unsustainable and a rebuilding of the rural economy, the livestock sector in particular, is an essential prerequisite to conservation of the Saiga antelope.

Cameo 2 - Serengeti, Tanzania

The main feature of the Serengeti ecosystem of northern Tanzania is an annual migration of large herds of ungulates, primarily wildebeest (*Connochaetes taurinus*). Over 2 million people to the west of Serengeti derive important benefits from Serengeti's wildlife, largely in the form of illegally hunted meat. The meat, hunted mainly by the resource-poor using largely traditional methods (snares, bow and arrow), is dried and sold locally or outside the area, at an estimated market value of about US$ 1 million per year (Campbell *et al.*, 2001). Growing human populations coupled with increasing household cash requirements, result in a situation where the 'safety net' provided by illegal hunting of wildlife constitutes one of the only support mechanisms and sources of cash for the poorest members of the community. Populations of many of the larger wildlife species have declined and hunters now travel greater distances and increase their hunting effort. A change in the age profile of hunters was recorded between 1992 and 2000, with a statistically significant increase in the 25 to 45 year age bracket – the main household income earners. The poorest households have few or no livestock, and there is a clear relationship between illegal hunting and lack of small stock in particular (poultry, sheep and goats). In these areas, hunting and utilisation of wildlife is not a long-term sustainable solution to the problem of rural poverty. Experiences of pro-poor tourism and community wildlife management projects, often seen as solutions in wildlife-rich areas, suggest that benefits accrue to the wider community, with few benefits directed at the poorest community members. Instead, poverty reduction and rural development may require integrated programmes with strong small stock components targeted at the poorest members of the communities – enabling and sustaining livestock ownership by the resource-poor.

Background

Resource-poor communities throughout the world experience a variety of interactions with wildlife species. For the purposes of this chapter, wildlife is defined as any wild animal, including mammals, birds and fish. In areas where resource-poor communities live adjacent to significant populations of wild herbivores there may be competition for grazing, predation of stock and transmission of disease to livestock. With growing numbers of people to feed, it is natural that planners are more oriented towards agricultural development than to the maintenance and establishment of natural reserves and conservation of wildlife. However, wildlife also provide a source of both food and income for the poor, particularly during times of hardship when crops may fail and domestic stock die.

Interactions between larger wildlife, livestock and livestock-keepers are necessarily limited to those areas where wildlife and livestock co-exist or live in reasonably close proximity. Worldwide, the numbers of resource-poor livestock-keepers are considerable. What is the potential scale of interaction, which livestock systems, and which groups of resource-poor livestock-keepers are most likely to be involved in these interactions? Thornton *et al.* (2002) showed that a majority of resource-poor livestock-keepers are to be found in mixed rain-fed production systems in arid/semi-arid and sub-humid/humid climatic zones in Sub-Saharan Africa and South Asia (see also Chapters 2 and 3). This chapter therefore focuses on interactions between wildlife and the rural poor of mixed systems in Sub-Saharan Africa, where significant populations of large wild ungulates still exist, where there are traditions of hunting and using wildlife, and where wildlife and wild areas remain important to the livelihoods of the rural poor. It is in these same areas that attempts at wildlife utilisation have been made. It is also worth noting that a significant part of the wildlife-rich areas in Sub-Saharan Africa is also home to the tsetse fly (*Glossina* spp.), vector of trypanosomiasis in domestic stock and of human sleeping sickness, and that the presence of tsetse is of some importance for wildlife conservation.

Interactions between wildlife and humans or livestock in different livestock production systems are varied and inevitably depend on individual circumstances. Nevertheless useful generalisations can be made. In livestock-only or range-based systems the predominant negative interactions are likely to be competition for grazing and browse, and potential disease transmission. Prins (2000) suggests there is little evidence of livestock numbers being reduced by competition with wildlife. On the other hand, wildlife numbers are negatively affected by livestock. Predation of domestic stock occurs but has declined due to reduced wildlife and predator populations. Direct benefits from wildlife include revenues from eco-tourism and commercial hunting, as well as the potential for wildlife ranching. Mixed systems include agro-pastoral and crop-based systems, with the latter characterised by greater human densities, more intensive production and few large wildlife. In crop-based systems, damage to standing and stored crops from pest species may be significant. Competition for grazing may occur near to protected areas, but favours livestock. Benefits include the use of wildlife as food and for cash (legally or illegally), and may also include tourism/eco-tourism. Landless or land-independent systems have little or no contact with large wildlife, whilst impact from pest species on stored feed may be significant. Compared with livestock-only systems, mixed rain-fed production systems support greater human populations. Where wildlife occurs, there is usually a steep gradient from high-density human settlement and low-density wildlife to low-density settlement and high-density wildlife.

Values of wildlife

Many wild species provide benefits to humans. In an economic context wild populations are often treated as economic goods, with utilisation through harvesting resulting in a net benefit. This is particularly true for wild fish populations, which constitute a large part of the diet of much of the world's population. Wild species may also result in economic loss, especially where resource-poor rural communities depend on crops and livestock for their livelihoods. A majority of contributions to livelihoods of the rural poor that have been attributed to livestock (Carney, 1998) can also be attributed to wildlife. These typically include a range of direct use, indirect use and non-use values, much of it non-marketed:

- a source of food and, in some instances, clothing or bedding
- an important source of cash income from the sale of bushmeat and other wildlife products
- a means of allowing the resource-poor to capture private benefits from common property resources
- a source of livelihood security by diversifying risk and buffering against low crop yields, livestock mortality and stock theft, particularly in drought-prone environments
- cultural well-being, and a variety of related functions, including use as medicines or rituals
- consumptive (trophy or sport hunting) and non-consumptive (photo-tourism or game viewing, and eco-tourism) tourism.

To this list one can add additional values of a wide variety of non-wildlife products that may also be directly gathered from the same natural environments. Examples include thatching grass, building poles, fuel wood, wild honey, and medicinal plants, to name but a few. It is therefore clear that there are potentially significant contributions from wild areas in general to the livelihoods and security of the rural poor. It is notable that the potential value of wild fish species is widely recognised and their exploitation for food and/or cash income is encouraged, whilst the exploitation of wild ungulates is usually strictly controlled, and under normal circumstances is not legally available as a resource to poor communities.

Biodiversity as a whole provides immense benefits to society (Koziell and Saunders, 2001; OECD, 2003) but they are difficult to quantify and largely ignored in national economic-accounting systems. As a result, markets undervalue biodiversity, thereby promoting its depletion, either directly or indirectly (Barbier, 1997; Tacconi, 2000). A major problem in considering the value of wildlife to the rural poor, and thus in making economic comparisons and policy decisions, lies in the exclusion of economic transactions from national statistics, as well as the techniques of valuation. Some exclusions from available statistics are deliberate, e.g. goods and services produced by women working at home are rarely included in national accounts. Other types of production are omitted because they are unpriced, unmarketed or unrecorded, or because producers or consumers have good reason to hide activities (e.g. illegal logging or hunting). Many of these activities are of particular importance to the poorest people, but the real value of wildlife and wild areas to resource-poor communities is overlooked. For example, estimates of crop damage are seldom compared with the benefits obtained from use of these same wildlife populations, irrespective of the legality of such utilisation. The consequences of crop damage are easily quantified, but hunting and the sale of wildlife products can result in significant unquantified benefits.

A similar problem is found in assessing the real value of rangelands. Typically, rangelands tend to be undervalued due to assessments that: i) are restricted to a specific sector, commonly livestock production; ii) biased towards a single marketed product, e.g. beef sales; and iii) exclude non-use values. The resulting undervaluation contributes towards poor rangeland management or a transformation to a monoculture, such as livestock alone, and leads to inappropriate policy recommendations. In some areas significant livestock farming subsidies make it more economic to keep cattle than to generate revenue from wildlife. In Botswana, Arntzen (1998) showed that livestock, wildlife and gathering, each make a significant contribution to the total direct use value, with hunting and gathering amounting to about a third of the total. Arntzen concluded that policies should pay greater attention to non-marketed products when assessing optimum uses for rangelands. Similarly, despite its often unlawful nature, DFID (2002) concluded that the bushmeat industry is important in securing the livelihoods of the rural poor, and that wildlife - poverty linkages are under-represented, in both poverty reduction strategies and country strategy papers.

In practice, economic utilisation of wildlife depends on utilitarian or aesthetic values, or both, and on policies regarding wildlife ownership and wildlife utilisation. Child (1995) discusses the history and evolution of wildlife utilisation in Zimbabwe and shows that large wild ungulate species are not significantly more efficient at converting grazing and browse into meat and hides than are livestock. In Zimbabwe, in areas where natural wildlife populations were dominated by a small number of species of bulk roughage feeders, these species have now been replaced by domestic cattle which have for centuries been bred for survival, with some meat production. Importantly, the behaviour of domestic stock enables them to be easily integrated into livestock production systems, unlike the great majority of wildlife species. However, the dressed weight of African wild animals is found to be 50 to 63 per cent of live weight, of which 2.5 per cent is typically fat, compared with 44 to 50 per cent of live weight in domestic stock, of which up to 40 per cent is fat (Ntiamoa-Baidu, 1997). In general, therefore, wild animals provide more lean meat than domestic animals. In addition, given an enabling policy environment, wildlife can be used in more ways than livestock. Their charismatic values allow them to be sold more than once through game viewing, providing an added-value that is clearly impossible with livestock-based production systems. These additional economic values can provide wildlife-based land uses, or integrated wildlife and livestock, with a competitive advantage over livestock-only systems. Child (1995) also demonstrated that there was little difference between costs of meat production using cattle or wildlife, but the latter became very competitive once safari hunting was introduced.

Problem animals

Problem animals and problem animal control have been a major focus of the wildlife - livestock and wildlife - agriculture interfaces. Indeed, many of Africa's wildlife management agencies started out with the primary aim of controlling problem animals, and even today many local communities in Africa view large wild mammals largely as a nuisance (Kiss, 1990). The extent and nature of the problem depends on the relevant production systems, as well as the wildlife species and frequency of contact. Crop and livestock damage includes direct consumption of crops and of grazing or browse, predation on livestock, rooting (e.g. by wild pigs), trampling, damage or flattening structures such as fences, and acting as carriers of weed species, parasites and diseases (Hone, 1994). Table 26.1 provides examples from the literature of types and costs of damage attributed to wildlife, much of which is caused by rodents. Wildlife damage to agriculture, by both rodents and larger wild herbivores, affects the production base of households. Such losses can make livelihoods that may already be insecure, become increasingly marginal in economic terms. Compared with rodents, large wildlife species generally result in smaller losses, particularly when considered on a national basis.

Table 26.1 Examples and cost estimates of problem animal damage

Locality	*Species*	*Damage*	*Reference*
China	Rodents	Loss of 15 million tonnes of cereals and vegetables	Stenseth *et al.*, 2003
East Africa	Multimammate rat, *Mastomys natalensis*	Loss of 5-15% annual maize crop: cost US$ 60 million. Up to 80% loss during outbreaks	Makundi *et al.*, 1999; Leirs, 2003; Skonhoft *et al.*, 2003
Kenya	Ground squirrel *Xerus* spp.	Loss of 9.7% planted maize seed	Stenseth *et al.*, 2003
Malawi	Large wildlife	All damage: estimated national cost US$ 17.3 million	Emerton, 1999a
Malawi	Large wildlife	Disease transmission: national cost US$ 27 million	Deodatus, 2000
Malawi	Predators	Predation of cattle: national cost < US$ 19,000	Deodatus, 2000
Morocco	Shaw's jird *Meriones shawi*	Loss of 40-70% cereal and vegetable crop during outbreaks	Stenseth *et al.*, 2003
Namibia	Large wildlife	All damage: US$ 757 per village in East Caprivi	Emerton, 1999a
Uganda	Large wildlife	All damage: national cost US$ 20 million	Emerton, 1999a

Pastoralists and wildlife have co-existed in African rangelands for hundreds of years. In the past, human and livestock populations were relatively small and widely dispersed, whilst livestock were managed to minimise predation and disease transmission risks by avoidance, for example, of particular areas or contact with certain species during specific periods. Today, however, competition for grazing and water resources is increasing, and the potential for conflicts between wildlife managers and livestock owners is growing, as both pastoralists and agro-pastoralists move into new areas and/ or live in the vicinity of protected areas (Boyd *et al.,* 1999; Prins, 2000). Factors driving these changes include population increase, an increasing urgency to protect the remaining wildlife and biodiversity resources and, perhaps most importantly, an expansion of cultivation into areas that formerly contained prime livestock-grazing resources, thereby reducing the total annual grazing resources available to livestock-keepers, and resulting in increased competition with livestock.

Pastoralist or transhumant societies typically view wildlife as predators, as a disease risk, or as competitors for scarce grazing resources or for water. Despite this, the costs attributed to predation are relatively low and, according to Grootenhuis (2000), the economic consequences of disease transmission to livestock from wildlife are generally overestimated. Although predation is no longer a significant threat, individual instances of predation, for example by lion or hyaena, can nevertheless be serious for the individual households involved.

Disease

Whenever wildlife and livestock share the same environment there are risks of disease transmission, and conflicts between livestock owners and wildlife managers are partly based on differing attitudes to disease control in livestock. In livestock-only systems (both ranching and pastoralism) the perceived risks to livestock of disease transmission from wildlife reservoirs has in the past led to large-scale eradication of larger wild herbivores over extensive areas in attempts at disease control (e.g. trypanosomiasis). However, there is little evidence that a reduction in wildlife has led to reduced incidence of disease or to lower costs of disease control in East Africa (Grootenhuis, 2000). It is clear that agents of disease are not confined to wild or domestic species, and transmission occurs in both directions. Disease transmission from livestock to livestock is more frequent, represents a far greater problem, and results in greater costs than wildlife to livestock disease transmission. Furthermore, disease problems are bi-directional at the wildlife/livestock interface, and livestock diseases may pass to wildlife and establish or re-establish wild reservoirs.

Some diseases are uncommon (or are rarely diagnosed); others are common but have little impact, whilst relatively few wild animal diseases are of major economic importance (Bourn and Blench, 1999). In practice, few diseases are transmitted between wild and domestic animals. Bengis *et al.* (2002) identify 12 diseases commonly transmitted from wildlife hosts to domestic livestock worldwide, four of which are identified as being of major economic importance. Similarly, Grootenhuis (2000) lists 19 diseases in Africa, of which four were considered of major economic importance: foot and mouth disease, rinderpest, East Coast fever, and trypanosomiasis, all of which have very effective domestic maintenance hosts (Table 26.2). According to Grootenhuis (2000), only three true wild maintenance and carrier hosts were identified from over 100 large wild animal species: warthog (African swine fever), wildebeest (malignant catarrhal fever), and bushbuck (bovine petechial fever).

Table 26.2 Examples of diseases transmissible between wild and domestic stock in Africa

Disease	*Domestic animals affected*	*Wild maintenance hosts*	*Typical domestic maintenance hosts*	*Control measures*
African swine fever	Pigs	Warthog, Bushpig	None	Avoid contact, resistant breeds
Anthrax	All livestock except poultry	None, abiotic soil phase (spores)	None	Surveillance, vaccination, cremation and burial of carcasses
Bovine petechial fever	Cattle	Bushbuck	None	Treatment, avoid risk zones
Brucellosis	Cattle (all live-stock except poultry)	African buffalo, hippopotamus, waterbuck	Cattle, goats	Avoid contact, vaccination, stamping out
East coast fever	Cattle	Buffalo	Cattle	Tick control, vaccination, treatment
Foot and mouth	Cattle, pigs, sheep, goats	African buffalo	Domestic ruminants, especially cattle	Vaccination, movement control
Malignant catarrhal fever	Cattle	Wildebeest	None	Avoid contact
Newcastle disease	Poultry	Wild birds	Chickens	Surveillance, vaccination
Rabies	All livestock except poultry	Carnivores, bats	Dogs	Vaccination of dogs
Rinderpest	Cattle	None	Cattle	Vaccination, movement control
Trypano-somiasis	Cattle, horses, pigs, sheep, goats, dogs	Ruminants and pigs	Ruminants, carnivores	Vector control, chemoprophy-laxis, treatment
Tuberculosis	Cattle	Buffalo and other wildlife species	Cattle	

Source: adapted from Grootenhuis (2000); Bengis *et al.* (2002); Perry *et al.* (2002)

Disease control is both possible and practical in domestic stock, but is usually impossible in wildlife. Furthermore, if control measures are poorly implemented in domestic stock, transmission to available wildlife reservoirs will inevitably occur. In such situations, the presence of disease in livestock will to a large extent mask the presence of wild reservoirs. It is only in countries with efficient veterinary services and where diseases amongst livestock are well controlled, that such wildlife reservoirs as do occur may become apparent. An important wildlife host is the African buffalo, but in their absence diseases are effectively maintained by cattle except in those situations where efficient control of disease in livestock exists. Rinderpest, for example, can be effectively controlled by vaccination of cattle, which results in the disappearance of this virus in wildlife (Bourn and Blench, 1999). The transmission of infectious agents from domestic disease reservoirs to sympatric wildlife is of particular concern (Daszak *et al.*, 2000). This spill-over effect can result in significant morbidity and mortality of wildlife in contact with infected livestock, and is of particular threat to endangered species, where it may lead to local population extinctions.

Community wildlife management

Control of the most important cattle diseases may not be cost-effective in some semi-arid rangelands (Grootenhuis, 2000), prompting a re-examination of current land-use pressures and development of more sustainable or more productive land-use options. Such analysis has already favoured the use of wildlife in many Southern African countries. These approaches, described for example by Child (1995), focus on the use of sparsely populated rangelands for wildlife or combined wildlife-livestock enterprises – typically on large-scale ranches or game conservancies (Barnes and de Jager, 1995) – with investment capital requirements far beyond the normal capacities of resource-poor livestock-keepers. Wildlife utilisation within a mixed crop and livestock smallholder farming system is also highly impractical for a variety of reasons, chief amongst which are that crops and wildlife are incompatible, and that commercial hunting, so far a major economic support of wildlife-oriented production, is both impractical and dangerous within a smallholder agricultural environment.

Community Wildlife Management (CWM) is an initiative that focuses on equipping resource users with the tools for sustainable use of available natural resource base in their jurisdictions. The premise of CWM is that communities will manage local resources in a sustainable manner if:

- they are assured of their ownership of the natural resources

- they are allowed to use the resources and/or benefit directly from others' use of them

- are given a reasonable level of control over management of the resources.

The first large-scale wildlife related CWM programme in Africa was established in Zimbabwe in the early 1980s, and is described by Child (1995). To protect wildlife, notably elephants, from excessive poaching, the government of Zimbabwe set up the CAMPFIRE programme (Communal Areas Management Programme For Indigenous Resources). Under this programme, authority over wildlife was given to Rural District Councils (RDCs). Conceived as an extension of game ranching to Zimbabwe's Communal Areas, the programme is of fundamental significance to natural resource conservation in the savannas of Africa. It has been emulated elsewhere, for example, the Luangwa Integrated Research and Development Project (LIRDP) and the Administrative Management Design for Community Based Wildlife management (ADMADE) in Zambia; and on communal land conservancies in Namibia (Child, 2000; Roe *et al.,* 2000). CAMPFIRE encourages sustainable trophy hunting of big game and revenue from fees paid by hunters goes to RDCs, a portion being distributed to lower administrative levels and/or individual households. Each RDC determines its own policy for the use and distribution of funds. Direct payments to households vary according to these policies and the availability of the most prized species of big game, e.g. elephant, buffalo, lion and leopard. Studies of selected wards show increases in wildlife populations and habitat retention and these were considered indictors of success (Getz *et al.,* 1999). The programme provides revenue that can make significant contributions to community livelihoods and provides motivation to conserve wildlife habitat and associated biodiversity.

Despite CAMPFIRE, local attitudes by smallholders in Zimbabwe towards wildlife generally remain negative (Campbell *et al.*, 1999). Roe *et al.* (2000) suggest that there are serious flaws in the implementation of these CWM models. Wildlife damage crops, are viewed as responsible for livestock mortality, and also pose dangers to human life. The CAMPFIRE programme is also seen as remote from individual aspirations, and as a government initiative, not a tool for empowerment. Much of the benefit goes towards community development initiatives, many previously funded by Government. Research shows that in the Sengwa area it is 5-10 times more lucrative from a household perspective to illegally hunt an animal than wait for the CAMPFIRE dividends (Campbell *et al.*, 2000). A study of LIRDP and ADMADE projects in Zambia (Gibson, 1999) concluded that illegal off-take of wildlife continued at pre-project levels, partly because individual returns from hunting greatly outweighed an individual's share in benefits from the projects. Such approaches are unlikely to be sustainable alongside mixed smallholder agriculture. In particular, where revenues depend on hunting, the lucrative wildlife species represent threats to smallholder agriculture and livestock. Nevertheless, there is evidence from elsewhere indicating positive ecological impacts from CWM and that it has enhanced the livelihood security of participating communities (Kothari *et al.*, 2000).

Wildlife – poverty linkages

Prevailing physical and socio-economic conditions in wildlife-rich areas generally mean that sources of employment, income and subsistence are relatively scarce and livelihoods are insecure. This is particularly so in the case of households in semi-arid areas engaged in mixed rain-fed production systems. Furthermore, settlements situated close to protected areas are typically less well-served by roads, infrastructure, markets and other services than settlements further away. As a result people tend to engage in a wide range of economic activities in search of secure livelihoods, and these impact on wildlife, e.g. by over-exploitation of resources, through both legal or illegal hunting, and secondary impacts resulting from land clearance for agriculture.

As a general rule, in mixed rain-fed production systems, land is a basic determinant of wealth, with livestock a valuable contributor, and the proportion of food needs derived from 'own crop production' increases with wealth. The smaller the area cultivated and fewer numbers of livestock managed, the poorer the household will be and the greater expenditure on food as a portion of household budget. Examples include: 70.4 per cent on food in Kitui district, Kenya (Barnett, 2000); and 40 per cent in Mchinji district, Malawi (Save the Children UK, 2001). The advantages of using cheaper or no-cost sources of protein are therefore considerable. Despite a lack of information on consumption of wildlife meat or 'bushmeat' and its total production, bushmeat is highly valued and a preferred animal protein in both rural and urban diets in many parts of Africa, and resource-poor people have a greater reliance on bushmeat than generally assumed (Ntiamoa-Baidu, 1997; DFID, 2002). In parts of East and Southern Africa bushmeat, either legally or illegally obtained, is regarded as the most important source of meat to households (Barnett, 2000). Critical factors include affordability relative to domestic sources, relative accessibility in wildlife-rich areas and, in Western Africa, cultural preferences.

In several areas bushmeat is now regarded as an important contributor to gross domestic product and national revenues (Barnett, 2000). Keita (1993) recorded a 10 per cent contribution of bushmeat to GDP in the Central African Republic. Bushmeat consumption in Côte d'Ivoire was estimated at 83,000 tonnes in 1990, valued at US$ 117 million (Feer, 1993) and three quarters of Liberia's meat production comes from bushmeat, with subsistence hunting yielding 105,000 tonnes of meat annually (Ntiamoa-Baidu, 1997). Near Korup National Park, Cameroon, hunting provided 56 per cent of total village income (Roe *et al.*, 2000). The extent and value of the bushmeat trade is also illustrated by the emergence of an illegal market for West African bushmeat in a number of European capitals with estimates of 10 tonnes a day passing through London's Heathrow airport, and 1.4 tonnes seized from a single flight (Bowen-Jones *et al.*, 2002). By contrast, in East and Southern Africa, the importance of bushmeat has been relatively poorly documented and its role in resource-poor households is less well understood, due to the largely illegal nature of consumptive utilisation of wildlife in these areas. However, Barnett (2000) underlines the importance of wildlife to many Eastern and Southern African societies. Despite the difficulty of obtaining data on what are essentially illegal activities, in Tanzania the utilisation of bushmeat was found to represent the largest economic value of wildlife, significantly greater than legal hunting, trophy values, or tourism (ITC and IUCN, 1998). DFID (2002) estimated that for forest and savanna systems, bushmeat is worth over US$ 1.5 billion a year as a source of food for the resource-poor of Sub-Saharan Africa alone. Declining wildlife populations will therefore present a major threat to livelihoods and food security. Furthermore, the value of wild products compares favourably with domestic livestock. In a study centred on Lake Mburo National Park, Uganda, Emerton (1999b) found that 61 per cent of the value was provided by livestock and 39 per cent by wild products, of which 41 per cent was from hunting. Since many activities were illegal, these figures may underestimate the real values of wildlife to households.

Illegal hunting has reduced wildlife populations of Tanzania's Serengeti ecosystem (Campbell and Borner, 1995). Reduced wildlife populations may in turn undermine local livelihoods partly dependent on this resource. Hofer *et al.* (2000), Campbell *et al.* (2001) and Loibooki *et al.* (2002) examined illegal hunting from the twin perspectives of conservation and livelihoods. The primary value of wildlife to resource-poor households was realised as dried meat sold within the local community or urban markets, and the funds used to purchase other goods and services. Cash was the most important contribution of wildlife to the rural poor, especially in times of hardship (e.g. poor rainfall, or stock theft). In these mixed crop-livestock production systems, hunting in wild areas represents a viable and profitable means for resource-poor people to generate cash incomes. For a villager, a day's hunting may produce an income equivalent to over 100 days of potential earnings through formal employment (Hofer *et al.*, 2000).

Participatory research and household surveys indicated that hunting was closely identified with the poorest people (Campbell and Loibooki, 2000). In particular, hunters had fewer livestock (or none) than the general village population. Information from arrested hunters demonstrated that the main reasons for hunting (75 per cent of arrests) were economic, rather than for food. Almost 80 per cent of the income from hunting was needed to pay taxes, village development contributions or levies, fees for education, and purchase of clothing (Table 26.3), clearly emphasising the need for cash. Based on recorded household consumption, the value of bushmeat in areas to the west of Serengeti was estimated at US$ 800,000 per annum. Including the likely trade to urban areas increased the value to US$ 987,000 per annum (Barnett, 2000). In strong contrast, it was estimated that legal community wildlife cropping schemes could, if extended to all villages north-west of Serengeti, reach a value of US$ 13,500 (Barrow *et al.*, 2000). Emerton and Mfunda (1999) estimated economic impacts of wildlife on landholders west of Serengeti to be roughly US$ 1 million per annum, but excluded benefits from illegal hunting of wildlife.

Table 26.3 Reasons given by hunters arrested in Serengeti National Park for needing a cash income provided by illegal hunting

Reason	*% of responses*
Taxes	32.8
Contributions/Levies, including schools	31.1
Clothes	15.1
Poverty	8.4
Debt	4.8
Medicine	3.5
Food or hunger	2.8
General purchasing needs	1.1
Other reasons	0.4

Source: Adapted from Campbell *et al.* (2001).

A number of factors were found to influence the probability of involvement in illegal hunting, itself strongly linked to poverty (Campbell *et al.*, 2001). The four most significant factors were:

- Distance to urban areas: Fewer hunters closer to urban areas and markets due to the increased income generating opportunities presented

- Distance to nearest main road: Fewer hunters closer to all-weather roads

- Livestock ownership: Livestock owners less likely to be involved in hunting. This effect was strongest for poultry, followed by sheep and goats. Cattle ownership had little or no effect

- Cash crops: Cash crop income reduced the necessity for hunting.

These results illustrate the importance of rural income diversification, and especially small stock ownership, to the reduction of rural poverty. Furthermore, the implications are that development initiatives must be specifically targeted at the poorest members of rural communities, many of whom have no livestock, but for whom its ownership is desirable. Encouraging livestock ownership, improved husbandry, cash crops and market access as parts of an integrated package of measures are likely to reduce poverty levels and to increase livelihood security. Without these combined measures, illegal hunting of wildlife to generate much needed cash will continue to provide a vital, although increasingly unsustainable, means of support for the poorest households.

Further reading

Campbell, B. M., Sithole, B., Frost, P., Getz, W., Fortmann, L., Cumming, D., du Toit, J. and Martin, R. 2000. CAMPFIRE experiences in Zimbabwe. *Science* 287(5450): 42.

Child, G. 1995. *Wildlife and people, the Zimbabwean success.* Wisdom Foundation, Harare, Zimbabwe and New York, USA.

Gibson, C.C. 1999. *Politicians and poachers: The political economy of wildlife policy in Africa.* Cambridge University Press, Cambridge, UK.

Kiss, A. (ed.). 1990. *Living with wildlife: Wildlife resource management with local participation in Africa.* World Bank Technical Paper No. 130, The World Bank, Washington, D.C., USA.

Kothari, A., Pathak, N. and Vania, F. 2000. *Where communities care: Community based wildlife and ecosystem management in South Asia.* Evaluating Eden Series, No. 3, International Institute for Environment and Development, London, UK.

Koziell, I. and Saunders, J. (ed.). 2001. *Living off biodiversity: Exploring livelihoods and biodiversity issues in natural resources management.* International Institute for Environment and Development, London, UK.

Ntiamoa-Baidu, Y. 1997. *Wildlife and food security in Africa.* FAO Conservation Guide 33, Food and Agriculture Organisation of the United Nations (FAO), Rome, Italy. http://www.fao.org/docrep/w7540e/w7540e00.htm

Prins, H.H.T., Grootenhuis, J.G. and Dolan, T.T. (ed.). 2000. *Wildlife conservation by sustainable use*. Kluwer Academic Publishers, The Netherlands and USA.

Roe, D., Mayers, J., Grieg-Gran, M., Kothari, A., Fabricius, C., and Hughes, R. 2000. *Evaluating Eden: Exploring the myths and realities of community based wildlife management*. Evaluating Eden Series, No. 8. International Institute for Environment and Development, London, UK.

Tacconi, L. 2000. *Biodiversity and ecological economics*. Earthscan Publications, London, UK.

Author addresses

Ken Campbell, Formerly of Natural Resources Institute (NRI), University of Greenwich at Medway, Chatham Maritime, Kent ME4 4TB, UK

Martin Loibooki, Tanzania National Parks, P.O. Box 3134, Arusha, Tanzania

27

Lessons learned and the way ahead

Christie Peacock, *Canagasaby Devendra*†, *Aichi Kitalyi*†, *John Best*‡, *Czech Conroy*‡

Introduction

Livestock have supported the livelihoods of humankind for thousands of years and will continue to do so for years to come. This book has shown the extraordinary diversity of livestock systems in the world and the problems commonly confronting livestock-keepers. It has also highlighted that in most developing countries livestock are kept by some of the poorest members of society and their improved husbandry and marketing has the potential to offer the resource-poor an important pathway out of poverty to greater wealth.

In countries that have experienced economic growth, a new generation of largely urban-dwellers has emerged with rising incomes and the desire to improve their diets through the consumption of increasing quantities of animal products. This widespread and increasing demand for animal products, the 'Livestock Revolution' (Delgado *et al.,* 1999; Owen *et al.*, 2004), presents a unique opportunity for the rural resource-poor to benefit from broader national economic growth, and for wealth to move from urban elites to the rural poor. How can this potential be unlocked? What must be done to ensure that the poor benefit from investment in livestock, and not just the wealthy?

In Chapter 3 it was argued that future efforts need to target priority agro-ecological zones and, within these, priority livestock systems in individual regions – the rationale being that these zones and systems offer the greatest opportunities for improved productivity and contribution from livestock, and for reducing poverty. Thornton *et al.* (2004) estimated that of the 600 million livestock-keepers globally, the mixed rain-fed systems contain the largest number (approximately 430 million) of resource-poor livestock-keepers. The critical regions are Sub-Saharan Africa and South Asia. Opportunities for increasing livestock productivity and livelihoods

---

* Principal author
† Co-author
‡ Consulting author

of the resource-poor in rain-fed environments world-wide were discussed in Chapter 3 and in Asia by Devendra (2000).

Key lessons from the past

It is important that lessons are learned from past livestock development interventions so that mistakes are not repeated, and the good practice that has emerged is built on and applied more widely.

A review of donor-funded livestock interventions concluded that in general they had performed poorly and had difficulty in catering to the needs of the poor, finding that wealthy livestock owners/investors had benefited most from external funding (LID, 1999). This poor performance, particularly in terms of poverty-reduction, was attributed to:

- An absence of a poverty-focus and specific targeting of the poor.
- Application of inappropriate technology to inappropriate species.
- Absence of service delivery to the poor.
- The capture of project benefits by the wealthy.

The underlying reasons identified for these failings are weaknesses in the institutional framework within which livestock interventions operate, such as:

- Under-performing public sector research and extension organisations, without a poverty-focus.
- Outdated and inhibiting national and international legislation (e.g. on issues related to animal health, land tenure, etc.).
- Weak private-sector development.
- Weak representation of the interests of the poor, particularly in policy formulation and reviews.

However there have been successes too in livestock development. When looking at these successes, certain key elements emerge as being essential:

- Empowered and enabled livestock-keepers.
- Access to services.
- A move towards market-orientation in livestock production, including adding value through processing.
- Investment in pro-poor livestock research.
- A strengthened legal and institutional framework to support pro-poor policies.

What can be done to counter the problems and build on the successful elements to support pro-poor strategies that empower and enable livestock-keepers to improve the production and returns from their livestock?

Key elements of successful livestock development

Empowered and enabled livestock-keepers

In order to achieve pro-poor economic growth it is imperative that resource-poor livestock-keepers are involved right from the start of any livestock intervention. In order to do this, resources must be allocated to identifying them, targeting them and planning, *with* them, appropriate interventions.

Targeting and wealth ranking

As indicated earlier, there is an increasing amount of information on where resource-poor livestock-keepers live, globally and nationally (cf. Chapter 2; Thornton *et al.*, 2004), and on their systems of production (cf. Chapter 3). This information can help direct investment into the most appropriate geographical locations. Once there, it is important to identify the poorest farmers. This can be done through a wealth-ranking exercise involving all members of the community, government and religious leaders, as well as men and women, old and young (cf. Chapter 4). Agreement should be reached among community members on their criteria for defining a resource-poor family - often livestock ownership itself is a key criterion, as well as landholding, quality of house, off-farm income, state of health etc. – and families ranked and targeted accordingly. Particular care needs to be taken in the process to ensure that the local elite do not bias decisions and the needs and concerns of the poorest families are incorporated. The needs of women- and child-headed households, or HIV/AIDS affected households, for example, are often of particular concern.

Participatory planning and management

The start of any development initiative is the time to ask fundamental questions about the situation to be improved and what constitutes an improvement. It is imperative that the existing situation is assessed by livestock-keepers themselves and that they actively participate in defining their problems and developing realistic solutions (Chapters 4 and 5).

In order to ensure the long-term sustainability of any livestock intervention it must be clear, over time, what roles will be played by livestock-keepers, farmer organisations, the government and the private sector. Livestock-keepers need either to be empowered to demand services for themselves, or enabled to supply their own services, through individual service providers or their own farmer organisations or groups.

Role and potential of farmer organisations and groups

There is no doubt that farmers acting as a group are stronger than when they act alone. Groups and organisations can serve a variety of purposes including:

- Mutual support and encouragement.
- Provision of non-formal micro-finance, through savings and credit schemes.
- Breeding services (e.g. sire services, AI, registration, etc.).
- Veterinary services (e.g. clinical, diagnostic, etc.).
- Feed/forage.
- Cost-effective input supply (e.g. feed, drugs, forage planting material, etc.).
- Technical support and training.
- Product collection, bulking and processing (e.g. milk, honey, etc.).
- Improve access to markets.
- Increase bargaining power and effective lobbying.

Cameo Case Study 1 is an example of a new farmer organisation, the Meru Goat Breeders Association, set up by very poor goat-keepers in Kenya. This organisation was also used to illustrate good practice in Chapters 12 and 19.

Cameo Case Study 1 - The Meru Goat Breeders Association

Christie Peacock[1]

The Meru Goat Breeders Association (MGBA) was formed to manage the breeding of dairy goats in two districts in Kenya. Its members were selected to benefit from acquiring goats through a credit scheme and were the poorest members of society, frequently poorly educated casual labourers. After four years the MGBA has become a hugely successful farmers' organisation able to manage the breeding programme (benefiting over 30,000 people), register and market breeding stock throughout Kenya and East Africa, and has recently started to provide members with technical advice as well as access to new breeding stock. The MGBA is a membership organisation run by a committee elected by members. Fees are charged for all services provided. The MGBA is now looking at ways in which value can be added to fresh goat milk though processing and marketing in Nairobi. Technical advice and inputs were provided by the NGO FARM-Africa and by staff of the Ministry of Rural Development, who will give the MGBA support in the future.

[1]Principal author

The interests of livestock-keepers are often poorly represented in governments and so their voices and concerns are not heard. As a result, livestock policies are often outdated and unhelpful, and limited investment is directed towards the sector. Farmer organisations can serve to represent their members and lobby government for better support. For example, an initiative on capacity building and strengthening of farmer organisations in East Africa led to the formation of the Agricultural Council of Uganda which has increased participation of farmers in policy and trade negotiations (RELMA, 2003).

Access to services

In order to improve production, it is important that livestock producers have access to reliable and affordable support services offering them access to knowledge and inputs, including credit and other financial services.

Input supply

Historically, in many developing countries, government extension and veterinary departments have provided services to livestock-keepers. These services were frequently subsidised to some extent and were often concentrated in the higher-potential districts, leaving marginalised livestock-keepers, such as pastoralists or the landless, under-served. The public sector reforms of the 1990s have led many of these public services to be cut back and in some cases withdrawn altogether. While the private sector has emerged in some countries to fill the gap, they typically provide services to wealthier livestock farmers such as dairy farmers and commercial poultry and pig keepers, leaving the poor even more marginalised from these vital services (Chapters 5 and 13).

In recent years new models of public-private service provision have developed which enable different actors to develop their strengths (Table 27.1).

Knowledge and information

Many livestock-keepers have kept their livestock for generations and have expert knowledge of their husbandry. Pastoralists are genuine livestock experts and have traditional treatments for animal diseases.

However livestock-keepers vary in their knowledge and so there is always a need to ensure that the best available information is accessible to them. Everyone learns best 'by doing' and there are various extension approaches used very successfully in livestock programmes, such as the Farmer Field School approach (Minjauw *et al.*, 2004).

In a recent study in Kenya it was found that farmers 'outside' a successful livestock project had acquired most of their information about the technology from churches and schools rather than government extension staff (Davis, 2004). There are many novel approaches to disseminating information that could be used in order to reach livestock-keepers, e.g. radio, at local markets, through religious leaders, schools, social groups etc. (cf. Chapter 6). Cameo Case Study 2 illustrates how two families in Bolivia achieved better livelihoods through practising improved livestock husbandry as a result of information gained from a CIAT project.

Table 27.1 Support services to livestock-keepers during the 1990s

	New roles played	*Problems*
Government	• Policy development and legislation • Regulation to ensure food safety, quality of inputs, compulsory vaccination, movement restrictions, etc. • Facilitation of farmer organisations, private sector, NGOs • Ensure issues of public benefit addressed	• Unreformed government • Attitude of staff • Inappropriate/lack of staff performance management and promotion criteria • Inflexible & bureaucratic • Unresponsive to needs of resource-poor • Political interference • Under-resourced to play new role • Fears of loss of status
Public sector research	• Pro-poor research strategy • Working on problems of all species not just large ruminants • Regular interaction with resource-poor livestock-keepers to understand problems and develop and test new technologies	• Lack of poverty focus • Inappropriate incentives and staff promotion criteria • Staff attitude
Private sector	• Innovative marketing and input supply • Some technical advice • Some new product development • Formal credit and insurance	• Risk averse • Few 'role models' • Lacking in long term vision of market potential • Lack of confidence in government
Farmer organisations	• Respond to farmers' needs • Bargaining power to purchase inputs and sell products • Provision of more flexible credit and financial products	• Potential to be taken over by elites • Political interference
Community groups	• Mutual and flexible support • Informal micro-finance for those ineligible for formal credit • Sharing and group management of valuable resources, e.g. breeding sire	• Political interference • Lack of appropriate legislation to enable them to develop

Cameo Case Study 2 - Empowered by knowledge and information: the story of two families in Bolivia

Rob Paterson[1]

Consuela and Eulalia live with their families in the community of El Salvador, a village of about 40 households. It is located in the area of Santa Rosa, about 120 km north-west of the city of Santa Cruz, in the tropical lowlands of Bolivia. Both families have similar farms, of about 50 ha each, but they use only simple hand tools and neither has enough capital to work more than two or three hectares at a time. Staple crops of maize and rice are grown every year, but each plot can be used only for a couple of years before it must be abandoned back to bush fallow, because of rapidly falling soil fertility and increasing problems with annual weeds. A new plot must then be cleared, either from forest or from bush regrowth, for the next cropping cycle. Consuela is from the nearby town of Portachuelo, while her husband, Benjamín, was born and bred in Santa Rosa. In contrast, Eulalia and Marco came to El Salvador from the highland Bolivian Department of Oruro soon after they were married some 15 years ago, to search for a new life in the sparsely populated tropical lowlands.

As well as the staple crops, Consuela and her family have always grown a wide range of fruits and vegetables, including citrus fruits, bananas, cassava, potatoes, lettuce, onions and tomatoes, while wild fruits are harvested from naturally occurring trees on the farm. They have several types of poultry, including chickens, ducks and guinea fowl, together with a couple of beef cows and a sow. The animals scavenge for what they can find to eat around the house, along the roadside and beside the stream that runs through the farm. Benjamín frequently hunts for meat in the forest and will sometimes bring back a live armadillo or a jochi from his hunting trips. These are fattened until they are slaughtered in time for a family celebration. The family has always been short of money, because off-farm casual work is hard to find and they sell farm produce, some grain or perhaps a steer, only when they have to pay school fees or medical bills. When this happens, Benjamín chooses what to sell and he administers the funds. Their whole operation is based on subsistence, but the farm produces a varied diet for all the family and after lunch, when the older children come home from school, there is usually time for leisurely and amusing conversation.

About two years ago, the family began to work with the Small Animals Project of CIAT, the nearby international agricultural research and development centre. Benjamín had little interest in the project when it started, because he did not see the importance of small animal species to the family, but Consuela and the children were enthusiastic. The first steps were to borrow a boar from a neighbouring community and to dose the sow against internal parasites. Following the advice of CIAT, Consuela planted some extra cassava and reserved some bananas for the sow, to provide a supplementary ration after farrowing. With the extra feed, the sow produced plenty of milk and the piglets grew big and strong. They were also dosed at two months of age to control worms and instead of losing half of the litter as usual, only two of the eight offspring died before weaning at about three months. The family slaughtered two of the pigs and gave a party for their friends in the community to celebrate Easter. Three more were sold in Portachuelo at a good price, while the best female was kept for breeding.

Benjamín was so pleased with the results that he gave Consuela the return from one of the piglets when it was sold. She spent part of the money buying wire netting. This was used with poles and thatch from the farm, to build a simple night shelter for the poultry. All the birds quickly adapted to laying their eggs in nest boxes placed in the shelter and this gave both the eggs and the hatchlings protection from the weather during the coldest part of the year and from attack by natural predators from the surrounding forest. Soon, there were more eggs and poultry meat than ever before on the table. The improvements in the family diet were there for all to see and Consuela quickly received the recognition that she deserved. With sufficient poultry to eat at least one bird each week, Benjamín was able to limit his trips into the forest, hunting only when he wanted to, rather than when meat was needed for the table. The local Mothers' Club visited the farm to see the changes that had been made and the whole family enjoyed their improved status in the community, as Consuela and the children were able to help their neighbours to design simple installations to shelter their small livestock.

Eulalia and her family have a fully market-oriented operation, although, because of its small scale, the surplus for sale has always been modest. As well as the staple crops grown by all of their neighbours, they produce water melons and oranges, which Eulalia sells from a roadside stall. They have a small area of planted pasture where they keep two cross-bred milking cows to make soft cheese, most of which Eulalia also sells to regular customers. As well as helping with the milking, the children provide minimal care to a small flock of scavenging chickens and cut fresh forage daily for a few guinea pigs that run free in the kitchen building. Despite coming from the highlands, the guinea pigs have few health problems in the tropics, but occasionally they are attacked by dogs from the neighbouring farms. When this happens, almost the whole flock can be killed in a single night. In common with most of his contemporaries from the highlands, Marco spends most of his time on the farm, tending to the crops and the cattle, while Eulalia handles all the sales and administers the family finances. Because they receive income from sales in almost all months of the year, she is able to plan the expenditure and seldom has to make forced sales to meet emergencies.

Eulalia watched with interest while Consuela built the shelter for her poultry and she was soon impressed by the obvious increase in the numbers of young birds. She bought some wire netting, using money from the sale of some of the farm produce and, with the help of her neighbour, she designed and built a night shelter for the chickens. There was some netting left over, which she used to make a small cage for the guinea pigs. She set this up on legs so that it was about a metre above the ground and out of the reach of dogs. Both shelters were thatched with palm fronds cut from the farm.

The guinea pigs settled easily into their new cage, where they suffered less disturbance than when they had lived in the kitchen. There was no obvious effect on productivity, but since the move there have been no more losses to dogs. This is important, because guinea pigs require little time for preparation and cooking and so are commonly eaten at times of land preparation and harvest, when the whole family is busy working in the field. As on Consuela's farm, the chickens prospered in their new environment and there was no need to carry out the regular morning search for eggs laid in the bushes during the night. Within a few months, there was a clear increase in the productivity of the flock, but Eulalia saw no need to increase the level of family consumption. Rather, she was now able to increase the family income through the sale of either eggs or live chickens, depending on her reading of the local market. She plans to save the extra money until she is able to buy some grafted seedlings of lemons, grapefruit and mandarins. When these trees start to bear fruit, there will be a larger variety of produce for her to sell at her roadside stall.

Small animal species can make a considerable contribution to the livelihoods of small-scale farmers, through either increased consumption or greater, regular income. In the Bolivian tropics, their potential has long been ignored by the local agricultural authorities and is only now being recognised.

[1]Principal author

Credit, insurance and livestock banks

Much 'successful' livestock development has relied on the provision of credit to enable poor people to acquire livestock or new breeding stock. This acquisition of livestock, for the first time, or to help re-stock people who have lost their animals, may be a first step out of poverty to a more secure and prosperous life.

Livestock credit can take many forms, from formal credit systems from a government institution or private bank with repayment in cash, to more informal credit schemes with repayment perhaps 'in-kind'. The less formal schemes are often promoted by NGOs and can be very effective at reaching the poor who would not be eligible for formal credit. These types of credit programmes generally rely on group management and peer pressure to ensure timely repayments. They can be self-administered by even illiterate people and the revolving fund managed in this way is 'inflation-proof' and can increase and multiply benefits very widely with repaid stock lent to new members. Very few of these group credit systems, with repayments 'in-kind', have been successfully adopted by the more formal institutions. There is scope for farmer organisations to manage revolving funds in this way for the benefit of their members. Credit and micro-credit schemes in Bangladesh have been particularly successful, as illustrated by Cameo Case Study 3.

Cameo Case Study 3 - Credit is crucial: the story of a landless livestock-keeper and her family in Bangladesh

David Barton[1], M Saadullah[2]

Dulali and her husband Amin live in Agbikromati village in Tangail District, Bangladesh. They have three children, all attending primary school. They own only eight decimals of land (0.032 ha) which makes up their homestead. Dulali cultivates vegetables in the homestead and keeps livestock, and Amin drives a hired rickshaw in the district town centre. Amin earns about Taka 20,000 (US$ 345) a year from his rickshaw work and Dulali about Taka 2,500 (US$ 43) a year from vegetables. They sought membership of a local NGO two years ago to access credit, to help them save so that they could improve their social standing and find a good husband for their daughter (and also afford the dowry that this husband would require). They joined a group of like-minded households (a condition of NGO membership) and demonstrated to the NGO their ability to save by lodging/saving Taka 100 with this organisation. Once this money had been saved they were eligible for a loan. The group is collectively responsible for repayment and any defaulting on repayment affects the credit worthiness of the group as a whole. In this way NGOs receive some guarantee of repayment (peer pressure) and are able to lend to poor households that have little or no physical assets or collateral.

Dulali first took a loan from an NGO in April 2000. The NGO provided two days of training on livestock rearing, health-care and first aid of livestock to help ensure the livestock enterprise was a success. The first loan was for Taka 3,000 and Dulali bought a young bull for fattening for Taka 2,800. The fattened bull was sold for Taka 8,000 11 months later. The sale of the bull was timed to coincide with a Muslim religious festival when many animals are slaughtered and market prices are high. The NGO required Dulali to pay weekly instalments of capital plus interest (Taka 75 per week for 50 weeks) on the loan, which were paid out of the earnings of Amin and from selling vegetables and chickens. The interest rate was around 20 per cent per annum which, although quite high, is considerably less than the interest charged by local moneylenders (often 20 per cent per month).

A second loan, Taka 5,000, was taken shortly afterwards and this money was used to buy another young bull for Taka 3,000 and books for their son (Taka 2,000). After one year the second bull was sold for around Taka 10,000. The family did not keep any account of the cost of fattening (forage, rice bran and veterinary services) so they are not sure of the exact profit they gained, however they are certain that they made money on the enterprise. A third loan (Taka 5,000) was taken for another young bull, which had not been fully repaid at the time of the interview (December, 2002). Part of this money (Taka 2,000) was used to buy a second-hand rickshaw for Amin, as previously he had to hire one on a daily basis. The family had not kept any livestock before their first loan, apart from a few chickens, but they intend to continue with this beef fattening business, as it has been successful.

As well as loan repayment, NGO members are also required to deposit savings on a weekly basis during the repayment period. Taka 10 per week is saved in this way, and their savings account has grown over the past 2 years to around Taka 1,000, and they receive a small amount of interest on this money. The family had never saved before they joined the NGO.

The additional income that livestock has provided has enabled the family to buy some furniture, improve their house by making a tin (corrugated iron) roof, constructing a tube-well for safe drinking water and building a latrine. Their daughter's education would have ceased without the extra income. They are now able to eat good-quality food like fish and meat, where previously they ate only vegetables. They believe that their social status has improved as neighbours comment on how their children are all in school. People now visit their house and they are proud to receive and greet their fellow villagers.

[1]Principal author
[2]Co-author

Livestock for many people represent their main, or only, asset and as such deserve protection, through good husbandry and veterinary care and, wherever possible, insurance. Livestock insurance has been offered only sporadically in livestock programmes, but offers a realistic opportunity to underpin the livelihoods of the poor, giving them the confidence to invest more in developing their valuable livestock assets.

Cameo Case Study 4, in South Africa, is an example of credit scheme in the form of a 'livestock bank' to help poor people invest in livestock.

Cameo Case Study 4 - Livestock 'banks' in South Africa

Herman Festus[1]

In the Northern Cape of South Africa many communities have benefited from the country's land reform programme, enabling them to have access to land for the first time.

Access to land, however, is merely the first step in a process of material benefits for the majority of land-reform beneficiaries. Access to finance, training, infrastructure, amongst others, remains an ellusive step for many. The financially-stronger beneficiaries quickly fill up the communally-owned land with their own livestock.

The FARM-Africa Northern Cape livestock banks now make it possible for the poorest beneficiaries to become livestock owners. Livestock ownership delivers two key benefits to these resource-poor farmers, namely:

- a precious income-generating asset for life, under very favourable terms and conditions
- an opportunity to share in the benefits of communal land ownership

Under the livestock bank, individual beneficiaries are granted loans in the form of ten female sheep or goats. The borrower is given a maximum of 30 months to repay ten young healthy females plus an extra four animals which could be either females or males. The extra four animals represent the interest on the loan, and would normally be sold for cash and deposited in the group's revolving fund for other projects. The repaid ten females are then given as a loan to a new beneficiary.

The group sets it own rules for the livestock bank, with monthly monitoring and evaluation of the scheme by a community livestock bank sub-committee.

Since the introduction of livestock banks in 2002, 236 households have benefited from this scheme.

[1]Principal author

Market-orientation and adding value: the keys to success?

The most successful livestock development programmes have enabled resource-poor livestock farmers to generate greater income from their valuable livestock assets through more stable and intensive market-focused production.

From subsistence to market orientation

In order for resource-poor livestock farmers to capture some of the benefits offered by a rising demand for livestock products, where the opportunity arises they need to move increasingly from a subsistence orientation to a more market orientation. Livestock are valuable assets for the poor and greater returns need to be generated from them. In order to do this, and sell an increasing quantity of livestock and livestock products, production needs to become more secure, reliable and, in most cases, intensified to generate more income per livestock unit over time.

Resource-poor livestock farmers often have barriers to accessing markets, including remoteness, poor infrastructure and poor market information. In some countries livestock product prices may be depressed by subsidised imports (cf. Chapter 8).

The Amul Model (Annand Milk Union Ltd.) in India is a successful case to cite (cf. Chapter 22), where organised markets and value adding has benefited resource-poor livestock-keepers. The Amul model has a unique farmer need-based integration between production, processing and marketing. This is built on a complementary and synergistic relationship between a large-scale modern dairy factory and large numbers of small-scale producers. A wide range of products, including milk powder, butter, cheese, chocolates and malted beverages, are produced (more information is available on www.nddb.coop/).

Livestock within social welfare

The pivotal role that livestock play in contributing to the security of the lives of the very poor has been noted elsewhere (especially Chapters 2 and 5).

However, consumption of animal products in developing countries is low. While people in developed countries obtain on average 27 per cent of their energy and 56 per cent of their protein from animal products, people in developing countries obtain only 11 and 26 per cent, respectively. Analysis of trends in consumption levels predicts a rapid growth of 2.8 and 3.3 per cent annually to 2020, for meat and milk in developing countries (Delgado *et al.*, 1999). However, this needs to be backed-up with awareness-raising campaigns. The school milk programme, supported by a number of government, non-government and multilateral development agencies, is one of the initiatives (Griffin, 2004). The programme is aiming at building a milk-drinking culture in the future generations.

As discussed in Chapter 7, milk, eggs and meat can provide protein as well as micro-nutrients. The vital role small quantities of animal products can play in healthy child development, as well as in adult health, is now well known. Vitamin A deficiency is widespread in the developing world and can cause night blindness.

In pastoral societies in Sub-Saharan Africa, consumption of animal products has gone down due to decreasing herd/flock sizes per household. Decreasing herd size coupled with diminishing grazing resources is a major cause of food insecurity in these areas. More effort in developing feeding strategies for these areas is a pre-requisite to increasing the livestock-keeper's welfare in pastoral communities. Pastoral communities in East Africa are potential sources of beef, if only more investments were put into infrastructure development.

Cameo Case Study 5 is an example from Ethiopia of the role of goats in food security.

Cameo Case Study 5 - Example from Ethiopia of role of goats in food security in very food insecure areas

Christie Peacock[1]

After years of war and regular droughts, and resulting famine, there were many impoverished, women-headed households on the edge of survival. These households typically had a small plot of land but no assets, not even a chicken, or realistic way of improving their lives. The British NGO FARM-Africa started a Dairy Goat Development Project to assist women and their families in this difficult situation. Two local female goats were provided on credit to each selected woman. Women were organised into small groups of 25-30 who were responsible for managing the repayment of the credit, which was re-paid in-kind by returning a kid to the group, who would select another woman to benefit. Women were trained in forage production, improved management and basic health-care and were given the opportunity to up-grade their goats through cross-breeding with Anglo-Nubian bucks. Two women per group were trained as community-based animal health-care workers (CBAHW) providing basic veterinary care for a small fee to members of the group and their neighbours.

Women quickly built up their herd of goats and in some cases were able to sell some goats to buy oxen for ploughing or a milking cow. During the severe drought of 1999-2000 households with livestock to sell were able to buy grain and survive the drought without resorting to food aid handouts. Selling one goat could buy enough grain to feed a family of five for two to three months.

[1]Principal author

Research issues

Pro-poor research agenda

Public funding to agricultural research relative to the contribution agriculture makes to national GDP is low in most developing countries (e.g. share of total agricultural GDP that developing countries invest in agricultural research is <1.0 per cent; in developed countries it is >2.5 per cent (Pardey and Bientema, 2001). This has led to a high dependency on donor funding, which has sometimes led to premature termination of research projects. At the same time, public-sector research organisations serving the needs of developing countries seldom adopt an explicit pro-poor research agenda. An analysis of research programmes will usually reveal a focus of research capacity on the needs of cattle or buffalo, to the exclusion of small ruminants, chickens and other 'minority' livestock, e.g. camels. While large ruminants may make a larger aggregate contribution to the formal national economy than smaller livestock, the crucial role played by smaller livestock in supporting poorer families should be recognised in research agendas.

There needs to be more investment in livestock research and development (R and D) from the standpoint of pro-poor agricultural growth. Past research investments in agricultural research have had major pay-offs, and clearly justify increased investments to strengthen future efforts (Pardey and Bientema, 2001).

There are many problems confronting smaller livestock that receive little research attention, e.g. sarcoptic mange in goats, developing an effective thermo-stable CCPP (contagious caprine pleuro-pneumonia) vaccine (cf. Chapter 19).

Undertaking systems R and D necessitates understanding many issues, such as the environment, natural resource management, crop-animal interactions and whole-farm analyses, in order to support the success of livestock farming activities and the real benefits to the poor. Informal training is part and parcel of the broader community-based participatory efforts and is a continuing one, but will need to be stressed.

Sustainable livestock farming necessitates understanding of the potential of livestock farming systems. There are two elements to this:

- an adequate knowledge of livestock farming systems research to provide improved understanding and support for livestock farming activities and societal issues

- a capacity to conduct systems R and D due to the complex interactions between, say, nutrition and genotype, reproduction and production system and the environment.

Links to other sectors

Livestock can compete with, or complement, other natural resource sectors, and any livestock intervention should be assessed for its impact on other sectors.

Crop production

There are many direct and indirect interactions between crop and livestock production, which need to be considered before any intervention in either enterprise. For example, the introduction of new, shorter varieties of crops may have an impact on by-product forage-feed supply. Conversely, the introduction of forage crops, particularly leguminous crops, into a cropping system can have a beneficial effect on the main crop, e.g. tree legumes (cf. Chapter 11). The intensification of livestock production will often mean that stock are zero-grazed/stall-fed or housed, enabling the easier collection of manure which is valuable for crop production.

The importance of the forestry sector in income and food security is growing in most developing countries. Furthermore, growing interests in low-external-input agriculture, both in terms of efficient use of natural resources and environment protection, have shown positive agro-forestry-livestock interactions. The Indian Grassland and Forage Research Institute, in collaboration with a sister institution, the Agroforestry Research Centre, has developed silvi-pasture management strategies, which can contribute to improving the husbandry of animals kept by the resource-poor (Pathak and Newaj, 2003).

Wildlife issues

Livestock often graze in areas used by wildlife species and they may compete for scarce grazing and water resources, as well as transmit diseases to each other (Sherman, 2002). In some countries wildlife can make a significant contribution to the national economy and be the basis for a tourist industry (cf. Chapter 26). Different countries have different approaches to wildlife conservation and management. Some conserve wildlife by strictly excluding both livestock and humans. Others may seek to find a way for livestock and wildlife to co-exist for their mutual benefit. As the human population expands, the competition between livestock and wildlife will only increase. This is an area where resource-poor livestock-keepers are likely to lose out unless their needs are considered at the highest level of policy formulation.

The environment

Expanding and intensifying livestock production can have both positive and negative impacts on the environment (cf. Chapter 9). Livestock have in recent years been considered harmful to the environment and have been accused of causing desertification, deforestation and pollution (LID, 1999). LID (1999) identified environmental concerns over livestock production, as well as competition for grain for human consumption and fears over the transmission of zoonotic diseases, as reasons for the recent lack of investment in livestock production. However, well-planned and -managed livestock development can have a very positive effect on the environment, for example through the growing of legumes for forage, the use of animal manure to improve the structure and nutrient content of poor soils, and the use of animal draught power rather than fossil-fuel power.

A legal and institutional framework to support pro-poor policies

There are many issues of policy at national and international levels that can affect resource-poor livestock-keepers directly or indirectly that will need to be modified to support pro-poor growth. These include:

Land and water rights

Livestock kept by resource-poor people often make use of communal grazing and water resources (cf. Chapter 3). However population growth is placing these communal resources under mounting pressure and in most countries they are shrinking in size. The legal recognition of traditional rights over grazing resources, in the face of competition from crop cultivators, wildlife reserves and other users, may be essential to secure the livelihoods of many livestock-keepers.

Veterinary legislation

In many countries current veterinary legislation may date back over half a century and often prevents the broader access of livestock producers to veterinary services. Much veterinary legislation is designed to regulate an exclusively state veterinary service and to preserve the professional status of qualified veterinarians. As state funding has declined, new models of veterinary service provision are being explored, including the use of trained livestock-keepers as community-based animal health-care workers (CBAHW) (cf. Chapters 5, 13). These 'para-professionals' represent a new cadre of animal-care workers and require new legislation to regulate them and so enable them to play an appropriate and supportive role within both state and private sector veterinary services.

Rights of women and children

In many countries the lack of rights of women over property, including livestock, can prevent them gaining access to formal credit and may mean that livestock are taken away from them on the death of their husband.

Children often play a vital role in tending grazing livestock during the day, which may prevent them from going to school. The enforcement of compulsory school attendance may cause a change in the way livestock are kept because of lack of labour to look after them during the day. In some societies boys may be allowed to go to school, leaving girls to look after the livestock.

Globalisation, international trade and animal health issues

The increasing global demand for animal products offers new opportunities for livestock-keepers in the developing world to access new international and regional markets (cf. Chapter 8). However, the sanitary and phyto-sanitary control measures imposed by countries in Europe and North America are very stringent and place huge demands on developing countries to attain high standards of disease surveillance and control (James, 2004; Upton and Otte, 2004). Some developing countries, such as Brazil and Thailand, have managed to penetrate developed-country markets and meet the standards required, but this is rare.

For many countries it is probably best to focus on national and regional trade in livestock and livestock products. There may still be many trade rules that will need to be negotiated to facilitate this and there will always be concern over disease control with any movement of livestock, e.g. Saudi Arabia banning live animal imports from East Africa because of concerns over Rift Valley fever.

The role of the government and public expenditure in pro-poor livestock development

Investment in agriculture and livestock by both donors and national governments has shrunk to less than half over the last 10 years (Eicher, 2003). Increasing proportions of public expenditure are being directed to the social services of education and health. This falling investment in productive sectors, such as agriculture and livestock, has led to agricultural growth in most countries barely keeping pace with population growth.

As far as Africa is concerned, the recent advent of the African Union and NEPAD give cause for optimism. As stated in the Foreword "The African Union Commission (AUC) has recognised as critical, the role of livestock in the socio-economic development of most farming communities in Africa. Recognising that institutional frameworks are essential, the AUC has established four Specialised Technical Offices on livestock in its Department of Rural Economy and Agriculture (AU-DREA). These, working in close collaboration with the New Partnership for Africa's Development (NEPAD) will implement livestock-focused poverty- and hunger-reduction strategies based on the coordination and harmonisation of policies for improved livestock production."

Ways forward – specific issues

Technical innovations to improve livestock production often cost money, thus there is a major constraint for resource-poor smallholders. However, the objective of this book is to raise livelihoods above the subsistence level, and inevitably, at some point, this will justify cash outlay to increase output of marketable livestock products. As indicated in earlier chapters, ways forward include considering the following:

- Need to plan ahead, both for feeding and other aspects of husbandry.
- Need to keep adequate records of major livestock events.
- Need to pool indigenous knowledge.

- Economic consequences of efficient and cheap delivery of veterinary services.
- Use of genetically-suitable animals coupled with amelioration of the environment and management.
- Development of AI services, both to increase the potential of livestock and remove the need for scrub bulls to be kept (these can be up to 20 per cent of the herd).
- Assessment of the future for mechanisation. Will draught animal power (DAP) remain a major aspect of African agriculture or will there be a spread of the developments in many parts of Asia?
- Impact of water harvesting (e.g. tied ridges for planting; collection of roof water – impact on supply and labour).
- Land tenure.
- Growing of improved forage.
- Conservation of forage.
- Making full use of locally-available novel feeds.

A vision for the future

Reducing global poverty, particularly in Sub-Saharan Africa and Asia, must be considered humankind's biggest challenge in the 21st century. It is clear that livestock-keepers represent some of the 'poorest of the poor' and that improving the production of their livestock is one of the best methods of securing their livelihoods and transforming their lives from one of poverty to relative prosperity. This book has attempted to show how this could be done. What is required now is the political will, investment and sustained personal commitment of all those involved.

Author addresses

Christie Peacock, FARM-Africa, 9-10 Southampton Place, London WC1A 2EA, UK

Canagasaby Devendra, Formerly of the International Livestock Research Institute (ILRI), P.O. Box 30709, Nairobi, Kenya. Current address: 130A Jalan Awan Jawa, 58200 Kuala Lumpur, Malaysia

Aichi Kitalyi, Regional Land Management Unit (RELMA) at World Agroforestry Centre (ICRAF), P.O. Box 30677, Nairobi 00100, Kenya

John Best, School of Agriculture, Policy and Development, University of Reading, Earley Gate P.O. Box 237, Reading RG6 6AR, UK

Czech Conroy, Natural Resources Institute (NRI), University of Greenwich at Medway, Chatham Maritime, Kent ME4 4TB, UK

Rob Paterson, Formerly of Natural Resources Institute (NRI), University of Greenwich at Medway, Chatham Maritime, Kent ME4 4TB, UK

David Barton, D Barton UK Ltd., Iyletts Farm, West Bourton, Gillingham, Dorset SP8 5PF, UK

M. Saadullah, Formerly of Department of Animal Science, Bangladesh Agricultural University, Mymensingh 2202, Bangladesh. Current address: 4/6 Maya Kanan, Sabujbagh, Dhaka, Bangladesh

Herman Festus, FARM-Africa NC Land Reform & Advocacy Programme, 4th Floor, Trust Bank Building, Jones Street, P.O. Box 2410, Kimberley 8300, N. Cape, South Africa

List of DFID Livestock Production Programme (LPP) research projects referred to in the book

R number	*Chapter*	*Project title*	*Research venue*
R5188	11	Improving the use of sorghum stover as ruminant feed in Ethiopia	Ethiopia
R5194	11 & 19	Tethering of small ruminants in Tanzania: purpose and implications	Tanzania
R5198	25	Feeding and management strategies for draught animals in Sub-Saharan Africa	Morocco, Niger, Zimbabwe, Ethiopia
R5926	25	Improving the productivity of draught animal power in Sub-Saharan Africa	Zimbabwe
R6166	25	Effects of feed quality and time of access to feed on feeding behaviour and nutrient intakes of tropical cattle and donkeys	Ethiopia, Zimbabwe
R6610	11	Introduction of fodder legumes into rice-based cropping systems and their use as supplements in straw-based rations for dairy cattle in Bangladesh	Bangladesh
R6619	11	Husbandry strategies for improving the sustainable utilisation of forages to increase profitable milk production from cows and goats on smallholder farms in Tanzania	Tanzania
R6953	11	Easing seasonal feed scarcity for small ruminants in semi-arid crop/ livestock systems through a process of participatory research	India
R6970	25	Improved management and use of draught animals in the Andean hill farming systems of Bolivia	Bolivia

R number	*Chapter*	*Project title*	*Research venue*
R6993	11 & 20	Effects of harvest and post-harvest practices on the production and nutritive value of maize and sorghum residues in Zimbabwe	Zimbabwe
R7010	11	The production of high quality silage from adapted forage and legume crops for the maintenance of dairy cow productivity on smallholder farms through the dry season in the semi-arid region of Zimbabwe	Zimbabwe
R7350	25	Use and management of donkeys by poor societies in peri-urban areas in Ethiopia	Ethiopia
R7351	11	Increasing the productivity in smallholder owned goats on Acacia thornveld	Zimbabwe
R7424	19	Can feeding locally available plant materials rich in tannins reduce parasitic burden in ruminants and hence improve their productivity?	Tanzania
R7425	6 & 11	Development, validation and promotion of appropriate extension messages and dissemination pathways	Kenya
R7432	11	Participatory development of community-based management plans for livestock feed resources in the semi-arid areas of Zimbabwe	Zimbabwe
R7524	11 & 16	The use of oilseed cake from small-scale processing operations for inclusion in rations for peri-urban poultry and small ruminant production	Zimbabwe
R7633	16	The use of alternative, tanniniferous, saponin and antioxidant containing materials as a means of improving the health and production of scavenging (desi) poultry	India

List of DFID LPP research projects

R number	*Chapter*	*Project title*	*Research venue*
R7855	11	Analysis, management and decision support for farmers' feeding strategies: Talking Pictures II	Tanzania, Kenya, India
R7955	11	Strategies for feeding smallholder dairy cattle in intensive maize forage production systems and implications for integrated pest management	Kenya
R8109	2 & 19	Using livestock to improve the livelihoods of landless and refugee-affected livestock keepers in Bangladesh and Nepal	Nepal & Bangladesh

Details of these projects can be found in the projects portfolio of the following website: http://www.lpp.uk.com. DFID's Livestock Production Programme is due to end in March 2006, so for information on where project information is stored post March 2006, please contact: info@nrint.co.uk.

References

Aarts, G. Sansoucy, R. and Levieux, G.P. 1990. *Guidelines for the manufacture and utilisation of molasses-urea blocks*. Animal Production and Health Division, Food and Agriculture Organisation of the United Nations (FAO), Rome, Italy.

Abate, A. and Abate, A.N. 1991. Wet season nutrient supply to lactating grade animals managed under different production systems. *East African Agricultural and Forestry Journal* 57: 33-39.

Abbott, J. 1993. *Agricultural and food marketing in developing countries: selected readings*. CABI Publishing, Wallingford, UK.

Abdullah, N., Ho, Y.W., Mahyuddin, M. and Jalaludin, S. 1990. Comparative studies of fibre digestion in cattle and buffaloes. In *Domestic buffalo production in Asia*, pp. 75-87. Panel Proceedings Series, International Atomic Energy Agency (IAEA), Vienna, Austria.

Abeygunawardena, H., Subasinghe, D.H.A., Jayatilaka, M.W.A.P., Perera, A.N.F. and Perera, B.M.A.O. 1996. Development of intensive buffalo management systems for smallholders in human settlement schemes in the dry zone of Sri Lanka. *Proceedings of the Second Asian Buffalo Association Congress,* pp. 63-75, Manila, the Philippines.

Abeygunawardena, H., Perera, B.M.A.O., Subasinghe, D.H.A. and Siriwardena, J.A. de S. 1998. Livestock research and development strategies aimed at smallholder farmers: A case for greater farmer participation. *Sri Lanka Veterinary Journal* 45: 1-14.

Acamovic T., Chandrasekaran, D., Natarajan, A., Anitha, K. and Sparks, N.H.C. 2004. Effects on production traits of feeding tannin-rich sorghum to desi poultry. Paper presented at *World's Poultry Congress*, Istanbul, Turkey, June, 8-13, 2004 (proceedings on CD).

Acamovic, T. and Brooker, J.D. 2005. Biochemistry of plant secondary metabolites and their effects in animals. *Proceedings of the Nutrition Society* 64: 1-10.

Acha, P.N. and Szyfres, B. 2003. *Zoonoses and communicable diseases common to animals and man. Three volumes.* Pan American Health Organisation, Washington D.C., USA.

Adu, E.K. 2003. Patterns of parturition and mortality in weaned greater cane rats (*Thryonomys swinderianus,* Temminck). *Tropical Animal Health and Production* 35: 425-431.

Adu, E.K., Alhassan, W.S. and Nelson, F.S. 1999. Smallholder farming of the greater cane rat, *Thryonomys swinderianus,* Temminck, in southern Ghana: A baseline survey of management practices. *Tropical Animal Health and Production* 31: 223-232.

Adu, E.K., Aning, K.G., Wallace, P.A. and Ocloo, T.O. 2000. Reproduction and mortality in a colony of captive greater cane rat, *Thryonomys swinderianus,* Temminck. *Tropical Animal Health and Production* 32: 11-17.

Afifi-Affat, K.A. 1998 Heifer-in-trust: a model for sustainable livestock development. *World Animal Review* 91: 13-20.

Adebambo, O.A., Ikeobi, C.O.N., Ozoje, M.O., Adenowo, J.A. and Osinowo, O.A. 1999. Colour variations and performance characteristics of the indigenous chicken of South-Western Nigeria. *Nigerian Journal of Animal Production* 26: 15-22.

AFRC (Agricultural and Food Research Council). 1993. *Energy and protein requirements of ruminants.* CABI Publishing, Wallingford, UK.

AFRC (Agricultural and Food Research Council). 1998. *The nutrition of goats. Technical Committee on Responses to Nutrients Report No. 10.* CABI Publishing, Wallingford, UK.

Ahuya, C.O. and Okeyo, A.M. 2002. Sustainable genetic improvement of goat meat and milk production in Kenya: A case of the Meru and Tharaka-Nithi Dairy and Animal Healthcare Community Based Breeding Programme. In *Proceedings of the Regional Consultative Workshop on Cattle and Goat Breeding Policy in East Africa*, United Kenya Club, April 15-17, 2002, Nairobi, Kenya.

Aidoo, K. 1999. The Saltpond hive. *Bees for Development Journal* 50: 6-7.

Alders, R.G. 2001. Sustainable control of Newcastle disease in rural areas. In *SADC planning workshop on Newcastle disease control in village chickens* (ed. R.G. Alders and P.B. Spradbrow), pp. 80-90. Proceedings of an International Workshop, Maputo, Mozambique, 6-9 March 2000. ACIAR Proceedings. No. 103.

Alexander, D.J., Bell, J.G. and Alders, R.G. 2004. *A technical review: Newcastle disease with special emphasis on its effect on village chickens. FAO Animal Production and Health Paper 161,* Food and Agriculture Organisation of the United Nations (FAO), Rome, Italy.

Al-Garib, S.O., Gielkens, A.L.J., Gruys, E. and Koch, G. 2003. Review of Newcastle disease virus with particular reference to immunity and vaccination. *World's Poultry Science Journal* 59: 185-200.

Animal Traction Network for Eastern and Southern Africa (ATNESA) website: www.atnesa.org

Anon. 1993. *Vetiver grass: the hedge against erosion.* The World Bank, Washington D.C., USA.

Anon. 2002. *Statistical book on livestock 2002.* Direktorat Jenderal Bina Produksi Peternakan. Departemen Pertanian R.I., Jakarta, Indonesia.

Anon. 2003. newsletter@amazonia.org.br, *Amazon News*, 17 July 2003. Reported in *Bees for Development Journal*, 2003, 69: 10.

Anta, E., Rivera, J.A., Galina, C.S., Porras, A. and Zarco, L. 1989. Análisis de la información publicada en México sobre eficiencia reproductiva de los bovinos. II. Parámetros reproductivos. *Veterinaria México* 20: 11-18.

ARC (Agricultural Research Council). 1980. *Nutrient requirements of ruminant livestock.* Commonwealth Agricultural Bureaux, CABI Publishing, Wallingford, UK.

Arntzen, J. 1998. *Economic valuation of communal rangelands in Botswana: a case study.* CREED Working Paper No. 17. International Institute for Environment and Development (IIED), London, UK, and Institute for Environmental Studies (IVM), Amsterdam, The Netherlands. http://www.biodiversityeconomics.org/pdf/topics-601-00.pdf

Ayantunde, A.A., Fernandez-Rivera, S. and McCrabb, G. (ed.). 2005. *Coping with feed scarcity in smallholder livestock systems in developing countries.* Animal Sciences Group, Wageningen UR, The Netherlands, University of Reading, Reading, UK, ETH (Swiss Federal Institute of Technology), Zurich, Switzerland, and ILRI (International Livestock Research Institute), Nairobi, Kenya.

Baratou, J. 1998. *Raising snails for food.* (Translated from French Herb, F. 1981. *Les escargots).* Illumination Press, Calistoga, California, USA.

Barbier, E.B. 1997. *Economic valuation of wetlands: A guide for policy makers and planners.* Ramsar Convention Bureau, Gland, Switzerland.

Barnes, J.I. and de Jager, J.L.V. 1995. *Economic and financial incentives for wildlife use on private land in Namibia and the implications for policy.* Directorate of Environmental Affairs, Ministry of Environment and Tourism, Namibia, Research Discussion Paper 8, September 1995. Windhoek, Namibia.

Barnes, R.D., Filer, D.L. and Milton, S.J. 1996. *Acacia karroo: monograph and annotated bibliography. Tropical Forestry Papers No. 32.* Oxford Forestry Institute, Oxford, UK.

Barnett, R. (ed.). 2000. *Food for thought: The utilisation of wild meat in Eastern and Southern Africa.* TRAFFIC/WWF/IUCN, Nairobi, Kenya.

Barrow, E., Gichohi, H. and Infield, M. 2000. *Rhetoric or reality? A review of community conservation policy and practice in East Africa. Evaluating Eden Series*, No. 5. International Institute for Environment and Development (IIED), London, UK.

Bath, G. and de Wet, I. 2000. *Sheep and goat diseases.* Tafelberg Press, Cape Town, South Africa

Baudelaire J.P. 1972. Water for livestock in semi-arid zones. *World Animal Review* 3:1-9.

Bayer, W. and Waters-Bayer, A. 1998. *Forage husbandry.* The Tropical Agriculturalist, Series Editors R. Coste and A.J. Smith. Macmillan Education Ltd., London, UK, in cooperation with Technical Centre for Agricultural and Rural Cooperation (CTA), Wageningen, The Netherlands.

Baxter, P.T.W. 1994. Pastoralists are people: why development for pastoralists not the development of pastoralism? *The Rural Extension Bulletin* No. 4: 3-8.

Beaugerie, L., Carbonnel, F., Carrat, F., Rached, A.A., Maslo, C., Gendre, J.P. 1998. Factors of weight loss in patients with HIV and chronic diarrhea. *J. Acquir. Immune. Defic. Syndr. Hum. Retrovirol.* 19: 34-9.

Behnke, R. 1994. Natural resource management in pastoral Africa *Development Policy Review* 12: 5-28.

Behnke, R., Scoones, I. and Kerven C. (ed.). 1993. *Range ecology at disequilibrium: New models of natural variability and pastoral adaptation in African savannas.* Overseas Development Institute (ODI), London, UK.

Bender, A. 1992. *Meat and meat products in human nutrition in developing countries. Food and Nutrition Paper No 53.* Food and Agriculture Organisation of the United Nations (FAO), Rome, Italy.

Bengis, R.G., Kock, R.A. and Fischer, J. 2002. Infectious animal diseases: the wildlife/livestock interface. *Scientific and Technical Review of the World Organisation for Animal Health. (Rev. sci. tech. Off. Int. Epiz.)* 21(1): 53-65. http://www.oie.int/eng/publicat/en_revue.htm.

Bennison, J.J. and Paterson, R.T. 1993. *Use of trees by livestock 3: Gliricidia.* Natural Resources Institute, Chatham, UK.

Ben Salem, H., Ben Salem, I., Nefzaoui, A. and Ben Said, M.S. 2003. Effect of PEG and olive cake feed blocks supply on feed intake, digestion, and health of goats given kermes oak (*Quercus coccifera*) foliage. *Animal Feed Science and Technology* 110: 45-59.

Ben Saleem, H., Nefzaoui, A. and Ben Said, L. 2004. Spineless cacti (*Opuntia ficus* f.*inemis*) and oldman saltbush (*Altriplex nummularia* L.) as alternative supplements for growing Barbarine lambs given straw-based diets. *Small Ruminant Research* 51: 65-73.

Berlo, D. 1964. *The process of communication. An introduction to theory and practice.* Rinehart Press, San Francisco, USA.

Betker, J. and Kutzbach, H-D. 1991. Role of donkeys in agricultural mechanisation in Niger - potential and limitations. In *Donkeys, mules and horses in tropical agricultural development* (ed. D. Fielding and R.A. Pearson), pp. 223-230. Proceedings of a colloquium held in Edinburgh, 3-6 September 1990, University of Edinburgh, Edinburgh, UK.

BHA (Brooke Hospital for Animals). 2003. *Brooke News,* Spring 2003.

Birthal, P.S. and Parthasarathy Rao P. 2002. Economic contribution of livestock sub-sector in India. In *Technologies for sustainable livestock production in India,* pp. 12-19. Proceedings of a workshop on documentation, adoption and impact of livestock technologies in India, 20-21 January, 2001, International Crops Research Institute for the Semi-Arid Tropics (ICRISAT) Centre, Patancheru, India.

Blake, R.W. and Nicholson, C.F. 2004. Livestock, land use change and environmental outcomes in the developing world. In *Responding to the Livestock Revolution – The role of globalisation and implications for poverty alleviation* (ed. E. Owen, T. Smith, M.A. Steele, S. Anderson, A.J. Duncan, M. Herrero, J.D. Leaver, C.K. Reynolds, J.I. Richards and J.C. Ku-Vera), pp. 133-153. British Society of Animal Science Publication No. 33, Nottingham University Press, Nottingham, UK.

Blench, R.M.B. 2001. " 'til the cows come home" Why conserve livestock diversity? *Living off biodiversity: exploring livelihoods and biodiversity issues in natural resource management* (ed. I. Koziell and J. Saunders). International Institute for Environment and Development (IIED), London, UK. (Chapter 9).

Blench, R. 2001. *You can't go home again: pastoralism in the new millennium.* ODI Report for the Food and Agricultural Organisation of the United Nations (FAO), Overseas Development Institute (ODI), London, UK. http://www.odi.org.uk/pdn/eps.pdf (Chapter 3).

Blench, R., Chapman, R. and Slaymaker, T. 2003. *A study of the role of livestock in poverty reduction strategy papers. FAO, Pro-poor Livestock Policy Initiative Working Paper No. 1.* Food and Agriculture Organisation of the United Nations (FAO), Rome, Italy.

BMC. 1986. *Annual Report 1985.* Botswana Meat Commission (BMC), Gaborone, Botswana.

Board on Science and Technology for International Development, National Research Council. 1991.

Microlivestock: Little-known small animals with a promising economic future. National Academy Press, Washington, D.C., USA.

Bourn, D. and Blench, R. (ed.). 1999. *Can livestock and wildlife co-exist? An interdisciplinary approach.* Overseas Development Institute (ODI), London, UK.

Bowen-Jones, E., Brown D. and Robinson E. 2002. *Assessment of the solution-orientated research needed to promote a more sustainable bushmeat trade in Central and West Africa.* Report produced for DEFRA (DETR) Wildlife and Countryside Directorate. Department for Environment, Food and Rural Affairs (DEFRA), UK. http://www.defra.gov.uk/wildlife-countryside/resprog/findings/bushmeat.pdf

Boyd, C., Blench, R., Bourn, D., Drake, L. and Stevenson, P. 1999. *Reconciling interests among wildlife, livestock and people in eastern Africa: a sustainable livelihoods approach. ODI Natural Resource Perspectives Number 45.* Overseas Development Institute (ODI), London. http://www.odi.org.uk/fpeg/publications/policybriefs/nrp/nrp-45.pdf

Braenkaert, R.D.S., Gaviria, L., Jallade, J. and Seiders, R.W. 2000. Transfer of technology in poultry production for developing countries. Paper presented at *World's Poultry Congress, Montreal, Canada, August 20-24, 2000* (proceedings on CD).

Bravo, P.W. 2003. *The reproductive process of South American camelids.* Tenino, WA, USA.

Bravo-Baumann, H. 2000. *Gender and livestock: Capitalisation of experiences on livestock projects and gender.* Working Document, Swiss Agency for Development and Cooperation, Bern, Switzerland.

Brockington, D. and Homewood, K. 1996. Wildlife, pastoralists and science: debates concerning Mkomazi Game Reserve, Tanzania. In *The lie of the land, challenging received wisdom on the African environment* (ed. M. Leach and R. Mearns), pp. 91-104. International Africa Institute, James Currey Limited and Heinemann, London, Oxford and Portsmouth, UK.

Brocklesby, D.W. and Sewell, M.M.H. 1990. *Handbook of animal diseases in the tropics. Fourth edition.* Bailliere Tindall, London, UK.

Bruggeman, H. 1994. *Pastoral women and livestock management: Examples from Northern Uganda and Central Chad. Issue Paper No. 50.* International Institute for Environment and Development, London, UK.

Bruinsma, J. (ed.). 2003. *World agriculture: towards 2015/2030 An FAO perspective.* Earthscan, London, UK, and Food and Agriculture Organisation of the United Nations (FAO), Rome, Italy.

Brumby, P. and Gryseels G. 1985. Stimulating milk production in milk deficit countries of Asia and Africa. *Milk production in developing countries* (ed. A.J. Smith), pp. 62-72. Centre for Tropical Veterinary Medicine, University of Edinburgh, Edinburgh, UK.

Buckland, R. and Guy, G. 2002. *Goose production. Animal Health and Production Paper No 154.* Food and Agriculture Organisation of the United Nations (FAO), Rome, Italy.

Bustinza, V. 1984. 'The camelidae of South America', In *The camelid. An all-purpose animal. Proceedings workshop, Khartoum, 1979, Volume 1,* pp. 112-143. Scandinavian Institute of African Studies, Uppsala, Sweden.

Calnek, B.W., Barnes, H.J., Beard, C.W., Reid, W.M. and Yoder Jr., H.W. 1991. *Diseases of poultry. Ninth edition.* Iowa State University Press, Iowa, USA.

Campbell, B.M., Sithole, B., Cavendish, W., Frost, P. and Mukamuri, B. 1999. *Peace parks: International aspirations vs. local agendas. IES Policy Briefs 2.* Institute of Environmental Studies (IES), University of Zimbabwe, Harare, Zimbabwe.

Campbell, B.M., Sithole, B., Frost, P., Getz, W., Fortmann, L., Cumming, D., du Toit, J. and Martin, R. 2000. CAMPFIRE experiences in Zimbabwe. *Science* 287(5450): 42.

Campbell, K.L.I. and Borner, M. 1995. Population trends and distribution of Serengeti herbivores: Implications for management. In *Serengeti II: research, management and conservation of an ecosystem.* (ed. A.R.E. Sinclair and P. Arcese), pp. 117-145. University of Chicago Press, Chicago, USA.

Campbell, K.L.I. and Loibooki, M. 2000. Game meat hunting in the Serengeti: a problem of sustainable livelihoods. *Conference on African Wildlife Management in the New Millennium*. College of African Wildlife Management, Mweka, Tanzania, 13-15 December 2000.

Campbell, K.L.I., Nelson, V. and Loibooki, M. 2001. *Sustainable use of wildland resources, ecological, economic and social interactions: An analysis of illegal hunting of wildlife in Serengeti National Park, Tanzania. Final Technical Report to DFID, Animal Health and Livestock Production Programmes*. Department for International Development (DFID), London, UK.

Carles, A.B. 1983. *Sheep production in the tropics*. Oxford University Press, Oxford, UK.

Carles, A.B. 1992. The non-medicinal prevention of livestock disease in African rangeland ecosystems. *Preventive Veterinary Medicine* 12: 165-173.

Carney, D. (ed.). 1998. *Sustainable rural livelihoods: What contribution can we make?* Department for International Development (DFID), London, UK.

CAST (Council for Agricultural Science and Technology). 1999. *Animal agriculture and global food supply. Task Force Report No*. 135. Ames, Iowa, USA.

Catley, A. 1997. *Non governmental organisations and the delivery of animal health services in developing countries: A discussion paper*. UK Department for International Development (DFID), London, UK.

Catley, A. 1999. *Methods on the move: a review of veterinary uses of participatory approaches and methods focusing on experiences in dryland Africa*. International Institute for Environment and Development (IIED), London, UK.

Catley, A., Blakeway, S. and Leyland, T. (ed.). 2002. *Community-based animal healthcare: A practical guide to improving primary veterinary services*. ITDG Publishing, London, UK.

Cernea, M. (ed.). 1991. *Putting people first: Sociological variables in rural development*. Oxford University Press, Oxford, UK, for the World Bank, New York, USA.

Chantalakhana, C. 1992. Genetics and breeding of swamp buffalo. *World animal science, Volume C6: Buffalo production* (ed. N.M. Tulloh and J.H.G. Holmes), pp. 95-109. Elsevier, Amsterdam, The Netherlands.

Charray, J., Humbert, J.M. and Levif, J. (Translated by Alan Leeson). 1992. *Manual of sheep production in the humid tropics of Africa*. CABI Publishing, Wallingford, UK, in collaboration with the Technical Centre for Agricultural and Rural Co-operation (CTA), Wageningen, The Netherlands.

Chigaru, P.R.N., Maundura, L. and Holness, D.H. 1981. Comparative growth, food conversion efficiency and carcass composition of indigenous and Large White pigs. *Zimbabwe Journal of Agricultural Research* 19: 31-36.

Child, G. 1995. *Wildlife and people, the Zimbabwean success*. Wisdom Foundation, Harare, Zimbabwe and New York, USA.

Child, B. 2000. Making wildlife pay: Converting wildlife's comparative advantage into real incentives for having wildlife in African savannas, Case studies from Zimbabwe and Zambia. In *Wildlife conservation by sustainable use* (ed. H.H.T. Prins, J.G. Grootenhuis and T.T. Dolan), pp. 335-388. Kluwer Academic Publishers, The Netherlands and USA.

Clauss, B. and Clauss, R. 1991. *Zambian beekeeping handbook*. Mission Press, Ndola, Zambia.

Cobbinah, J.R. 1994. *Snail farming in West Africa: a practical guide*. Technical Centre for Agriculture and Rural Co-operation (CTA), Wageningen, The Netherlands.

Cockrill, W.R. (ed.). 1974. *The husbandry and health of the domestic buffalo*. Food and Agriculture Organisation of the United Nations (FAO), Rome, Italy.

Codex Alimentarius. 1989. *Codex standards for sugars (honey). Supplement 2 to Volume III*. Food and Agriculture Organisation of the United Nations (FAO), Rome, Italy.

Coetzer, J.A.W. and Tustin, R.C. (ed.). 2004. *Infectious diseases of livestock. Three volumes*. Oxford University Press, Cape Town, South Africa.

Cohen, J. and Uphoff, N. 1977. *Rural development participation: Concepts and measures for project design, implementation and evaluation*. Cornell University, Ithaca, USA.

Cole, H.H. and Garrett, W.N. (ed.). 1980. *Animal agriculture. The biology, husbandry and use of domestic animals*. W.H. Freeman and Company, San Francisco, USA.

Collins, P. and Solomon, G. (ed.). 1999. *Proceedings of the First Caribbean Beekeeping Congress*, Tobago 1998. Tobago Apicultural Society, Trinidad and Tobago.

Colombatto, D., Mould, F.L., Bhat, M.K. and Owen, E. 2003. Use of fibrolytic enzymes to improve the nutritive value of ruminants' diets. A biochemical and *in vitro* rumen degradation assessment. *Animal Feed Science and Technology* 107: 201-209.

Condorena, N. 1980. Algunos índices de producción de la alpaca bajo el esquema de esquila anual establecido en 'La Raya'. *Rev. Inv. Pec. (IVITA) Uni. Nac. M.S. Marcos* 5(1): 50-54.

Conroy, C. 2005. *Participatory livestock research: a guide.* ITDG Publishing, London, UK.

Conroy, C. and Rangnekar, D. 1999. Livestock and the poor in rural India with particular reference to goat keeping. Paper presented at the DSA Annual Conference, September 12-13, 1999, University of Bath, Bath, UK.

Conroy, C., Sparks, N., Joshi, A.L. and Chandrasekaran, D. 2004a Getting the villagers' perspective on poultry-keeping. *Proceedings of the 3rd National Seminar on Rural Poultry for Adverse Environment,* held on 24th and 25th February 2004 at Veterinary College and Research Institute, Namakkal, Tamil Nadu, India.

Conroy, C., Sparks, N., Acamovic, T., Joshi, A.L. and Chandrasekaran, D. 2004b. *Improving the productivity of traditional scavenging poultry systems: constraints and solutions. INFPD Newsletter,* 14, 10-15.

Cook, B.G., Pengelly, B.C., Brown, S.D., Donnelly, J.L., Eagles, D.A., Franco, M.A., Hanson, J., Mullen, B.F., Partridge, I.J., Peters, M. and Schultze-Kraft, R. 2005. *Tropical forages: an interactive selection tool.* [CD-ROM]. CSIRO, DPI&F(Qld), CIAT and ILRI, Brisbane, Australia.

Coppock, D.L. 1994. *The Borana Plateau of Southern Ethiopia: synthesis of pastoral research, development and change 1898-1991*. International Livestock Centre for Africa (ILCA), Addis Ababa, Ethiopia.

Crane, E. 1999. *The world history of beekeeping and honey hunting*. Duckworth, London, UK.

Crawford, R.J.M. and Coppe, P. 1990. *The technology of traditional milk products in developing countries. Animal Production and Health Paper No 85.* Food and Agriculture Organisation of the United Nations (FAO), Rome, Italy.

Creswell, D.C. and Gunawan, B. 1982. *Indigenous chickens in Indonesia: Production characteristic in an improved environment. Report No.2*, pp. 9-14, Research Institute for Animal Production, Indonesia.

Cronje, P. 2000. *Ruminant physiology: Digestion, metabolism, growth and reproduction. Proceedings of the Ninth International Symposium on Ruminant Physiology,* CABI Publishing, Wallingford, UK.

CSO. 1984. *Annual Report*. Central Statistical Office (CSO), Ministry of Internal Trade, Addis Ababa, Ethiopia.

Cullison, A.E. 1979. *Feeds and feeding. Second edition*. Prentice-Hall Company, Reston Publishing Company Inc., Reston, USA.

Curry J., Huss-Ashmore R., Perry B. and Mukhebi, A. 1996. A framework for the analysis of gender, intra-household dynamics, and livestock disease control with examples from Uasin Gishu District, Kenya. *Human Ecology* 24 (2): 161-189.

Czerkawski, J,W. and Breckenridge, G. 1977. Design and development of a long-term rumen simulation technique (RUSITEC). *British Journal of Nutrition* 38: 371-384.

Daszak, P., Cunningham, A.A. and Hyatt, A.D. 2000. Emerging infectious diseases of wildlife - threats to biodiversity and human health. *Science* 287: 443-449.

Davis, K. 2004. *Technology dissemination among small-scale farmers in Meru Central District of Kenya: Impact of group participation. Doctoral thesis*. University of Florida, Gainesville, USA.

de Haan, C. (ed.). 2004. *Veterinary institutions in the developing world: current status and future needs. World Organisation for Animal Health Scientific and Technical Review* 23: 414 pp.

de Haan, C., Steinfeld, H. and Blackburn, H. 1997. *Livestock and the environment: finding a balance.*

WRENmedia, Fressingfield, UK.

de Haan, C., Schillhorn van Veen, T., Brandenburg, B., Gauthier, J., Le Gall, F., Mearns, R. and Simeon, M. 2001. *Livestock development: Implications for rural poverty, the environment and global food security*. The World Bank, Washington D.C., USA.

D'Haese, M., Vink, N., Van Huylenbroeck, G., Bostyn, F. and Kirsten, J. 2003. *Local institutional innovation and pro-poor agricultural growth: The case of small-woolgrowers' associations in South Africa.* Garant, Antwerpen-Apeldoorn, The Netherlands.

de Jong, R. 1996. *Dairy stock development and milk production with smallholders. PhD thesis*, Wageningen University, Wageningen, The Netherlands.

de Lasson, A. and Dolberg, F. 1985. The causal effect of landholding on livestock. *Quarterly Journal of International Agriculture* 24 (no. 4): 339-354.

de Leeuw, P.N., Grandin, B.E. and Bekure, S. 1991. Introduction to the Kenyan rangelands and Kajiado District. In *ILCA Systems Study 4: Maasai herding - An analysis of the livestock production system of Maasai pastoralists in eastern Kajiado District, Kenya* (ed. S. Bekure, P.N. de Leeuw, B.E. Grandin and P.J.H. Neate), pp. 7-19. International Livestock Centre for Africa (ILCA), Addis Ababa, Ethiopia.

Delgado, C., Rosegrant, M., Steinfeld, H., Ehui, S. and Courbois, C. 1999. *Livestock to 2020: the next food revolution. Food, Agriculture, and the Environment Discussion Paper No. 28,* International Food Policy Research Institute (IFPRI), Washington D.C., USA.

Deodatus, F. 2000. Wildlife damage in rural areas with emphasis on Malawi. In *Wildlife conservation by sustainable use* (ed. H.H.T. Prins, J.G. Grootenhuis and T.T. Dolan), pp. 115-140. Kluwer Academic Publishers, The Netherlands and USA.

Devendra, C. 1987. Herbivores in the arid and wet tropics. In *The nutrition of herbivores* (ed. J.B. Hacker and J.H. Ternouth), pp. 23-46. Academic Press, Sydney, Australia.

Devendra, C. (ed.). 1988. *Non-conventional feed resources and fibrous agricultural residues. Strategies for expanded utilisation.* International Development Research Centre (IDRC), Ottawa, Canada, and Indian Council of Agricultural Research, New Delhi, India.

Devendra, C. 1995. Mixed farming and intensification of animal production in Asia. In *Livestock strategies for low income countries* (ed. R.T. Wilson, S. Ehui and S. Mack). *Proceedings of the joint FAO/ILRI Roundtable on Livestock Development Strategies for Low Income Countries, Addis Ababa, Ethiopia, 27 February-2 March 1995,* pp. 133-144. Food and Agriculture Organisation of the United Nations (FAO)/International Livestock Research Institute (ILRI), Nairobi, Kenya.

Devendra, C. 1998. Improvement of small ruminant production systems in rain-fed agro-ecological zones of Asia. *Annals of Arid Zones* 37: 215-232.

Devendra, C. 2000. Animal production and rain-fed agriculture in Asia: potential opportunities for productivity enhancement. *Outlook on Agriculture* 29: 161-175.

Devendra, C. 2002. Crop-animal systems in Asia: future perspectives. *Agricultural Systems* 71: 179-186. (Chapter 3)

Devendra, C. 2002. Small ruminants: imperatives for productivity enhancement and rural growth. *Asian-Australian Journal of Animal Science* 14: 1483-1496. (Chapter 19)

Devendra, C. and Burns, M. 1983. *Goat production in the tropics. Second edition.* Commonwealth Agricultural Bureaux, CABI Publishing, Wallingford, UK.

Devendra, C. and Fuller, M.F. 1979. *Pig production in the tropics*. Oxford University Press, Oxford, UK.

Devendra C. and McLeroy G. B. 1982. *Goat and sheep production in the tropics.* Intermediate Tropical Agriculture Series. Longman Scientific and Technical, Harlow, UK.

Devendra, C., Thomas, D., Jabbar, M.A. and Kudo, H. 1997. *Improvement of livestock production in crop-animal systems in rain-fed agro-ecological zones of South-East Asia.* International Livestock Research Institute (ILRI), Nairobi, Kenya.

Devendra, C., Thomas, D., Jabbar, M.A. and Zerbini, E. 2000. *Improvement of livestock production in crop-animal systems in agro-ecological zones of South Asia*. International Livestock Research Institute (ILRI), Nairobi, Kenya.

Devendra, C. and Chantalakhana, C. 2002. Animals, poor people and food insecurity –opportunities for improved livelihoods through natural resource management. *Outlook on Agric*ulture 31: 161-175.

Devendra, C. and Haenlein, G.F. 2002. Goat breeds. In *Encyclopaedia of dairy sciences* (ed. R. Roginski, J.W. Fuquay and P.F. Fox), pp. 498-495. Academic Press, Oxford, UK.

Devendra, C. and Thomas, D. 2002. Crop-animal interactions in mixed farming systems in Asia. *Agricultural Systems* 71: 27-40.

Devendra, C. and Pezo, D. 2004. Crop - animal systems in Asia and Latin America: Characteristics, opportunities for productivity enhancement and emerging challenges, including comparisons with West Africa. In *Proceedings of International Conference on sustainable crop-livestock production for improved livelihoods and natural resource management in West Africa* (ed. T.O. Williams, S.A. Tarawali, P. Hiernaux and S. Fernandez-Reivera), pp. 123-159. International Livestock Research Institute (ILRI), Nairobi, Kenya.

DFID (UK Department for International Development). 1999. *Sustainable Livelihoods Guidance Sheets: No.1 Introduction*. DFID, London, UK. Available at: http://www.livelihoods.org/info/guidance_sheets_pdfs/section1.pdf

DFID (UK Department for International Development). 2000. *Sustainable livelihoods guidance sheets*. Institute of Development Studies, London, UK. http://www.livelihoods.org.

DFID. 2002. *Working to reduce poverty. Highlights from DFID's Departmental Report 2002*. UK Department for International Development (DFID), London, UK. (Chapter 1)

DFID. 2002. *Wildlife and poverty study*. DFID Livestock and Wildlife Advisory Group, Department for International Development (DFID), London, UK. Available at http://www.dfid.gov.uk/. (Chapter 26)

DFID (UK Department for International Development). 2005. *Millennium development goals*. http://www.dfid.gov.uk/mdg/

Díaz, G.S., Galina, C.S., Basurto, C.H. and Ochoa, G.P. 2002. Efecto de la progesterona natural con o sin la adición de benzoato de estradiol sobre la presentación de celo, ovulación y gestación en animales tipo *Bos indicus* en el trópico mexicano. *Archivos de Medicina Veterinaria* 34: 235-244.

Din, M., Madan, J. and Srivastava, P.K. 1999. Technico-economic analysis of different modes of short (up to 5 km) and medium haulage distance (5-20 km) comparing truck, tractor trailer and animal carts (bullock, buffalo, camel, mule and donkey). *Draught Animal News* 30: 10-11.

Dixon J., Gulliver, A. and Gibbon, D. 2001. *Farming systems and poverty – improving farmers' livelihoods in a changing world* (ed. M. Hall). Food and Agriculture Organisation of the United Nations (FAO), Rome, Italy. http//www.fao-org./DOCRP/ Y1860E/y1860e00.htm

Dorji, T., Roder, W. and Sijium, Y. 2003. Disease in the yak. In *The yak. Second edition* (ed. G. Wiener, Han Jianlin and Long Ruijun), pp. 221-236. Food and Agriculture Organisation of the United Nations (FAO), Bangkok, Thailand.

Dorward, A.R. and Anderson, S. 2002. Understanding small stock as livelihood assets: indicators for facilitating technology development and dissemination. *Report on review and planning workshop 12th to 14th August 2002*, pp. 4-7, Imperial College, Wye, UK.

Draught Animal News. A biannual publication, website: www.vet.ed.ac.uk/ctvm

Duckham, A.N. and Masefield, G.B. 1970. *Farming systems of the world*. Chatto and Windus, London, UK.

Ebozoje, M.O. and Ikeobi, C.O.N. 1995. Productive performance and the occurrence of major genes in the Nigerian local chicken. *Nigerian Journal of Genetics* 10: 67-77.

Ehui, S. 1999. A review of the contribution of livestock to food security, poverty alleviation and environmental sustainability in Sub-Saharan Africa. *UNEP Industry and Environment, April-September 1999*, 37.

Eicher, K. 2003. Flashback: Fifty years of donor aid to African agriculture. *Conference Paper 16. IFPRI*

Conference 'Successes in African Agriculture', Pretoria, South Africa. International Food Policy Research Institute (IFPRI), Washington, D.C., USA.

Emerton, L. 1999a. The nature of benefits and the benefits of nature: Why wildlife conservation has not economically benefited communities in Africa. *Community conservation research in Africa: Principles and comparative practice. Paper No. 9.* Institute for Development Policy and Management, University of Manchester, Manchester, UK.

Emerton, L. 1999b. Balancing the Opportunity Costs of Wildlife Conservation for Communities around Lake Mburo National Park, Uganda. Community Conservation Research in Africa: Principles and Comparative Practice. *Evaluating Eden Series, Discussion Paper No. 5*. International Institute for Environment and Development (IIED), London, UK.

Emerton, L. and Mfunda, I. 1999. *Making wildlife economically viable for communities living around the Western Serengeti, Tanzania. Evaluating Eden Series, No. 1.* International Institute for Environment and Development (IIED), London, UK.

England, B.G., Foote, W.C., Cardozo, A., Matthews, D.H. and Riera, S. 1971. Oestrus and mating behaviour in the llama (*Lama glama*). *Animal Behaviour* 19: 722-726.

English, P.R., Burgess, G., Cockran, R.S. and Dunn, J. 1992. *Stockmanship, improving the care of the pig and other livestock*. Farming Press, Ipswich, UK.

Ernst, W.H.O., Kuiters, A.T., Nelissen, H.J.M. and Tolsma, D.J. 1991. Seasonal variation in phenolics in several savannah tree species in Botswana. *Acta Botanica Neelandica* 40: 63-74.

Escobar, A. 1985. *Encountering development: The making and unmaking of the third world.* Princeton University Press, Princeton, USA.

Eusebio, J.A. 1980. *Pig production in the tropics*. Longman, Harlow, UK.

Evans, J.O., Simpkin, S.P. and Atkins, D. 1995. *Camel keeping in Kenya. Range management handbook of Kenya, Volume III, 8.* Republic of Kenya, Ministry of Agriculture, Livestock Development and Marketing, Nairobi, Kenya.

Ewbank, R., Kim-Madslien, F. and Hart, C.B. 1999. *Management and welfare of farm animals. Fourth edition*. Universities Federation for Animal Welfare, The Old School, Brewhouse Hill, Wheathampstead AL4 8AN, UK.

Fadare, S.O. 2003. *Bees for Development Journal* 66: 13.

Fafchamps, M., Udry, C. and Czukas, K. 1998. Drought and saving in West Africa: are livestock a buffer stock? *Journal of Development Economics* 55: 273-305.

Falvey, L. 1999. The future for smallholder dairying. In *Smallholder dairying in the tropics* (ed. L. Falvey and C. Chantalakhana), pp. 417-432. International Livestock Research Institute (ILRI), Nairobi, Kenya.

Falvey, L. and Chantalakhana, C. (ed.). 1999. *Smallholder dairying in the tropics*. International Livestock Research Institute (ILRI), Nairobi, Kenya.

FAO. 1986. *Better farming series: Farming snails. Economic and Social Development Series Nos.33 and 34*. Food and Agriculture Organisation of the United Nations (FAO), Rome, Italy.

FAO. 1990. *The technology of traditional milk products in developing countries. Animal Health and Production Paper No 85.* Food and Agriculture Organisation of the United Nations (FAO), Rome, Italy.

FAO. 1996. *Production yearbook. Volume 50.* Food and Agriculture Organisation of the United Nations (FAO), Rome, Italy.

FAO. 1998. *Domestic Animal Diversity Information System*. Food and Agriculture Organisation of the United Nations (FAO), Rome, Italy. http://dad.fao.org/en/Home.htm

FAO (Food and Agriculture Organisation of the United Nations). 1999. *Poverty alleviation and food security in Asia: Role of livestock.* FAO, Regional Office for Asia and the Pacific, Bangkok, Thailand In http://www.fao.org. (Chapter 5)

FAO (Food and Agriculture Organisation of the United Nations), 1999. *The rabbit research program.* http://www.fao.org . (Chapter 17)

FAO. 2001. *Mixed crop-livestock farming: A review of traditional technologies based on literature and field experience. FAO Animal Production and Health Paper 152.* Food and Agriculture Organisation of the United Nations (FAO), Rome, Italy. Full document available at: http://www.fao.org/DOCREP/004/Y0501E/Y0501E00.HTM. (Chapter 23)

FAO. 2001. *FAO Yearbook: Production, Volume 55.* Food and Agriculture Organisation of the United Nations (FAO), Rome, Italy. (Chapter 21)

FAO. 2002. *Quarterly Bulletin of Statistics 2002.* Food and Agricultural Organisation of the United Nations (FAO), Rome, Italy. (Chapter 18)

FAO (Food and Agriculture Organisation of the United Nations). 2002. www.fao.org/ag/againfo/resources FAOSTAT. (Chapter 25)

FAO. 2003. *Livestock, Environment and Development (LEAD) Digital Library. CD-ROM.* Livestock, Environment and Development (LEAD) Initiative, Animal Production and Health Division, Food and Agriculture Organisation of the United Nations (FAO), Rome, Italy. (Chapter 9)

FAO. 2003. *Egg marketing: a guide for the production and sale of eggs. Agricultural Services Bulletin No. 150.* Food and Agriculture Organisation of the United Nations (FAO), Rome, Italy. (Chapter 7)

FAO. 2003. *Rabbit breeding raises income in Egypt.* Food and Agriculture Organisation of the United Nations (FAO), Rome, Italy. (Chapter 17)

FAO (Food and Agriculture Organisation of the United Nations). 2003. *Pro-poor Livestock Policy Initiative.* http://www.fao.org/ag/againfo/projects/en/pplp/home.html. (Chapter 2)

FAO/APHCA. 1988. *Non-conventional feed resources in Asia and the Pacific. Third edition.* Food and Agriculture Organisation (FAO) of the United Nations, Regional Animal Production and Health Commission (APHCA), Bangkok, Thailand.

FAO/ILRI. 1999. *Farmers, their animals and the environment.* CD-ROM. Food and Agriculture Organisation of the United Nations (FAO), Rome, Italy, and International Livestock Centre for Africa (ILRI), Nairobi, Kenya. http://www.virtualcentre.org/en/enl/vol1n2/read.htm

FAO/PNUMA. 1985. 'Manejo de la fauna silvestre y desarrollo rural. Información sobre siete especies de América Latina y el Caribe'('Management of the wild fauna and rural development. Information about seven species of Latin America and the Caribbe'), pp.118-151. Actas Taller, Food and Agriculture Organisation of the United Nations (FAO) and Programa de las Naciones Unidas para el Medio Ambiente (PNUMA), Lima, Peru.

FAOSTAT. 2003. *FAO Statistical database* at http://faostat.fao.org. Accessed February 2004. (Chapters 1, 3, 8, 19, 23)

FARM-Africa. 1996. *Goat types of Ethiopia and Eritrea. Physical description and management systems.* Joint publication FARM-Africa, London, UK, and International Livestock Centre for Africa (ILRI), Nairobi, Kenya.

Fattah, K.A. 1999. *Poultry as a tool in poverty eradication and promotion of gender equality.* Proceedings of a Workshop. http://www.husdyr.kvl.dk

Feeding standards for Australian livestock. Ruminants. 1990. CSIRO Publications, East Melbourne, Victoria, Australia.

Feer, F. 1993. The potential for sustainable hunting and rearing of game in Tropical forests. In *Tropical forests, people and food* (ed. C.M. Hladik, A. Hladik, O.F. Linares, A. Semple and M. Hadley), pp. 691-708. The Parthenon Publishing Group, Paris, France.

Fernandez-Baca, S. (ed.). 1991. *Avances y perspectivas del conocimiento de los camélidos sudamericanos. (Advances and perspectives of the knowledge on South American camelids),* Food and Agriculture Organisation of the United Nations (FAO), Santiago, Chile.

Field, C.R. 2005. *Where there is no development agency. A manual for pastoralists and their promoters. With special reference to the arid zone of the Greater Horn of Africa,* Natural Resources International, Aylesford, Kent, UK.

Fielding, D. 1991. *Rabbits.* The Tropical Agriculturalist, Series Editors R. Coste and A.J. Smith. Macmillan Education Ltd., London, UK, in cooperation with Technical Centre for Agricultural and Rural Cooperation (CTA), Wageningen, The Netherlands.

Forbes, J.M. 1995. *Voluntary food intake and diet selection in farm animals.* CABI Publishing, Wallingford, UK.

Fowler, M.E. 1993. *Improving Andean sheep and alpacca production. Recommendations from a decade of research in Peru.* Small Ruminant Collaborative Research Project, University of California, USA. Published by University of Missouri-Columbia, USA.

Galal, S., Boyazoglu, J. and Hammond, K. (ed.). 2000. *Workshop on Developing Breeding Strategies for Lower Input Animal Production Environments.* Bella, Italy, 22-25 September 1999. ICAR Technical Series, No.3. (available online at www.icar.org)

Galaz, J.L., and González, G. (ed.). 2001. Conservación y manejo de la vicuña en Sudamérica. (Conservation and management of the vicuña in South America). *Actas del I Seminario Internacional: Aprovechamiento de la fibra de vicuña en los Andes de Argentina, Bolivia, Chile y Perú (Proceedings of the First International Seminar: Utilisation of the vicuna fibre in the Andes of Argentina, Bolivia, Chile and Peru)*, November 1996, Arica, Chile.

Galina, C.S. and Arthur, G.H. 1989. Review of cattle reproduction in the tropics. Part 2. Parturition and calving intervals. *Animal Breeding Abstracts* 57: 679-686.

Galina, C.S., Orihuela, T.A. and Rubio, I. 1996. Behavioural trends affecting oestrous detection in Zebu cattle. *Animal Reproduction Science* 42: 465-470.

Garcia, L. O. and Restreppo, J.I.R. 1995. *Multi-nutrient block handbook. Better farming series 45.* Food and Agriculture Organisation of the United National (FAO), Rome, Italy.

Garforth, C. 2004. Knowledge management and dissemination for livestock production: global opportunities and local constraints. In *Responding to the Livestock Revolution – The role of globalisation and implications for poverty alleviation* (ed. E. Owen, T. Smith, M.A. Steele, S. Anderson, A.J. Duncan, M. Herrero, J.D. Leaver, C.K. Reynolds, J.I. Richards and J.C. Ku-Vera), pp. 287-298. British Society of Animal Science Publication No. 33, Nottingham University Press, Nottingham, UK.

Gatenby, R.M. 1986. *Sheep production in the tropics and sub-tropics.* Longman, Scientific and Technical, Harlow, UK.

Gatenby, R.M. 2002. *Sheep.* The Tropical Agriculturalist, Series Editors R. Coste and A.J. Smith. Macmillan Education Ltd., London, UK, in cooperation with Technical Centre for Agricultural and Rural Cooperation (CTA), Wageningen, The Netherlands.

Gemeda, T., Zerbini, E., Wold, A.G. and Demissie, D. 1995. Effects of draught work on performance and metabolism of crossbred cows. 1. Effect of work and diet on body-weight change, body condition, lactation and productivity. *Animal Science* 60: 361-367.

Getz, W.M., Fortmann, L., Cumming, D., du Toit, J., Hilty, J., Martin, R., Murphree, M., Owen-Smith, N., Starfield, A.M. and Westphal, M.I. 1999. Sustaining natural and human capital: villagers and scientists. *Science* 283(5409): 1855-1856.

Ghirotti, M. 1999. Making better use of animal resources in a rapidly urbanising world: a professional challenge. *World Animal Review* 92: 2-14.

Gibson, C.C. 1999. *Politicians and poachers: The political economy of wildlife policy in Africa.* Cambridge University Press, Cambridge, UK.

Gilchrist, P., Sinurat, A., Basuno, E., Hamid, H. 1994. *Village Chicken Smallholder Production and Marketing Project. Report.* Indonesian International Animal Science Research And Development Foundation (INI ANSREDEF), Bogor, Indonesia.

Givens, D.I., Owen, E., Axford, R.F.E. and Omed, H.M. (ed.). 2000. *Forage evaluation in ruminant nutrition.* CABI Publishing, Wallingford, UK.

Gohl, B. 1981. *Tropical feeds. Feed information summaries and nutritive values. FAO Animal Production and Health Series No. 12.* Food and Agriculture Organisation of the United Nations (FAO), Rome, Italy. (now available on line from FAO)

González, B., Bas, F., Tala, C., Iriarte, A. (ed.). 2000. Manejo Sustentable de la Vicuña y el Guanaco (Sustainable management of the vicuña and guanaco), *Actas del Seminario Internacional Pontifica*

Universidad Catolica de Chile, November 1998. Servicio Agrícola Ganadero, Pontificia Universidad Católica de Chile y Fundación para la Innovación Agraria, Santiago, Chile.

Grandin, B.E. 1991. The Maasai: Socio-historical context and group ranches. In *ILCA Systems Study 4: Maasai herding - An analysis of the livestock production system of Maasai pastoralists in eastern Kajiado District, Kenya* (ed. S. Bekure, P.N. de Leeuw, B.E. Grandin and P.J.H. Neate), pp. 21-39. International Livestock Centre for Africa (ILCA), Addis Ababa, Ethiopia.

Griffin, M. 2004. Issues on the development of school milk. Paper presented at *School Milk Workshop, FAO Intergovernmental Group on Meat and Dairy Trade. Winnipeg, Canada, 17- 19 June, 2004*. Food and Agriculture Organisation of the United Nations (FAO), Rome, Italy.

Grimes, S.E. 2002. *A basic laboratory manual for the small-scale production of and testing of 1-2 Newcastle disease vaccine*. FAO RAP publication 2002/22, Food and Agriculture Organisation of the United Nations (FAO), Rome, Italy.

Grootenhuis, J.G. 2000. Wildlife, livestock and animal disease reservoirs. In *Wildlife conservation by sustainable use* (ed. H.H.T. Prins, J.G. Grootenhuis and T.T. Dolan), pp. 81-113. Kluwer Academic Publishers, The Netherlands and USA.

Gryseels, G., Groenewold, J. P. and Kassam, A. 1997. *The Technical Advisory Committee database for quantitative analysis of CGIAR priorities and strategies*. TAC Secretariat, Food and Agriculture Organisation of the United Nations (FAO), Rome, Italy.

Gupta, N.P., Patni, P.C. and Sugumar, S. 1989. Properties and processing of camel hair in India. *Indian Textile Journal* 99: 180-188.

Gutteridge, R.C. and Shelton, H.M. (ed.). 1998. *Forage tree legumes in tropical agriculture*. The Tropical Grassland Society of Australia Inc., St Lucia, Queensland, Australia. http://www.tropicalgrasslands.asn.au/

Hadley Centre. 2004. *Climate change predictions from the Hadley Centre*. UK Meteorological Office, Hadley Centre, Exeter, Devon, UK. http://www.met-office.gov.uk/research/hadleycentre/models/modeldata.html

Hadrill, D. 2002. *Horse healthcare – A manual for animal health workers and owners*. ITDG Publishing, London, UK.

Hall, S.J.G. and Clutton-Brock, J. 1989. *Two hundred years of British farm livestock*. British Museum. London, UK.

Hammond, J.A., Fielding, D. and Bishop, S.C. 1997. Prospects for plant anthelmintic in tropical veterinary medicine. *Veterinary Research Communications* 21: 213-228.

Hansen, J.W. and Perry, B.D. 1994. *The epidemiology, diagnosis and control of helminth parasites of ruminants: A handbook*. International Laboratory for Research on Animal Diseases (ILRAD), Nairobi, Kenya.

Hardin, G. 1968. "The tragedy of the commons." *Science* 162: 1243-1248.

Heffernan, C. 2000. *The socio-economic impact of restocking destitute pastoralists: A case study from Kenya. PhD thesis*. The University of Reading, Reading, UK.

Heffernan, C. and Sidahmed, A. 1998. Issues in the delivery of veterinary services to the rural poor. *Paper presented at the Conference on the Delivery of Veterinary Services to the Poor, June, 1998, at The University of Reading*. University of Reading, Reading, UK.

Heffernan, C. and Misturelli, F. 2000. *The delivery of veterinary services to the poor: Findings from Kenya.* Report for the UK Department for International Development (DFID) Animal Health Programme, University of Edinburgh, Edinburgh, UK.

Heffernan, C., Misturelli, F. and Nielsen, L. 2001a. *Restocking and poverty alleviation: Perceptions and realities of livestock keeping among poor pastoralists in Kenya.* Report for the UK Department for International Development (DFID) Livestock Production Programme, NR International, Aylesford, Kent, UK.

Heffernan, C., Nielsen, L. and Misturelli, F. 2001b. *Restocking pastoralists: A manual.* Report for the UK Department for International Development (DFID) Livestock Production Programme, NR

International, Aylesford, Kent, UK.

Hill, D.H. 1988. *Cattle and buffalo meat production in the tropics.* Intermediate Tropical Agriculture Series, General Editor W.J.A. Payne. Longman Scientific and Technical, Harlow, Essex, UK.

Hodasi, J.K.M. 1975. Preliminary studies on the feeding and burrowing habits of *Achatina achatina*. *Ghana Journal of Science* 15: 193-199.

Hofer, H., Campbell, K.L.I., East, M.L. and Huish, S.A. 2000. Modelling the spatial distribution of the economic costs and benefits of illegal game meat hunting in the Serengeti. *Natural Resource Modelling* 13(1): 151-177.

Hoffmann, R., Otte, K., Ponce, C. and Ríos, M. 1983. *El manejo de la vicuña silvestre.* (The management of the wild vicuña). Deutsche Gesellshaft für Technische Zusammenarbeit, Eschborn. Two volumes.

Holden S., Ashley S. and Bazeley P. 1996. *Improving the delivery of animal health services in developing countries: a literature review*. Livestock in Development (LID), Crewkerne, UK.

Holland, J. and Blackburn, J. 1998. *Whose voice: Participatory research and policy change.* ITDG Publishing, London, UK.

Holness, D.H. 1985. Pig slurry for fish production. *Farming World (Zimbabwe)*, November 1985, pp. 3-7.

Holness, D.H. 1991. *Pigs*. The Tropical Agriculturalist, Series Editors R. Coste and A.J. Smith. Macmillan Education Ltd., London, UK, in cooperation with Technical Centre for Agricultural and Rural Cooperation (CTA), Wageningen, The Netherlands.

Homewood, K.M. and Rodgers, W.A. 1991. *Maasailand ecology: pastoralist development and wildlife conservation in Ngorongoro, Tanzania*. Cambridge University Press, Cambridge, UK.

Hone, J. 1994. *An analysis of vertebrate pest control*. Cambridge University Press, Cambridge, UK.

Humphrey, C. and Sneath, D. (ed.). 1996. *Culture and environment in inner Asia. Volume 1 The pastoral economy and the environment.* White Horse Press, Cambridge, UK.

Humphreys, L.R. 1987. *Tropical pastures and fodder crops. Second edition.* Intermediate Tropical Agriculture Series, Longman Scientific and Technical Group UK Limited, Harlow, UK.

Hunter, A. 1994. *Animal health. Volume 2 Specific diseases*. The Tropical Agriculturalist, Series Editors R. Coste and A.J. Smith. Macmillan Education Ltd., London, UK, in cooperation with Technical Centre for Agricultural and Rural Cooperation (CTA), Wageningen, The Netherlands.

Huntington, J.A. and Givens, D.I. 1995. The *in situ* technique for studying the rumen degradation of feeds: a review of the procedure. *Nutrition Abstracts and Reviews (series B)* 65: 63-93.

Huque, Q.M.E. 1989. Village poultry production system in Bangladesh. *Proceeding of the 3rd National conference of Bangladesh Animal Husbandry Association* held on December 23-24, Dhaka, Bangladesh.

IDL (In Development Livestock) Group. 2003. *Community animal health workers: Threat or opportunity.* IDL, Crewkerne, UK.

Ikeobi, C.O.N. and Oladotun, O.A. 1998. Visible genetic profiling of single comb and head spurs in the Nigerian local chicken. *Proceedings of 3rd Annual Conference of the Animal Science Association of Nigeria* (ed. A.D Ologhobo and E.A. Iyayi), pp. 14-17.

Ikeobi, C.O.N., Ozoje, M.O., Adebambo,O.A. and Adenowo, J.A. 1998. Modifier genes and their effects in the Nigerian local chicken: Ptiploidy and comb type. *Proceedings of 6th World Congress on Genetics Applied to Livestock Production, Armidale, Australia*, 24: 318-321.

Ikeobi, C.O.N. and Godwin, V.A. 1999. Presence of polydactyly gene in the Nigerian local chicken. *Tropical Journal of Animal Science* 1: 57-65.

Ikeobi, C.O.N., Ozoje, M.O., Adebambo, O.A. and Adenowo J.A. 2001. Frequencies of feet feathering and comb type genes in the Nigerian local chicken. *Pertanika Journal of Tropical Agricultural Science* 24: 147-150.

ILCA. 1988. *International Livestock Centre for Africa. Annual Report*. International Livestock Centre for Africa (ILCA), Addis Ababa, Ethiopia.

ILRI. 2000. *ILRI strategy to 2010. Making the livestock revolution work for the poor.* International

Livestock Research Institute (ILRI), Nairobi, Kenya.
ILRI. 2003. *The livestock revolution*. International Livestock Research Institute (ILRI), Nairobi, Kenya.
ILRI (International Livestock Research Institute). 2004. Unpublished data.
Inns, F. 2003. Integrated harness and implement design – a key factor in developing improved equipment for animal draught tillage operations. In *Working animals in agriculture and transport. A collection of some current research and development observations* (ed. R.A. Pearson, P. Lhoste, M. Saastamoinen and W Martin-Rosset), pp. 145-163. EAAP Technical Series No 6, European Association of Animal Production, Wageningen Press, The Netherlands.
IPCC. 1995. *Greenhouse gas inventory reference manual: IPCC guidelines for national greenhouse gas inventories*. United Nations Environment Programme, the Organisation for Economic Co-operation and Development, the International Energy Agency and the Intergovernmental Panel on Climate Change (IPCC), Bracknell, UK.
IPCC. 2001. *Climate Change 2001: Working group I: The scientific basis*. Intergovernmental Panel on Climate Change (IPCC). Cambridge University Press, Cambridge, UK.
ITC and IUCN. 1998. *Report on development and promotion of wildlife utilisation*. Ministry of Lands, Natural Resources and Tourism, Dar es Salaam, Tanzania.
Jahnke, H.E. 1982. *Livestock production systems and livestock development in Tropical Africa*. Kieler Wissenschaftsverlag Vauk, Kiel, Germany.
Jainudeen, M.R. and Hafez, E.S.E. 2000. Cattle and buffalo. *Reproduction in farm animals, Seventh edition* (ed. B. Hafez and E.S.E. Hafez), pp. 159-171. Lea and Febiger, Philadelphia, USA.
James, A.D. 2004. Disease and biosecurity constraints to trade in animal products. In *Responding to the Livestock Revolution – The role of globalisation and implications for poverty alleviation* (ed. E. Owen, T. Smith, M.A. Steele, S. Anderson, A.J. Duncan, M. Herrero, J.D. Leaver, C.K. Reynolds, J.I. Richards and J.C. Ku-Vera), pp. 85-90. British Society of Animal Science Publication No. 33, Nottingham University Press, Nottingham, UK.
James, A.D. and Carles, A.B. 1996. Measuring the productivity of grazing and foraging livestock. *Agricultural Systems* 52: 271-291.
Janssen, W., Sanint, L.R., Rivas, L. and Henry, G. 1991. CIAT commodity factfile revisited: Indicators of present and future importance. In *CIAT (Centro Internacional de Agricultura Tropical), in the 1990s and beyond: A Strategic Plan,* pp. 15-50. International Centre for Tropical Agriculture (CIAT), Cali, Colombia.
Jessup, D.A. Lance, W.R. 1982. What veterinarians should know about South American camelids. *Calif. Vet.* 11: 12-19.
Jianlin, Han. 2003. Molecular and cytogenetics in yak - a scientific basis for breeding and evidence for phylogeny. In *The yak. Second edition* (ed. G. Wiener, Han Jianlin and Long Ruijun), pp. 415-435. Food and Agriculture Organisation of the United Nations (FAO), Bangkok, Thailand.
Jianlin, Han, Richard, C., Hanotte, O., McVeigh, C. and Rege, J.E.O. (ed.). 2002. *Proceedings of the Third International Congress on Yak held in Lhasa, P.R. China, 4-9 September 2000*. International Livestock Research Institute (ILRI), Nairobi, Kenya.
Johnson, D. 1969. The nature of nomadism; a comparative study of pastoral migrations in South western Asia and Northern Africa. *Department of Geography Research Paper 118*, University of Chicago, Chicago, USA.
Joshi, D.D., Awasthi, B.D. and Sharma, M. 1999. *An assessment of yak cheese factories in Nepal*. National Zoonoses and Food Research Centre, Kathmandu, Nepal.
Kadiyala, S. and Gillespie, S. 2004. Rethinking food aid to fight AIDS. *Food Nutrition Bulletin* 25 (3): 271-282.
Kaimowitz, D. 1995. *Livestock and deforestation in central America*. Environment and Production Technology Division Discussion Paper No. 9, International Food Policy Research Institute (IFPRI), Washington D.C., USA and Instituto Interamericano de Cooperación para la Agricultura, Coronado, Costa Rica.

Kalita, N., Sarma, D., Talukdar, J.K., Barua, N. and Ahmed, N. 2004a. Comparative performance of Khaki Campbell ducks, desi ducks and their reciprocal crosses for certain economic traits in rural conditions. *World's Poultry Science Journal* 60: 349-356.

Kalita, N., Saikia, N.D., Barua, N. and Talukdar, J.K. 2004b. Economic housing design and stocking densities for rural poultry production in Assam. *World's Poultry Science Journal* 60: 356- 366.

Katzina, E.V. 2003. Yak in other countries with a long tradition of yak keeping - Buryatia. In *The yak. Second edition* (ed. G. Wiener, Han Jianlin and Long Ruijun), pp. 291-299. Food and Agriculture Organisation of the United Nations (FAO), Bangkok, Thailand.

Kaumbutho, P.G. 2003. Recent developments in the role of equines in transport: experiences from Africa. In *Fourth International Colloquium on Working Equines* (ed. R.A. Pearson, D. Fielding and D. Tabbaa), pp. 7-18. SPANA, London, UK.

Keita J.D. 1993. Non-wood forest products in Africa: an overview. In *Non-wood forest products - A regional expert consultation for English-speaking African countries*, Commonwealth Science Council and Food and Agriculture Organisation of the United Nations (FAO), in co-operation with Ministry of Tourism, Natural Resources and Environment, Tanzania. Series Number CSC(94)AGR-21. *FAO Technical Paper 306*, Food and Agriculture Organisation of the United Nations (FAO), Rome, Italy.

Kerven, C., Alimaev, I.I., Behnke, R., Davidson, G., Franchois, L., Malmakov, N., Mathijs, E., Smailov, A., Temirbekov, S. and Wright, I. 2003. Retraction and expansion of flock mobility in Central Asia: costs and consequences. *Proceedings III Rangeland Congress 'Rangelands in the new millennium'*, pp. 543-556. Durban, South Africa.

Khan, A.G. 1983. Improvement of Desi birds (Part II). *Poultry Adviser* 16: 53-61.

Kindness, H., Sikosana, J.L.N., Mlambo, V. and Morton, J.F. 1999. *Socio-economic surveys of goat keeping in Matobo and Bubi Districts*. Report No. 2451, Natural Resources Institute (NRI), Chatham Maritime, Kent, UK.

King, J.M., Parsons, D.J., Turnpenny, J.R., Nyangaga, J.R., Bakari, P. and Wathes, C.M. 2005. Ceiling to milk yield on Kenya smallholdings requires rethink of dairy development policy. *Proceedings of the British Society of Animal Science 2005*, p. 25.

Kingston, D.J. and Creswell, D.C. 1982. *Indigenous chickens in Indonesia: Population and production characteristics in five villages in West Java. Report No.2*, pp. 3-8. Research Institute for Animal Production, Bogor, Indonesia.

Kiss, A. (ed.). 1990. *Living with wildlife: Wildlife resource management with local participation in Africa. World Bank Technical Paper No. 130,* Washington, D.C., USA.

Kitalyi, A.J. 1998. *Village chicken production systems in rural Africa. Household food security and gender issues. FAO Animal Production and Health Paper 142*. Food and Agriculture Organisation of the United Nations (FAO), Rome, Italy.

Kitalyi, A., Miano, D., Mwebaza, S. and Wambugu, C. 2005. *More forage, more milk. Forage production for small-scale zero grazing systems. Technical handbook No. 33*. Regional Land Management Unit (RELMA in ICRAF)/World Agroforestry Centre, Nairobi, Kenya.

Kohls, R.L. and Uhl, J.N. 1998. *Marketing of agricultural products. Eighth edition*. Prentice-Hall, Upper Saddle River, New Jersey, USA.

Kon, S.K. 1972. *Milk and milk products in human nutrition. Nutritional Studies No 27*. Food and Agriculture Organisation of the United Nations (FAO), Rome, Italy.

Kossila, V.L. 1984. Global review of the potential of crop residues as animal feed. In *Better utilisation of crop residues and by-products in animal feeding: research guidelines 1. State of knowledge* (ed. T.R. Preston, V.L. Kossila, J. Goodwin and S.B. Reed), *Proceedings of the FAO/ILCA Expert Consultation 5–9 March 1984, ILCA Headquarters, Addis Ababa. FAO Animal Production and Health Paper 50.* Food and Agriculture Organisation of the United Nations (FAO), Rome, Italy.

Kothari, A., Pathak, N. and Vania, F. 2000. *Where communities care: Community based wildlife and ecosystem management in South Asia. Evaluating Eden Series, No. 3*. International Institute for Environment and Development (IIED), London, UK.

Koziell, I. and Saunders, J. (ed.). 2001. *Living off biodiversity: Exploring livelihoods and biodiversity issues in natural resources management.* International Institute for Environment and Development (IIED), London, UK.

Krell, R. 1996. *Value added products from beekeeping. Agricultural Services Bulletin No 124.* Food and Agriculture Organisation of the United Nations (FAO), Rome, Italy.

Kristensen, E., Larsen, C.E.S., Kyvsgaard, N., Madsen, J. and Hendriksen, J. 1999. Livestock Production: the twenty first century's food revolution. In *Poultry as a tool in poverty eradication and promotion of gender equality.* Proceedings of a workshop, March 22-26, Tune Landsboskole, Denmark. At: http://www.husdyr.kvl.dk/htm/php/tune99/index2.htm (December, 2004).

Krueger, R. and Casey, M. 2000. *Focus groups: A practical guide for applied research.* Sage Publication, London, UK.

Latham, M.C. 1997. *Human nutrition in the developing world. Food and Nutrition Series No 29.* Food and Agriculture Organisation of the United Nations (FAO), Rome, Italy.

Lawrence, P.R. and Pearson, R.A. 1998. *Feeding standards for cattle used for work.* Centre for Tropical Veterinary Medicine, University of Edinburgh, Edinburgh, UK. *See also* http://www.vet.ed.ac.uk/ctvm

LEAD (Livestock, Environment and Development). 2003. *Livestock and environment toolbox.* http://lead.Virtualcenter.org/en/dec/toolbox/

Lebas, F., Coudert, P., de Rocambeau, H. and Thébault, R.G. 1997. *The rabbit, husbandry, health and production. FAO Animal Production and Health Series No. 21.* Food and Agriculture Organisation of the United Nations (FAO), Rome, Italy. http://www.fao.org/docrep/t1690E/t1690e00.htm

Lee, Y-w. 1999. *Silk reeling and testing manual. Agricultural Services Bulletin No. 136.* Food and Agriculture Organisation of the United Nations (FAO), Rome, Italy.

Leeuwis, C. (with contributions from van den Ban, A.). 2004. *Communication for rural innovation: Rethinking agricultural extension. Third edition.* Blackwell Publishing, Oxford, UK.

Leirs, H. 2003. Management of rodents in crops: the Pied Piper and his orchestra. In *Rats, mice and people: rodent biology and management* (ed. G.R. Singleton, L.A. Hinds, C.J. Krebs and D.M. Spratt), pp. 183-190. ACIAR, Canberra, Australia.

Lekasi, J.K., Tanner, J.C., Kimani, S.K. and Harris, P.J.C. 2001. *Managing manure to sustain smallholder livelihoods in the east African highlands.* Kenya Agricultural Research Institute (KARI), International Livestock Research Institute (ILRI), Nairobi, and Department for International Development (DFID), UK.

Lemma Gizachew. 1993. Comparison of legumes, hay, urea and noug cake as protein supplements to Horro sheep fed on tef straw. In P*roceedings of the fourth national livestock improvement conference held in Addis Ababa, Ethiopia. 13-15th November 1991,* pp. 211-215. Institute of Agricultural Research, Addis Ababa, Ethiopia.

Leng, R.A. 1995. Appropriate technologies for field investigations in ruminant livestock nutrition in developing countries. In *Agricultural science for biodiversity and sustainability in developing countries* (ed. F. Dolberg and P.H. Petersen), Proceedings of a workshop organised by Danish Agricultural and Rural Development Advisers Forum, 3-7 April 1995 at Tune Landboskole, Denmark.

Leng, R.A., Preston, T.R., Sansoucy, R. and Kunju, P.J.G. 1992. Multi-nutrient blocks as a strategic supplement for ruminants. *Harnessing Biotechnology to Animal Production and Health* 67.

Leonard, D.K. 1993. Structural reform of the veterinary profession in Africa and the new institutional economics. *Development and Change* 24: 227-267.

Leonard, D.K. (ed.). 1999. *Africa's changing markets for health and veterinary services: new institutional issues.* Palgrave, Basingstoke, UK.

LID. 1999. *Livestock in poverty focused development.* Livestock in Development (LID), Crewkerne, UK.

Livestock Development Group (LDG). 2002. *Livestock services and the poor.* Report for The Global Initiative on Livestock Services and the Poor, International Fund for Agricultural Development (IFAD), Rome, Italy.

Livestock Development Group (LDG). 2003a. *Poverty and participation: An analysis of bias in participatory methods*. Livestock Development Group (LDG), The University of Reading, Reading, UK.

Livestock Development Group (LDG). 2003b. *The livestock and poverty assessment methodology: A toolkit for practitioners*. Livestock Development Group (LDG), The University of Reading, Reading, UK.

Livingstone, D and Livingstone, C. 1865. *Narrative of an Expedition to the Zambezi and its tributaries 1858 to 1864*. London. Cited in Mason, I.L. and Maule, J.P. 1960. *The indigenous livestock of Eastern and Southern Africa*. Commonwealth Agricultural Bureaux, Publication No. 14, CABI Publishing, Wallingford, UK.

Loibooki, M., Hofer, H., Campbell, K.L.I. and East, M.L. 2002. Bushmeat hunting by communities adjacent to the Serengeti National Park, Tanzania: the importance of livestock ownership and alternative sources of protein and income. *Environment and Conservation* 29(3): 391-398.

LPP (Livestock Production Programme). 2000. *Annual Report 1999-2000*. NR International, Aylesford, Kent, UK.

Lukefahr, S.D. and Goldman, M. 1985. A technical assessment of production and economic aspects of small-scale rabbit farming in Cameroon. *Journal of Applied Rabbit Research* 8: 126-135.

Lukefar, S. and Preston, T. 1999. Human development through livestock projects: Alternate global approaches for the next millennium. *World Animal Review* 93: 24-25.

Magash, A. 2003. Yak in other countries with a long tradition of yak keeping - Mongolia. In *The yak. Second edition* (ed. G. Wiener, Han Jianlin and Long Ruijun), pp. 306-315. Food and Agriculture Organisation of the United Nations (FAO), Bangkok, Thailand.

Magnum, W. 2001. Top-bar hives in the USA. *Bees for Development Journal* 58: 3-5.

Mahadevan, P. 1992. Distribution, ecology and adaptation. *World animal science, Volume C6: Buffalo production* (ed. N.M. Tulloh and J.H.G. Holmes), pp. 1-12. Elsevier, Amsterdam, The Netherlands.

Makundi, R. H., Oguge, N. O. and Mwanjabe, P.S. 1999. Rodent pest management in East Africa – an ecological approach. In *Ecologically-based management of rodent pests* (ed. G.R. Singleton, L.A. Hinds, H. Leirs and Z. Zhang), pp. 460-476. ACIAR, Canberra, Australia.

Mallia, J.G. 1999. Observations on family poultry units in parts of Central America and sustainable development opportunities. *Livestock Research for Rural Development* 11. At: http://www. Cipav.org.co/lrrd/

Manyuchi, B., Ncube, S. and Smith, T. 1990. Optimising crop residue utilisation in Zimbabwe. *British Society of Animal Production* 50: 589.

Manyuchi, B., Smith, T. and Mikayiri, S. 1991. Effect of dry season feeding on the growth of Mashona steers of two ages kept on natural pasture in the subsequent dry season. *Zimbabwe Journal of Agricultural Research* 30: 105-116.

Marake, M., Mokuku, C., Majoro, M. and Mokitimi, N. 1998. *A preliminary report and literature review for Lesotho*. INCO-DC Project No. ERBIC18CT970162: Management and policy options for the sustainable development of communal rangelands and their communities in southern Africa. Collaborating partners institution, National University of Lesotho, Lesotho. http://www.maposda.net/

Mariadas, P. 2000. *Nadakkaum Panam Nattukozhigal, II edition* (*Desi birds - Walking money*). Nivey Comprint, Trichy 620 001, India.

Martin, M. 2001. *The impact of community animal health services on farmers in low-income countries: A literature review*. VETAID, Edinburgh, UK.

Mason, I.L. 1996. *A world dictionary of livestock breeds, types and varieties*. CABI Publishing, Wallingford, UK.

Mason, I.L. and Maule, J.P. 1960. *The indigenous livestock of Eastern and Southern Africa*. Commonwealth Agricultural Bureaux, Publication No 14, CABI Publishing, Wallingford, UK.

Mathewman, R.W. in collaboration with Chabeuf, N. 1993. *Dairying*. The Tropical Agriculturalist, Series Editors R. Coste and A.J. Smith. Macmillan Education Ltd., London, UK, in cooperation with Technical Centre for Agricultural and Rural Cooperation (CTA), Wageningen, The Netherlands.

Mathias, E., Rangnekar, D.V. and McCorkle, C.M. 1999. *Ethenoveterinary medicine, alternatives for livestock production.* BAIF Development Research Foundation, Pune, India.

Mauricio, R.M., Mould, F.L., Dhanoa, M.S., Owen, E., Channa, K.S. and Theodorou, M.K. 1999. A semi-automated *in vitro* gas production technique for ruminant feedstuff evaluation. *Animal Feed Science and Technology* 79: 321-330.

McAinsh, C.V., Kusina, J., Madsen, J and Nyoni, O. 2004. Traditional chicken production in Zimbabwe. *World's Poultry Science Journal* 60: 233-246.

McClean, M. and Ramsey, M. 2001. *A basic course for village animal health workers in Cambodia.* International Fund for Agricultural Development (IFAD), Rome, Italy.

McCorkle, C.M. (ed.). 1990. *Improving Andean sheep and alpaca production. Recommendations from a decade of research in Peru.* Small Ruminant Collaborative Research Support Project, University of California, USA. Published by University of Missouri-Columbia, USA.

McCorkle, C.M., Mathias, E. and Schillhorn van Veen, T.W. (ed.). 1996. *Ethnoveterinary research and development.* Intermediate Technology Publications, London, UK.

McDonald, P., Edwards, R.A., Greenhalgh, J.F.D. and Morgan, C.A. 2002. *Animal nutrition. Sixth edition.* Prentice Hall (Pearson Education), London, UK.

McDonald, P., Henderson, A.R. and Heron, S.J.E. 1991. *The Biochemistry of silage.* Chalcombe Publications, Lincoln, UK.

McIntire, J., Bourzat, D. and Pingali, P. 1992. *Crop-livestock interaction in Sub-Saharan Africa.* The World Bank. Washington, D.C., USA.

McLeod, A., Saadullah, M., Jayaswal, M.L., Best, J., Barton, D., Rymer, C., Regmi, B.N. and Noor, T.R. 2002. Using livestock to improve the livelihoods of landless and refugee-affected livestock keepers in Bangladesh and Nepal. In. *Responding to the increasing global demand for animal products. Programme and summaries.* An international conference organised by the British Society of Animal Science, American Society of Animal Science and Mexican Society of Animal Production, 12-15 November 2002, Merida, Yucatan, Mexico.

McNitt, J.I., Patton, N.M., Lukefahr, S.D. and Cheeke, P.R. 2000. *Rabbit production. Eighth edition.* The Interstate Printers and Publishers Inc., Danville, IL, USA.

MDG (Millennium Development Goals). 2003. *Millennium Development Goals.* http://www.developmentgoals.org. Accessed January 2004.

Mdoe, N.S.Y., Mla, G.I. and Urio, N.A. 1992. Comparing costs associated with the utilisation of crop residues and planted pastures in smallholder production systems in the highlands of Hai District, Tanzania. In *The complementarity of feed resources for animal production in Africa* (ed. J.E.S. Stares, A.N. Said and J.A. Kategile), pp. 385-392. Proceedings of the joint feed resources networks workshop held in Gabarone, Botswana, 4-8 March 1991, International Livestock Centre for Africa (ILCA), Addis Ababa, Ethiopia.

Meherez, A.Z., Ørskov, E.R. and McDonald, I. 1977. Rates of rumen fermentation in relation to ammonia concentration. *British Journal of Nutrition* 38: 437-443.

Mendes, L. 1988. *Private and communal land tenure in Morocco's Western High Atlas Mountains: complements, not ideological opposites. ODI Pastoral Development Network Paper 26a.* Overseas Development Institute (ODI), London, UK. http://www.odi.org.uk/pdn/papers26a.html

Menke, K.H., Raab, L., Salewski, A., Stringab, H., Fritz, D. and Scheider, W. 1979. The estimation of the digestibility and metabolisable energy content of ruminant feeding stuffs from the gas production when they are incubated with rumen liquor *in vitro. Journal of Agricultural Science, Cambridge* 93: 217-222.

Menzi, H. 2001. Minimising environmental impacts of livestock production through good manure and nutrient management. *http://lead.virtualcentre.org/apps/cams/user/default.asp?idf_cams_forum=4&idf_cams_topic=382&lang=en*

Methu, J.N., Owen, E., Abate, A., Mwangi, D.M. and Tanner, J.C. 1996. Smallholder dairying in central Kenya highlands: Practices in the utilisation of maize stover as a feed resource. In *Focus on agricultural research for sustainable development in a changing economic environment, 5th Biennial KARI Scientific Conference*, pp. 243-251. Kenya Agricultural Research Institute (KARI), Nairobi, Kenya.

Mhere, O., Ncube, S. and Matshe, F. 1995. Yield, quality and persistence of Elephant grass (*Pennisetum purpureum*, K. Schum) and its inter-specific hybrids with Pearl millet (*Pennisetum glaucum*, L., R.BR) as affected by irrigation and soil type. *Annual Report, Division of Livestock and Pastures 1994-95*, pp. 124-128. Department of Research and Specialist Services, Harare, Zimbabwe.

Miller, D. 2001. *Sustainable development of mountain rangelands in Central Asia: an update from Kyrgyzan Republic*. http//www.mtnforum.org/ resources/milld01a.htm

Milner-Gulland, E.J., Kholodova, M.V., Bekenov, A.B., Bukreeva, O.M., Grachev, I.A., Amgalan, L., and Lushchekina, A.A. 2001. Dramatic declines in saiga antelope populations. *Oryx* 35(4): 340-345.

Milner-Gulland, E.J., Bukreeva, O.M., Coulson, T., Lushchekina, A.A., Kholodova, M.V., Bekenov, A.B. and Grachev, I.A. 2003. Reproductive collapse in saiga antelope harems. *Nature* 432: 135.

Ministry of Agriculture. 1997. *Beekeeping in Botswana (Beekeeping handbook 4th edition)*. Ministry of Agriculture, Gaborone, Botswana.

Minjauw, B., Muriuki, H.G. and Romney, D.L. 2004. Development of Farm Field School methodology for smallholder dairy farmers in Kenya. In *Responding to the Livestock Revolution – The role of globalisation and implications for poverty alleviation* (ed. E. Owen, T. Smith, M.A. Steele, S. Anderson, A.J. Duncan, M. Herrero, J.D. Leaver, C.K. Reynolds, J.I. Richards and J.C. Ku-Vera), pp. 299-313. British Society of Animal Science Publication No. 33, Nottingham University Press, Nottingham, UK.

Minson, D.J. 1990. *Forage in ruminant nutrition*. Academic Press Inc., San Diego, New York, Boston, London, Sydney, Tokyo and Toronto.

Minson, D.J. and McLeod, M.N. 1972. *The in vitro technique. Its modification for estimating digestibility of large numbers of tropical pasture samples. Division of Tropical Pastures Technical Paper, 8*, pp. 1-15. CSIRO, Australia.

Misra, A.K. and Pandey, A.S. 2000. Seasonality of bullock power use in rain-fed areas. *Draught Animal News* 32: 11-13.

Molina, J.I. 2003. *Aceptación de la técnica de transferencia de embriones bovinos en productores adscritos al programa para el mejoramiento genético de la ganadería del estado de Chiapas*. Facultad de Medicina Veterinaria y Zootecnia, Universidad Nacional Autónoma de México, México, D.F., Mexico.

Morse, R.A. and Calderone, N. 2000. The value of honeybees as pollinators of US crops in 2000. *Bee Culture* 128: 3.

Morton, J. 1988. Sakanab: greetings and information among the Northern Beja. *Africa* 58: 423-436.

Morton, J. 1990. *Aspects of labour in an agro-pastoral economy. ODI Pastoral Development Network Paper 30b*. Overseas Development Institute (ODI), London, UK. http://www.odi.org.uk/pdn/papers/paper30b.html

Morton, J. 1994. Pastoralism in Pakistan: concepts unrecognised and opportunities missed. *Rural Extension Bulletin* No. 4, pp. 33-36.

Morton, J. and Meadows, N. 2000. Pastoralism and sustainable livelihoods: an emerging agenda. *NRI Policy Series No. 11*. Natural Resources Institute, Chatham, UK.

Morton, J. and Barton, D. 2002. Destocking as a drought mitigation strategy: clarifying rationales and answering critiques. *Disasters* 26(3): 213-228.

Moss, A. 1993. *Methane: global warming and production by animals*. Chalcombe Publications, Lincoln, UK.

Moyo, S. 1996. *The productivity of indigenous and exotic beef breeds and their crosses at Matopos, Zimbabwe. PhD thesis*, University of Pretoria, Pretoria, Republic of South Africa.

Msangi, B.S.J. 2001. *Studies of smallholder dairying along the coast of Tanzania with special reference to influence of feeding and supplementation on reproduction and lactation in crossbred cows. PhD thesis*, University of Reading, Reading, UK.

Mtengeti, E.J., Urio, N.A. and Mlay, G.I. 1992. Intensive fodder gardens for increasing fodder availability for smallholder dairy production in Hai District, Tanzania. In *The complementarity of feed resources for animal production in Africa* (ed. J.E.S. Stares, A.N. Said and J.A. Kategile), pp. 129-133. Proceedings of the joint feed resources networks workshop held in Gaborone, Botswana, 4-8 March 1991, International Livestock Centre for Africa (ILCA), Addis Ababa, Ethiopia.

Muchaal, P.K. 2002. *Urban agriculture and zoonoses in West Africa: an assessment of the potential impact on public health. Report 35, Cities Feeding People Series*. International Development Research Centre (IDRC), Ottawa, Canada.

Muinga, R.W., Topps, J.H., Rooke, J.A. and Thorpe, W. 1995. The effect of supplementation with *Leucaena leucocephala* and maize bran on voluntary food intake, digestibility, live weight and milk yield of *Bos indicus*Bos taurus* dairy cows and rumen fermentation in steers offered *Pennisetum purpureum ad libitum* in the semi-humid tropics. *Animal Science* 60: 13-23.

Muirhead, M.R. and Alexander, T.J.L. 1999. *Managing pig health and the treatment of disease*. Nottingham University Press, Nottingham, UK.

Mupeta, B., Coker, R. and E Zaranyika. 2003. *The use of oilseed cake from smallholder processing operations for inclusion in rations for poultry production.* LPP Project R 7524

Muralimanohar, B. 2004. Diseases affecting village poultry and their prevention. *Proceedings of the 3rd National Seminar on Rural Poultry for Adverse Environment,* held on 24th and 25th February 2004 at Veterinary College and Research Institute, Namakkal, Tamil Nadu, India.

Muriuki, H., Omore, A., Hooton, N., Waithaka, M.H. Staal, S.J. and Odhiambo, P. 2003. *The policy environment in the Kenya dairy sub-sector: A review. SDP Research and Development Report No 2.* Smallholder Dairy (R&D) Project, International Livestock Research Institute (ILRI), Nairobi, Kenya.

Murphy, S. and Allen, L. 1996. A greater intake of animal products could improve the micronutrient status and development of children in East Africa. In *Proceedings of East Africa Livestock Assessment Workshop.* SR-CRSP, Davis, California, USA.

Murrell, K.D. 2003. International Action Planning Workshop on *Taenia solium* cysticercosis/taeniosis with special focus on eastern and southern Africa. *Acta Tropica* 87:1-189.

NAADS (National Agricultural Advisory Services). 2000. *Master document of the NAADS Task Force and Joint Donor Group*, Ministry of Agriculture, Animal Industry and Fisheries, Entebbe, Uganda.

Narayan, D., Patel, R., Schafft, K., Rademacher, A. and Koch-Schulte, S. 2000. *Can anyone hear us?* The International Bank for Reconstruction and Development/The World Bank, Washington D.C., USA, Oxford University Press, New York, USA.

Natarajan, A., Anitha, K., Chandrasekaran, D. and Sparks, N. 2004. Improving the health of desi poultry in the study area. *Proceedings of the 3rd National Seminar on Rural Poultry for Adverse Environment,* held on 24th and 25th February 2004 at Veterinary College and Research Institute, Namakkal, Tamil Nadu, India.

Nations, J.D. and Komer, D.T. 1987. Rainforests and the hamburger society. *The Ecologist* 17(4/5): 161-67.

Ncube, S. and Mpofu, D. 1994. The nutritive value of wild fruits and their use as supplements to veldt hay. *Zimbabwe Journal of Agricultural Research* 32: 71-77.

Ndikumana, J., Stuth, J., Kamidi, R., Ossiya, S., Marambii, R. and Hamlett, P. 2000. *Coping mechanisms and their efficacy in a disaster-prone pastoral system of the Greater Horn of Africa. Effects of the 1995-97 drought and 1997-98 El Nino rains and the responses of pastoralists and livestock.* ILRI Project Report. ASARECA (Animal Agricultural Research Network), Nairobi, Kenya; Global Livestock-Collaborative Research Support Programme Livestock Early Warning System, College Station, Texas, USA, and International Livestock Research Institute (ILRI), Nairobi, Kenya.

Nell, A.J. (ed.). 1998. *Livestock and the environment: Proceedings of an international conference organised by the World Bank and the Food and Agriculture Organisation of the United Nations (FAO)*. International Agriculture Centre, Wageningen, The Netherlands.

Nelson, N. and Wright, S. (ed.). 1995. *Power and participatory development.* ITDG Publishing, London, UK.

Nengomasha, E.M., Pearson, R.A. and Smith T. 1999a. The donkey as a draught power resource in smallholder farming in semi-arid western Zimbabwe: 1. Live weight and feed and water requirements. *Animal Science* 69: 297-304.

Nengomasha, E.M., Pearson, R.A. and Smith T. 1999b. The donkey as a draught power resource in smallholder farming in semi-arid western Zimbabwe: 2. Performance compared with that of cattle when ploughing on different soil types using two plough types. *Animal Science* 69: 305-312.

Nestel, B. 1984. Animal production systems in different regions. In *Development of animal production systems. World animal science*, A2, (ed. B. Nestel), pp.131-140. Elsevier Scientific Publishing Company, Amsterdam, The Netherlands.

Ngere, L.O. 1973. Size and growth rate of the West African Dwarf sheep and a new breed, the Nungua Black Head of Ghana. *Ghana Journal of Agricultural Science* 6: 113-117.

Nicholson, C.F., Blake, R.W., Reid, R.S. and Schelhas, J. 2001. Environmental impacts of livestock in the developing world. *Environment* 43: 7-17.

Nicholson, C.F., Mwangi, L., Staal, S.J. and Thornton, P.K. 2003. *Dairy cow ownership and child nutritional status in Kenya.* Department of Applied Economics and Management, College of Agriculture and Life Sciences, Cornell University, Ithaca, NY, USA.

Nielsen, L. 2004. *Motivation and livestock-based livelihoods: Findings from Kenya, Bolivia and India. PhD thesis*. The University of Reading, Reading, UK.

Nitis, I.M., Lana, K., Sukhantan, W., Suarna, M and Putra, S. 1990. The concept and development of the three strata forage system. In *Shrubs and tree fodders for farm animals* (ed. C. Devendra), pp. 92-102. International Development Research Centre, IDRC-276e., Ottawa, Canada.

North, D.C. 1990. *Institutions, institutional change and economic performance*. Cambridge University Press, Cambridge, UK.

North, D.C. 1995. The new institutional economics and Third World development. In *New institutional economics and Third World development* (ed. J. Harris, J. Hunter and C.M. Lewis). Routledge, London, UK.

Norval, R.A.I., Perry, B.D. and Young, A.S. 1992. *The epidemiology of Theileriosis in Africa*. Academic Press, London, UK.

Novoa, M.C. 1981. La conservacion de especas nativas en America Latina [The conservation of native species in Latin America]. *Animal genetic resources conservation and management. Proceedings of an FAO/UNDP Technical Consultation*, pp. 349-362. Food and Agriculture Organisation of the United Nations (FAO), Rome, Italy.

Ntenga, G.M. and Mugongo, B.T. 1991. *Honey hunters and beekeepers: beekeeping in Babati District, Tanzania.* Swedish University of Agricultural Science, Uppsala, Sweden.

Ntiamoa-Baidu, Y. 1997. *Wildlife and food security in Africa. FAO Conservation Guide 33*. Food and Agriculture Organisation of the United Nations (FAO), Rome, Italy. http://www.fao.org/docrep/w7540e/w7540e00.htm

Nutrient requirements of goats; Angora, dairy, and meat goats in temperate and tropical countries. 1981. National Academy Press, Washington D.C., USA.

Nutrient requirements of poultry. Ninth revised edition. 1994. National Academy Press, Washington D.C., USA.

Nutrient requirements for swine. Tenth revised edition. 1998. National Academy Press, Washington D.C., USA.

NWRC. 1997. *Low productivity in East African beekeeping*. Njiro Wildlife Research Centre, Arusha, Tanzania.

Oakley, R. 1998. *Experiences with community-based livestock worker programmes, methodologies and impact: A literature review.* Veterinary Epidemiology Economic Research Unit (VEERU), The University of Reading, Reading, UK.

O'Donovan, P.B. 1984. Compensatory growth in cattle and sheep. *Nutrition Abstracts and Reviews, Series B* 54: 389-410.

OECD. 2003. *Harnessing markets for biodiversity: Toward conservation and sustainable use*. OECD Publications, Paris, France. http://www1.oecd.org/publications/e-book/9703031E.pdf.

Ogle, B. and Preston, T.R. 2004. Ecological impacts of sustainable systems for smallholders in Vietnam. In *Responding to the Livestock Revolution – The role of globalisation and implications for poverty alleviation* (ed. E. Owen, T. Smith, M.A. Steele, S. Anderson, A.J. Duncan, M. Herrero, J.D. Leaver, C.K. Reynolds, J.I. Richards and J.C. Ku-Vera), pp. 179-190. British Society of Animal Science Publication No. 33, Nottingham University Press, Nottingham, UK.

Ørskov, E.R. 1993. *Reality in rural development aid, with emphasis on livestock.* Rowett Research Services Ltd., Bucksburn, Aberdeen AB2 9SB, UK.

Ørskov, E.R. 1998. *The feeding of ruminants. Second edition***.** Chalcombe Publications, Lincoln, UK.

Ørskov, E.R. and McDonald, I. 1979. The estimation of protein degradability in the rumen from incubation measurements weighted according to rate of passage. *Journal of Agricultural Science, Cambridge* 92: 499-503.

Osafo, E.L.K., Owen, E., Said, A.N., Gill, M. and Sherington, J. 1997. Effects of amount offered and chopping on intake and selection of sorghum stover by Ethiopian sheep and cattle. *Animal Science* 65: 55-62.

Osuji, P.O. 1994. Nutritional and anti-nutritional values of multipurpose trees used in agroforestry systems. In *Agroforestry and animal production for human welfare* (ed. J.W. Copeland, A. Djajanegara and M. Sobrani). *Proceedings of an international symposium held in association with the 7th AAAP Animal Science Congress*, Bali, Indonesia, 11-16 July, 1994. *ACIAR Proceeding No. 55*: 82-88.

Otieno-Oruko, L., Upton, M. and McLeod, A. 2000. Restructuring of animal health services in Kenya: Constraints, prospects and options. *Development Policy Review* 18 (2): 123-138.

Otte, J. and Chilonda, P. 2003. Classification of cattle and small ruminant production systems in Sub-Saharan Africa. *Outlook on Agriculture* 32: 183-190.

Owen, E. and Jayasuriya, M.C.N. 1989a. Recent developments in chemical treatment of roughages and their relevance to animal production in developing countries. In *Feeding strategies for improving productivity of ruminant livestock in developing countries,* pp. 205-230. International Atomic Energy Agency (IAEA), Vienna, Austria.

Owen, E. and Jayasuriya, M.C.N. 1989b. Use of crop residues as animal feeds in developing countries. *Research and Development in Agriculture* 6: 129-138.

Owen, E., Smith, T., Steele, M.A., Anderson, S., Duncan, A.J., Herrero, M., Leaver, J.D., Reynolds, C.K., Richards, J.I. and Ku-Vera, J.C. (ed.). 2004. *Responding to the Livestock Revolution – The role of globalisation and implications for poverty alleviation.* British Society of Animal Science Publication No. 33, Nottingham University Press, Nottingham, UK.

Pardey, P.G. and Bientema, N.M. 2001. *Slow magic. Agricultural R and D a century after Mendel.* International Food Policy Research Institute (IFPRI), Washington D.C., USA.

Paterson, D. and Palmer, M. 1991. *The status of animals. Ethics, education and welfare.* CABI Publishing (on behalf of Humane Education Foundation), Wallingford, UK.

Paterson, R.T. 1994. *Use of trees by livestock 8: Calliandra*. Natural Resources Institute, Chatham, UK.

Paterson, R.T. and Clinch, N.J.L. 1993. *Use of trees by livestock 6: Cassia*. Natural Resources Institute, Chatham, UK.

Paterson, R.T., Joaquín, N., Chamón, K. and Palomino, E. 2001. The productivity of small animal species in small-scale mixed farming systems in subtropical Bolivia. *Tropical Animal Health and Production* 33: 1-14.

Paterson, R.T. and Rojas, F. 2002. Small animal species in the livelihoods of small-scale farmers in Tropical Bolivia. In *Responding to the increasing global demand for animal products. Programme and summaries.* An international conference organised by the British Society of Animal Science, American Society of Animal Science and Mexican Society of Animal Production, 12-15 November 2002, Merida, Yucatan, Mexico.

Pathak, P.S. and Newaj, R. 2003. *Agroforestry potentials and opportunities.* Agorbios and Indian Society of Agrofrestry, India.

Payne, W.J.A. 1990. *An introduction to animal husbandry in the tropics. Fourth edition.* Longman Scientific and Technical, Harlow, Essex CM20 2JE, UK. Co-published in the United States by John Wiley & Sons, New York, USA.

Payne, W.J.A. and Wilson, R.T. 1999. *An introduction to animal husbandry in the Tropics. Fifth Edition.* Blackwell Science, Ltd., London, UK.

Peacock, C.P. 1996. *Improving goat production in the tropics. A manual for development workers.* Oxfam/FARM-Africa, Oxford, UK.

Pearson, L.J. and Dutson, T.R. (ed.). 1990. *Meat and health. Advances in meat research, Volume 6.* Elsevier Applied Science, Amsterdam, The Netherlands.

Pearson, R.A. and Dijkman, J.T. 1994. Nutritional implications of work in draught animals. *Proceedings of the Nutrition Society* 53: 169-179.

Pearson, R.A., Muirhead, R.H. and Archibald, R.F. 2001. The effect of forage quality and level of feeding on digestibility and gastrointestinal transit time of oat straw and alfalfa given to ponies and donkeys. *British Journal of Nutrition* 85: 599-606.

Pearson, R.A., Alemayehu, M., Tesfaye, A., Allan, E.F., Smith, D.G. and Asfaw, M. 2002. *Use and management of donkeys in peri-urban areas of Ethiopia. Draught Animal Power Technical Report 5.* EARO/CTVM, Division of Animal Health and Welfare, University of Edinburgh, Easter Bush Veterinary Centre, Roslin, Midlothian EH25 9RG, UK.

Pearson, R.A., Lhoste, P., Saastamoinen, M. and Martin-Rosset, W. (ed.). 2003a. *Working animals in agriculture and transport. A collection of some current research and development observations.* EAAP Technical Series No 6, European Association of Animal Production, Wageningen Press, Wageningen, The Netherlands.

Pearson, R.A., Simalenga, T.E. and Krecek, R.C. 2003b. *Harnessing and hitching donkeys, horses and mules for work.* Centre for Tropical Veterinary Medicine, Division of Animal Health and Welfare, University of Edinburgh, Easter Bush Veterinary Centre, Roslin, Midlothian EH25 9RG, UK.

Peeler, E.J. and Omore, A.O. 1997. *Manual of livestock production systems in Kenya.* Kenya Agricultural Research Institute (KARI)/UK Department for International Development (DFID), National Agricultural Research Project, Kikuyu, Kenya.

Perera, A.N.F. *et al.* 2001. Impact of UMMB in small scale dairy farms in Central Province of Sri Lanka. *Proceedings of the 53rd Annual Convention of the Sri Lanka Veterinary Association*, Kandy, Sri Lanka.

Perera, B.M.A.O. 1999a. Reproduction in water buffalo: comparative aspects and implications for management. *Journal of Reproduction and Fertility, Supplement* 54: 157-168.

Perera, B.M.A.O. 1999b. Management of reproduction. *Smallholder dairying in the tropics* (ed. L. Falvey and C. Chantalakhana), pp. 241-264. International Livestock Research Institute (ILRI), Nairobi, Kenya.

Perera, B.M.A.O. 1994. Current buffalo production systems and future strategies for improvement. *Proceedings of the fourth World Buffalo Congress*, pp. 27-38. Sao Paulo, Brazil.

Perera, B.M.A.O., de Silva, L.N.A., Kuruwita, V.Y. and Karunaratne, A.M. 1987. Post-partum ovarian activity, uterine involution and fertility in indigenous buffalo at a selected village location in Sri Lanka. *Animal Reproduction Science* 14: 115-127.

Perez, R. 1997. *Feeding pigs in the tropics. FAO Animal Production and Health Paper 132.* Food and Agriculture Organisation of the United Nations (FAO), Rome, Italy.

Pérez, R., Valenzuela, S., Merino, V., Cabezas, I., Garcia, M., Bou, R. and Ortiz, P. 1996. Energetic requirements and physiological adaptation of draught horses to ploughing work. *Animal Science*

63: 343-351.

Permin, A. 1997. *Helminths and helminthosis in poultry with special emphasis on* Ascardia galli *in chickens. PhD thesis*, The Royal Veterinary and Agricultural University, Copenhagen, Denmark.

Perry, B.D. (ed). 1999. The Economics of animal disease control. *World Organisation for Animal Health Scientific and Technical Review, Special Edition* 18, (2).

Perry, B.D., McDermott, J.J. and Randolph, T.F. 2001. Can epidemiology and economics make a meaningful contribution to national animal disease control? *Preventive Veterinary Medicine* 48: 231-260.

Perry, B.D., Randolph, T.F., McDermott, J.J., Sones, K.R. and Thornton, P.K. 2002. *Investing in animal health research to alleviate poverty*. International Livestock Research Institute (ILRI), Nairobi, Kenya. Available at www.dfid.gov.uk and www.ilri.org.

Perry, B.D., Randolph, T.F., McDermott, J.J. and Sones, K.R. 2003. Pathways out of poverty: A novel typology of animal diseases and their impacts. In *Proceedings of the 10th International Symposium for Veterinary Epidemiology and Economics (ISVEE)*, Vina del Mar, Chile, 17-21 November 2003. Compact disk.

Perry, B.D., McDermott, J.J. and Randolph, T.F. 2004. Control of infectious diseases: making appropriate decisions in different epidemiological and socio-economic conditions. In *Infectious diseases of livestock, Volume 1* (ed. J.A.W. Coetzer and R.C. Tustin), pp. 178-224. Oxford University Press, Cape Town, South Africa.

Peters, D., Tinh, N.T. and Thuy, T.T. 2000. Improving pig feed in Vietnam. *Urban Agriculture Magazine* 1 (2): 37-38.

Petrie, O.J. 1995. *Harvesting of textile animal fibres. Agricultural Services Bulletin No 122.* Food and Agriculture Organisation of the United Nations (FAO), Rome, Italy

Phillips, C.J.C. (ed.). 1989. *New techniques in cattle production.* Butterworths, London, UK. Now Nottingham University Press, Nottingham, UK.

Phororo, D.R. and Letuka, P.P. 1993. *A review and analysis of land tenure in Lesotho, with recommendations for reform.* United Nations Development Programme (UNDP), Maseru, Lesotho.

Pillay, T.V.R. 1993. *Aquaculture principles and practices*. Fishing News Books, Blackwell Scientific Publications, Oxford, UK.

Pisulewski, P.M., Okorie, A.U., Buttery, P.J., Haresign, W.H. and Lewis, D. 1981, Ammonia concentration and protein synthesis in the rumen. *Journal of the Science of Food and Agriculture* 32: 759-766.

Porter, V. 1996. *Goats of the World.* Farming Press, Ipswich, UK.

Prabakaran, R. 2003. *Good practices in planning and management of integrated commercial poultry production in South Asia. FAO Animal Production and Health Paper 159.* Food and Agriculture Organisation of the United Nations (FAO), Rome, Italy.

Preston, T.R. and Leng, R.A. 1984. Supplementation of diets based on fibrous residues and by-products. In *Straw and other fibrous by-products as feeds* (ed. F. Sundstøl and E. Owen), pp. 371-413. Elsevier, Amsterdam, The Netherlands.

Preston, T.R. and Leng, R.A. 1987. *Matching ruminant production systems with available resources in the tropics and sub-tropics.* Penambul Books, Armidale, Australia.

Preston, T.R. and Murgeitio, E. 1992. *Strategy for sustainable livestock production in the tropics*. Centro para la Investigacion en Sistemas Sostenibles de Produccion Agropecuaria (CIPAV), AA20591 Cali, Colombia; Swedish Agency for Research Cooperation with Developing Countries (SAREC), P.O. Box 16140, S-103 23, Stockholm, Sweden.

Prins, H.T. 2000. Competition between wildlife and livestock in Africa. In *Wildlife conservation by sustainable use* (ed. H.H.T. Prins, J.G. Grootenhuis and T.T. Dolan), pp. 51-80. Kluwer Academic Publishers, The Netherlands and USA.

Prins, H.H.T., Grootenhuis, J.G. and Dolan, T.T. (ed.). 2000. *Wildlife conservation by sustainable use*. Kluwer Academic Publishers, The Netherlands and USA.

Putt, S.N.H., Shaw, A.P.M., Woods, A.J., Tyler, L. and James, A.D. 1987. *Veterinary epidemiology and*

economics in Africa. International Livestock Centre for Africa (ILCA). Addis Ababa, Ethiopia.
Rahnema, M. and Bawtree, V. (ed.). 1997. *The post-development reader.* Zed Books, London, UK.
Rajaguru, A.S.B. 1973. Effect of rubber seed meal on the performance of mature chicken. *R.R.I. S.L. Bulletin (Sri Lanka)* 8: 39-45.
Rangnekar, D. and Thorpe, W. (ed.). 2002. *Smallholder dairy production and marketing - opportunities and constraints*. Proceedings of a south-south workshop held at NDDB, Anand, India, 13-16 March 2001. National Dairy Development Board (NDDB) Anand, India and International Livestock Research Institute (ILRI), Nairobi, Kenya.
Ranjhan, S.K. 1998. *Textbook on buffalo production. Fourth edition*. Viskas Publishing House, New Delhi, India.
Reid, R.S. and Ellis, J.E. 1995. Livestock-mediated tree regeneration: impacts of pastoralists on dry tropical woodlands. *Ecological Applications* 5: 978-992.
RELMA. 2003. Regional Land Management Unit RELMA/Sida, Nairobi, Kenya.
Rico, N.E. and Rivas, V.C. 2000. *Manual Sobre Manejo de Cuyes. Second edition*. Editora Gráfica Soliz, Cochabamba, Bolivia.
Rischkowsky, B., Thomson E.F., Shnayien, R. and King, J.K. 2003. Mixed farming systems in transition: the case of five villages along a rainfall gradient in north-West Syria. *Experimental Agriculture* 40: 1-18.
Rivera, W.M. and Zijp, W. 2002. *Contracting for agricultural extension. International case studies and emerging practices*. CABI Publishing, Wallingford, UK, and New York, USA.
Roe, D., Mayers, J., Grieg-Gran, M., Kothari, A., Fabricius, C., and Hughes, R. 2000. *Evaluating Eden: Exploring the myths and realities of community based wildlife management. Evaluating Eden Series, No. 8*. International Institute for Environment and Development (IIED), London, UK.
Rogers, E.M. and Kincaid, D. 1981. *Communication networks. Toward a new paradigm for research.* The Free Press, New York, USA, and London, UK.
Röling, N. 1988. *Extension science. Information systems in agricultural development*. Cambridge University Press, Cambridge, UK.
Romney, D.L. and Gill, M. 2000. Intake of forages. In *Forage evaluation in ruminant nutrition* (ed. D.I. Givens, E. Owen, R.F.E. Axford and H.M. Omed), pp. 43-62. CABI Publishing, Wallingford, UK.
Romney, D., Kaitho, R., Biwott, J., Wambugu, M., Chege, L., Omore, A., Staal S., Wanjohi, P. and Thorpe, W. 2000. Technology development and field testing: access to credit to allow smallholder dairy farmers in central Kenya to reallocate concentrates during lactation. *Paper presented at the 3rd All Africa Conference on Animal Agriculture and 11th Conference of the Egyptian Society of Animal Production, 6-9 November, 2000, Alexandria, Egypt.*
Romney, D., Utiger, C., Kaitho, R., Thorne, P., Wokabi, A., Njoroge, L., Chege, L., Kirui, J., Kamotho, D. and Staal, S. 2004. Effect of intensification on feed management of dairy cows in the Central Highlands of Kenya. In *Responding to the Livestock Revolution – The role of globalisation and implications for poverty alleviation* (ed. E. Owen, T. Smith, M.A. Steele, S. Anderson, A.J. Duncan, M. Herrero, J.D. Leaver, C.K. Reynolds, J.I. Richards and J.C. Ku-Vera), pp. 287-298. British Society of Animal Science Publication No. 33, Nottingham University Press, Nottingham, UK.
Rose, S.P. 1997. *Principles of poultry science*. CABI Publishing, Wallingford, UK.
Roubik, D. 1995. *Pollination of cultivated plants in the tropics*. Food and Agriculture Organisation of the United Nations (FAO), Rome, Italy.
Roubik, D. 2002. The value of bees to the coffee harvest. *Nature* 417: 708.
Ruijun Long. 2003a. Alpine rangeland ecosystems and their management in the Qinghai-Tibetan Plateau. In *The yak. Second edition* (ed. G. Wiener, Han Jianlin and Long Ruijun), pp. 359-388. Food and Agriculture Organisation of the United Nations (FAO), Bangkok, Thailand.
Ruijun Long. 2003b. Yak nutrition - a scientific basis. In *The yak. Second edition* (ed. G. Wiener, Han Jianlin and Long Ruijun), pp. 389-414. Food and Agriculture Organisation of the United Nations (FAO), Bangkok, Thailand.

Ruthenberg, H. 1980. *Farming systems in the tropics. Third edition.* Clarendon Press, Oxford, UK.

Sadia M Ahmed, Hassan Mohamed Ali and Amina Mohamoud Warsame. 2001. *Survey on the state of pastoralism in Somaliland.* PENHA and ICD, London, UK.

Said, R., Bryant, M.J. and Msechu, J.K.K. 2003. The survival, growth and carcass characteristics of crossbred beef cattle in Tanzania. *Tropical Animal Health and Production* 35: 441-454.

Sakho, K. 1999. Sustainability in Senegal: the Vautier hive. *Bees for Development Journal* 51: 3-5.

Sandford, S. 1983. *Management of pastoral development in the Third World.* John Wiley and Sons, Chichester, UK, in association with Overseas Development Institute (ODI), London, UK.

Sansoucy, R, Aarts, G. and Leng, R.A. 1988. Molasses-urea blocks as nutrient supplements for ruminants. In *Sugarcane as a feed. FAO Animal Production and Health Paper No. 72,* pp. 263-279. Food and Agriculture Organisation of the United Nations (FAO), Rome, Italy.

Save the Children UK. 2001. *Final report on the findings of a household economy assessment and training in Mchinji District, Malawi, October 2001.* The Food Security and Livelihoods Unit and The Malawi Country Programme, Save the Children, London, UK.

Sawyerr, L.C. 1995. *Short notes on practised snail farming.* CDS Press, Accra, Ghana.

Schaller, G.B. 1998. Wild yak. In *Wildlife of the Tibetan Steppe*, pp. 125-142. University of Chicago Press, Chicago, USA.

Schantz, P.M., Cruz M., Pawlowski, Z. and Sarti, E. 1993. Potential eradicability of taeniasis and cysticercosis. *Bulletin of the Pan American Health Organisation* 27:397-403.

Scherf, B.D. (ed.). 2000. *World watch list for domestic animal diversity.* Food and Agriculture Organisation of the United Nations (FAO), Rome, Italy.

Schiere, H. and van der Hock, R. 2001. *Livestock keeping in urban areas – a review of traditional technologies based on literature and field experiences. FAO Animal Production and Health Paper 151.* Food and Agriculture Organisation of the United Nations (FAO), Rome, Italy.

Schrage, R. and Yewadan, L.T. 1999. *Raising grasscutters.* Deutsche Gesellschaft fur Technische Zusammenarbeit (GTZ) GmbH, Eschborn, Germany.

Schwartz, H.J. 1992. Productive performance and productivity of dromedaries (*Camelus dromedarius*). *Animal Research and Development* 35: 86-98.

Schwartz, H.J. and Dioli, M. 1992. *The one-humped Camel (*Camelus dromedarius*) in Eastern Africa: A pictorial guide to diseases, health care and management.* Verlag Josef Margraf, Weikersheim, Germany.

Scoones, I. and Wolmer, W. 2002. *Pathways of change: crops, livestock and livelihoods in Mali, Ethiopia and Zimbabwe.* James Currey, Oxford, UK and Heinemann, Portsmouth NH., USA.

Scott, G.J. (ed.). 1995. *Prices, products and people: analysing agricultural markets in developing countries.* Rienner, London, UK.

Sebastian, L., Mudgal, V.D. and Nair, P.G. 1970. Comparative efficiency of milk production by Sahiwal cattle and Murrah buffalo. *Journal of Animal Science* 30: 253-256.

Secretariat on the Convention on Biological Diversity. 2001-2004. *Convention on Biological Diversity. Convention Text.* United Nations Environment Programme (UNEP). www.unep-ucmc.org/

Seré, C. and Steinfeld, H. 1996. *World livestock production systems; current status, issues and trends. FAO Animal Production and Health Paper No. 127.* Food and Agriculture Organisation of the United Nations (FAO), Rome, Italy.

Shanawany, M.M. 1995. Recent developments in ostrich farming. *World Animal Review* 83: 8 pages. Also available at www.fao.org/ag/AGA/AGAP/WAR/Contents.htm

Sharma, V.P., Kohler-Rollefson, I. and Morton, J. 2003. *Pastoralism in India: A scoping study.* Indian Institute of Management, League for Pastoral Peoples and Natural Resources Institute, Chatham, UK, for UK Department for International Development (DFID) Livestock Production Programme.

Sheldon, C. 1988. *Raising snails.* Special Reference Briefs (National Agricultural Library SRB 88-04), United States Department of Agriculture (USDA), Beltsville, Maryland, USA.

Sherman, D. 2002. *Tending animals in the global village. A guide to international veterinary medicine.* Lippincott, Williams and Wilkins, Philadelphia, USA.

Shindey, D.N. and Pathan, R.K. 2002. *Scavenging Poultry Report, April 2001-March 2002*. BAIF Development Research Foundation, Pune, India.

Shukla, R.K. and Brahmankar, S.D. 1999. *Impact evaluation of Operation Flood on the rural dairy sector.* National Council of Applied Economic Research (NCAER). New Delhi, India (http://www.ncaer.com/)

Sims, B.G. and O'Neill, D.H. 2003. Aspects of work animal use in semi-arid farming systems. In *Working animals in agriculture and transport. A collection of some current research and development observations* (ed. R.A. Pearson, P. Lhoste, M. Saastamoinen and W Martin-Rosset), pp. 39-50. EAAP Technical Series No 6, European Association of Animal Production, Wageningen Press, The Netherlands.

Sinclair, A.R.E. and Fryxell, J.M. 1985. The Sahel of Africa: ecology of a disaster. *Canadian Journal of Zoology* 63: 987-994.

Singh, B.B. and Tarawali, S.A. 1997. Cowpea and its improvement: Key to sustainable mixed crop/livestock farming systems in West Africa. In *Crop residues in sustainable mixed crop/livestock farming systems* (ed. C. Renard), pp. 79-100. CABI Publishing, Wallingford, UK.

Singh, G. and Prabhakar, S. 2002. Taenia solium *cysticercosis: from basic to clinical science*. CABI Publishing, Wallingford, UK.

Singh, I. and Dhanda, O.P. (ed.). 2003. *Buffalo for food security and rural employment. Proceedings of fourth Asian Buffalo Congress, Volume 1 (Lead papers).* New Delhi, India.

Sinurat, A.P., Santoso, J. E., Sumanto, Murtisari, T. and Wibowo, B. 1992. *Improving productivity of village chickens held by smallholders by farming system approach* (in Indonesian). *Ilmu dan Peternakan* 5: 73-77.

Sinurat, A.P., Gilchrist, P., Hamid, H. and Basuno, E. 1998 Diseases of village poultry. *Proceedings. 10th World Poultry Association Congress*, pp. 19-25. Sydney, Australia.

Skonhoft A., Stenseth, N.C., Leirs, H., Andreassen, H.P. and Mulungu, L.S.A. 2003. *The bioeconomics of controlling an African rodent pest species.* Working Paper Series, 5/2003, Department of Economics, Norwegian University of Science and Technology, Trondheim, Norway. http://www.svt.ntnu.no/iso/wp/wp.htm

Skunmun, P. 2000. Buffalo as a draught animal within smallholder farming systems in Asia. *Proceedings of the third Asian Buffalo Congress on the changing role of the buffalo in the New Millennium in Asia*, pp. 63-71. National Science Foundation Press, Colombo, Sri Lanka.

Smallholder Dairy Project. 1999. *Longitudinal recording in three dairy production systems*. Unpublished data. International Livestock Research Institute (ILRI), Nairobi, Kenya.

Smirnov, D.A. *et al.* 1990. Meat yield and meat quality in yaks. *Sel'skokhozyaistvennykh Nauk Im. V.I. Lenina* (Soviet Agricultural Sciences), No. 1: 46-49.

Smith, A.J. (ed.). 1976. *Beef production in developing countries*. Centre for Tropical Veterinary Medicine, University of Edinburgh, Edinburgh, UK.

Smith, A.J. 2001. *Poultry.* The Tropical Agriculturist, Series Editors R. Coste and A.J. Smith. Macmillan Education Ltd., London, UK, in cooperation with Technical Centre for Agriculture and Rural Cooperation (CTA), Wageningen, The Netherlands.

Smith, D.G. 1999. *The impact of traditional African grazing systems on the feed intake and behaviour of cattle and donkeys. PhD thesis,* University of Edinburgh, Edinburgh, UK.

Smith, F.G. 2003. *Beekeeping in the tropics (Reprint).* Northern Bee Books, Mytholmroyd, UK.

Smith, M.C. and Sherman, D.M. 1994. *Goat medicine*. Lea and Febiger, Philadelphia, USA.

Smith, T. 2002. *On-farm treatment of straws and stover with urea. International Atomic Energy Agency (IAEA) TECDOC-1294*, pp. 15-20. International Atomic Energy Agency (IAEA), Vienna, Austria.

Smith, T., Manyuchi, B. and Mikayiri, S. 1990. Legume supplementation of maize stover. In *Utilisation of research results on forage and agricultural by-product materials as animal feed resources in*

Africa (ed. B.H. Dzowela, A.N. Said, A. Wendem-Agenehu and J.A. Kategile). *Proceedings of the first joint PANESA/ARNAB workshop, Lilongwe, Malawi, 5-9 December 1989*, pp. 302-320. PANESA/ARNAB, Addis Ababa, Ethiopia.

Sohai, M.A. 1983. The role of the Arabian camel (*Camelus dromedarius*) in animal production. *World Review of Animal Production* 19 (3): 37-40.

Sonaiya, E.B. 2000. *Issues in family poultry development research.* Publication of International Network for Family Poultry Development. Department of Animal Science, Faculty of Agriculture, Obafemi Awolowo University, Ife-Ife, Nigeria.

Sonaiya, E.B. 2004. Direct assessment of nutrient resources in free-range and scavenging systems. *World's Poultry Science Journal* 60: 523-535.

Sparks, N.H.C. and Shindey, D.N. 2004. Management of the hatching egg. *Proceedings of the 3rd National Seminar on Rural Poultry for Adverse Environment,* held on 24th and 25th February 2004 at Veterinary College and Research Institute, Namakkal, Tamil Nadu, India.

Spedding, C.R.W. 1975. *The biology of agricultural systems.* Academic Press, London, UK.

Spradbrow, P.B. 1999. Epidemiology of Newcastle disease and the economics of its control. In *Poultry as a tool in poverty eradication and promotion of gender equality. Proceedings of a workshop*, March 22-26., Tune Landsboskole, Denmark. At: http://www.husdyr.kvl.dk/htm/php/tune99/index2.htm (December, 2004).

Staal, S.J., Owango, M., Muriuki, H., Kenyanjui, M., Lukuyu, B., Njoroge, L., Njubi, D., Baltenweck, I., Musembi, F., Bwana, O., Muriuki, K., Gichungu, G., Omore, A., Kenyanjui, M., Njubi, D. and Thorpe, W. 2001. *Dairy systems characterisation of the Nairobi milk shed: Application of spatial and household analysis*. Kenya Agricultural Research Institute (KARI)/Ministry of Agriculture and Rural Development (MoARD)/International Livestock Research Institute (ILRI), Nairobi, Kenya.

Stack, J.A., Bell, S.J., Burke, P.A., Forse, R.A. 1996. High-energy, high-protein, oral, liquid, nutrition supplementation in patients with HIV infection: effect on weight status in relation to incidence of secondary infection. *Journal of the American Dietetic Association* 96: 337-41.

Steele, M. 1996. *Goats*. The Tropical Agriculturalist, Series Editors, R. Coste and A.J. Smith. Macmillan Education Ltd., London, UK in cooperation with Technical Centre for Agricultural and Rural Cooperation (CTA), Wageningen, The Netherlands.

Steinfeld, H., de Haan, C. and Blackburn, H. 1997. *Livestock and the environment: issues and options*: WRENmedia, Fressingfield, Eye, Suffolk, IP21 5SA, UK. http://www.fao.org/ag/aga/LSPA/LXEHTML/Default.htm

Stenseth, N.C., Leirs, H., Skonhoft, A., Davis, S.A., Pech, R.P., Andreassen, H.P., Singleton, G.R., Lima, M., Machang'u, R.M., Makundi, R.H., Zhang, Z., Brown, P.B., Dazhao Shii and Xinrong Wan. 2003. Mice and rats: the dynamics and bioeconomics of agricultural rodents pests. *Frontiers in Ecology and the Environment* 1(7): 367-375.

Stickney, R.R. 2000. *Encyclopedia of aquaculture*. John Wiley & Sons, New York, USA.

Stobbs, T.H. and Thompson, P.A.C. 1975. Milk production from tropical pastures. *World Animal Review* 13: 27-31.

Sumar, J. 1977. Algunos índices de producción en la llama. *Anales de la I Reunión de la Asociación Peruna de Produción Animal (APPA) y I Symposio sobre producción de leche en el país*, pp. 31-32.

Sumar, J. and Garcia, M. 1986. 'Fisiología de reproducción de la alpaca' ('Reproductive physiology of the alpaca'). In *Nuclear and Related Techniques in Animal Production and Health. Proceedings of Symposium, Vienna, 1986*, pp. 149-177. International Atomic Energy Agency (IAEA), Vienna, Austria.

Sundstøl, F. and Owen, E. 1993. *Urine – A wasted, renewable natural resource. Proceedings of a workshop: NORAGRIC Occasional Papers Series C,* 67, NORAGRIC, As, Norway.

Sundstøl, F. and Owen, E. (ed.). 1984. *Straw and other fibrous by-products as feed.* Elsevier, Amsterdam, The Netherlands.

Sundstøl, F., Fox, D.G. and Tveitnes, S. 1995. Farm animals in sustainable agriculture. In *Second*

International Conference on Increasing Animal Production with Local Resources, Zhanjiang, China.

Surai, P.F. 2002. *Natural antioxidants in avian nutrition and reproduction.* Nottingham University Press, Nottingham, UK.

Swai, E.S. 2002. *Epidemiological studies of tickborne diseases in small-scale dairy farming in Tanzania. PhD thesis*. University of Reading, Reading, UK.

Swallow B.M. 2000. *Impacts of trypanosomiasis on African agriculture. PAAT Technical and Scientific Series 2*. Food and Agriculture Organisation of the United Nations (FAO). Rome, Italy.

Swallow, B.M. and Bromley, D.W. 1998. *Institutions, governance and incentives in common property regimes for African rangelands. Livestock Policy Analysis Brief.* International Livestock Research Institute (ILRI), Nairobi, Kenya. http://www.ilri.cgiar.org/research/proj6/pol_br01.cfm

Swallow, D. 2003. Genetics of lactose persistence and lactose intolerance. *Annual Review of Genetics* 37: 197-219.

Swift, J. 1988. *Major issues in pastoral development with special emphasis on selected African countries.* Food and Agriculture Organisation of the United Nations (FAO), Rome, Italy.

Syrstad, O. 1990. A genetic interpretation of results obtained in *Bos indicus* x *Bos taurus* crossbreeding for milk production. *Proceedings of the 4th World Congress on Genetics Applied to Livestock Production, Edinburgh, 1990,* XIV, pp. 195-198.

TAC (Technical Advisory Committee). 1994. *Review of CGIAR priorities and strategies*. TAC Secretariat. Food and Agriculture Organisation of the United Nations (FAO), Rome, Italy.

Tacconi, L. 2000. *Biodiversity and ecological economics*. Earthscan Publications, London, UK.

Taneja, V.K. 2000. Cattle breeding programmes in India. In *Workshop on Developing Breeding Strategies for Lower Input Animal Production Environment* (ed. S. Galal, J. Boyazoglu and K. Hammond), pp. 445-454, Bella, Italy 22-25 September, 1999. ICAR Technical Series, No.3.

Tanner, J.C., Reed, J.D. and Owen, E. 1990. The nutritive value of fruits (pods with seeds) from four *Acacia* spp. compared with extracted noug (*Guizotia abyssinica*) meal as supplements to maize stover for Ethiopian highland sheep. *Animal Production* 51: 127-133.

Theodorou, M.K. and France, J. (ed.). 2000. *Feeding systems and feed evaluation models*. CABI Publishing, Wallingford, UK.

Thomas, C. (ed). 2004. *Feed into milk*. Nottingham University Press, Nottingham, UK.

Thomson, E.F. 1987. *Feeding systems and sheep husbandry in the barley belt of Syria*. International Centre for Research in the Dry Areas (ICARDA), Aleppo, ICARDA-106 En., Syria.

Thorne, P.J. 1998. *DRASTIC. A dairy rationing system for the tropics, evaluation version for Windows 3.1*. Natural Resources Institute (NRI), Chatham, UK.

Thornton, P.K., Kruska, P.L. Henninger, N., Kristjanson, P.M., Reid, R.S., Atieno, F., Odero, A.N. and Ndegwa, T. 2002. *Mapping poverty and livestock in the developing world*. International Livestock Research Institute (ILRI), Nairobi, Kenya. Available at www.dfid.gov.uk and www.ilri.org.

Thornton, P.K., Kristiansen, P.M., Kreskas, R.L. and Reid, R.S. 2004. Mapping livestock and poverty. In *Responding to the Livestock Revolution – The role of globalisation and implications for poverty alleviation* (ed. E. Owen, T. Smith, M.A. Steele, S. Anderson, A.J. Duncan, M. Herrero, J.D. Leaver, C.K. Reynolds, J.I. Richards and J.C. Ku-Vera), pp. 37-50. British Society of Animal Science Publication No. 33, Nottingham University Press, Nottingham, UK.

Thorpe, W., Muriuki, H.G., Omore, A., Owango, M.O. and Staal, S. 2000. *Dairy development in Kenya: The past, the present and the future*. Paper presented at the Annual Symposium of the Animal Production Society of Kenya (APSK). Theme: Challenges to Animal Production in this Millennium, KARI Headquarters, Nairobi, March 22-23, 2000.

Tilley, J.M.A. and Terry, R.A. 1963. A two stage technique for *in vitro* digestion of forage crops. *Journal of the British Grassland Society* 18: 104-111.

Tinsley, R.L. 2004. *Developing smallholder agriculture – a global perspective*. AgBe Publishing, 1050 Brussels, Belgium.

Titterton, M. and Bareeba, F. B. 2000. Grass and legume silages in the tropics. In *Silage making in the tropics with particular emphasis on smallholders. Proceedings of the FAO Electronic Conference on Tropical Silage 1 September - 15 December 1999. FAO Plant Production and Protection Paper No. 161*, 43-50. 38. Food and Agriculture Organisation (FAO), Rome, Italy.

Topps, J.H. 1992. Chemical composition and use of legume shrubs and trees as fodder for livestock in the tropics. *Journal of Agricultural Science, Cambridge* 118: 1-8.

Torres, H. 1984. *Distribution and conservation of the Vicuña (Vicugna vicugna), Special Report N° 1*. International Union for Conservation of Nature and Natural Resources, Gland, Switzerland.

Tulloh, N.M. and Holmes, J.H.G. (ed.). 1992. *World animal science, Volume. C6: Buffalo production*. Elsevier, Amsterdam, The Netherlands. (Contents page available at: http://www.elsevier.com/wps/find/bookdescription.cws_home/522234/description#description).

Tuszyñski, W.B., Diakowska, E.A.A. and Hall, N.S. 1983. *Solar energy in small-scale milk collection and processing. Animal Health and Production Paper No 39*. Food and Agriculture Organisation of the United Nations (FAO), Rome, Italy.

UK Government. 2000. *Eliminating world poverty: Making globalisation work for the poor. White Paper on international development*. Presented to Parliament by the Secretary of State for International Development, by Command of Her Majesty, December 2000.

UNDP (United Nations Development Programme). 1996. *Urban agriculture, food, jobs and sustainable cities. Publication series for habitat II*. United Nations Development Programme (UNDP), New York, USA.

UNDP (United Nations Development Programme). 1997. *Human development report 1997: poverty from a human development perspective*. Oxford University Press, Oxford, UK.

UNEP and UNFCCC. 2002. *Climate change information kit*. United Nations Environment Programme (UNEP) and United Nations Framework Convention on Climate Change (UNFCCC). http://unfccc.int/resource/iuckit/index.html

UNEPCA (Unidad Ejecutora Proyecto Camélidos). 1999. *Camélidos. La ganadería del futuro* (Camelids. The future's husbandry), Ministerio de Asuntos Campesinos, Indígenas y Agropecuarios. La Paz, Bolivia.

UNEPCA (Unidad Ejecutora Proyecto Camélidos). 2002. *Camélidos. La ganadería del futuro*. (Camelids. The future's husbandry). CD ROM UNEPCA, Oruro, Bolivia. http://www.infoagro.gov.bo/camelidos/panorama.htm

Upton, M. and Otte, J. 2004. The impact of trade agreements on livestock producers. In *Responding to the Livestock Revolution – The role of globalisation and implications for poverty alleviation* (ed. E. Owen, T. Smith, M.A. Steele, S. Anderson, A.J. Duncan, M. Herrero, J.D. Leaver, C.K. Reynolds, J.I. Richards and J.C. Ku-Vera), pp. 67-83. British Society of Animal Science Publication No. 33, Nottingham University Press, Nottingham, UK.

Vale, W.G. 1994a. Collection, processing and deep freezing of buffalo semen. *Buffalo Journal Supplement* 2: 65-82.

Vale, W.G. 1994b. Reproductive management of water buffalo under Amazon conditions. *Buffalo Journal* 10: 85-90.

Vale, W.G. 1996. The buffalo production in the Amazon Valley. *International Symposium on buffalo products, EAAP Publication* No. 82, pp. 99-116. Wageningen Press, Wageningen, The Netherlands.

Vall, E. 1996. Capacités de travail, comportement à l'effort et réponses physiologiques du zébu, de l'âne et du cheval au Nord-Cameroun. *Thèse de Doctorat, ENSAM, Montpellier,* France.

Van der Westhuysen, J.M. 1982. Mohair as a textile fibre. *Proceedings of the Third International Conference on Goat Production and Disease*, pp. 264 -267. University of Arizona, Tucson, USA.

Van Soest, P.J. 1994 *Nutritional ecology of the ruminant. Second edition.* Cornell University Press, Ithaca, New York, USA.

Van Ufford, P.Q. and Bos, A.K. 1996. Distorted beef markets and regional livestock trade in West Africa: experience from the Central West African Corridor. *Tijds.Sociaalwetenschapp. Onderzoek Landbouw* 11 (1): 5-19.

Van Zyl, A., Meyer, A.J. and van der Merwe, M. 1999. The influence of fibre in the diet on growth rates and the digestibility of nutrients in the greater cane rat (*Thryonomys swinderianus*). *Comparative Biochemistry and Physiology A. Molecular, Integrative, Physiology* 123: 129-135.

Vivanco, W., Cardenas, H., Bindon, B. 1985. 'Relación entre la duración de la cópula y momento de ovulación en alpacas' ('Relationship between the copulation duration and the moment of ovulation in alpaca'). In *Libro de Resúmenes V Convención Internacional Sobre Camélidos Sudamericanos*, Cuzco, Perú.

Wagner, J.E. and Manning, P.J. (ed.). 1976. *The biology of the guinea pig*. Academic Press, New York, USA.

Walters, J.R. 1981. Peri-urban piggeries in Papua New Guinea. In *Intensive animal production in developing countries* (ed. A.J. Smith and R.G. Gunn), pp. 275-278. British Society of Animal Production, Occasional Publication No. 4.

Wanjaiya, J.K. and Pope, C.A. 1985. Alternative income and protein sources for rural communities: Prospects for the rabbit in East Africa. *Journal of Applied Rabbit Research* 8: 19-22.

Wengraf, T. 2001. *Qualitative research interviewing*. Sage Publications, London, UK.

WHO. 1996. *Investing in health research and development. Report of the Ad Hoc Committee on Health Research Relating to Future Intervention Options*. World Health Organisation (WHO), Geneva, Switzerland.

Wiener, G. 1994. *Animal breeding*. The Tropical Agriculturalist, Series Editors R. Coste and A.J. Smith. Macmillan Education Ltd., London, UK, in cooperation with Technical Centre for Agricultural and Rural Cooperation (CTA), Wageningen, The Netherlands.

Wiener, G. 2003. Yak in non-traditional environments - North America. In *The yak. Second edition* (ed. G. Wiener, Han Jianlin and Long Ruijun), pp. 337-346. Food and Agriculture Organisation of the United Nations (FAO), Bangkok, Thailand.

Wiener, G., Jianlin, Han and Ruijun, Long (ed.). 2003. *The yak. Second edition*. Food and Agriculture Organisation of the United Nations (FAO), Bangkok, Thailand.

Wilkins, J.V. and Martinez, L. 1983. Bolivia: an investigation of sow productivity in humid lowland villages. *World Animal Review* 47: 15-18.

Willis, M.B. 1998. *Dalton's introduction to practical animal breeding. Fourth edition*. Blackwell Science, London, UK.

Wilson, R.T. 1976. The Ostrich, *Struthio camelus*, in Darfur, Republic of Sudan. *Bulletin of the British Ornithologists Club* 96: 123-125.

Wilson, R.T. 1992. Goat and sheep skin and fibre production in selected Sub-Saharan African countries. *Small Ruminant Research* 8: 13-29.

Wilson, R.T. 1995. *Livestock production systems*. The Tropical Agriculturalist, Series Editors R. Coste and A.J. Smith. Macmillan Education Ltd., London, UK, in cooperation with Technical Centre for Agricultural and Rural Cooperation (CTA), Wageningen, The Netherlands.

Wilson, R.T. 2000. The use and value of animal power in Zimbabwe. *Draught Animal News* 33: 13-24.

Wilson, R.T. 2003. The environmental ecology of oxen used for draught power. *Agriculture, Ecosystems and the Environment* 97: 21-37.

Wilson, R.T., de Leeuw, P.N. and de Haan, C. 1983. *Recherches sur les systèmes en zones arides: Résultats preliminaries (Rapport de recherché No. 5)*. International Livestock Centre for Africa (ILCA), Addis Ababa, Ethiopia.

World Bank. 1996. *The World Bank participation sourcebook*. The World Bank, Washington D.C., USA. In http://www.worldbank.org

World Bank. 2001. *World Development Report 2000/2001. Attacking poverty*. Oxford University Press, New York, USA.

World Bank. 2003. *Livestock and poverty reduction*. http://lnweb18.worldbank.org/ESSD/ardext.nsf/26ByDocName/LivestockAnimalResourcesLivestockPoverty

Wu Ning. 2003. Social, cultural and economic context of yak production. In *The yak. Second edition* (ed. G. Wiener, Han Jianlin and Long Ruijun), pp. 347-358. Food and Agriculture Organisation of the United Nations (FAO), Bangkok, Thailand.

Yang Rongzhen, Han Xingtai and Luo Xiaolin (ed.). 1997. *Proceedings of the Second International Congress on Yak held in Xining, P.R. China, 1-6 September 1997*. Qinghai People's Publishing House, Xining, P.R. China.

Zerbini, E. and Alemu Gebre Wold. 1999. In *Feeding dairy cows for draught in smallholder dairying in the tropics* (ed. L. Falvey and C. Chantalakhana), pp. 133-156. International Livestock Research Institute (ILRI), Nairobi, Kenya. Also available at: www.ssdairy.org/AdditionalRes/Smhdairy/chap8.html

Zhang Rongchang, Han Jianlin and Wu Jianpin (ed.). 1994. *Proceedings of the First International Congress on Yak held in Lanzhou, P.R. China, 4-9 September 1994*. Supplement of *Journal of Gansu Agricultural University*, Lanzhou, P.R. China.

Index